Introduction to Microprocessor-Based Systems Design

Giuliano Donzellini · Andrea Mattia Garavagno ·
Luca Oneto

Introduction
to Microprocessor-Based
Systems Design

Giuliano Donzellini
DITEN
University of Genoa
Genova, Italy

Andrea Mattia Garavagno
DITEN
University of Genoa
Genova, Italy

Luca Oneto
DIBRIS
University of Genoa
Genova, Italy

ISBN 978-3-030-87346-2   ISBN 978-3-030-87344-8   (eBook)
https://doi.org/10.1007/978-3-030-87344-8

To my wife Melina and my sons Sara and Paolo
for their presence and infinite patience.
To my Father and my Mother, always in my thoughts.

Giuliano Donzellini

To my grandparents who couldn't witness this happy moment.
To my parents for always being there.
To my friends for their presence despite my absences,
and for listening to many of my ideas.

Andrea Mattia Garavagno

To Irene.
Hoping to become the father she deserves.

Luca Oneto

Foreword of Prof. Donatella Sciuto

Writing a university textbook is no easy task. It requires a broad knowledge of the material, an ability to explain it clearly and to inspire students to want to learn more. Compared to commonly used tools such as lecture notes or slides, a textbook requires far greater attention to detail, meticulous planning of the approach, the order of the subjects and the level of detail for each part.

What a university textbook adds to other teaching tools is a critical approach and reasoning. This entails not only meticulously listing data or technical aspects but relating them to each other to develop a method in order to challenge preconceived ideas. It is precisely the method, the ability to analyze, which studying should provide us, not just a set of ideas, for which there are other useful and complementary tools. Textbooks, however, educate the mind and require a keen understanding of the material and ability to analyze and summarize. Like a good teacher, they also inspire questioning, the basis of any valid scientific principle.

In fact, this textbook stems from the considerable experience of its authors who have taught this material and developed a flexible methodology over the years. They have developed a flexible educational tool that allows for the simulation of logic networks and the emulation of a system based on an educational processor, used in this text as an example to explain the basic concepts of a programmable digital system architecture. This book deals with computers from the perspective of their logical structure and that of its language. It provides enough ideas and examples for students to learn to program directly using machine language. It emphasizes the conceptual, technological and structural aspects, that is the hardware architecture of programmable digital systems.

The book takes a bottom-up approach; the broad first chapter shows how to "build" the architecture of a simple 8-bit processor step-by-step, detailing every internal mechanism. The aim of the chapter is to give the fundamentals of microcontroller design and also to make clear that design choices have an effect on how the machine performs and what it can do from the programmer's point of view. The book then introduces another 8-bit microprocessor, the DMC8, inspired by the Z80-CPU. This processor is the heart of the rest of the book, which deals with the issue from the perspective of initially having to integrate the processor in a system and then programming it to perform the required functions.

The book offers a "learning by doing" approach with very detailed examples that help the reader develop analyzing and summarizing skills. The strong connection between the book's content and the *Deeds* (Digital Electronics Education and Design Suite) environment makes this possible. The *Deeds* environment provides the simulation tools so students can master the concepts, work through the examples and check how they work in a hands-on way.

Chapter 2 deals with the integration of the processor with the memory and input/output subsystems by means of the bus. It also introduces the individual components, the hardware elements required in the programming model and the steps required to execute simple instructions. This chapter shows the dynamic interaction among the different components through bus signals. The chapter concludes with a presentation of the *Deeds* module that allows for the emulation of the workings of the complete microcomputer system introduced previously. A description of the steps from writing the assembly code to the execution are provided.

Chapter 3 goes into greater detail on processor instructions, specifically for the DMC8 and subprogram management mechanisms, while Chapter 4 develops the subject of interfacing with input/output devices both in terms of synchronization and data transfer.

All the chapters come with detailed examples and exercises with solutions to help students understand microprocessors system architecture and their own programming techniques. These skills are key to improving computer designers' and programmers' understanding of what software applications can do in relation to the characteristics of the underlying hardware.

Chapter 5 deals with implementation on programmable components such as FPGAs of systems based on the processor presented in the textbook, creating prototypes to check one's own design on FPGA boards supported by the *Deeds* environment and FPGA company-produced development tools. This is a process of experimentation which is a good ending to this book and brings the reader through the different phases of application design, so that they may check what they have learned over the previous chapters on a physical system.

This book will certainly be a useful aid for those teaching introductory applied microcontroller architecture courses. It provides students with detailed, practical knowledge on how to create and design these devices, while enabling them to set up workshops where they can carry out projects on their own.

There is a good balance between theory and practice in the book, which provides students with tools related to content and methodology on the one hand, and requires them to apply them on the other. If theory allows us to apply an abstract analysis to a problem from the right distance, practice brings us close to concrete scientific data. They go hand in hand.

This is a trend that will generally be more and more present in technical engineering, which is open to a theoretical-creative approach in robustly automated environments. This is a step toward an approach that is free from the recursive mental processes that we have learned to delegate to machines.

This will make a difference in the educational and professional experience of engineers who must develop technologies, each component of which follows an ethical approach and truly sustainable developmental principles that serve social needs.

Preface of the Authors

This textbook is the natural successor to *"Introduction to Digital Systems Design"*, by the same authors[1], which is a recommended reading.

In this book in fact, the authors offer the most open and general approach to microprocessor systems design and programming. By starting from the simplest examples and ending with medium-complexity problems, readers will acquire the theoretical-practical bases to enable them to later extend their programming knowledge to other types of microprocessors.

With this guiding aim, the text provides a brief reference on general computer concepts, then dives into the subject of "how to design" a processor using an initial "problem-solving" approach, based on knowledge of logic networks.

Starting from a simple design idea, we will build a computing network from scratch and develop it step by step into a small processor with limited, basic features, which can be programmed to perform simple tasks.

This process teaches students not only the basic architectural elements found in all microprocessors, but also how to program them. First, programs will be written in "machine language" and then, with the help of mnemonic code, in "assembly language". The resulting basic computing network will be able to take decisions, a feature at the root of all microprocessors.

At the end of Chapter 1, interested students can use the large amount of material developed to continue working independently on processor design.

After the basic concepts are introduced in Chapter 1, the aim and perspective change as of Chapter 2. Here we deal with a complete microcomputer, not in terms of a logic-gate-by-logic-gate analysis of the processor's architecture, as in Chapter 1, but on a slightly more abstract level.

---

[1] G.Donzellini, L.Oneto, D.Ponta, and D.Anguita, *"Introduction to Digital Systems Design"*, Springer, ©2019, ISBN 978-3-319-92803-6.

The focal point is the functional aspects of the microprocessor and the elements connected to it. There is a particular focus on its set of instructions and their relationship with the system components that the programmer should manage directly.

The explanation takes on two perspectives, often simultaneously. The first is that of systems engineers, who use their electronics skills to treat microprocessors as components (as supplied by the producer) to integrate them into a more complex system. The second is that of programmers who use their computer science and electronics skills to fulfill the required specifications, while making the most of the hardware available.

Chapter 3 deals in detail with microprocessor instructions from a programmer's perspective, and adds numerous examples to fully understand their behaviour. Subjects like "call and return" instructions for manage subprograms, "delay loops" and other programming techniques common in microprocessor systems.

Chapter 4 deals with so-called "interrupts". This technique, which is used in all types of microprocessors, enables the interruption of a program's normal execution flow, and the execution of another if the system components directly command it.

Thus, we will see how to interface a microprocessor system with other devices, paying special attention to the way two or more systems interact by using parallel and serial connections.

In some cases, we will use input/output components expressly designed for our purposes. This way, students will be well positioned to understand the basic concepts without spending too much time on commercial components, which while highly configurable, are often very complex, hard to deal with and ill-adapted for educational use.

In Chapter 5, the focus shifts to tests and trials on real systems. Technology has made a wide array of programmable components called FPGA[2] available. There are also myriad prototype boards based on them called "FPGA boards." These days, we can discard the idea of a system made of prefabricated components connected together since re-programmable hardware is available.

FPGA components have made it possible to quickly produce project prototypes, thus saving time and production costs. On one single FPGA chip, we can build a system including a microprocessor, memory and any accessory device we would like (within the limits imposed by the FPGA hardware).

Aside from the theoretical elements, this book offers a large number of examples and exercises with solutions to help students hone and solidify their

---

[2] Field Programmable Gate Array

understanding. They can derive great educational benefits from using the *Deeds* simulation tool[3] (Digital Electronics Education and Design Suite), developed by one of the authors, Giuliano Donzellini, to support Engineering and Computer Science students in their learning and lab work.

The close connection with *Deeds* is a strong point that renders this book unique in that schematics, programs and exercises, from the simplest to the most complex, were created with *Deeds* and are immediately available online for simulation. *Deeds* covers all the main facets of digital systems projects, including combination and sequential networks, finite state machines, user-defined components and especially microprocessor systems suitable for "assembly language" programmable "embedded" applications.

The environment supports a microprocessor created for educational purposes, the DMC8. In devising the DMC8 the creators used a technically and historically relevant 8-bit microprocessor as a model, the Z80-CPU by Zilog, but it also has a compatibility mode with the earlier I8080 by Intel. The DMC8 maintains much of the architecture of the Z80-CPU and the I8080 but its structure is simplified and a few elements have been added to bring it closer to more recent microprocessors. The DMC8 can be programmed by using the assembly language of either of the processors it derives from.

Deeds was developed with an educational and semi-professional purpose in mind so it needed to be very user-friendly while at the same time usable for complex projects. The main differences between *Deeds* and a professional simulator are the simple, direct interface and the vast collection of educational and project materials. *Deeds* is a continuously evolving, "living" system; updates are periodically available to improve the existing tools and to add new ones. The same is true of the educational materials.

The transition toward FPGA devices happens because one can export an entire *Deeds*-created and -simulated project into a professional tool and then actually test it on the hardware. *Deeds* makes it possible to avoid the complexity of the whole process, which is normally inevitable in a specific professional software. Thus access to these devices is immediate and intuitive.

---

[3] https://www.digitalelectronicsdeeds.com

Teaching Objectives

Based on the authors' experience, the whole book including the *Deeds* project exercises and simulations can be used in an introductory microprocessor systems course.

Below is a schema of the contents of each chapter. Subjects that could be omitted without sacrificing course continuity are commented with notes in *italics*:

1. Introduction to programmable computing networks
   → A general introduction to microprocessors
      The next five sections of the chapter can be omitted if the course is mainly on programming:
   → Design of a programmable computing network
   → Sequencing, microinstructions and microprograms
   → Jumps, loops and decisions
   → Input and output ports
   → Constants, variables and read/write memory

2. A system based on the DMC8 microprocessor
   → The DMC8 microprocessor
   → Bus signals and timing
   → Input/output and memory subsystems
   → Introduction to Deeds-McE

3. Programming the DMC8
   → Introduction to assembly language programming
   → Addressing modes
   → Types of instructions
   → Subprograms and the Stack area
   → Programming examples

4. Interfacing with external devices
   → Managing communication with external devices
   → Hardware-supported handshake
   → Polling
      If the course does not deal with interrupts, the next four sections in the chapter can be omitted.
   → Interrupt techniques
   → Using vectored interrupts
   → Interrupt timers
   → Examples of programming and interfacing

5. Microprocessor systems on FPGA

> *If the course does not include laboratory activities, or does not deal with FPGAs, the entire chapter may be omitted.*

→   Introduction to FPGAs
→   The architecture of FPGA components
→   FPGA development tools
→   The FPGA boards used in the examples
→   Microprocessor system prototypes on FPGA
→   Project examples

How to Use the Book

The strong connection between this book and the *Deeds* environment should encourage the reader to use it along with the simulation tools to actively test out the concepts and procedures in the textbook examples.

Another benefit is that readers get solutions to all the system design and programming exercises. Learning by doing helps students progressively build their analytical and organizational skills, which is the aim.

Digital Contents of the Book

This textbook alternates between theory and practice (examples, exercises and solutions). All the examples and exercises were created with the *Deeds* simulator, which is available at this address:

https://www.digitalelectronicsdeeds.com

The site describes the simulator and gives instructions on how to download and use it (using Windows or other operating systems with the appropriate virtual machines). The simulator is to be used *locally* so it does not require constant internet connection.

The site also has accessory material, *Deeds* schematics and programs, related to all the figures and examples in the book. Finally, the site offers all the materials needed to do the exercises and test the solutions with *Deeds*.

The site is set up with the same chapter, section and subsection titles as the textbook itself, so it is easy to use. In the future, the site will host any updates, corrections or improvements that will be made to the book.

Contents

1

Introduction to programmable computing networks

Abstract After a brief opening to the microprocessors' world, this chapter will gradually introduce the idea of building a digital programmable network able to solve different problems. The reader will be led through the basics of microprocessors, introducing step by step the microcomputer architecture's cornerstones. Several concepts, such as sequencer, ALU, registers, RAM and ROM memories, and input/output ports will be presented. At the same time, the reader will learn how to program the network exploiting, for the first time, machine codes, mnemonic codes, assembly language, and microprogramming. At the end of the chapter, the reader will be able to understand the components of more complete microprocessors.

1.1 A general introduction to microprocessors

Microprocessors are one of the most commonly used devices in digital electronics. They include the fundamental parts of a digital computer, integrated into a single component. They are the central element in personal computers, tablets, smartphones and printers, but are found in many other devices such as satellites, cars, industrial plants or wherever a large number of arithmetical-logical calculations are needed.

Digital computers help us to solve Physics problems such as how to calculate the trajectory of a missile or asteroid, or even that of an object in a virtual environment like a video game. They help solve monitoring problems such as calculating the maneuvers a satellite must execute to stay in geostationary orbit, or what a robotic arm must execute to cross the shortest distance to grasp an object. Essentially, they help with any problem that requires numerous calculations to solve and decisions based on the results.

1.1.1 A brief history of microprocessors

The production of microprocessors is possible today because of the development of low-cost, integrated circuit technology on silicon LSI and VLSI[1] chips in the 1970s. These chips made it possible to create very small (a couple mm per side) yet sufficiently complex integrated circuits. One of the best-known integration technologies is the SGT (Self-aligned Gate Transistor), which is still largely in use in microprocessor production today.

The first microprocessor made with this technology was called 4004, and was sold by Intel in 1971 under the direction of Federico Faggin[2]. The microprocessor was very limited; it worked with only one 4-bit unit of calculation. At that time, however, the novelty was that all its circuits were included on a single integrated component, reducing the cost of production.

Although the product was still very new, the potential commercial success of this type of electronic device was clear. The relatively low cost of production together with product's versatility made it the ideal component for an economy of scale.

The first commercially successful microprocessor was the Intel 8080, which could perform 8-bit arithmetic and logical operations back in 1974. In those same years other big companies designed and commercialized their microprocessors: the Z80 (Zilog), the 6800 (Motorola) and the 6502 (MOS Technology). In the ensuing years, the great potential of microprocessors, together with their low cost brought about the age of the first Personal Computers like the first APPLE II (based on the 6502) and the IBM PC, initially based on the 8088, a low-cost version of the 8086 made for industrial applications.

Over the next decades, microprocessors became ever more complex and efficient. Some examples are the Motorola 68000 (and its successors); the Intel 80286, 80386 and 80486; the whole Pentium series and the Core I7, just to name a few commercial successes. ARM microprocessors, by "Advanced Risc Machines", can be found in most smartphones today (for example the ARM Cortex-A8 chip).

---

[1] Large Scale Integration and Very Large Scale Integration. LSI has 100 to 9,999 logic gates in a single component; VLSI has from 10,000 into the millions and beyond.

[2] Federico Faggin (class of 1941) is an Italian physicist, inventor and entrepreneur who became a US citizen in 1968. At Intel, he was responsible for developing the 4004, 8008, 4040 and 8080 microprocessors. He also worked in integrated circuit technology. He developed MOS (Metal Oxide Semiconductor) technology, a prime example of which is the SGT (Self-aligned Gate Transistor), that came with the first microprocessors with connectable memory. In 1974, he resigned from Intel and founded Zilog, a company exclusively dedicated to microprocessors, and created the famous Z80-CPU. Later, he co-founded the Synaptics company (that created the first touchpads and touch-screens).

Depending upon the type, today's microprocessors can consist of hundreds of thousands or even billions of transistors integrated on one single silicon (or another semiconductor material) chip.

1.1.2 Types of microcomputers

Microprocessors and microcomputers are not the same. Simply put, microprocessors, also known as Central Processing Units (CPUs), are the central elements of microcomputers. In fact as we will see in the following, a microcomputer is a complete system that includes the microprocessor plus other devices: memory, data input/output circuits, etc. Personal computers, tablets and smartphones are all examples of microcomputers.

When all the devices of a microcomputer are integrated on a single device, it is called a single-chip microcomputer. A specific type of single-chip microcomputer is called a microcontroller. As the name suggests, a microcontroller is designed for control applications. In addition to the basic elements of all microcomputers, we find specialized input/output modules designed to control industrial plants and/or dedicated systems such as timers, Analog to Digital Converters (ADCs) and Digital to Analog Converters (DACs).

A microcomputer inserted within a more complex system is "embedded". In this type of application, the user sees and uses the whole system without directly noticing any microcomputers within it. In everyday life, for example, we find microcomputers and microcontrollers used in televisions, multimedia apparatus, home appliances, cars and video and music players.

Another type of specialized microcomputer is the Digital Signal Processor (DSP), expressly designed and optimized for signals engineering with digital techniques. Consider audio signals engineering in multimedia systems, for example the sound cards in personal computers, musical instruments, music workstations or motion picture production.

1.1.3 Microcomputers and systems

A standard microcomputer inserted into a system performs the following functions:

- acquires external data
- stores data
- processes data, produces and stores results
- based on those results, takes action within the system

Consider, for example, a microcomputer used in industrial plant control. The data are acquired through sensors or transducers in the plant (speed sensors, pressure sensors, temperature sensors, etc.); after the data are stored and processed, the microcomputer will use those results to take action in the plant through specific actuators (motors, electromagnets, hydraulic pistons, solenoid valves, electric brakes, etc.).

In principle, one could design a complex digital sequential network based on logic gates and basic memory units instead of a microcomputer, and it could perform the same functions (a "wired system"). That said, this type of system would be so complex it would barely be manageable in the design phase or in maintenance. It would be "dedicated" specifically to a particular application and its operation algorithm[3] would not be modifiable a posteriori without re-designing and rebuilding it because the way it works depends on the specific connection between its logic devices.

Aside from the aforementioned technical reasons, the production of a dedi-cated circuit can only be justified by an economy of scale or the lack of an alternative component. In any case, only very large companies can afford to produce dedicated circuits.

A microcomputer control, however, has numerous advantages; it allows for far greater efficiency and flexibility. The sequence of operations to compute is ob-tained by the designer through a program written in a language the machine understands (this refers to a programmed system). The system's architec-ture was devised to make it possible to change the sequence of operations to carry out without changing the circuits that execute them (the hardware), but rather by changing only the program (the software).

One advantage of microcomputers is precisely their general, modular type of architecture, which is adaptable to specific applications by choosing the right devices and writing the right programs.

Often, the same commercial system can be used in different applications by simply changing the program that determines how it functions. Furthermore, the standard, modular architecture generally makes it possible to keep design costs low since the designers and programmers reuse the same functionalities for a wide range of projects.

1.1.4 The basic structure of a generic computer

One of the basic features of a digital computer is its ability to execute ordered sequences of instructions rather quickly. A list of instructions is called a pro-gram and it is executed through a special internal structure made expressly to carry out instructions.

---

[3] An algorithm is a process that provides a solution to a problem in a finite number of well-defined steps. The term "algorithm" comes from the transcription into the Latin alphabet of "al-Khuwārizmī", habitational surname of Persian mathemati-cian, Muhammad ibn Mūsa, who lived in the 9th century AD. He was among the first to describe the concept of an ordered procedure to solve problems. In the theory of information, a problem is defined as computable if it can be described with an algorithm.

An instruction is an order sent to a machine that identifies a specific and limited sequence of basic operations that the machine will carry out. A computer is designed to execute a finite set of instructions. For example, one instruction might be to copy a group of bits from one memory element to another; a second instruction might be to perform an addition or to increase a number by one, etc.

Aside from the base function of carrying out mathematical calculations, another important feature of computers is that they can make decisions based on data processing results. These decisions translate into the execution of different tasks depending on the decision that was made.

The most commonly used architectures in computer production today are the Von Neumann[4] and the Harvard[5]. The Von Neumann architecture has been in use since the dawn of the computer, that is since the 1940s. It was devised for maximum circuit simplicity and can be broken down into three blocks, from a logical point of view (see the figure below).

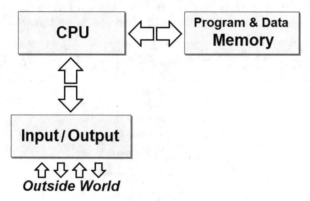

On the upper left hand side we see the Central Processing Unit (CPU); on the right, the Memory subsystem (Memory); while the lower block is the Input/Output subsystem, which allows the computer to receive data from the outside world and return the processing result to it.

---

[4] John Von Neumann (1903-1957), Hungarian mathematician, physicist and scientist, who taught Mathematical Physics at Princeton University (USA). In 1945 he published the First Draft of a Report on the EDVAC (Electronic Discrete Variable Automatic Calculator), one of the first electronic digital computers based on the binary number system, which he helped develop as a consultant. The architecture is named after him although it was initially devised by designers John Mauchly and John Presper Eckert (University of Pennsylvania), who also invented the previous ENIAC (Electronic Numerical Integrator and Computer) based on the decimal system.

[5] The term refers to the architecture of one of the first digital, electromechanical computers, the Harvard Mark I (1943), designed by Howard Hathaway Aiken (1900-1973), at Harvard University (USA).

A Von Neumann machine is based on the following base criteria:

— there is only one processing unit (the CPU)
— only one instruction is executed at a time (within the CPU)
— the memory contains both the data that the computer works on and the programs that process that data.

The CPU plays a fundamental role, in that its job is to take the instructions that make up the program from the memory and execute them one by one. The calculation of one instruction can correspond to one basic calculation, an addition for example, or an exchange of data with the memory or the outside world.

The whole system will execute only the tasks required by the program (and by extension, the programmer who wrote it), no more, no less. It does nothing autonomously, nothing that doesn't derive from the individual instructions.

The Harvard architecture is a variation on that of Von Neumann, and it differs on the last criterion. Its data and programs reside in different memory subsystems and are separately accessible (see the figure below).

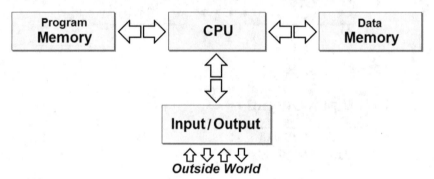

In the Von Neumann machine, the CPU is required to execute memory access instructions in sequence, whereas the Harvard architecture makes it possible to get program instructions and access data at the same time. The more expensive Harvard architecture is cost-effective when the size of the data is similar to that of the programs.

The most commercially available computers are based on the Von Neumann architecture. Some microcontrollers and digital signal processors use the Harvard model.

1.1.5 The common bus connection

A microcomputer based on Von Neumann criteria is normally organized around one single common bus (see the figure below). A bus is a bundle of parallel wires shared among the different system units. Its name is related to the *bus* of public transportation and in fact, these wires transport information from one unit to another.

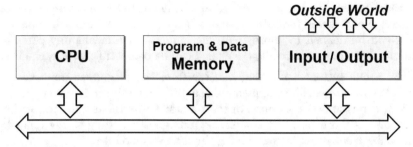

This type of connection is used mainly because it is economical and modular. The figure shows the three blocks mentioned previously. Communication happens among them through the bundle of wires (the bus) that runs through them. Only one unit at a time can use these wires to send information on the bus, while it is possible for the other units to receive it simultaneously.

The unit that transmits data at any point is called the "talker", while the receivers are "listeners". The unit that manages who can use the bus wires at any point is the "master" (or "arbiter"), while the others who obey are the "slaves". There are "multi-master" systems, but here we will limit ourselves to systems with the CPU as the only master.

Let's look at the advantages and disadvantages of a bus connection. All the subunits of the system are connected by the same bundle of wires making the bus a shared resource. Clearly, all the devices sharing the same bundle of wires may pose a functional limitation: if two subunits are engaged in dialog, the bus is unavailable to any other subunit.

If we connect the individual units through reserved connections we get a potentially much quicker "star" system (a fully connected system). The figure below shows a generic star system.

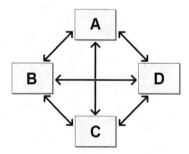

Nonetheless, this is not in principle, a severe limitation for a microcomputer with Von Neumann architecture since only one operation will be executed at a time. What is more, a bus connection is far more economical than a star system, which would require many more wires for the dedicated connections.

An advantage of the bus connection is that it is modular. For example, one could add new memory or input/output elements if needed, just by connecting them to the bus without having to add dedicated connections. Another advantage of it being modular is that a bus can be standardized, that is its specs can be defined by more than one builder. This means any computer add-ons that are commercially developed can be compatible with each other.

Lastly, a bus system has added benefits from the diagnostic point of view for both the project development phase and maintaining a system in use. In fact, a bus is easier to observe than a star system; one can troubleshoot malfunctions more easily by connecting a diagnostic system to the bus wires and seeing what is exchanged among the different modules.

1.2 Design of a programmable computing network

The introduction on general terms and concepts of the world of microprocessors is now completed. Then reader can begin step by step to learn the
fundamental components and mechanisms that make up microprocessors and
allow them to run.

In this chapter, we will take the point of view of digital network designers who
must build a programmable computing network from components and their
own basic skills. We begin with a simple, introductory design specification
and make it more and more versatile until a sort of precursor to a 8-bit
microprocessor is reached.

This is a bottom-up approach where readers (helped by the authors) take the
role of a designer who needs to create a programmable network that can solve
a vast array of problems, without having to redesign the same circuit each
time.

As of Chapter 2, we will analyze the complete and tested architecture of a
microprocessor. The point of view adopted will be different; the goal will no
longer be to study the logical working of the network in minute detail. This
time, the explanation will focus on programming techniques and on the design
of systems based on microprocessors intended as component.

1.2.1 The design specification: a dedicated computing network

Let's imagine we have been asked to design
a digital network that can calculate the average of four whole positive numbers (OP3, OP2,
OP1, OP0), less than 64 (6 significant bits).

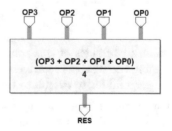

The result RES will also be a 6-bit whole number (see the figure at the right).

1.2.1.1 Combinational solution

Imagine a purely combinational circuit that calculates the average by first
adding the four operands inputted through a cascade of adders (see the following figure).

We can divide by 4 through a scaled output connection (scaling an output
two positions to the right reduces the weight of each bit by a factor of 4). In
the calculation sample in the figure, we get 5 as an average of 3, 2, 9 and 6.

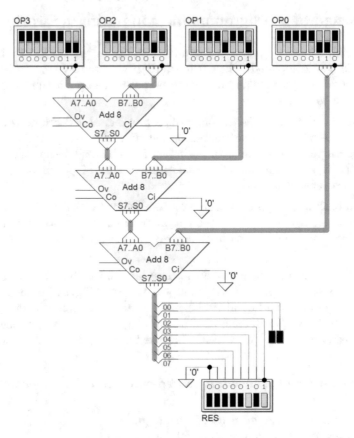

Some advantages of this solution are that it doesn't require a clock and processing the result is relatively quick (it takes at most 3 times the time it takes for each adder to calculate the sum).

On the other hand, the network is difficult to reconfigure. For example, if one needed to add more operands in the input or raise the number of bits of the operands, the circuit would need to be redesigned and remade, which would add to costs.

If we needed to have one single network model that was able to be reused in a wide range of cases, we could design an oversized circuit equipped with a large number of operands and more bits.

Nevertheless, this would be underutilized in many applications and we would be using a network that would be too expensive for the specifications. For other applications, it might not be enough and this would require us to redesign it. If the level of complexity is decided a priori, the network will always be very specialized.

1.2.1.2 Sequential solution

To get a more versatile and reconfigurable solution, we can assume we'll use one single adder with a parallel register. With only one adder available, we must execute sums one by one as we memorize the partial results in the register.

The clearest advantage of this approach is that the number of operands to add is independent of the number of adders in the circuit. The figure below shows the circuit schematic.

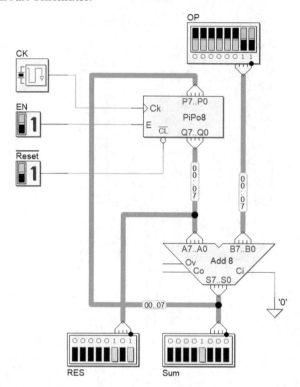

To add an operand, we show its value at input OP and enable the register for one clock cycle CK (by activating EN). Clearly, first we have to clear the register so that we add zero to the first operand.

Doing all these operations manually is obviously cumbersome so we will need to add a "controller" to our arithmetic network (the "datapath") to automatically sequence the operations. It is also inconvenient to read the operand on only one input. Also the network still cannot execute the final division.

Adding these observations to the initial specifications, we can complete our datapath schematic, as shown in the figure below.

As we can see, the operand to add at a particular moment (OP3, OP2, OP1 or OP0) is chosen by a 4-channel multiplexer[6].

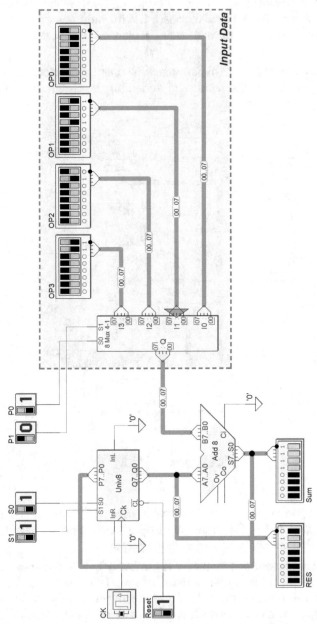

---

[6] Notice that if we wanted to raise the number of operands, we shouldn't change the arithmetic part but rather add a larger multiplexer. This is valid until the summation of operands no longer exceeds the maximum number expressible by the arithmetic part.

To make the schematic easier to understand the data acquisition subsystem in the input is enclosed in a frame marked "Input Data". Channel selection is controlled by P1 and P0. At the point of the simulation captured in the figure, lines P1 = '0' and P0 = '1' make it so that operand OP1 is routed by the channel selected by the multiplexer at the adder input. See the arrow in the figure.

The parallel register has been replaced by a "universal register", whose functioning is controlled by lines S1 and S0. This way, we can order the register to shift its contents two positions to the right at the end of the addition operations. The table below shows the functions obtained by the universal register depending on the values of S1 and S0.

S1	S0	Function
0	0	Content does not change
0	1	"Right" shift
1	0	"Left" shift
1	1	Parallel loading

In the simulation in the figure, we can see that the number '00000110' is in the register. The adder adds this number to the one from operand OP1, '00001001', since it was chosen by the setting of P1 and P0. We can see the result in the output: '00001111'.

At the next rising edge of the clock CK, S1 and S0 being set to '1' means that the output from the adder will be memorized in the register. The new partial result will replace the previous one.

After defining the network datapath, now we need to take care of the automatic sequencing of lines S1, S0, P1 and P0. So let's add a Finite State Machine (FSM) controller, which gives us the definitive schematic. See the following figure.

Moreover, the network has an added END output that comes from the controller and signals the end of a calculation.

In summary, the universal register is used in this architecture to memorize the intermediate results of a sum and to double shift them to the right.

From now on, we will refer to this register as "Accumulator", since it can memorize (accumulate) the partial results of calculations. Also, the output that visualizes the content will be called ACC.

The FSM functions as a system "Controller", also known as "Sequencer". That is it produces the correct sequence of signals to send to the other components of the "Datapath" that process the data to execute the task (in our case, calculating the average).

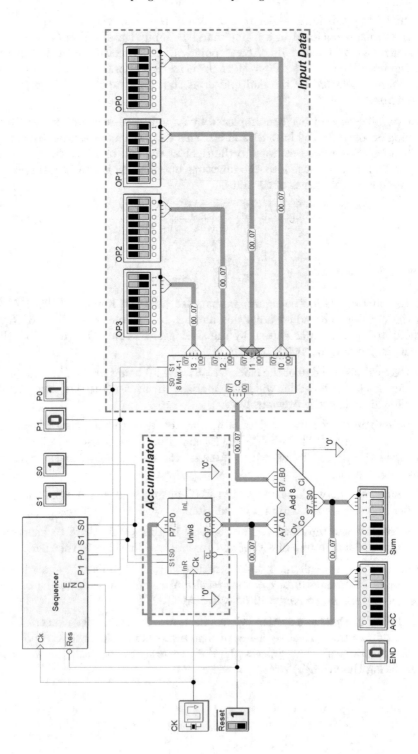

1.2.1.3 The sequencer algorithm

The Algorithmic State Machine (ASM) chart at the right describes how the algorithm of the sequencer works.

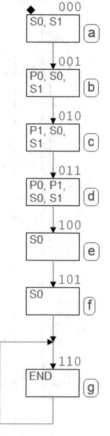

In state (a) the lines S1 (='1') and S0 (='1') order to load the data available on the adder output into the accumulator. It is the same as the sum of operand OP0 (selected by the multiplexer setting P1='0', P0='0') and the content of the register itself, which was cleared by the. $\overline{\text{Reset}}$ before (in other words, the value of the operand is loaded in register OP0).

In state (b) the register is ordered (S1='1', S0='1') to load the value from the adder, which is the same as the sum of operand OP1 (selected by the multiplexer setting P1='0', P0='1') and the register itself, which as we have seen is the same as operand OP0.

In state (c) the accumulator (S1='1', S0='1') is called on again to memorize the value generated by the adder. In this case, the sum of operand OP2 (choosing P1='1', P0='0') is the current content of the register (the sum of operands OP0 and OP1).

In state (d), we maintain S1='1' and S0='1', and order the register one last time to load the output from the adder, which is the same as the sum of operand OP3 (choosing P1='1', P0='1') and the content of the register itself, that is the sum of operands OP0, OP1 and OP2. After this final operation, the total sum of of the 4 operands will be in the register.

In states (e) and (f) we order the register to right shift its content twice and to insert a zero to the left side thus dividing by four.

In state (g) the END output is activated to indicate the "end of process", and the FSM stands by until it is reset.

1.2.1.4 Simulation

Timing diagram simulation allows us to examine the evolution of signals as of the initialization of the system. See the figure below.

We can see the change of control signals S1, S0, P1 and P0 at each rising edge of the clock CK. The result of the (Sum) is loaded onto the register at edges 2, 3, 4 and 5. On edges 6 and 7, the contents right shift. Finally, the END output indicates that computing has ended (RIS = 5).

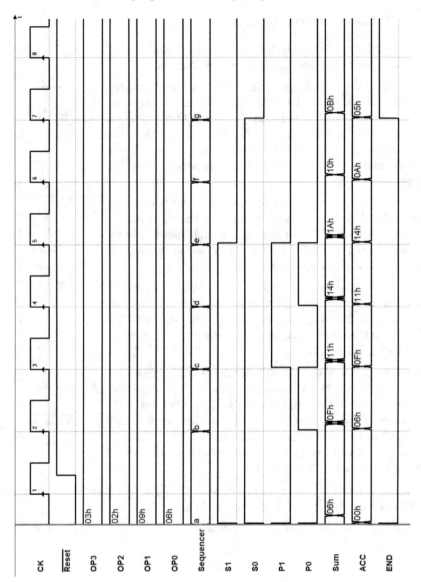

1.2.2 Counter and ROM memory based sequencer

A much more general type of sequencer is made up of a counter addressing a ROM memory (see Appendix A.1) that contains the sequence of control signals to emit. See the figure below.

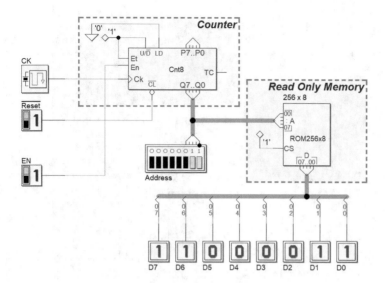

This type has a great advantage: there is no need to resynthesize the FSM each time we want to change the sequence to execute. All that is needed is to change the bits in the ROM.

In other words, the "physical" hardware of our circuit can remain the same (it won't need to be redesigned or rebuilt); we only need to change the contents of the memory.

For our network, we have chosen to use an 8-bit binary, universal counter that counts forward, starting from zero when it is reset. Line EN lets us start and stop the count.

With 8 address bits we can read the 256 ROM locations. Memory addresses also have 8 bits so we can connect at most 8 bits of control line to the datapath (denoted in the figure as D7, D6, ..., D0).

As an example, in the figure above, the counter has reached number "00000011", during the simulation. The memory gives the value "11000011", which is contained in the location at that address.

So let's replace the FSM with the new sequencer and get the schematic in the figure below. Only 5 of the 8 output lines have been used from the ROM, P1 and P0 at bits D7 and D6, S1 and S0 at bits D1 and D0. We will only use D4, D3 and D2 in the next example.

The datapath is identical to the one in the previous network except for a small modification on line END. As we can see, line END was brought back to the counter by a NOT, so that when it is activated, it will make it possible to disable the counter and stop the sequence.

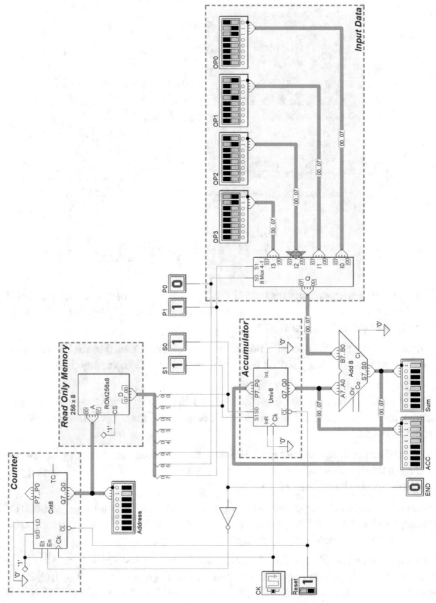

The ROM contains the ordered sequence of control signals to send to the components. In this version, we want it to be identical to the one generated by the previous circuit's FSM.

When the system is reset, the counter is initialized to zero so the memory will send the commands that are held at the zero address. So, this location must contain the signals emitted by the previous version's FSM in state (a), the reset state. At the next rising edge of the clock the counter increments by one and sends out the contents of the second location, which should contain the signals that the FSM emits in the second state...and so on for the rest of the locations.

By following this approach, we complete the programming of the memory locations and get the result shown in the table below. Next to each ROM location, we see the corresponding state of the previous FSM. We are not interested in the other memory locations so they can contain any code since our network doesn't address them.

		ROM Contents								FSM
Address	(Hex)	P1	P0	END	x	x	x	S1	S0	State
00000000	00h	0	0	0	0	0	0	1	1	(a)
00000001	01h	0	1	0	0	0	0	1	1	(b)
00000010	02h	1	0	0	0	0	0	1	1	(c)
00000011	03h	1	1	0	0	0	0	1	1	(d)
00000100	04h	0	0	0	0	0	0	0	1	(e)
00000101	05h	0	0	0	0	0	0	0	1	(f)
00000110	06h	0	0	1	0	0	0	0	0	(g)
00000111	07h	...	...	...	...	...	...	...	...	...
00001000	08h	...	...	...	...	...	...	...	...	...

Our network is growing more complex but also more versatile. By introducing ROM memory in the sequencer, we have made the behavior of the circuit more easily redefinable.

1.2.3 Extending computing possibilities

Having only the adder available doesn't offer many degrees of freedom in using the system, but when new components are introduced, the circuit's possibilities begin to get more interesting. Consider the option of replacing the adder with a combinational network that can execute different operations like adding, subtracting, logical operations, etc.

If we add computing options, we can define much more versatile operation sequences. Logic networks that can execute different mathematical and logical operations are called Arithmetic Logic Units (ALUs). The table to the right shows a possible set of executable functions.

Function
A+B
A - B
A and B (bitwise)
A or B (bitwise)
not A
not B
A
B

Let's design an ALU that carries out this set of functions. The figure below shows our result.

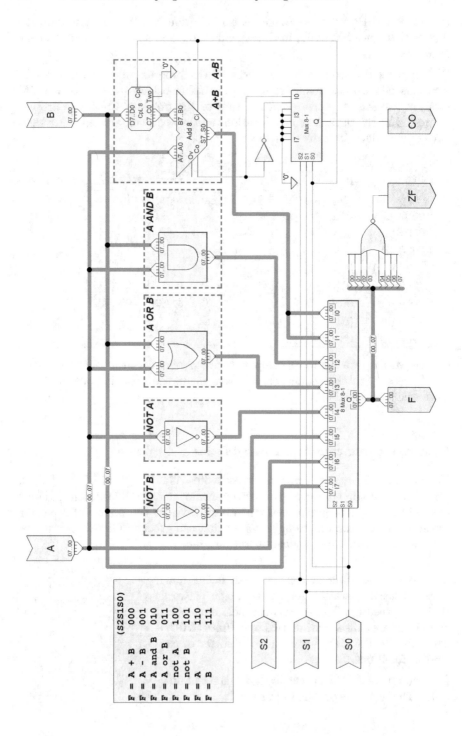

The schematic is made up of arithmetic and logic circuits each of which executes one of the functions listed above on the input operands. The individual circuits' outputs merge into an 8-channel selector that allows us to choose the operation we want at output F in function of inputs S2, S1 and S0.

Notice that the network is purely combinational. The table below shows the ALU's functions depending on its control inputs S2, S1 and S0.

S2	S1	S0	Function
0	0	0	A+B
0	0	1	A - B
0	1	0	A and B (bitwise)
0	1	1	A or B (bitwise)
1	0	0	not A
1	0	1	not B
1	1	0	A
1	1	1	B

In the schematic, we find a complementer component across from the adder. When we subtract A-B, this is calculated as $A+C_2(B)$. Two's complement is calculated by inverting all the bits of operand B and adding '1'. We add '1' by setting the adder's input (Ci) equal to '1', in the case of subtraction.

There are also two auxiliary outputs called ZF (Zero Flag) and CO (Carry Output). ZF signals when the result is zero, while CO activates whenever the circuit generates a carry after adding (or a borrow request in the case of subtraction).

The auxiliary outputs are called "Flags" and are very useful when evaluating the result of processing, as we will see ahead. For example, when subtracting, we can establish if the two operands are equal and if not, which is greater than the other.

We can do this by assessing the value of the flags after the operation. See the table on the right. If the zero flag ZF is activated, the operands are equal. If not, the carry output CO is active only if A < B. This convention is used in Intel processors, for example.

Flags		
Result	CO	ZF
A = B	0	1
A > B	0	0
A < B	1	0

In commercial processors, there are also other types of flags such as the Parity Flag PF that indicates if the number of bits at '1' in the ALU result is even or odd. Here, we will only consider the basic flags.

1.2.4 ALU-based computing networks

The figure below shows a very similar network to the one shown above, but
this one has an ALU rather than a simple adder.

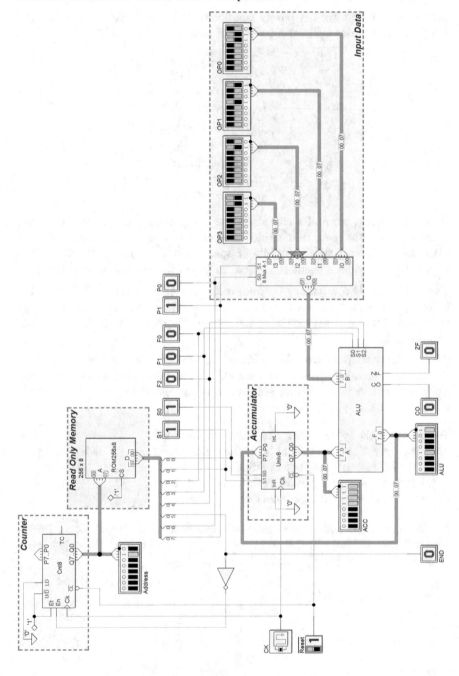

Also, the schematic differs from the previous one by the fact that the ALU set-up wires are connected to the ROM output lines. The table below shows a summary of the connections between the 8 ROM lines and the control lines used in the datapath.

D_{07}	D_{06}	D_{05}	D_{04}	D_{03}	D_{02}	D_{01}	D_{00}
P1	P0	END	F2	F1	F0	S1	S0

The new network can calculate the average of the four values with the same bit setup as in the previous version's ROM, with no modifications.

ROM Contents

Address	P1	P0	END	F2	F1	F0	S1	S0	(Hex)
00h	0	0	0	0	0	0	1	1	03h
01h	0	1	0	0	0	0	1	1	43h
02h	1	0	0	0	0	0	1	1	83h
03h	1	1	0	0	0	0	1	1	C3h
04h	0	0	0	0	0	0	0	1	01h
05h	0	0	0	0	0	0	0	1	01h
06h	0	0	1	0	0	0	0	0	20h

In the previous version in fact, we had forced the unused lines to zero whereas now they (here renamed F2, F1 e F0) control the ALU's function and the configuration '000' corresponds precisely to the function of the sum of the two operands.

1.2.4.1 Another computing example

Here, let's recap what our logic network is capable of in its current configuration:

1. carry out up to 256 operations in sequence
2. carry out 8 arithmetical and logical operations, such as adding, subtracting and bitwise (AND, OR, NOT) logical operations
3. generate ZF and CO flags to give the user an evaluation of the operands and the result (the result of these evaluations is not yet reusable by the network itself in this version).

Now, let's define another computing example. We want a sequence of operations that evaluates if the result of the sum of operands OP0 and OP1 is greater than equal to or lesser than operand OP2.

The execution of the sequence can be divided into 3 steps:

1. adding operand OP0 to operand OP1
2. subtracting operand OP2 from the previous result, generating flags ZF and CO, and stopping the computing sequence
3. evaluating the flags.

Our network can carry out the first two steps if the ROM memory is programmed appropriately. The last step must be carried out by an end user or another logic circuit.

— P1 and P0: govern the selection of input operands OP3..OP0
— END: is used to stop the sequencing
— F2, F1, F0: define the operation of the ALU;
— S1, S0: control the function of the universal register.

Carrying out step 1) means: selecting operand OP0 through the multiplexer, setting P1='0', P0='0'; commanding the ALU to copy the operand OP0 to its output (F=B) and therefore setting F2='1', F1='1' and F0='1'; commanding the accumulator to update its contents with the result coming from the ALU (in other words, operand OP0), setting S1='1' and S0='1'. Obviously, we must set END='0'. Thus we end the coding of the first location of ROM memory:

P1	P0	END	F2	F1	F0	S1	S0	Description
0	0	0	1	1	1	1	1	A ← OP0

Then, we select operand OP1 making P1='0' and P0='1'; command the ALU to add the contents of the accumulator to operand OP1, defining F2='0', F1='0' and F0='0'; and force the update of the content of the accumulator with the result from the ALU (S1='1', S0='1'). The next ROM location will then have the following values:

P1	P0	END	F2	F1	F0	S1	S0	Description
0	1	0	0	0	0	1	1	A ← (A + OP1)

At step 2), to subtract operand OP2 from the result in the accumulator we must command the multiplexer to select operand OP2 (P1='1', P0='0') and to the ALU to subtract (F2='0', F1='0', F0='1'). Below we see the contents of the third location:

P1	P0	END	F2	F1	F0	S1	S0	Description
1	0	1	0	0	1	0	0	(A - OP2)

To stop the sequence at the end of this operation, we set END=1.

Note that the result in the accumulator is not set to load (S1='0', S0='0'). This is because the operands from the subtraction at the ALU input must remain stable until we can see the value taken by the flags. The table below shows the final evaluation (step 3).

Final Evaluation

Result	CO	ZF
(OP0 + OP1) = OP2	0	1
(OP0 + OP1) > OP2	0	0
(OP0 + OP1) < OP2	1	0

Here below is a summary of the codes to insert in the ROM.

ROM Contents

ROM Address (Hex)	P1	P0	END	F2	F1	F0	S1	S0	Description
00h	0	0	0	1	1	1	1	1	A ← OP0
01h	0	1	0	0	0	0	1	1	A ← (A + OP1)
02h	1	0	1	0	0	1	0	0	(A - OP2)

1.2.5 The "instructions"

Since the machine carries out the operations that we put into the memory, from now on we will call the operations the network can do "instructions".

Defining the instructions in the ROM bit by bit can be a long, boring job easily subject to errors. To make this easier, we'll introduce mnemonic codes that we'll associate to every possible instruction.

This way it will be easier to define the contents of the ROM. We will simply express the algorithm that we want to perform in terms of mnemonic codes, which can be easily translated into the corresponding binary code (also known as "machine code").

The writing phase of the mnemonic codes will be supported by a text editor, and the translation into machine code will be done later.

Going back to the example above, shifting operand OP0 into register A was done using the following code:

P1	P0	END	F2	F1	F0	S1	S0	Description
0	0	0	1	1	1	1	1	A ← OP0

This machine code could be associated to this mnemonic:

$$\text{IN} \quad \text{A,OP0}$$

where the IN refers to "Input", getting an operand from outside; the A, which stands for accumulator register, is the destination of the operand, and the OP0 after the comma represents the source of the operand. This use of mnemonic codes is common in technical manuals for microprocessors.

Naturally we need to define a similar code for the remaining operands. See below:

$$\text{IN} \quad \text{A,OP0}$$
$$\text{IN} \quad \text{A,OP1}$$
$$\text{IN} \quad \text{A,OP2}$$
$$\text{IN} \quad \text{A,OP3}$$

These four codes can be succinctly expressed by the following code:

$$\text{IN} \quad \text{A,P}$$

where P represents one of the four possible inputs. Coding bits by using these four mnemonics is quicker if we indicate bits p_1 and p_0 that control the multiplexer:

Mnemonic Code	P1	P0	END	F2	F1	F0	S1	S0
IN A,P	p_1	p_0	0	1	1	1	1	1

The same reasoning can be applied to addition: a good code to represent addition is the word ADD. In our network, the only possible destination of the result is the accumulator register (this goes for all the operations the ALU carries out) so A will be the second part of the mnemonic code.

Given that there are 4 possible operands for this instruction as well, the last part of the mnemonic code, after the comma, is P. So we get this mnemonic:

$$\text{ADD A,P}$$

represented by the binary code below:

Mnemonic Code	P1	P0	END	F2	F1	F0	S1	S0
ADD A,P	p_1	p_0	0	0	0	0	1	1

Following the same reasoning, we'll associate a code for each of the operations the ALU carries out, giving us the following mnemonics. Please note that if the operand coincides with the destination of the result, only the destination is indicated.

Mnemonic Code	P1	P0	END	F2	F1	F0	S1	S0	Description
IN A,P	p_1	p_0	0	1	1	1	1	1	A ← P
ADD A,P	p_1	p_0	0	0	0	0	1	1	A ← (A + P)
SUB A,P	p_1	p_0	0	0	0	1	1	1	A ← (A - P)
AND A,P	p_1	p_0	0	0	1	0	1	1	A ← (A and P)
OR A,P	p_1	p_0	0	0	1	1	1	1	A ← (A or P)
NOT A	0	0	0	1	0	0	1	1	A ← NOT(A)

Also note that when the NOT A instruction is being executed the system ignores the first two bits. For simplicity's sake, we have decided to set it to 0, but any other value would have worked as well (this method will also be adopted in the following).

What we have examined so far are not the only instructions that the network can execute. Consider for example right shifting the contents of the accumulator by placing a 0 to the left, indicated in the row of bits below:

Mnemonic Code	P1	P0	END	F2	F1	F0	S1	S0
SRL A	0	0	0	0	0	0	0	1

The first 6 bits could take any value since the network ignores them while carrying out the instruction.

This operation is very useful for dividing by two and can be denoted with the initials SRL which indicate Shift Right Logic.

As for the shift instructions, the word "Logic" is usually used to indicate inserting a zero on the left. The word "Arithmetic" however, is used in relation to right shifting when the most significant bit itself (which represents the sign) is inserted on the left[7]. The result of the shifting operation is done in the accumulator and the only possible operand is contained there, so we have chosen the mnemonic SRL A.

Similarly, we will use the mnemonic SLL A to denote the left shift instruction, that will enter a '0' on the right:

Mnemonic Code	P1	P0	END	F2	F1	F0	S1	S0
SLL A	0	0	0	0	0	0	1	0

---

[7] Since our circuit can only insert zeroes in the serial inputs (InR and InL) of the universal register, arithmetic shifting operations will not appear here. In any case, it is useful to understand this distinction.

It is useful to formalize the operation that leaves the result in the accumulator unaltered. Let's call this operation NOP ("No Operation"). See the corresponding configuration of bits below:

Mnemonic Code	P1	P0	END	F2	F1	F0	S1	S0
NOP	0	0	0	0	0	0	0	0

Notice that the first 2 bits and those related to the ALU could take any other value since they have no influence on the contents of register A.

Finally, we have already used the END bit to stop the execution. A variation on the NOP is shown below, in which the network does nothing, but the sequencer also stops; we'll call it HALT.

Mnemonic Code	P1	P0	END	F2	F1	F0	S1	S0
HALT	0	0	1	0	0	0	0	0

The following table sums up the mnemonic codes:

Mnemonic Code	P1	P0	END	F2	F1	F0	S1	S0	Description
IN A,P	p_1	p_0	0	1	1	1	1	1	$A \leftarrow P$
ADD A,P	p_1	p_0	0	0	0	0	1	1	$A \leftarrow (A + P)$
SUB A,P	p_1	p_0	0	0	0	1	1	1	$A \leftarrow (A - P)$
AND A,P	p_1	p_0	0	0	1	0	1	1	$A \leftarrow (A$ and $P)$
OR A,P	p_1	p_0	0	0	1	1	1	1	$A \leftarrow (A$ or $P)$
NOT A	0	0	0	1	0	0	1	1	$A \leftarrow NOT(A)$
SRL A	0	0	0	0	0	0	0	1	$A \leftarrow (A$ div $2)$
SLL A	0	0	0	0	0	0	1	0	$A \leftarrow (A$ per $2)$
NOP	0	0	0	0	0	0	0	0	No Operation
HALT	0	0	1	0	0	0	0	0	Sequencer Halt

As an example, let's look at the table below showing the algorithm (discussed previously) that calculated the average of numbers, and juxtaposing its machine code with the mnemonic equivalent.

Clearly, the mnemonic code offers greater comprehension and readability.

Mnemonic Code		Machine Code							
ADD	A,OP0	0	0	0	0	0	0	1	1
ADD	A,OP1	0	1	0	0	0	0	1	1
ADD	A,OP2	1	0	0	0	0	0	1	1
ADD	A,OP3	1	1	0	0	0	0	1	1
SRL	A	0	0	0	0	0	0	0	1
SRL	A	0	0	0	0	0	0	0	1
HALT		0	0	1	0	0	0	0	0

The full table of instructions with all the variants and the network schematic are available in Appendix B.1 (the network is called Mp8A).

1.2.6 "Program", "programming" and other important terms

We have seen that instructions in mnemonic code are useful to denote the basic operations a machine is asked to carry out, in a reasonably readable format. As of now, we will call the sequence of instructions describing a sequence of operations (an algorithm) to execute, a "program".

We will "program the system" in two phases: in phase one we write a text file with a sequence of instructions in mnemonic code that describes the algorithm that we want to obtain. In phase two, we translate the mnemonic code into machine code to insert into the ROM. Taking into account its function, the ROM is usually called Program Memory.

As mentioned above, phase one is supported by a text editor, while phase two requires the use of a mnemonic code-machine code translator (or "compiler") (this type of application is called an "assembler").

As we have seen before, the instruction address to be executed is generated by the counter inside the network sequencer. As of now, that counter will be called by its proper name, "Program Counter" (PC), in that it indexes the individual program instructions, so that they are executed in order, one by one. In technical literature, the term Program Counter is not the only way to refer to this; the term "Instruction Pointer" (IP) is also in use.

1.3 Sequencing, microinstructions and microprograms

In this section we will change the sequencer so that our processing system will be able to execute more complex tasks. We will introduce new elements such as the microinstructions, the microprogram memory and the instruction pipeline.

1.3.1 A more compact sequencer

The network designed in Section 1.2 can be developed further. So far, it can execute only about 10 instructions. It has yet to gain the ability to carry out different sequences of instructions based on intermediate calculation results, which is fundamental for any processing system. For example, in the control system of a heating plant, if the sensor measures a temperature lower than what is desired, a sequence must be carried out to turn the heater on, if the temperature is already at the right level, no action is taken.

Furthermore, to make the processing network more versatile, we will add many elements to the datapath. The network and the number of lines required to control it will become more complex. With the current sequencing structure, introducing a new line of control requires adding another bit to the machine codes.

Please note that, when writing a program, we will use the same codes again and again and as they grow in size, we will need much more program memory. Given these necessary changes, it might be useful to rethink the system's current sequencing network (see the figure below) and add an instruction coding system.

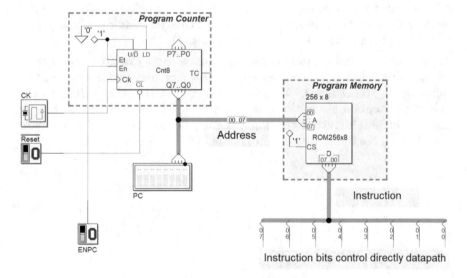

Instruction bits control directly datapath

In the figure above, we have separated the sequencer from the rest of the network. Our first goal is to review its structure to be able to use the most compact machine codes. In this current configuration, the Program Counter (PC) addresses the Program Memory (PM). In a combinational way, the PM returns the instruction code found at the address. The address the PC generates is incremented on the active edge of the clock CK if the enable signal of the program counter (ENPC) has been activated.

In this structure, the instruction code is represented by the set of bits required to directly control the elements of the datapath. As said before, if we keep this setting and raise the elements in the datapath, the number of bits per memory location raises correspondingly. We need to achieve code compaction of the individual instruction codes by choosing the appropriate criteria.

The figure below shows a different instruction decoding network. As you can see there is one more element added vis-a-vis the previous structure: a new ROM memory at the bottom right-hand side.

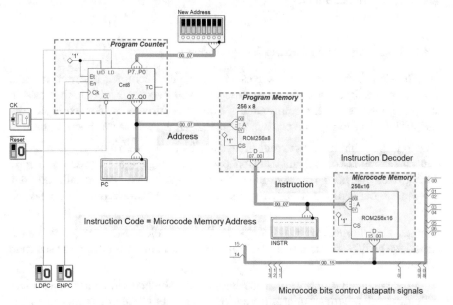

In this new version, the machine code supplied by the PM is no longer defined as the combination of control signals to activate in order to carry out instructions, but as an address that selects a location stored in the second ROM. This is what contains the control signals to activate.

From now on, we'll call the set of control signals related to a certain instruction a "Microcode", and the ROM containing it "Microcode Memory", (MCM'). The network designer will have previously memorized all the microcodes in the MCM. The machine codes for the instructions themselves (those that the programmer will insert in the PM) will actually be the microcode addresses.

Obviously, the network designer needs to have documented all the available instructions along with their codes for the programmer to use.

Since a memory location address doesn't depend on the length of its content, this new version will help us make the length of the machine code independent of the number of control lines. In our example the MCM has 256 locations at 16-bits each. This means we can control up to 16 datapath control lines even though we use only 8-bit instruction codes.

Notice that the MCM acts as an instruction decoder and so it could be replaced by a specially designed combinatorial network. The ROM memory-based solution, however, is more practical since it allows us to express the microcodes explicitly.

Lastly, a new control line (LDPC) has been added, and is connected to the PC load command (LD). It will be used in Section 1.4 to load a new value inside the PC, as a way to vary the order of instructions execution. For now, however, it remains inactive.

1.3.2 The microprogrammed sequencer

The solution discussed here resolves the problem of compact instruction code but still has a limitation. Some instructions will need multiple clock cycles to complete, because of the component's timing requirements. For example, the instructions to modify the contents of a PC, as we will see in Section 1.4.

This is impossible in the current structure; just one control signal configuration corresponds to each instruction code, and is generated over one single clock cycle. We need to make it so they are generated sequentially so that different signal configurations are presented to the network in successive clock cycles.

To make this happen, we make a change that will allow us to carry out the instructions as if each one was a small program made up of multiple consecutive microcodes.

To get a better distinction of what happens in the general context, let's call individual microcodes "microinstructions" and sets of microinstructions that define the behavior of an instruction "microprograms". We call this new structure a "microprogrammed sequencer".

The change we need to make consists of adding another counter before the previously introduced MCM. We get a structure where the instruction machine code taken from the PM is loaded in the new counter rather than to address directly the MCM.

Let's call this new element "Microprogram Counter" (MPC), and the ROM previously added, that from now on will contain the microprograms "Microprogram Memory" (MPM). See the following figure.

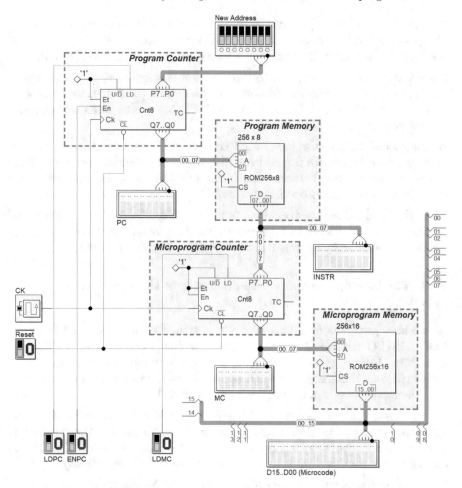

In this new sequencer, the machine code no longer corresponds to an individual microcode address but represents the address of the first microinstruction of the microprogram to execute. After the MPC counter addresses the MPM's first microinstruction, it is incremented so that it targets the microprogram's next microinstructions. The upper part of the network, made up of the PC and the PM supplies the instructions. The lower part (MPC and MPM) carries out the corresponding microprograms.

A certain instruction's microprogram is definitively formed in the design phase and is never changed again unless there are later corrections to make or an expansion to do.

We can control all the elements of a calculation network with MPM outputs $D_{15}..D_0$. From a functional perspective, the MPM acts on the network in the same way as the MCM seen previously.

As visible in the lower part of the schematic, the MPC counter is forced to always count forward unless the LDMC line forces it to load. When it is active, the instruction code in the program memory output is loaded in the counter.

If we always ordered the MPC to load, we would simply get the same functioning as the previous network. Once the instruction code is loaded, it is still possible to deactivate LDMC and get the MPC counter to increment at each successive rising edge of the clock.

As mentioned before, the instruction code in the program memory of this new structure represents the address that the microprogram (corresponding to the instruction itself) starts from. In the simplest cases, the microprogram is made up of only one microinstruction.

The MPC and MPM together do a "sequential decoding" of the instruction retrieved from the PM in that they make all the control signals necessary for execution available in the order dictated by the microinstructions.

1.3.3 The microprogrammed sequencer and the computing network

As we have seen, the functions of the new sequencer are controlled by lines LDPC, ENPC and LDMC. We will make it so that the sequencer itself manages them. Let's connect the microprogram memory outputs so that the microinstructions themselves decide if and when to increment or load the PC, load the instruction in the MPC or force it forward.

Let's consider the calculation network examined in Section 1.2, and replace its sequencer with a microprogrammed one. Here, we've divided the schematic into two parts with a dotted red line to make it easier to read. The whole schematic is shown in Appendix B.2.2 (the network is called Mp8B).

The next page shows the part with the datapath with some minor adaptations on the connections to the new control lines. The page after next shows the part with the microprogrammed sequencer where line LDMC has been connected to bit D15 of the MPM and line ENPC to bit D14 so that the microcode can directly force the PC forward (or not) by means of the ENPC, and/or load or advance the MPC by means of the LDMC.

For now, we don't need to reload the PC with a new value, which is why we've eliminated line LDPC from the schematic making the counter load command inactive (we will use this line next, in Section 1.4). Notice that there are control lines that haven't been used yet but they will be in the networks coming up next.

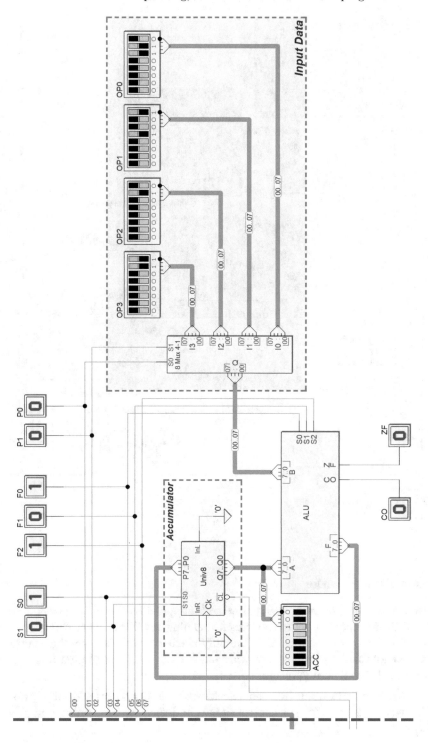

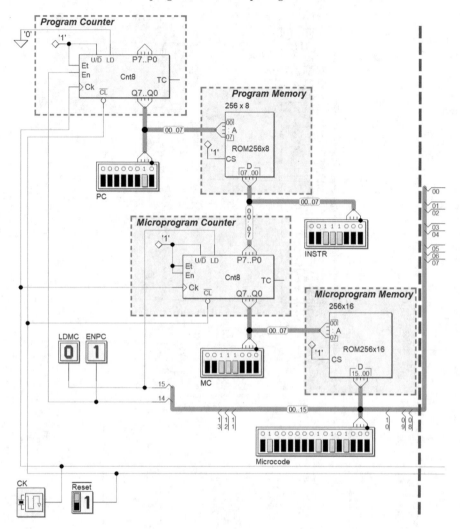

1.3.4 How it works

As we have seen, once the instruction code supplied by the PM is loaded in the MPC, it becomes the source address for reading the MPM. Therefore, in order to set out a specific instruction, we only need to write the microinstructions in that location and in the following ones. Each microinstruction must have a tailored signal configuration.

In order to fully examine how sequencers work, let's look at the Mp8B network. The table below shows the control signals and their corresponding microcode bit words.

D_{15}	D_{14}	D_{13}	D_{12}	D_{11}	D_{10}	D_{09}	D_{08}	D_{07}	D_{06}	D_{05}	D_{04}	D_{03}	D_{02}	D_{01}	D_{00}
LDMC	ENPC	-	-	-	-	-	-	F2	F1	F0	S1	S0	P1	P0	-

Below is a summary of the control lines and the tasks they carry out:

LDMC	'1': MPC loading; '0': MPC increment
ENPC	'1': PC increment
F2,F1,F0	ALU function select
S1,S0	Accumulator register function select
P1,P0	Input multiplexer channel select

The table below shows an example of instruction code (keep in mind that the codes in the left hand column are only examples). We see microprograms made up of one or more microinstructions.

Microprogram Memory

Instruction Code	Address	Microprogram
1Ah	1Ah	1st Microinstruction
	1Bh	2nd Microinstruction
1Ch	1Ch	1st Microinstruction
1Dh	1Dh	1st Microinstruction
	1Eh	2nd Microinstruction
	1Fh	3rd Microinstruction

The examples above are illustrations of the topics discussed so far. We can refer to instructions simply by indicating the address where the first micro-program microinstruction is memorized.

This means that the instruction machine code (the microprogram address) works as an identifier of the elementary operations that must be actually executed. These operations can even be quite complex but they remain within the logic of the sequencer itself and not directly visible to the programmer.

The microinstruction designer will have to manage the MPC load line LDMC, which is generated by the microprogram memory. Line LDMC will have to be activated at the last microinstruction of each microprogram so that the next instruction is loaded in the MPC and the network is continually active.

Let's examine the timing diagram in the figure below, where we see LDMC is active between CK clock edges 2 and 3. The instruction retrieved from the PM is loaded in the MPC at edge 3.

The PC was incremented beforehand so that it targets the current instruction.

Since the instruction machine code is nothing more than the address supplied to the MPM, this will output the corresponding control signals after edge 3 and before edge 4 (remember that ROM memory is a combinational network and it produces outputs after a simple delay).

As we have seen, microinstructions contain the combination of control signals that need to be active at a given time so that the computing logic will produce the desired results.

The results will be memorized only at the next active edge of the clock (4 in the figure), so if it is made up of only one microinstruction, we can assume that the instruction has been executed on that edge.

For example, the result of an addition produced by the ALU is memorized in the accumulator at this moment.

Let's take the case of an instruction made up of multiple microinstructions that will take multiple clock cycles to execute.

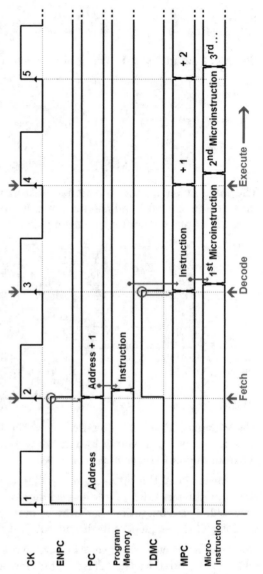

We need to make the address in the MPC advance progressively so that the microprogram memory produces the next microinstructions one by one in the output. To allow the MPC to advance at each clock cycle we only need to make sure that it hasn't been commanded to load (LDMC = '0') in the current microinstruction.

Briefly, the Fetch cycle happens at clock edge 2, the Decode cycle at edge 3 and the Execute cycle as of edge 4.

This cyclical, continual succession of events is called an instruction cycle. See the figure below.

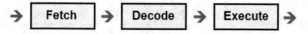

1.3.5 Executing a sequence of instructions

Consider an instruction (A) and the following instruction (B). To ensure the two instructions are executed one right after the other, B's first microinstruction must be executed in the clock cycle immediately following A's last microinstruction.

Considering the timed sequence of operations examined above, a preliminary solution could be to activate ENPC (to force the next instruction to be taken) in the second to the last instruction and activate LDMC in the last one.

Therefore, a standard microprogram as far as these two lines are concerned will be in the format shown below. The other lines, which are insignificant for this issue, are shown in a generic form here.

Generic Instruction			
LDMC	ENPC	D13..D0	
0	0	xxxxxxxxxxxxxx	1st Microinstruction
0	0	xxxxxxxxxxxxxx	2nd Microinstruction
0	0	...	...
0	1	xxxxxxxxxxxxxx	$(N-1)^{th}$ Microinstruction
1	0	xxxxxxxxxxxxxx	N^{th} Microinstruction

This solution is not optimal because it requires the microprograms to have at least two microinstructions. If we only needed one microinstruction to execute a certain instruction, we would still have to add a second one so that command LDMC could be activated. See below. This would waste memory and execution time.

LDMC	ENPC	D13..D0	
0	1	xxxxxxxxxxxxxx	1st Microinstruction
1	0	xxxxxxxxxxxxxx	2nd Microinstruction

Notice that this problem would disappear if the PC targeted the next instruction while the network executed the first. To achieve this we need to think about activating ENPC differently. We will activate it not just one microinstruction before as was done above, but a whole instruction before.

Since each instruction is followed by the next, we can activate ENPC in the current instruction's last microinstruction knowing that the PC won't increment by the very next instruction, but by the one after it.

In sum, for any instruction, we activate both LDMC and ENPC in the last microinstruction as shown in the table below.

	Generic Instruction		
LDMC	ENPC	D13..D0	
0	0	xxxxxxxxxxxxxx	1st Microinstruction
0	0	xxxxxxxxxxxxxx	2nd Microinstruction
0	0	...	...
0	0	xxxxxxxxxxxxxx	$(N-1)^{th}$ Microinstruction
1	1	xxxxxxxxxxxxxx	N^{th} Microinstruction

On the next edge of the clock, activating LDMC makes the instruction available in the program memory load in the MPC, while activating ENPC increments the PC. However, this new value won't be used until the next fetch because the instruction loaded in the MPC now is the one that was taken with the previous PC value.

1.3.6 Executing the first instruction during start up

We have just dealt with how to make instruction execution continuous, assuming the current instruction is already being executed. This section deals with how to execute the first instruction in the program memory when the system is reset.

When the MPC is reset, it targets microprogram memory location zero, so the microinstruction in that location will be executed at the first rising edge of the clock after reset.

Similarly, when the PC is reset it targets microprogram memory location zero, so the instruction in that location is ready to be loaded in the MPC and then decoded in the next instruction cycle.

The microinstruction in microprogram memory location zero should only order the MPC to load the current instruction, at the same time enabling the PC to increment and then prepare to execute the next instruction (control bits D13..D0 are all set at zero).

Notice that whatever is set in microprogram memory location zero corresponds to the NOP (No Operation) instruction. The following table shows its microprogram.

Mnemonic	Hex	LDMC	ENPC	-	-	-	-	-	-	-	-	-	S1	S0	P1	P0	-
NOP	00h	1	1	0	0	0	0	0	0	0	0	0	0	0	0	0	0

From the programmer's point of view, this does nothing except increment the PC. As it is located at MPM address zero, the instruction machine code NOP is the address itself, 00h.

1.3.7 The "instruction pipeline"

This new way to manage the sequencer allows us to do the following things simultaneously (in the same clock cycle):

1. fetch the next instruction
2. decode the instruction fetched in the previous cycle
3. execute the instruction decoded in the cycle before the previous one

In the literature, this mechanism is called the "instruction pipeline" and it is now used in many microprocessors. Theoretically it makes it possible to execute an instruction in one clock cycle (if the instructions are made up of one microinstruction each). A thorough treatment of this subject is beyond the scope of this book.

Here, we will simply point out that our sequencer allows for a three-stage instruction pipeline (Fetch-Decode-Execute). The figure below shows the pipeline executing a sequence of instructions[8] made up of one microinstruction each.

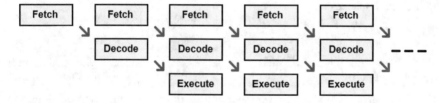

1.3.8 Defining microprograms

As we have mentioned, the network designer sets the microprogram memory up with the microcodes for all the instructions that our network can execute. The complete list of instructions for this network and their corresponding microprograms can be found in Appendix B.2.1).

Remember that this is done only once in the design phase of the processor. The programmer's role is to set up the program memory with the sequence of instructions that the algorithm may be required to carry out.

This separation of duties allows the system programmer to ignore the internal workings of the processor and concentrate on the logic of the program.

---

[8] The order of execution of the instructions is usually referred as the "control flow".

For practice, let's attempt to build the microprograms of a few instructions (even though this has already been done). For simplicity's sake, let's choose instructions we already know because they were used in the programming example of calculating the average of four operands (see Page 28, in Section 1.2.5). We wrote:

$$
\begin{array}{ll}
\text{ADD} & \text{A,OP0} \\
\text{ADD} & \text{A,OP1} \\
\text{ADD} & \text{A,OP2} \\
\text{ADD} & \text{A,OP3} \\
\text{SRL} & \text{A} \\
\text{SRL} & \text{A} \\
\text{HALT} &
\end{array}
$$

Let's examine the instructions used in the example one by one and define the microprograms. We'll assign a value to each control bit based on the operations to carry out, like we did in Section 1.2.5 starting on Page 25.

For the ADD instruction, we need to set LDMC and ENPC to '1' to activate sequencing. Then let's set all the unused bits to '0'. Let's also set F2F1F0 to '000' to force the arithmetic logic unit to perform the addition. The accumulator register has to store the result so we set S1S0 to '11'. The operand should be retrieved from the channel 0 of the multiplexer so let's set P1P0 to '00'. Here is the result:

Mnemonic	Hex	LDMC	ENPC	-	-	-	-	-	-	F2	F1	F0	S1	S0	P1	P0	-
ADD A,OP0	04h	1	1	0	0	0	0	0	0	0	0	0	1	1	0	0	0

The designer loaded this microinstruction at microprogram memory location 04h so this is the instruction code the programmer will use.

Let's build the microcode of the other three variations of ADD in a very similar way. We'll change only the multiplexer selection bits, as shown in the tables below. These microinstructions were defined continuously in the microprogram memory so the machine codes for ADD A,OP1 and the following ones are in order as 05h, 06h and 07h.

Mnemonic	Hex	LDMC	ENPC	-	-	-	-	-	-	F2	F1	F0	S1	S0	P1	P0	-
ADD A,OP1	05h	1	1	0	0	0	0	0	0	0	0	0	1	1	0	1	0

Mnemonic	Hex	LDMC	ENPC	-	-	-	-	-	-	F2	F1	F0	S1	S0	P1	P0	-
ADD A,OP2	06h	1	1	0	0	0	0	0	0	0	0	0	1	1	1	0	0

Mnemonic	Hex	LDMC	ENPC	-	-	-	-	-	-	F2	F1	F0	S1	S0	P1	P0	-
ADD A,OP3	07h	1	1	0	0	0	0	0	0	0	0	0	0	1	1	1	1 0

For the SRL A instruction, we, of course, set LDMC and ENPC to '1', then we set all the bits unrelated to the operation to '0'. However, we need to force the register to right shift its contents so we set S1S0 = '01'. We get:

Mnemonic	Hex	LDMC	ENPC	-	-	-	-	-	-	F2	F1	F0	S1	S0	P1	P0	-
SRL A	20h	1	1	0	0	0	0	0	0	0	0	0	0	1	0	0	0

This instruction machine code is 20h because this is the microprogram memory address where the designer loaded this microcode.

The HALT instruction, introduced in the previous chapter, stops the processor. For HALT, we load in the MPC a special variant of the NOP instruction that does not increment the PC.

Starting from the NOP microcode, we force line ENPC to zero, and give HALT the next microcode, placing it at address 01h, where the only enabled line is LDMC and all other controls are at '0':

Mnemonic	Hex	LDMC	ENPC	-	-	-	-	-	-	F2	F1	F0	S1	S0	P1	P0	-
HALT	01h	1	0	0	0	0	0	0	0	0	0	0	0	0	0	0	0

If the processor executes machine code 01h from HALT, we disable the raising of the PC in the next clock cycle. In this clock cycle, the PC has been incremented by 1 by the previous instruction's microprogram. Now it targets the next machine code. Since we cannot disable the loading of MPC, that next machine code will be loaded regardless.

We can overcome this problem by inserting an identical code after the first HALT code. This way, when the second HALT code is executed, it definitively blocks the processor because that second code continues to load in the MPC at every clock cycle. The PC doesn't increment and only a system reset can start execution again.

Since we need two consecutive 01h codes to stop our processor, we will make things shorter by indicating one single HALT instruction that corresponds to a pair of consecutive 01h codes.

1.3.9 Rewriting the program of the average of four operands

The work we've just done for practice was actually the job of the microprogram designer and was definitely finished in the network design phase. Here we simply want to program the network to calculate the average of four values.

All we need to do to achieve this is consult the table of instructions (see Appendix B.2.1), and insert into the program memory the codes that translate our program into machine language, as shown below.

Mnemonic Code		Machine Code
ADD	A,OP0	04h
ADD	A,OP1	05h
ADD	A,OP2	06h
ADD	A,OP3	07h
SRL	A	20h
SRL	A	20h
HALT		01h 01h

Let's use a timing simulation session to trace the progress of this program's execution inside our network. The figure on the opposite page shows that the PC targets program memory location 00h in the first clock cycle. This location contains code 04h, the machine code for ADD A,OP0 (the Fetch cycle).

At the first cycle (as of edge 1), the pipeline starts, as described in section 1.3.6. In the next cycle (as of edge 2), the instruction code is loaded in the MPC and starts the Decode cycle. In this cycle microinstruction ADD A,OP0 activates all the necessary signals, including those that order the accumulator to store a new value.

At clock edge 3 the result of the addition in the register is loaded (the Execute phase). The arrows show the three cycles in succession.

Similarly, in all the following cycles, the instruction machine codes leave the program memory (PM) one by one since the PC increments (ENPC='1') for each clock cycle.

Note that ENPC and LDMC are set at '1' in all the microinstructions related to the instructions we use, except for HALT where ENPC='0'.

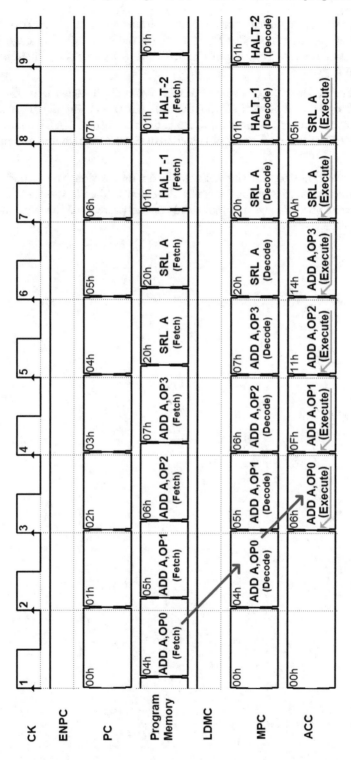

Finally, program memory location 06h contains the first of the two identical HALT instruction codes. The PC is incremented once again on edge 8 of the clock by the microinstruction of the last SRL A instruction, which is now in the decode cycle.

After edge 8, the first of the two HALT instruction codes is in the decode cycle. ENPC is set at '0', so the PC will not increment at the next edge. LDMC at '1' orders the next machine code (the second HALT code) to load in the MPC. As of now, none of the control signals change and the processor stops.

In conclusion, notice that the last calculation instruction (the second SRL A) is executed on edge 8, right when the second HALT code is retrieved. And it is then that you can see the final result of the average of the 4 values in the accumulator.

1.4 Jumps, loops and decisions

In this section, we are going to expand the capabilities of our system. Some modifications to the microprogrammed sequencer make it possible to: a) execute instructions that allow for changing the program execution sequence (or "control flow"), b) execute the same sequence of instructions multiple times, and c) make decisions on the tasks to carry out.

1.4.1 Loops and jump instructions

Our network is not yet able to repeat the same sequence of instructions multiple times. Also, as we alluded to in the previous chapter, it does not have the tools yet to choose which of the available sequences to execute based on intermediate computing results[9].

These two capabilities are at the basis of modern computing systems. The aim of this chapter is to change our network so that it can:

— cyclically carry a sequence of instructions.
— decide which sequence of instructions to execute based on intermediate computing results.

Underpinning these new actions is the need to vary the control flow of the program by interrupting one sequence and beginning another.

The first change we will make is to add the ability for our network to repeat a sequence of instructions. Remember that so far we have only been able to execute a single sequence that ended only by stopping the processor.

Consider introducing an instruction that can change the current flow of execution and make it restart from any other instruction. This would make it possible to write a sequence of instructions only once and to cyclically repeat its execution infinitely.

We have seen that the PC addresses the next instruction to be executed. By varying the PC's contents, we can change the program's flow of execution and make it restart from another instruction. In computer jargon, changing the flow of execution is referred to as jumping to a new sequence of instructions.

The following figure uses arrows to identify the changes as A, B, C and D. We introduce a connection (arrow 'A' in the figure) between the PM and the pre-load input of the PC.

---

[9] In other words, the networks examined so far are not able to execute operations like the classic if - then - else construct, or loops that result from the for, or while - do constructs that are common in programming language.

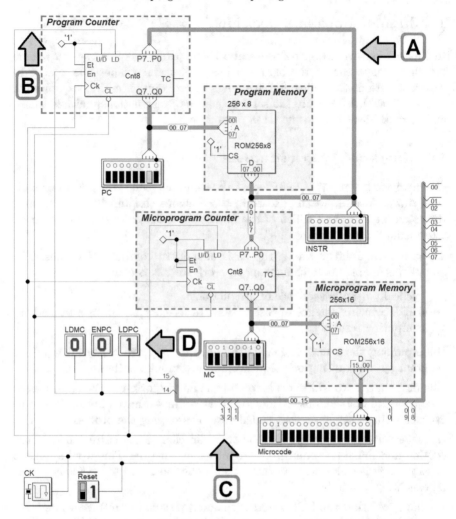

With this new change, the output of the PM is now connected not only with the MPC but also with the PC, forming a sort of feedback. We can change the content of the PC by retrieving from the PM itself the address we want to jump to and loading it in the PC.

The PC's pre-load command (arrow 'B') is now connected to line D13, an MPM output (arrow 'C'). This control line was introduced in the previous chapter with the name LDPC but it was not used (arrow 'D').

The microinstruction designer can now order the PC to be loaded. As seen above, the value to load comes from the PM. We want to write the address to jump to in the location immediately after the one containing the jump instruction.

To execute the jump, we need to activate LDPC, at the appropriate time, to command the loading of the address in the PC.

Let's give a new mnemonic code JP (Jump) to represent the new instruction. Its operand is the address we want to jump to. The format is as follows:

$$\text{JP} \quad <\text{address}>$$

The microprogram related to this can be found in the table below. The first microinstruction commands the instruction address that we want to jump to, which is already in the PM output thanks to the pipeline, to load in the PC (LDPC = '1'). The second commands the instruction we have jumped to load in the MPC (LDMC = '1') and the PC to increment so that the next instruction can be retrieved (ENPC = '1'), so that the pipeline can restart. All the other control signals are kept inactive.

LDMC	ENPC	LDPC	D12..D0	
0	0	1	0000000000000	1st Microinstruction
1	1	0	0000000000000	2nd Microinstruction

The instruction requires two clock cycles to execute: one to retrieve the instruction we want to jump to from the memory and another to execute that instruction, i.e. to complete the jump.

Below we see an example: our program creates an "infinite loop" where the sequence made up of the second and third instructions is cyclically repeated indefinitely:

	Mnemonic Code		Address	Machine Code	
	IN	A,OP0	00h	1Ch	
LOOP:	ADD	A,OP1	01h	05h	
	JP	LOOP	02h	22h	01h

The first instruction loads the value of input OP0 in the accumulator register. The second instruction is to add operand OP1 to it. A label followed by a colon has been placed in front of the mnemonic code for the instruction.

This helps identify an instruction, or a sequence that begins with that instruction. This label allows us to write the jump instruction more legibly and in a more versatile way on the row below. In full, the JP LOOP instruction reads like a command: "jump to the instruction with the label LOOP". The assembler will replace the argument of JP with the address of the instruction labeled as LOOP, that is 01h.

As the table shows, the mnemonics correspond to machine codes 1Ch, 05h, 22h and 01h. 01h is the address to jump to and corresponds to the label LOOP. It is placed at address 03h, immediately after the code of JP (22h).

The following timing diagram shows the sequence of the operations (where we have set OP0 = 06h and OP1 = 09h).

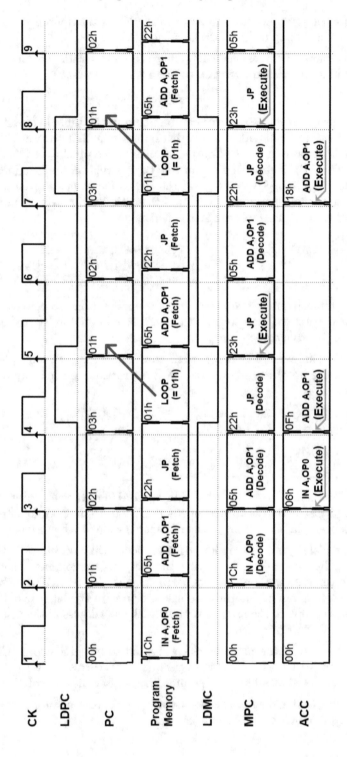

The red arrows highlight the transfer of the address we want to jump to from the program memory to the PC. Because of that transfer, the pipeline stops for one cycle. This happens at clock positive edges 5 and 8, where LDPC is high. The figure also shows the Fetch, Decode and Execute cycles for all the instructions.

The program memory output shows jump instruction code 22h between edges 3 and 4. On edge 4, that code is loaded in the MPC so it enters the Decode cycle. Between edges 4 and 5, the first jump microinstruction orders the PC to load the contents of the program memory output (i.e. the address to jump to, 01h in this case). It also orders the MPC to increment in order to execute the second microinstruction in the next clock cycle.

On edge 5, the PC loads the new address 01h, and so between fronts 5 and 6 we can read the instruction code found at the new address, ADD A,OP1, from the program memory. Meanwhile on the same clock cycle, the second microinstruction orders the new instruction to load in the MPC and so everything proceeds as normal. The jump has been executed.

1.4.2 Decisions and conditional jump instructions

We have seen that the jump instruction allows us to change the order of execution of the instructions set in memory as well as to define infinite loops. Now we want to introduce a similar jump instruction that will allow the network to make decisions based on the results of a previous calculation.

It will allow for example to define counter-based loops (that is sequences of instructions to execute a certain number of times) or loops that are repeated until a certain condition is reached.

In the previous example of the infinite loop, we continually incremented the accumulator with no concern for its content. Now we are going to load a number (taken from an input) in the accumulator and decrement it at each repetition of the loop. This time, however, we want to stop when the register content reaches zero, reload the initial number and repeat the process. The graph below shows the number in the accumulator and its progress over time.

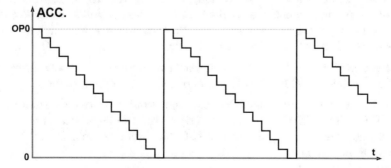

The program in mnemonic code looks like this:

```
START:   IN    A,OP0
LOOP:    SUB   A,OP1
         JP    NZ, LOOP
         JP    START
```

We added a label at the beginning so that the last instruction (JP START) makes everything repeat from the top, creating an infinite cycle. ADD was replaced by SUB, which subtracts the OP1 input operand (set to 1) from the accumulator, and re-saves the result in the accumulator itself.

What is new is the JP NZ,LOOP instruction, which we call a "conditional jump", in that the jump operation depends on a condition, the one written just before the LOOP label. In this case, the condition concerns the fact that the last calculation produced a non-zero result (NZ = "Not Zero") at the ALU output.

In natural language, we can express the meaning of this instruction thus: "Jump to the instruction labeled as LOOP if the result of the last calculation is not zero".

In our example, we have defined two "nested" loops (one inside the other). The outer one is labeled as START and repeats everything from the top. The inner one re-executes the calculation only if the result of the operation is non-zero. At some point, the result reaches zero. Following this condition, we don't want the jump to be executed, but rather the very next instruction in the program.

1.4.3 The FLAG register

There is a problem with the implementation of our network. As we saw in the previous chapter, the ALU generates outputs ZF and CO. ZF signals if the result of the operation being executed is zero (this activates ZF). The CO, however, activates whenever the circuit generates a carry (or a borrow).

The issue is that these signals no longer refer to the last logical-arithmetic operation executed when the conditional jump instruction is being decoded. Therefore, they must be memorized the moment they are produced, i.e. in the clock cycle when the ALU executes the operation.

To memorize their value, we can introduce two flip-flops. This way we can use the flags' values to affect the execution of the following instructions.

As we see in the figure below, the ALU signal outputs are connected to two E-PET flip-flops. The enable inputs (E) of the flip-flops are connected together and connected to the new EFLG line, coming from bit D10 of the microprogram memory.

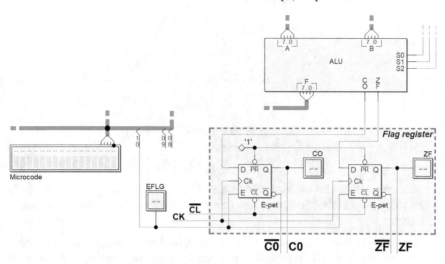

The microcode will enable the memorization (or lack thereof) of the flags. It will also allow for the memorization during the correct clock cycle and only for those operations that require it (arithmetic and logical operations).

In the most complex systems there can be a greater variety of flags. The set of flip-flops that memorize the ALU signals is normally bundled in a register called the "Flag Register".

In sum, the addition of the flag register allows us to keep their values even once the operation that generated them is finished.

1.4.4 Controlling jump conditions

Let's review our network schematic. The load enable of the PC was connected (with the name of LDPC) to microcode line D13. Remember that the activation of LDPC causes the loading of the jump address, coming from the program memory, into the PC.

Now we need to generate the LDPC line in function of the type of jump requested. It has to be imposed by the microcode in function of the values of the flags. We will use not only microcode line D13 but also lines D12 and D11. For our purposes, they have been renamed J2, J1 and J0, respectively.

The table below shows all the possible types of jumps. For the moment, some combinations are not being used as they are reserved for possible future expansions.

J_2	J_1	J_0	Selected Operation
0	0	0	No Jump
0	0	1	Unconditional Jump
0	1	0	Not Used
0	1	1	Not Used
1	0	0	Jump if ZF = '1'
1	0	1	Jump if ZF = '0'
1	1	0	Jump if CO = '1'
1	1	1	Jump if CO = '0'

As you can see in the table, according to the combination of J2, J1 and J0, we can choose whether to jump, not jump or condition the jump on the value of a flag.

So, let's add a network based on a multiplexer that can carry out the functions defined by this table. The complete schematic of the network can be found in Appendix B.3.2 under the name Mp8C.

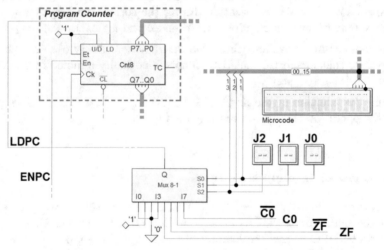

If the combination requested by J2J1J0 is '000', the multiplexer forces LDPC to '0' (this is the case for all the other non-jump instructions). To get an unconditional jump, the J2J1J0 lines have to be set to '001'. This is the same as activating LDPC in any case, regardless of the value of the flags.

In the other four cases, LDPC is directly connected to the flag requested (direct or negated) so there will be a jump only if the flag requests it. For example if we make J2J1J0 = '101', the LDPC will equal the negated ZF, as in the case of instruction "JP NZ, LOOP" at the start of this section. We get a jump if ZF is inactive, that is if the result of the previous operation is non-zero.

Now let's look at the carry flag, CO. If we make J2J1J0 = '110', the load command for the PC will be active only if the flag is. The mnemonic code is "JP C,LOOP". The jump will be executed if the previous calculation has generated a carry.

The mnemonic codes for any jump instructions that are possible in our system are shown in the following table. We can denote the address both in numerical and in symbol form, by using a label as in the examples below. (It is the assembler's job to replace the labels with the real addresses).

Mnemonic Code		Operation
JP	<address>	Unconditional Jump
JP	Z,<address>	Jump if Zero
JP	NZ,<address>	Jump if Not Zero
JP	C,<address>	Jump if Carry
JP	NC,<address>	Jump if Not Carry

The microprogram for the unconditional jump instruction is as follows:

Mnemonic	Hex	LDMC	ENPC	J2	J1	J0	EFLG	-	-	F2	F1	F0	S1	S0	P1	P0	-
JP <address>	22h	0	0	0	0	1	0	0	0	0	0	0	0	0	0	0	0
	+1	1	1	0	0	0	0	0	0	0	0	0	0	0	0	0	0

In the first microinstruction, lines J2J1J0 = '001' order the PC to be loaded with the instruction to jump to. In the second, LDMC orders the MPC to load the instruction we are jumping to, while the ENPC forces the PC to be incremented to retrieve the next instruction. The other control signals are not needed so they are left inactive.

The table below shows the conditional jump instructions and their microprograms.

Mnemonic	Hex	LDMC	ENPC	J2	J1	J0	EFLG	-	-	F2	F1	F0	S1	S0	P1	P0	-
JP Z,<address>	24h	0	1	1	0	0	0	0	0	0	0	0	0	0	0	0	0
	+1	1	1	0	0	0	0	0	0	0	0	0	0	0	0	0	0
JP NZ,<address>	26h	0	1	1	0	1	0	0	0	0	0	0	0	0	0	0	0
	+1	1	1	0	0	0	0	0	0	0	0	0	0	0	0	0	0
JP C,<address>	28h	0	1	1	1	0	0	0	0	0	0	0	0	0	0	0	0
	+1	1	1	0	0	0	0	0	0	0	0	0	0	0	0	0	0
JP NC,<address>	2Ah	0	1	1	1	1	0	0	0	0	0	0	0	0	0	0	0
	+1	1	1	0	0	0	0	0	0	0	0	0	0	0	0	0	0

In the first microinstruction, bits J2, J1 and J0 indicate the flag to use to condition the jump. Meanwhile, ENPC = '1' enables incrementing the PC by one, which will happen if the jump condition is not reached.

The second microinstruction orders the MPC to be loaded and the PC to be incremented so it always stays one instruction ahead (for the pipeline to work). All the other datapath controls are kept inactive in both microinstructions.

If the flag has permitted the jump, the instruction we've jumped to will be loaded in the MPC register. Otherwise, if the flag has not permitted the jump, the instruction that follows the conditional jump will be loaded.

To thoroughly understand the logic of the conditional jump mechanism we must focus on the state of the sequencer. In the same clock cycle where the first microinstruction is executed, the program memory output supplies the address where we want to jump. Since we cannot know at that moment whether the jump will be executed or not, we must consider both possibilities. Let's look at the following test program:

Mnemonic Code			Address	Machine Code	
	IN	A,OP0	00h	1Ch	
LOOP:	SUB	A,OP1	01h	09h	
	JP	NZ, LOOP	02h	26h	01h
NEXT:	NOP		04h	00h	
	NOP		05h	00h	
	...		06h	...	

Suppose that we read number 2 from input OP0 while OP1 is set at 1. When the first SUB is executed, the value of the accumulator goes from 2 to 1. The result is "non zero", so the jump is executed and goes to LOOP.

Let's use the timing diagram of the signals used here to analyze what happens in the sequencer. See the figure on the next page.

We see that machine code 26h from JP NZ,LOOP is retrieved from the program memory during the clock cycle between edges 3 and 4. In the next cycle, (between edges 4 and 5) the instruction is decoded.

The logic takes into account that ZF is '0', and consequently the load command for LDPC is activated (to jump) and ENPC is also activated (to retrieve the next instruction).

The PC ignores the ENPC enable because the load command for LDPC has priority over the enable due to the way the counter functions. The jump address (on edge 5) is available on the program memory output and is loaded in the PC.

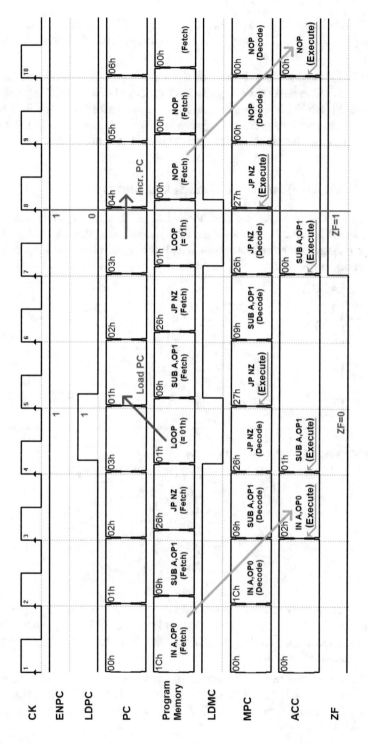

In the cycle between edges 5 and 6 the instruction machine code of where we jumped to is read in the program memory. Therefore, this is the next instruction to be decoded.

When SUB is executed for the second time, the accumulator is reset to zero (on edge 7). ZF='0' means that the jump isn't executed and LDPC isn't activated. Even so, the ENPC is active, so the PC is incremented on edge 8. Then the instruction immediately after the jump is retrieved (here, it's NOP) and the order set out by the pipeline (fetch, decode and execute) recommences.

1.4.5 Example: How to use conditional jumps

In the previous example, we first loaded a number in the accumulator and decremented it until it reached zero and started the process again. Now we want to change the numerical sequence that is generated following the triangular trend in the figure.

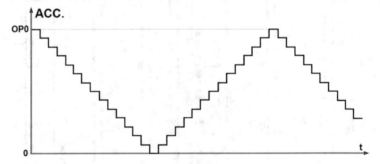

The program in mnemonic code looks like this:

```
                IN      A,OP0
        LDOWN:  SUB     A,OP1
                JP      NZ, LDOWN
        LUP:    ADD     A,OP1
                CP      A,OP0
                JP      NZ, LUP
                JP      LDOWN
```

The first three instructions are the same as in the previous example, except that the label is now called LDOWN (abbreviation of Loop Down). The first instruction takes the number we'll start with from input OP0 and puts it in the accumulator register.

The next two instructions bring about the progressive reduction of the content of the register until it reaches zero (we set input OP1 to 1). When the register goes to zero, instruction JP NZ,LDOWN stops jumping and we move to the next instruction.

At the LUP (Loop Up) label, we find the ADD instruction, which adds the constant read from input OP1 to the accumulator, which is set to 1. We want to repeat this increment until conditional jump JP NZ,LUP gets us to the initial value again. Then we will restart by using JP LDOWN to jump to the second instruction, to decrement the accumulator again.

Nonetheless, the flag mechanism can't help us directly because neither of the two available flags activate when we reach the value OP0 (the value is not zero nor has the increment produced a carry). Do we have to add a magnitude comparator to the ALU? As we can see in the mnemonic code, we have inserted a CP instruction ("Compare") which is meant to compare the accumulator with the OP0 input.

To achieve the same result previously, we subtracted (SUB) one operand from the other and discussed how to observe the flags to evaluate if the two operands were equal or if one was larger than the other (see Section 1.2.4.1, on Page 25). Now, we save the flags for a later use, so a subtraction operation seems the perfect choice to do an evaluation through conditional jumps.

If the two operands are different, the result of the subtraction will be non-zero so instruction JP NZ,LUP will continue to jump backwards to instruction ADD, labeled as LUP. Nevertheless, our case requires the accumulator to keep its content after the compare, as every time the cycle repeats we increment it by one. Unfortunately, a SUB changes the content of the accumulator at the moment it saves the result of the subtraction.

We can solve this problem by introducing the compare instruction CP, which works like a SUB instruction but updates only the flags based on the result. The CP instruction doesn't save the result of the subtraction in the accumulator so its content remains intact.

Let's compare the SUB, ADD and CP microprograms.

Mnemonic	Hex	LDMC	ENPC	J2	J1	J0	EFLG	-	-	F2	F1	F0	S1	S0	P1	P0	-
ADD A,OP0	04h	1	1	0	0	0	1	0	0	0	0	0	1	1	0	0	0

Mnemonic	Hex	LDMC	ENPC	J2	J1	J0	EFLG	-	-	F2	F1	F0	S1	S0	P1	P0	-
SUB A,OP0	08h	1	1	0	0	0	1	0	0	0	0	1	1	1	0	0	0

Mnemonic	Hex	LDMC	ENPC	J2	J1	J0	EFLG	-	-	F2	F1	F0	S1	S0	P1	P0	-
CP A,OP0	30h	1	1	0	0	0	1	0	0	0	0	1	0	0	0	0	0

As we can see, all the microprograms of ADD, SUB and CP activate the line EFLG (as well as all the other arithmetic and logical instructions defined for this network), which orders the memorization of the flags. ADD and SUB order the register to store the result from the ALU (setting S1S0 = '11'), while CP calculates the result but doesn't save it (S1S0 = '00').

Appendix B.3.1 has the complete list of instructions and the microprograms related to the network we have developed so far (Mp8C), as well as the complete schematic.

By consulting this list, we can translate the proposed program into machine code and get the following:

Mnemonic Code			Address	Machine Code	
	IN	A,OP0	00h	1Ch	
LDOWN:	SUB	A,OP1	01h	09h	
	JP	NZ, LDOWN	02h	26h	01h
LUP:	ADD	A,OP1	04h	05h	
	CP	A,OP0	05h	30h	
	JP	NZ, LUP	06h	26h	04h
	JP	LDOWN	08h	22h	01h

1.5 Input and output ports

This section introduces the base concepts surrounding connecting the processor to external devices through the use of input and output ports. [10]

1.5.1 Input ports

An input port means a connection between the processing system and the outside, which allows it to receive data to process. We have encountered input ports since the beginning of this book and we have used them in examples to get the values of the operands to use in our calculations. We have also used them to acquire constants, due to the lack of other methods available.

Let's review the structure we used before in the following figure. A multiplexer allows us to choose which port we wish to read from, within the limits imposed by bits P1 and P0. This kind of port is called "parallel" since it allows us to read a whole group of bits (8 in our case) in one single operation.

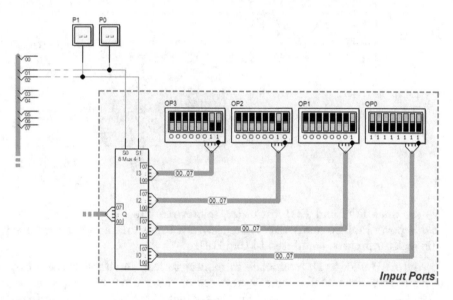

We have seen that the information selected can be copied in the accumulator by following the paths allowed within the network. To do this we can use the instruction IN A, <port>, which is translated to microcode terms in the line P1 and P0 settings.

---

[10] Port in the maritime sense of the term. A port where ships dock to load and unload cargo is a metaphor for data that enter and results that exit.

1.5.2 Output ports

In the examples from the previous section, we've always retrieved results by reading the accumulator for simplicity's sake. In a real processor, however, the accumulator is not visible from the outside. The internal part of the calculation is always markedly separated from the part that produces the results.

Therefore, it is necessary to introduce something that will make it possible to communicate with the outside. This is what output ports are for; they make the results available to the outside through the use of parallel registers. One benefit is that they make it possible to keep the results permanently legible while the whole network continues processing other data.

The two parallel output ports that we added to our system are made up of registers that are synchronized by the clock and controlled by the microprogram memory like all the other elements of the datapath. See the figure below.

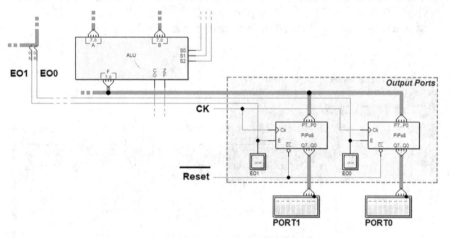

We see lines EO0 and EO1 (in order, microcode bits D08 and D09), that enable ports PORT0 and PORT1, respectively, to load data in the registers. The register input is connected to the ALU.

So, the ALU allows the microcode to request to load in either of these two registers regardless of the number produced or transferred. In our processor, we will restrict ourselves to copy only the content of the accumulator to these ports. The complete network schematic, including the output ports is available in Appendix B.4.2 under the name Mp8D.

To manage the output ports, we add a new instruction:

OUT   <port>,A

Where <port> identifies the output port chosen by the programmer. This instruction loads the accumulator content onto the selected port by passing it through the ALU.

These two instructions' microprograms are made up of only one microinstruction each, where the new control lines EO0 and EO1 appear. The ALU is set up to copy input A (F2F1F0='110') to the output.

Mnemonic	Hex	LDMC	ENPC	J2	J1	J0	EFLG	EO1	EO0	F2	F1	F0	S1	S0	P1	P0	-
OUT PORT0,A	34h	1	1	0	0	0	0	0	1	1	1	0	0	0	0	0	0

Mnemonic	Hex	LDMC	ENPC	J2	J1	J0	EFLG	EO1	EO0	F2	F1	F0	S1	S0	P1	P0	-
OUT PORT1,A	35h	1	1	0	0	0	0	1	0	1	1	0	0	0	0	0	0

The value copied to the selected output port will remain there until a new OUT instruction is performed on it.

In Appendix B.4.1 you will find the list of instructions and the microprograms related to the network we have developed so far.

1.5.3 How to use ports

Below is a basic example:

```
LOOP:    IN      A,OP0
         OUT     PORT0,A
         JP      LOOP
```

where the number in input port OP0 is first copied in the accumulator and then in turn is copied in output port PORT0 without being processed. The unconditioned jump instruction JP LOOP means that this pair of instructions is repeated infinitely. Once it is translated into machine language, we have:

Mnemonic Code			Address	Machine Code	
LOOP:	IN	A,OP0	00h	1Ch	
	OUT	PORT0,A	01h	34h	
	JP	LOOP	02h	22h	00h

1.5.3.1 Generating a periodic triangular waveform

Let's return to an example we studied in the previous section, where a sequence of increasing/decreasing binary numbers was generated in the accumulator and repeated infinitely. As you can see in the following figure, the plot of this sequence of values over the time recalls a triangular waveform.

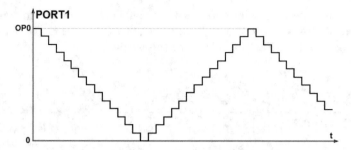

In this new version of the program we want to use port PORT1 to generate the values. Remember what we discussed before, that the accumulator in a real system is never directly accessible from the outside.

Therefore, if we want to generate a number sequence, we must copy them, value by value to an output port. The listing below shows the program's mnemonic code rewritten with the OUT instructions added where the content of the accumulator is changed.

```
                IN      A,OP0
                OUT     PORT1,A
        LDOWN:  SUB     A,OP1
                OUT     PORT1,A
                JP      NZ, LDOWN
        LUP:    ADD     A,OP1
                OUT     PORT1,A
                CP      A,OP0
                JP      NZ, LUP
                JP      LDOWN
```

The first two instructions initialize the content of the accumulator with the starting value (read from port OP0) and then copy it to port PORT1. This is followed by a loop that decrements the accumulator at every repetition. This is achieved by subtracting from it the number read in port OP1 (set at 1).

At each decrement the new value is calculated and copied to the output port. The loop ends only when the accumulator reaches zero. At this point, instruction JP NZ, LDOWN stops jumping to the LDOWN row, so the execution continues with the next instruction.

On this row of the program we find a loop with a similar structure to the previous one. Every time this repeats, we increment the accumulator and update the output port with the new value.

In this case, however, the loop's end condition is not the accumulator reaching zero but comparing (with instruction CP) the value of the accumulator with the final number of the count (which has to be the same as the starting number and is read again at the OP0 port). When they are equal, the loop is finished and we move on to the instruction in the position after JP NZ, LUP, that is the unconditioned jump JP LDOWN. The program goes back and repeats the sequence of the two loops infinitely.

After the translation of the mnemonic code we get the following machine code. Now, we can use the simulator to test the functionality of the program.

	Mnemonic Code		Address	Machine Code	
	IN	A,OP0	00h	1Ch	
	OUT	PORT1,A	01h	35h	
LDOWN:	SUB	A,OP1	02h	09h	
	OUT	PORT1,A	03h	35h	
	JP	NZ, LDOWN	04h	26h	02h
LUP:	ADD	A,OP1	06h	05h	
	OUT	PORT1,A	07h	35h	
	CP	A,OP0	08h	30h	
	JP	NZ, LUP	09h	26h	06h
	JP	LDOWN	0Bh	22h	02h

1.5.3.2 Generating a periodic trapezoidal waveform

This example shows us another use of loops: generating a delay. As in the previous case, we want to generate a periodic sequence of values, but here the arrangement will look trapezoidal. See below.

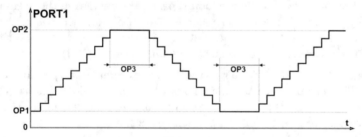

As you can see, there is a pause between the rising part and the descending part, where we keep the last value that was generated for a duration proportional to the value read at input port OP3.

Also, notice that the sequence that was generated stays between two values, which are read at input ports OP1 and OP2. The following list shows a sample program that generates this sequence.

We see four loops one after the other, closed in a single external loop that repeats the whole sequence infinitely. Unlike the previous example, here both loops that increment the value (LOOPUP) and decrement it (LOOPDN) use the CP instruction to evaluate the loop end condition.

```
START:      IN      A,OP1
            OUT     PORT1,A
LOOPUP:     ADD     A,OP0
            OUT     PORT1,A
            CP      A,OP2
            JP      NZ, LOOPUP
            IN      A,OP3
PAUSEHI:    SUB     A,OP0
            JP      NZ, PAUSEHI
            IN      A,OP2
LOOPDN:     SUB     A,OP0
            OUT     PORT1,A
            CP      A,OP1
            JP      NZ, LOOPDN
            IN      A,OP3
PAUSELO:    SUB     A,OP0
            JP      NZ, PAUSELO
            JP      START
```

The pauses, obtained by a delay loop, are inserted between the two loops. We will explore the delay loop concept in detail further on. For now, let's look in our example at the loop that repeats from label PAUSEHI.

Instruction IN A,OP3 loads the value of OP3 in the accumulator before entering the first loop. Each time the loop repeats, the content of the accumulator lowers by one until it reaches zero. It is easy to verify that the two instructions that repeat are executed in 3 clock cycles. The loop is repeated OP3 times so its execution time is proportional to this number (3 x OP3).

Note that after this delay loop is executed, the accumulator's previous content is lost. This means when we move forward, we must reload the values from the input ports to resume output generation. After the assembler translates the program, we get the following machine codes:

Mnemonic Code			Address	Machine Code	
START:	IN	A,OP1	00h	1Dh	
	OUT	PORT1,A	01h	35h	
LOOPUP:	ADD	A,OP0	02h	04h	
	OUT	PORT1,A	03h	35h	
	CP	A,OP2	04h	32h	
	JP	NZ, LOOPUP	05h	26h	02h
	IN	A,OP3	07h	1Fh	
PAUSEHI:	SUB	A,OP0	08h	08h	
	JP	NZ, PAUSEHI	09h	26h	08h

(cont.)

	IN	A,OP2	0Bh	1Eh	
LOOPDN:	SUB	A,OP0	0Ch	08h	
	OUT	PORT1,A	0Dh	35h	
	CP	A,OP1	0Eh	31h	
	JP	NZ, LOOPDN	0Fh	26h	0Ch
	IN	A,OP3	11h	1Fh	
PAUSELO:	SUB	A,OP0	12h	08h	
	JP	NZ, PAUSELO	13h	26h	12h
	JP	START	15h	22h	00h

1.5.3.3 Generating signals with the PWM technique

"Pulse Width Modulation" (PWM) is a very common technique used to generate a definable medium voltage on a line. A succession of fixed period, variable length pulses is generated on the line. The average voltage generated depends on the ratio between the duration of the high level and the period.

In this example, we use our system to generate two PWM signals, whose average values are proportionate to the number set in the input.

A PWM signal is retrieved from the bit in position 0 of port PORT0, and its average value is proportionate to the value read on input port OP1. The other PWM signal is generated on bit 0 of PORT1 and depends on input OP2.

The figure below shows the operating principle. The program cyclically increments the number in the accumulator from zero to the maximum representable value (255) generating a sort of staircase.

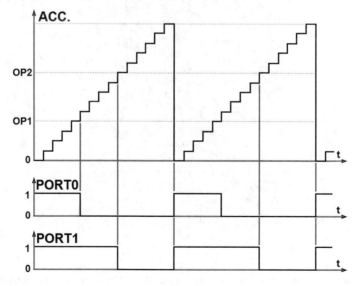

At the start of the staircase, bit 0 in both ports is set to '1'. At each increment, the program compares the new number with the one set on the corresponding input. When they are equal, bit 0 in the corresponding port goes to zero. As a result, we get a 2-level pulse signal on bit 0 in both ports. Its duration is proportional to the number set on the corresponding input (as per specifications).

The following listing shows the mnemonic code for the program of the PWM generator. Remember that ports OP0 and OP3 are used to keep constants 0 and 1 available, respectively.

The program's main loop begins and repeats on the first instruction, labeled as START. At the beginning of the sequence, bit 0 of ports PORT0 and PORT1 are both brought high, while the accumulator is brought to zero.

At the LOOP line, the value in the accumulator is incremented. At lines TEST1, TEST2 and TEST3 that value is compared to OP1, OP2 and zero, respectively[11].

If the first compare produces a positive result, bit 0 of PORT0 is set to zero. Similarly, bit 0 of PORT1 is set to zero if the second compare requires it, while the last compare checks if the accumulator count has returned to zero. If it has, we start again from the top and bring bit 0 of both ports high again.

```
START:    IN      A,OP3
          OUT     PORT0,A
          OUT     PORT1,A
          IN      A,OP0
LOOP:     ADD     A,OP3
TEST1:    CP      A,OP1
          JP      NZ, TEST2
          IN      A,OP0
          OUT     PORT0,A
          IN      A,OP1
TEST2:    CP      A,OP2
          JP      NZ, TEST3
          IN      A,OP0
          OUT     PORT1,A
          IN      A,OP2
TEST3:    CP      A,OP0
          JP      NZ, LOOP
          JP      START
```

---

[11] Note how similar this evaluation sequence is to the switch-case-case construct found in many programming languages.

The limitations of this architecture based on one only register, the accumulator, make it so that on every positive compare we need to clear the accumulator just to write zero in the output ports.

If no corrections are made, this operation loses the memory of the number we are at on the staircase. Luckily, all we need to do in our case is this: take the value (that the compare found equal to the number that was in the accumulator) from the corresponding input port and reload it.

Finally we assemble the program and get the machine codes as follows:

Mnemonic Code			Address	Machine Code	
START:	IN	A,OP3	00h	1Fh	
	OUT	PORT0,A	01h	34h	
	OUT	PORT1,A	02h	35h	
	IN	A,OP0	03h	1Ch	
LOOP:	ADD	A,OP3	04h	07h	
TEST1:	CP	A,OP1	05h	31h	
	JP	NZ, TEST2	06h	26h	0Bh
	IN	A,OP0	08h	1Ch	
	OUT	PORT0,A	09h	34h	
	IN	A,OP1	0Ah	1Dh	
TEST2:	CP	A,OP2	0Bh	32h	
	JP	NZ, TEST3	0Ch	26h	11h
	IN	A,OP0	0Eh	1Ch	
	OUT	PORT1,A	0Fh	35h	
	IN	A,OP2	10h	1Eh	
TEST3:	CP	A,OP0	11h	30h	
	JP	NZ, LOOP	12h	26h	04h
	JP	START	14h	22h	00h

1.6 Constants, variables and read/write memory

In this section, we'll complete the design of our processor for educational purposes by adding support for managing constants and a small RAM memory bank for managing variables.

1.6.1 Constants

The term "constant" means a number or value that a program uses for its calculations, which is defined when the program is written and remains unchangeable during the program's execution. We have already used constants in previous sections. We retrieved them from input ports since we had no other means to provide these values to our programs.

The use of ports for this purpose is quite limited. This is because we would like to use ports to acquire numbers from the outside to process and because quite often we need very many constants.

It would be useful to memorize the constants inside the program itself and be able to load them in the accumulator. To do that, we will add a new instruction:

$$LD \quad A,<const>$$

where LD is the abbreviation for Load, and operand <const> is a generic 8-bit constant to copy in the accumulator. For example, to load constant 01h in the register, we write:

$$LD \quad A,01h$$

But where should we memorize this constant? As we did for jump instructions (see the figure below), it is convenient to put the instruction operand in the memory location immediately after that of its machine code. This way, we take advantage of the opportunity to increment the PC to target the constant following the instruction.

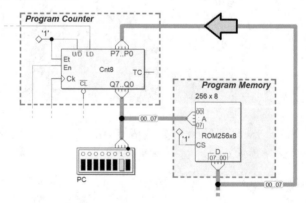

With the jump instructions, we added a connection between the program memory output and the pre-load inputs of the PC. The green arrow shows the path of the jump address.

With constants, we need to bring the program memory output toward the accumulator, but we cannot do that directly since the accumulator is connected to the ALU output.

The first solution that comes to mind is to add a multiplexer to choose where to retrieve the value to load in the accumulator. This is shown in the figure below where the arrows highlight the two paths.

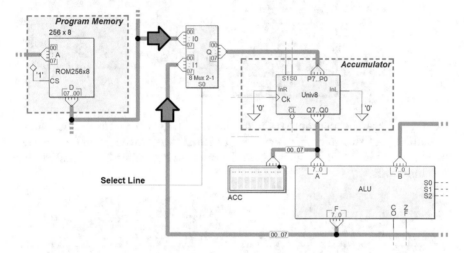

Under the control of the Select Line, we can choose to direct the ALU output (as before) or the program memory output toward the accumulator. Clearly, the selection line should originate with the microcode so that the microinstructions can control it.

This simple choice from the point of view of the paths is incompatible with what we are adding to our processor in this section. It would force us to add a bit to the microcode but the only line available (D0) will soon be used for other things.

Adding other bits would require a larger microprogram memory. Although this is technically feasible, let's remember the educational scope of the network we are building. We want to limit the network complexity.

This means we should compromise and eliminate something to make space for the new connection. Let's take advantage of the processor's existing data paths and sacrifice an input port (OP3). This isn't ideal but if we consider that we have always used at least one port to read a constant in the previous examples, this is not an unreasonable sacrifice, seeing that, in its place, we could put many constants in the program memory.

The following figure shows there is no OP3 port. Instead, the program memory input has been connected to the multiplexer input. The arrows show the path that the constant follows before being loaded in the accumulator.

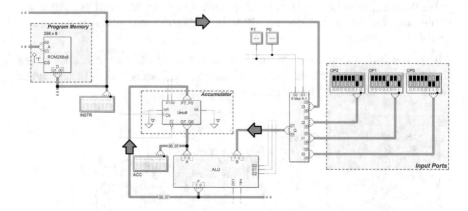

Below, we see the microprogram of the LD A,<const> instruction, which is made up of two microinstructions. The first was placed at microprogram memory address 3Dh so, as we have learned, this number is the instruction machine code.

Mnemonic	Hex	LDMC	ENPC	J2	J1	J0	EFLG	EO1	EO0	F2	F1	F0	S1	S0	P1	P0	-
LD A,<const>	3Dh	0	1	0	0	0	0	0	0	1	1	1	1	1	1	1	0
	+1	1	1	0	0	0	0	0	0	0	0	0	0	0	0	0	0

In order to activate this path, F2F1F0 should be set at '111' in the first microinstruction so that operand B is copied at the ALU output. Also, P1P0 should be at '11' to select multiplexer channel 3. Finally, we should force the number in the register to load with S1S0 = '11'.

When these signals are active the PC is targeting the machine code's next location, which contains the constant that will be loaded in the accumulator (on the very next rising edge of the clock). LDMC is set at '0', so that the second microinstruction can be read. ENPC = '1' so that the PC is brought forward to retrieve the next instruction in the next clock cycle.

The second microinstruction forces the PC to load in the MPC, bringing it forward to target the next instruction.

1.6.2 Immediate addressing instructions

The term "addressing" refers to the mode used to retrieve data to process. This category of instruction uses "immediate" addressing to retrieve a number found immediately after the instruction machine code.

Once we have changed the connection to multiplexer input I3, we are forced to review all the instructions that use that input. So far, none of these instructions involves reading a constant from the memory program. For example, if we tried to execute ADD A,OP3 now, it would add the contents of A with the machine code of the next instruction in the program memory. This does not interest us. Instead, let's add the instruction:

$$\text{ADD}\quad \text{A,<const>}$$

In the example below, the programmer asks to add constant 27h to the accumulator. The constant is placed directly after the machine code.

Mnemonic	Machine Code
ADD A,27h	40h 27h

This instruction has machine code 40h, in that it is allocated in the MPM at that address. The microprogram below shows that the first microinstruction orders (F2F1F0 = '000' and P1,P0 = '11') the sum of the content of the accumulator with the value from the microprogram memory, aka the constant that follows the machine code.

Remember that the sequencing pipeline makes it that the PC always targets the next memory location after the instruction being executed.

Mnemonic	Hex	LDMC	ENPC	J2	J1	J0	EFLG	EO1	EO0	F2	F1	F0	S1	S0	P1	P0	-
ADD A,<const>	40h	0	1	0	0	0	1	0	0	0	0	0	1	1	1	1	0
	+1	1	1	0	0	0	0	0	0	0	0	0	0	0	0	0	0

After the sum, flag register (EFLG = '1') is updated. Also, the PC is forced to increment (ENPC = '1') to keep the pipeline active, and the result of the sum (S1,S0 = '11') is memorized in the accumulator. The second microinstruction simply commands LDMPC = '1' and ENPC = '1' so the next instruction can be executed.

Let's apply the immediate addressing mode to define instructions SUB, AND, OR and CP as well. We get:

Mnemonic	Hex	LDMC	ENPC	J2	J1	J0	EFLG	EO1	EO0	F2	F1	F0	S1	S0	P1	P0	-
SUB A,<const>	42h	0	1	0	0	0	1	0	0	0	0	1	1	1	1	1	0
	+1	1	1	0	0	0	0	0	0	0	0	0	0	0	0	0	0
AND A,<const>	44h	0	1	0	0	0	1	0	0	0	1	0	1	1	1	1	0
	+1	1	1	0	0	0	0	0	0	0	0	0	0	0	0	0	0
OR A,<const>	46h	0	1	0	0	0	1	0	0	0	1	1	1	1	1	1	0
	+1	1	1	0	0	0	0	0	0	0	0	0	0	0	0	0	0
CP A,<const>	4Ah	0	1	0	0	0	1	0	0	0	0	1	0	0	1	1	0
	+1	1	1	0	0	0	0	0	0	0	0	0	0	0	0	0	0

Thanks to immediate address instructions, we no longer have to dedicate input ports to the acquisition of constants. Consider the example in Section 1.5.3.1 where we needed a constant 01h on input port OP1 to be able to execute the increasing and decreasing of the content of the accumulator.

Now that we have immediate address instructions at our disposal, let's revisit previous examples and substitute each increment (ADD A,OP1) with ADD A,01h and every decrement (SUB A,OP1) with SUB A,01h. This frees us from the constraint of needing constants on input ports.

1.6.3 Variables

The term "variable" refers to a container that preserves its content (in our case, binary code that can represent a number, ASCII character, etc.). This concept reflects a generic register. As the processor is working, the program changes the content of a variable, for example to memorize the intermediate results of operations, as we have seen with the accumulator.

Still, having only one register available is insufficient for most of the programs that we will write. Take for example the calculation of a bitwise logical expression like this:

$$(OP0 \cdot OP1) + \overline{(OP0 + OP1)}$$

where we must first execute the calculation of the first term, memorize it somewhere, execute the calculation of the second term and finally calculate the logical sum of the two results.

We could add a set of registers like the accumulator to the processor. Many commercial processors have a certain number of registers inside. Our choice, however relies on using a read/write memory of a comparable size to the rest of the network, which will allow it to be read and written by the processor's calculation unit.

1.6.4 Read/write memory (RAM)

The term "read write memory" refers to a system that can memorize numbers in an organized way. For historical reasons, this type of memory is called RAM (Random Access Memory. See Appendix A.2 for more details).

By principle, we can conceive of RAM as a set of equal parallel registers, each of which can be loaded (written) with a number and then re-read when we want to fetch the number. Each of these registers can be considered the physical manifestation of a variable, able to contain a binary code.

RAM can contain a high number of registers that from now on we will call memory locations (or cells). The number of locations can range from a few to hundreds in small components and there are systems that can memorize several billion locations.

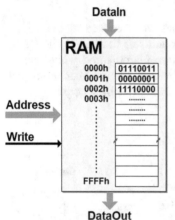

Except for specific cases, RAM allow us to write or read one location at a time. Each cell is identified by a number called an address (as with ROM).

The figure on the left shows a simplified representation of the main "static" RAM connections. See Appendix A.2.

The inside of the memory is shown in an idealized form, as a table. Each block of the table is identified by its address.

Each location preserves the number written within it or an indeterminate value if nothing has been written in it yet.

To memorize a number in RAM, we must first decide which location to choose among all the possibilities. To do this, we must provide the RAM with the location address we want on the Address lines. Then we must submit the number to the RAM's DataIn inputs and then order it to be written (here a generic Write command is represented).

When we want to re-read the number in a previously written location, we must provide its address to the memory and then retrieve the contents from the DataOut outputs.

1.6.5 RAM read/write instructions

Now let's write the instructions we will need to use RAM. They look like this:

$$\text{LD} \quad (\text{<address>}),\text{A}$$
$$\text{LD} \quad \text{A},(\text{<address>})$$

As before, LD is the abbreviation for Load, and operand <address> is the 8-bit address to use to access the memory. LD (<address>),A copies the content of the accumulator and writes it in the memory location specified in parentheses. LD A,(<address>) reads the content of the memory location in parentheses and copies it in the accumulator.

In the example below, we load constant 3Fh in the accumulator and then write it in memory address 45h:

$$\text{LD} \quad \text{A,3Fh}$$

$$\text{LD} \quad \text{(45h),A}$$

Then after using the accumulator for other calculations, we can retrieve the value that was saved in the memory by writing:

$$\text{LD} \quad \text{A,(45h)}$$

Now let's adopt the same process to some cases we've looked at before, regarding where to memorize the operand and the address to send to the RAM. As before, we count on the fact that the PC was incremented when the instruction was decoded. So we indicate the address of the RAM memory cell that we want to read or write, immediately after the machine code.

1.6.6 The RAM and the processor

We add a RAM to our processor so that the partial results of our calculations can be temporarily memorized in it. To do this we must manage the various data paths and necessary control signals.

We will use a component from the simulator library, a RAM with 256 locations at 8 bits each. See the figure to the right. This is a synchronous read/write RAM.

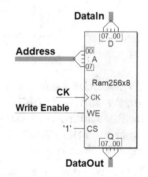

To write, first we introduce the address of the cell on the Address line, and the number to write in the DataIn inputs. If the Write Enable (WE) line is active, the number will be memorized at the next rising edge of the clock CK. As for reading, first we route the memory location, then its content is available in the output as of the next active edge of the clock. Clearly, to execute the read/write functions, the CS (Chip Select) line must be activated.

The following figure shows the writing path. In our processor we should retrieve the number to write from the ALU output (yellow arrows) to avoid changing existing paths.

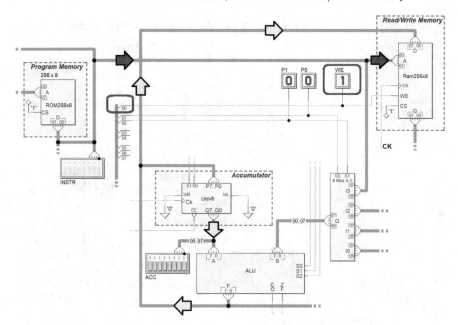

We don't need special changes for the address (red arrows) since we've decided to retrieve it from the program memory as mentioned above. As with jumps and constants, it will be in the location after that of the machine code.

The figure also highlights write enable line WE which we have connected to microcode line D0.

Thanks to the connections that we have set so far, we can now define the microprogram for the write instruction in the RAM.

This instruction is coded as 38h, we're placing the microprogram at this MPM address. The first microinstruction (see the following table) sets F2F1F0 at '110' so that the accumulator is copied at the ALU output and then it orders the writing by activating WE.

Mnemonic	Hex	LDMC	ENPC	J2	J1	J0	EFLG	EO1	EO0	F2	F1	F0	S1	S0	P1	P0	WE
LD (<address>), A	38h	0	1	0	0	0	0	0	0	1	1	0	0	0	0	0	1
	+1	1	1	0	0	0	0	0	0	0	0	0	0	0	0	0	0

When these signals are active, the PC targets the location holding the address, which is used by the memory when it writes (at the very next rising edge of the clock). Here LDMC = '0' to read the second microinstruction, and ENPC = '1' to force the PC to increment, to retrieve the next instruction.

As in all the previous cases we've studied, the second microinstruction forces the next instruction to load in the MPC to bring the PC forward one increment to prepare to retrieve the next instruction and keep the pipeline active.

In the figure below, the arrows show the path of the data being read.

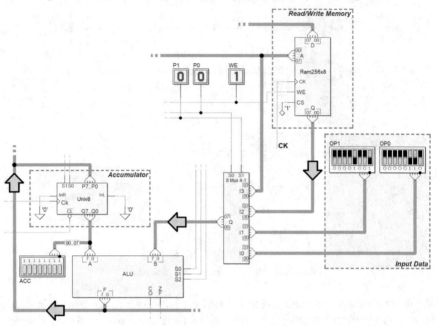

So that the number in the RAM output is loaded in the accumulator, it needs to follow a dedicated path. For the sake of consistency with previous circuits, let's limit the network complexity and use the multiplexer we already have rather than adding a new one and sacrificing another input port (OP2). Once we have set the read paths, then we can define the microprogram for the write instruction in the RAM. It is placed at address 3Ah, so this is its machine code. The following table shows the microprogram.

Mnemonic	Hex	LDMC	ENPC	J2	J1	J0	EFLG	EO1	EO0	F2	F1	F0	S1	S0	P1	P0	WE
LD A, (<address>)	3Ah	0	1	0	0	0	0	0	0	0	0	0	0	0	0	0	0
	+1	1	1	0	0	0	0	0	0	1	1	1	1	1	1	0	0

The first microinstruction only commands the PC to increase by one increment to keep the pipeline mechanism going. This instruction is submitted to the datapath on the same clock cycle as the location address to read is made available to the RAM. This is because the PC is targeting the location that contains the instruction operand. Through synchronous read, the RAM makes the number available after the active edge of the clock.

The second microinstruction sets F2F1F0 = '111' to let operand B pass through the ALU, and 'P1P0 = 10' so that the multiplexer copies the content of the RAM location. The number from the RAM is then loaded in the accumulator on the rising edge of the clock because signals S1 and S0 are both set at '1'. This microinstruction activates lines LDMC and ENPC to take care of loading the next instruction in the MPC and to bring the PC forward one increment as in previous cases.

1.6.7 Instructions with direct addressing

The instructions introduced here take advantage of the "direct addressing" mode to read and write numbers in the memory. Direct addressing means explicitly indicating the memory location that contains the operand by placing the address immediately after the instruction machine code.

The last modifications have changed the nature of the connection of multiplexer input I2. Now we need to rethink all the instructions that address that input, as we had to do for input I3.

At the moment, none of the instructions that work on multiplexer input I2 address the RAM memory through an address retrieved from the program memory. If we tried to execute ADD A,OP2, we would add the content in the ROM location targeted by the next instruction to the value in the accumulator. This doesn't make much sense.

Instead, let's introduce instruction ADD A,(address) whose machine code is 4Ch. This instruction specifies the RAM memory location address that we want to access immediately after its machine code. This way, the programmer can decide which memory cell to retrieve the second operand from.
The following table shows the microprogram of the instruction.

Mnemonic	Hex	LDMC	ENPC	J2	J1	J0	EFLG	EO1	EO0	F2	F1	F0	S1	S0	P1	P0	WE
ADD A,(<address>)	4Ch	0	1	0	0	0	0	0	0	0	0	0	0	0	0	0	0
	+1	1	1	0	0	0	1	0	0	0	0	0	1	1	1	0	0

As we can see, the first microinstruction simply orders the PC to increment so that the pipeline remains active.

On the next rising edge of the clock, a register inside the RAM memorizes the address from the program memory, i.e. the one we want to access. (Remember that because of the pipeline, the PC always targets the memory cell right after the one with the instruction being executed). After the memory propagation time, the content of the addressed memory cell will be available at the RAM output.

So, with the second microinstruction, all we need to do is prepare the datapath to get the sum of the content of the accumulator and the current RAM output. All we need to do to get this result is select RAM output (P1P0 = '10'), request the ALU to execute the sum (F2F1F0 = '000') and set S1S0 = '11' for it to be memorized. Lines LDMC and ENPC are activated so that the next instruction can be executed.

Similar reasoning is applied to set up instructions SUB, AND, OR, CP, IN and OUT with direct addressing, giving us:

Mnemonic	Hex	LDMC	ENPC	J2	J1	J0	EFLG	EO1	EO0	F2	F1	F0	S1	S0	P1	P0	WE
SUB A,(<address>)	4Eh	0	1	0	0	0	0	0	0	0	0	0	0	0	0	0	0
	+1	1	1	0	0	0	1	0	0	0	0	1	1	1	1	0	0
AND A,(<address>)	50h	0	1	0	0	0	0	0	0	0	0	0	0	0	0	0	0
	+1	1	1	0	0	0	1	0	0	0	1	0	1	1	1	0	0
OR A,(<address>)	52h	0	1	0	0	0	0	0	0	0	0	0	0	0	0	0	0
	+1	1	1	0	0	0	1	0	0	0	1	1	1	1	1	0	0
IN (<address>),OP0	56h	0	1	0	0	0	0	0	0	1	1	1	0	0	0	0	1
	+1	1	1	0	0	0	0	0	0	0	0	0	0	0	0	0	0
IN (<address>),OP1	58h	0	1	0	0	0	0	0	0	1	1	1	0	0	0	1	1
	+1	1	1	0	0	0	0	0	0	0	0	0	0	0	0	0	0
CP A,(<address>)	5Ah	0	1	0	0	0	0	0	0	0	0	0	0	0	0	0	0
	+1	1	1	0	0	0	1	0	0	0	0	1	0	0	1	0	0
OUT PORT0,(<address>)	5Ch	0	1	0	0	0	0	0	0	0	0	0	0	0	0	0	0
	+1	1	1	0	0	0	0	0	1	1	1	1	0	0	1	0	0
OUT PORT1,(<address>)	5Eh	0	1	0	0	0	0	0	0	0	0	0	0	0	0	0	0
	+1	1	1	0	0	0	0	1	0	1	1	1	0	0	1	0	0

Appendix B.5.1 has the list of instructions and microprograms for the final version Mp8E of the processor to consult. Appendix B.5.2 has its complete schematic.

1.6.8 Use of the Mp8E network: examples

The following are programming examples for the Mp8E network, which use RAM memory and immediate constants.

1.6.8.1 Calculating a logical expression

We want to write a program that calculates the logical expression that appeared at the beginning of Section 1.6.3, and give the result on output PORT0:

$$PORT0 = (OP0 \cdot OP1) + \overline{(OP0 + OP1)}$$

As mentioned before, this expression cannot be calculated by using the only accumulator we have. Even if we try to simplify by using theorems of Boolean algebra, we will still need to use a temporary variable to memorize at least one intermediate calculation result.

Therefore, we'll dedicate RAM location 00h for this task. The following is the mnemonic code and the corresponding machine code:

Mnemonic Code		Address	Machine Code	
IN	A,OP0	00h	1Ch	
AND	A,OP1	01h	0Dh	
LD	(00h),A	02h	38h	00h
IN	A,OP0	04h	1Ch	
OR	A,OP1	05h	11h	
NOT	A	06h	14h	
OR	A,(00h)	07h	52h	00h
OUT	PORT0,A	09h	34h	
HALT		09h	01h	01h

As you can see, after the execution of the bitwise AND between the two operands, the partial result is saved in the RAM. Then the second operation is executed (bitwise OR, negated) and the value obtained is placed in OR along with that of the RAM, giving us the final result.

1.6.8.2 Calculation of a mathematical expression

The following expression has been assigned to calculate:

$$PORT0 = OP0/2^{OP1}$$

Notice that dividing a number by the power of two means executing a number of right shifts that is equal to the exponent to the power of 2, inserting zeroes at the left. Also, OP1 should be set to a value lesser than or equal to 8 (for larger values, the result is forced to zero, whatever the value of OP0 may be). What follows is one possible solution:

```
              IN      A,OP0
              LD      (00h),A
              IN      A,OP1
              LD      (01h),A
              OR      A,00h
      LOOP:   JP      Z,EXIT
              LD      A,(00h)
              SRL     A
                      (cont.)
```

```
                   LD      (00h),A
                   LD      A,(01h)
                   SUB     A,01h
                   LD      (01h),A
                   JP      LOOP
          EXIT:    LD      A,(00h)
                   OUT     PORT0,A
                   HALT
```

First, the values of the two input ports are acquired by the program and memorized in two variables. We load OP0, the dividend, in the first variable. Then, this variable will contain the partial (and also the final) results of the calculations. The value of port OP1 (the exponent of the divisor) is loaded in the second variable, but it will be decremented to keep count of the number of right shifts necessary to get the result.

Before entering in the loop started by the LOOP label, the number of required shifts (which is still in A) must be checked. The instruction OR is used to make sure this number is not zero (executing an OR with zero does not change the value of A, but it updates the flags).

If the number of shifts to execute is zero, there is no division to execute and so we jump to EXIT. There, we update the output port with the initial value of the dividend and then place the CPU in the HALT state.

Otherwise, we divide the dividend by two by shifting it one position to the right (the rest of the division is lost). The variable containing the dividend is retrieved from the memory and brought to the accumulator where its content is shifted to the right. Then, the new value is updated in the memory.

Then, the variable containing the number of shifts is lowered by one. The process to achieve this is as follows: the processor reads the content of the variable from the memory location, subtracts constant 1 and updates the same location with the new value.

This value is equal to the number of shifts that still have to be executed. The subtraction has updated the zero flag. This will allow us to see whether that number is indeed zero.

The unconditioned jump brings us back to the LOOP label where the loop repeats if the number of shifts hasn't reached zero yet. If it has, we jump to EXIT where the final value of the calculation is brought to output port PORT0 and then the processor stops.

Moving to machine code, we get the following result:

Mnemonic Code		Address	Machine Code	
	IN A,OP0	00h	1Ch	
	LD (00h),A	01h	38h	00h
	IN A,OP1	03h	1Dh	
	LD (01h),A	04h	38h	01h
	OR A,00h	06h	46h	00h
LOOP:	JP Z,EXIT	08h	24h	17h
	LD A,(00h)	0Ah	3Ah	00h
	SRL A	0Ch	20h	
	LD (00h),A	0Dh	38h	00h
	LD A,(01h)	0Fh	3Ah	01h
	SUB A,01h	11h	42h	01h
	LD (01h),A	13h	38h	01h
	JP LOOP	15h	22h	08h
EXIT:	LD A,(00h)	17h	3Ah	00h
	OUT PORT0,A	19h	34h	
	HALT	1Ah	01h	01h

1.6.8.3 Generating the samples of a sinusoidal wave

This example will show a program that can generate samples of any kind of wave, including sinusoidal form waves on port PORT0.

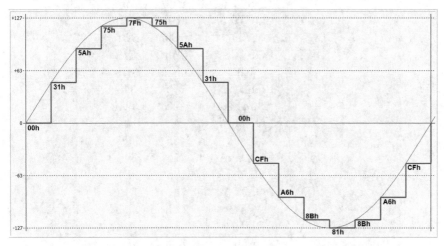

Just by looking at the figure above, we can intuit how the sinusoid samples were calculated, 8-bit integer approximations of the ideal waveform.

There are 16 values within an interval of +127..-127 (rendered in hexadecimals in the figure).

If we have samples representing a whole waveform period, we can memorize the constants in the program and then transfer them to output port PORT0. Below is an example of how the program should look for any waveform:

```
START:    LD      A,Sample1
          OUT     PORT0,A
          LD      A,Sample2
          OUT     PORT0,A
                .          .
                .          .
          LD      A,SampleN
          OUT     PORT0,A
          JP      START
```

For the sinusoid waveform in the previous figure, we get the following program (along with the corresponding machine code):

Mnemonic Code			Address	Machine Code	
START:	LD	A,00h	00h	3Dh	00h
	OUT	PORT0,A	02h	34h	
	LD	A,31h	03h	3Dh	31h
	OUT	PORT0,A	05h	34h	
	LD	A,5Ah	06h	3Dh	5Ah
	OUT	PORT0,A	08h	34h	
	LD	A, 75h	09h	3Dh	75h
	OUT	PORT0,A	0Bh	34h	
	LD	A, 7Fh	0Ch	3Dh	7Fh
	OUT	PORT0,A	0Eh	34h	
	LD	A, 75h	0Fh	3Dh	75h
	OUT	PORT0,A	11h	34h	
	LD	A, 5Ah	12h	3Dh	5Ah
	OUT	PORT0,A	14h	34h	
	LD	A, 31h	15h	3Dh	31h
	OUT	PORT0,A	17h	34h	
	LD	A, 00h	18h	3Dh	00h
	OUT	PORT0,A	1Ah	34h	
	LD	A, CFh	1Bh	3Dh	CFh
	OUT	PORT0,A	1Dh	34h	
	LD	A, A6h	1Eh	3Dh	A6h
	OUT	PORT0,A	20h	34h	
				(cont.)	

Mnemonic Code		Address	Machine Code	
LD	A, 8B	21h	3Dh	8Bh
OUT	PORT0,A	23h	34h	
LD	A, 81h	24h	3Dh	81h
OUT	PORT0,A	26h	34h	
LD	A, 8B	27h	3Dh	8Bh
OUT	PORT0,A	29h	34h	
LD	A, A6h	2Ah	3Dh	A6h
OUT	PORT0,A	2Ch	34h	
LD	A, CFh	2Dh	3Dh	CFh
OUT	PORT0,A	2Fh	34h	
JP	START	30h	22h	00h

The samples have been designed to be ideally visualized by a "Digital To Analog Converter" (DAC)[12]. The Deeds simulator offers several types of virtual DAC components that can graphically visualize the resulting analog waveform.

1.6.9 Final considerations on the processor developed here

The Deeds simulator website has the circuit schematics of all the networks we have studied here. We can use it to test all the programs we've developed. We can also broaden our horizons and design new systems or expansions based on the processors we have designed so far.

For example, some of the exercises at the end of this chapter ask you to design new instructions. In fact, the processor we designed here can execute more complex instructions than those presented so far.

Among the types of processors examined at the beginning of the chapter, our processor is clearly inspired by the Harvard architecture because it separates the program memory accesses from those of the RAM memory. We could continue the process of extending the processor's capabilities, making it ever more powerful until it becomes a modern microprocessor.

Nevertheless, proceeding with the approach from this chapter would make the network ever more complex, diverting us from one of the goals of this book: preparing the reader to efficiently program any processor from its functional specifications. This is why the authors believe that the treatment of this topic should conclude here.

---

[12] Digital To Analog Converters are electronic circuits that can transform a sequence of numbers (a "digital signal") into an "analog signal", whose value is proportional to the digital signal in the input.

As of the next chapter, we'll discard the perspective of the microprocessor designer and analyze the architecture of a complete 8-bit microprocessor, inspired by a processor that really existed and supported by complete development instruments. This time we will take the perspective of someone who wants to use the microprocessor as a component to design electronic systems. We will no longer pay attention to the minute details of how the processor functions in its interior.

We will learn to construct a small microcomputer around it (Chapter 2) and to program it in assembly language (from Chapter 3 on). Finally, we will expand the system hardware by experimenting with some microprocessor "interfacing" techniques with numerous types of devices.

1.7 Exercises

The digital content pages of the book on the Deeds simulator website have outlines of the schematics, diagrams and/or programs to complete for each exercise. Those same web pages also have the files for the solutions, so that students can check their work.

1.7.1 Dedicated computing networks

1. Design a combinational network that can multiply an unsigned 6-bit integer by 3.

2. Given two (6-bit unsigned) integers A and B, design a combinational network that can calculate the expression $((A + B)/2) + 5$. The result must also be a 6-bit integer. This means any fractional part must be cut from the number.

3. Given (8-bit unsigned) integers A, B and C, design a combinational network that can calculate the expression $(A/2+B/4+C/8)$. The result must also be an 8-bit integer. This means any fractional part must be cut from the number.

4. Define the ASM chart of the datapath sequencer on Page 14, that make it possible to calculate the following expressions.
   a) $3 \cdot (OP0) + 2 \cdot (OP3)$
   b) $3/2 \cdot (OP1) + 3 \cdot (OP2)$

5. Program the ROM memory of the network on Page 18, so that it can calculate the following expressions.
   a) $3 \cdot (OP0) + 2 \cdot (OP3)$
   b) $3/2 \cdot (OP1) + 3 \cdot (OP2)$

6. Add parity flag P to the ALU on Page 20. It should be calculated according to logic: $P = '1'$ if the result has an odd number of bits at '1', $P = '0'$ if that number is even.

7. Using the ALU on Page 20 as a reference, design a combinational component with two inputs: ZF and CO, and three outputs: A=B, A>B and A<B. The component should evaluate the flags supplied by the ALU and determine if A=B, A>B or A<B.

1.7.2 Programmable computing networks

1. Program the ROM memory of the Mp8A network on Page 22 and in Appendix B on Page 587, so that it is able to calculate the expressions indicated.

We suggest reasoning first in mnemonic code and using the table in Appendix B on Page 585. It is acceptable to truncate the fractional part in the calculation.

a) NOT [(OP2) AND (OP3) OR (OP1)]
b) 3/2 · (OP2) + (OP1)
c) [(OP0) + 2 · (OP1) − (OP2) + 3/2 · (OP3)]/4

2. Program the Mp8C network (see Section 1.3.2 and Appendix B.3) so that the accumulator assumes the timing sequence represented in the figures. If you need to use a constant, we suggest dedicating an unused input port to acquire it. For example, if you need to increment the value of the accumulator, you can assume you have a constant 1 available at input port OP1.

a) Create the following sequence in the accumulator.

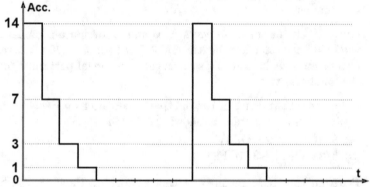

Note: you can freely choose the duration of the values and that of the interval between the two sequences; note the relation between the values in the figure (overlooking the remainders, 7 = 14 : 2, 3 = 7 : 2, 1 = 3 : 2...).

b) Create the following sequence in the accumulator.

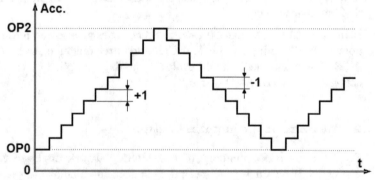

Note: you can freely choose the duration of the values.

3. Program the Mp8D network (see Section 1.5 and Appendix B.4) so that port PORT0 assumes the timing sequence represented in the figures.

a) Create the following sequence:

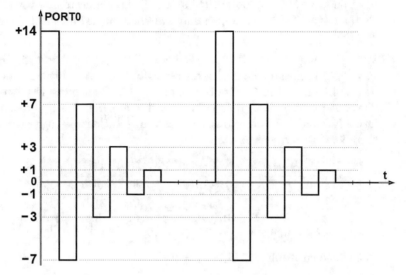

Note: you can freely choose the duration of the values and that of the time interval between the two sequences. Also, input ports can provide the necessary constants.

The relation between the values in the figure, overlooking the remainders, is as follows: $-7 = -(14 : 2)$, $-3 = -(7 : 2)$, $-1 = -(3 : 2)$...

b) Create the following sequence:

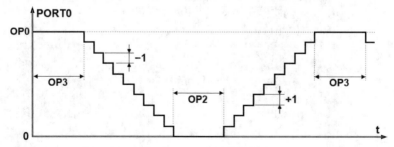

Note: assume OP0, OP2 and OP3 are different from zero. If you need to acquire constant 1, it can be set on port OP1.

4. Write a program for the Mp8E network (refer to Appendix B.5 on Page 596) that acts as a timer counting the time elapsed from reset. Once the value at port OP0 is read, the program should decrement the content of the accumulator until it reaches zero. When the count ends, we must enter an infinite loop where all the LEDs connected to output ports PORT0 and PORT1 are repeatedly switched on and off until the system is reset.

5. Write a program for the Mp8E network (refer to Appendix B.5 on Page 596) that can calculate a bitwise EXOR between memory locations 01h and 02h, and memorize the result in 00h. Remember that the ALU does not have the EXOR function but this function can be broken down into AND and OR operations. To test your solution, we suggest inserting the following segment of code.

Mnemonic	Comment	Address	Machine Code	
LD A,55h	;01010101b	00h	3Dh	55h
LD (01h),A		02h	38h	01h
LD A,AAh	;10101010b	04h	3Dh	AAh
LD (02h),A		06h	38h	02h

The expected result: RAM(00h) = FFh.

6. Write a program for the Mp8E network (refer to Appendix B.5 on Page 596) that can add two 16-bit integers. The variables containing the addends will be memorized in the RAM memory as of locations 02h and 04h. The result should be memorized as of location 00h. Any carry generated in the add should be reported by writing FFh in RAM memory location 06h. If there is no carry, the value saved in location 06h should be 00h. To test your solution, we suggest inserting the following segment of code.

Mnemonic	Comment	Address	Machine Code	
LD A,FFh	;first number	00h	3Dh	FFh
LD (02h),A		02h	38h	02h
LD A,FFh		04h	3Dh	FFh
LD (03h),A		06h	38h	03h
LD A,01h	;second number	08h	3Dh	01h
LD (04h),A		0Ah	38h	04h
LD A,00h		0Ch	3Dh	00h
LD (05h),A		0Eh	38h	05h

Expected results: RAM(00h) = 00h, RAM(01h) = 00h, RAM(06h) = FFh.

7. Write a program for the Mp8E network (refer to Appendix B.5 on Page 596) that can emulate the function of a pre-settable 8-bit syn-

chronous up/down counter. Output port PORT0 represents the state of the counter; we can read the current value of the count there.

On input port OP0 we find inputs CK, EN, UP and LD. When EN is high, it enables the counter. When EN is low the state of the counter is preserved and the other inputs are ignored.

When UP is set at '1', it commands the counter to count up (+1), otherwise it counts down (-1). When LD is set at '1', it commands the value at input port OP1 to load in the state of the counter.

The inputs should only be evaluated on the positive edge of the clock CK. The command priority is as follows: EN, LD, UP. The following table represents the connections of port OP0.

7	6	5	4	3	2	1	0
nc	nc	nc	nc	CK	LD	UP	EN

1.7.3 Microprogramming new instructions

In the exercises below, you will be asked to change the microprogram memory of the Mp8E network (see Section 1.6 and Appendix B.5 on Page 596) with the aim to add new instructions.

1. The mnemonic code below represents an instruction that makes it possible to load constant <const> directly in the RAM memory at the <address> indicated.

$$\text{LD} \quad (<\text{address}>), <\text{const}>$$

Use memory address 60h to define the microprogram associated to that instruction. In the program memory, the instruction machine code (60h) is followed by the constant and then by the RAM location address where we want to memorize the constant.

During the loading process, the previous value in the accumulator is overwritten by the constant. To test your solution, we suggest running the following program.

Mnemonic		Address	Machine code		
LD	(02h),01h	00h	60h	01h	02h
HALT		03h	01h	01h	

The expected result: RAM(02h) = 01h.

2. The following mnemonic code represents an instruction that compares the content of the RAM memory cell <address> with the value of the constant <const>.

$$\text{CP} \quad (\text{<address>}), \text{<const>}$$

Define the microprogram associated to that instruction as of memory address 63h. In the program memory, the instruction machine code (63h) is followed by the constant then by the location address holding the value to compare with that constant.

During the compare process, the previous value on the accumulator is overwritten by the constant. To test your solution, we suggest running the following program.

Mnemonic		Address	Machine code		
LD	A,01h	00h	3Dh	01h	
LD	(02h),A	02h	38h	02h	
CP	(02h),01h	04h	63h	01h	02h
HALT		07h	01h	01h	

The expected result: active ZF.

3. The following mnemonic code represents an instruction that compares the content of the RAM memory cell pointed by the first <address1> with the content of the RAM memory cell pointed by the second <address2>.

$$\text{CP} \quad (\text{<address1>}),(\text{<address2>})$$

Define the microprogram associated with that instruction at memory address 66h. In the program memory, the machine code for instruction (66h) is followed by the first <address1> and then by the second <address2>. Before executing the comparison, it is necessary to load the content of the RAM location (pointed by the first address) in the accumulator. To test your solution, we suggest running the following program.

Mnemonic		Address	Machine code		
LD	A,01h	00h	3Dh	01h	
LD	(02h),A	02h	38h	02h	
LD	A,01h	04h	3Dh	01h	
LD	(03h),A	06h	38h	03h	
CP	(02h),(03h)	08h	66h	02h	03h
HALT		0Bh	01h	01h	

The expected result: active ZF.

4. The following mnemonic code represents an instruction that adds the content of the RAM memory cell pointed by the first <address1> with the content of the RAM memory cell pointed by the second <address2>, saving the result in the accumulator.

$$\text{ADD} \quad (\text{<address1>}),(\text{<address2>})$$

Define the microprogram associated to that instruction at memory address 69h. In the program memory, the machine code for instruction (69h) is followed by the first <address1> and then by the second <address2>. We suggest using the following program to test your solution.

Mnemonic		Address	Machine code		
LD	A,01h	00h	3Dh	01h	
LD	(02h),A	02h	38h	02h	
LD	A,FFh	04h	3Dh	FFh	
LD	(03h),A	06h	38h	03h	
ADD	(02h),(03h)	08h	69h	02h	03h
HALT		0Bh	01h	01h	

The expected result: A = 00h, ZF = 1, CO = 1.

5. The following mnemonic code represents an instruction that adds the content of the RAM memory cell pointed by the first <address1> with the content of the RAM memory cell pointed by the second <address2>. It then saves the result in the RAM location pointed by the third <address3>.

$$\text{ADD} \quad (<address1>),(<address2>),(<address3>)$$

Define the microprogram associated with that instruction at memory address 6Ch. In the program memory, the instruction machine code (6Ch) is followed by <address1>, <address2> and <address3> one after the other. In the add process, the previous value in the accumulator is overwritten. We suggest using the following program to test your solution.

Mnemonic		Address	Machine code			
LD	A,01h	00h	3Dh	01h		
LD	(02h),A	02h	38h	02h		
LD	A,FFh	04h	3Dh	FFh		
LD	(03h),A	06h	38h	03h		
ADD	(02h),(03h),(04h)	08h	6Ch	02h	03h	04h
HALT		0Ch	01h	01h		

The expected result: RAM(04h) = 00h, ZF = 1, CO = 1.

6. Following the same criteria as those in the previous exercise, define the microprograms associated to the instructions listed in the following table and allocated at the address indicated.

Mnemonic		Address
SUB	(<address1>),(<address2>),(<address3>)	71h
AND	(<address1>),(<address2>),(<address3>)	76h
OR	(<address1>),(<address2>),(<address3>)	7Bh

We suggest testing your solutions with the following test programs.

Mnemonic		Address	Machine code			
LD	A,01h	00h	3Dh	01h		
LD	(02h),A	02h	38h	02h		
LD	A,02h	04h	3Dh	02h		
LD	(03h),A	06h	38h	03h		
SUB	(02h),(03h),(04h)	08h	71h	02h	03h	04h
HALT		0Ch	01h	01h		

The expected result: RAM(04h) = FFh, CO = 1.

Mnemonic		Address	Machine code			
LD	A,FFh	00h	3Dh	FFh		
LD	(02h),A	02h	38h	02h		
LD	A,55h	04h	3Dh	55h		
LD	(03h),A	06h	38h	03h		
AND	(02h),(03h),(04h)	08h	76h	02h	03h	04h
HALT		0Ch	01h	01h		

The expected result: RAM(04h) = 55h.

Mnemonic		Address	Machine code			
LD	A,00h	00h	3Dh	00h		
LD	(02h),A	02h	38h	02h		
LD	A,55h	04h	3Dh	55h		
LD	(03h),A	06h	38h	03h		
OR	(02h),(03h),(04h)	08h	7Bh	02h	03h	04h
HALT		0Ch	01h	01h		

The expected result: RAM(04h) = 55h.

7. The following mnemonic code represents an instruction that right shifts the content of the <address> memory location and inserts zeroes at the left.

$$\text{SRL} \quad (\text{<address>})$$

Define the microprogram associated to that instruction at memory address 80h. In the program memory, the instruction machine code (80h) is followed by the <address>. Only the accumulator can right shift its content so its previous value will be overwritten. We suggest using the following program to test your solution.

Mnemonic		Address	Machine code	
LD	A,01h	00h	3Dh	01h
LD	(02h),A	02h	38h	02h
SRL	(02h)	04h	80h	02h
HALT		06h	01h	01h

The expected result: RAM(02h) = 00h, ZF = 1.

8. The following mnemonic code represents an instruction that left shifts the content of the memory location pointed by <address> and inserts zeroes at the right.

$$\text{SLL} \quad (\text{<address>})$$

Define the microprogram associated to that instruction at memory address 85h. In the program memory, the instruction machine code (85h) is followed by the <address>. Only the accumulator can left shift its content so its previous value will be overwritten. We suggest using the following program to test your solution.

Mnemonic		Address	Machine code	
LD	A,80h	00h	3Dh	80h
LD	(02h),A	02h	38h	02h
SLL	(02h)	04h	85h	02h
HALT		06h	01h	01h

The expected result: RAM(02h) = 00h, ZF = 1.

1.8 Solutions

1.8.1 Dedicated computing networks

1.

Multiplying by three is the same as adding the same number three times. In base ten, the highest expressible 6-bit number is $2^6 - 1 = 63$. So, the lowest number of bits that are needed to express the result is $\lceil \log_2(63 \cdot 3) \rceil = 8$.

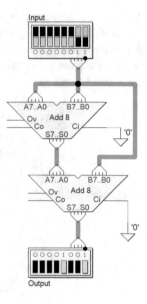

2.

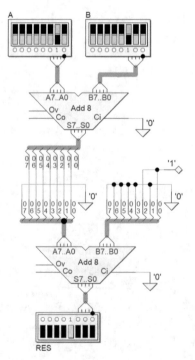

After adding A and B, we right shift the result by one bit and we add it to constant 5.

3.

Let's shift operands A, B and C before adding them. Shifting is required to divide them by the amount assigned.

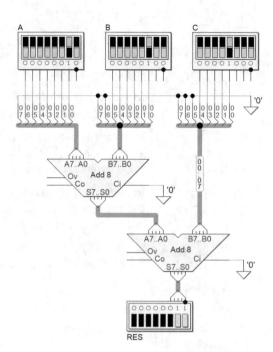

4. a) Calculating expression $3 \cdot (OP0) + 2 \cdot (OP3)$.

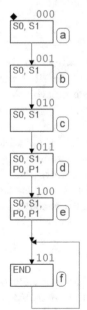

In state (a), OP0 is added to the content of the accumulator (zero). In other words, OP0 is imported into the accumulator.

In states (b) and (c), OP0 is again added to the content of the accumulator, giving us $3 \cdot (OP0)$.

In states (d) and (e), OP3 is added twice to the content of the accumulator, giving us the calculation of the expression that was called for.

Finally in state (f), the network is stopped. To execute a new calculation, the network must be reset.

b) Calculation of expression $3/2 \cdot (OP1) + 3 \cdot (OP2)$.

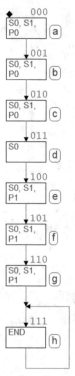

In state (a), OP1 is added to the content of the accumulator (zero). In other words, OP1 is imported into the accumulator.

In states (b) and (c) OP1 is again added to the content of the accumulator, giving us $3 \cdot (OP1)$.

In state (d) the accumulator is ordered to right shift, giving us $3/2 \cdot (OP1)$. Remember that the shift operation inserts a zero at the left.

In states (e), (f) and (g) OP2 is added three times to the content of the accumulator, giving us the calculation of the expression that was called for.

In state (h) the network is stopped. To execute a new calculation, the network must be reset.

5. a) Calculation of expression $3 \cdot (OP0) + 2 \cdot (OP3)$.

Address	Content
00h	0000.0011
01h	0000.0011
02h	0000.0011
03h	1100.0011
04h	1100.0011
05h	0010.0000

At address 00h, OP0 is added to the content of the accumulator (zero). In other words OP0 is imported into the accumulator. At the next two locations, OP0 is added again to the content of the accumulator, giving us $3 \cdot (OP0)$. At locations 03h and 04h, OP3 is added twice to the content of the accumulator, giving us the calculation of the expression that was called for. Finally, the network is stopped at the last ROM location. To execute a new calculation, the network must be reset.

b) Calculating expression $3/2 \cdot (OP1) + 3 \cdot (OP2)$.

At address 00h, OP1 is added to the content of the accumulator, which has a value of 00h after the reset (see the following table). At addresses 01h and 02h, OP1 is again added to the content of the accumulator, giving us $3 \cdot (OP1)$.

Address	Content
00h	0100.0011
01h	0100.0011
02h	0100.0011
03h	0000.0001
04h	1000.0011
05h	1000.0011
06h	1000.0011
07h	0010.0000

At 03h the accumulator is ordered to right shift, giving us $3/2 \cdot$ (OP1). Remember that the shift operation inserts a zero at the left. From address 04h to 06h, OP2 is added to the content of the accumulator three times, giving us the calculation of the expression that was called for. At location 07h the network is stopped. To execute a new calculation, we must reset the network.

6. The logical function EXOR is also called an "odd function" because it generates a 1 only if the number of 1's at the input is odd.

We get: $P = F_7 \oplus F_6 \oplus F_5 \oplus F_4 \oplus F_3 \oplus F_2 \oplus F_1 \oplus F_0$.

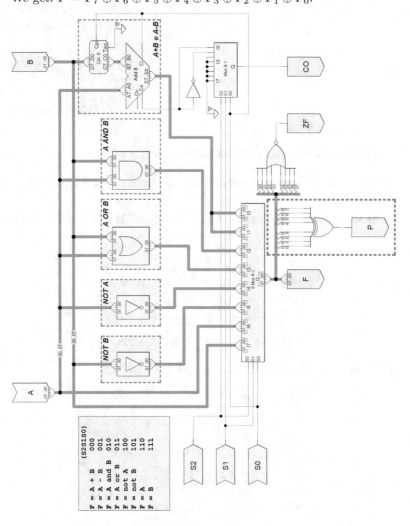

7. By analyzing the Flags table below, (also available on Page 21), we can derive the network seen on the right.

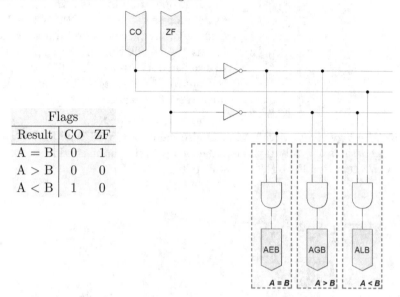

Flags		
Result	CO	ZF
A = B	0	1
A > B	0	0
A < B	1	0

1.8.2 Programmable computing networks

1. The following are solutions expressed in terms of mnemonic codes and machine codes.

a) Calculating the expression: NOT [(OP2) AND (OP3) OR (OP1)]

Mnemonic		Address	Machine code
IN	A,OP2	00h	9Fh
AND	A,OP3	01h	CBh
OR	A,OP1	02h	4Fh
NOT	A	03h	13h
HALT		04h	20h

b) Calculating the expression: $3/2 \cdot (OP2) + (OP1)$

Mnemonic		Address	Machine code
IN	A,OP2	00h	9Fh
ADD	A,OP2	01h	83h
ADD	A,OP2	02h	83h
SRL	A	03h	01h
ADD	A,OP1	04h	43h
HALT		05h	20h

c) Calculating the expression: $1/4 \cdot [(OP0) + 2 \cdot (OP1) - (OP2) + 3/2 \cdot (OP3)]$

Mnemonic		Address	Machine code
IN	A,OP3	00h	DFh
ADD	A,OP3	01h	C3h
ADD	A,OP3	02h	C3h
SRL	A	03h	01h
SUB	A,OP2	04h	87h
ADD	A,OP1	05h	43h
ADD	A,OP1	06h	43h
ADD	A,OP0	07h	03h
SRL	A	08h	01h
SRL	A	09h	01h
HALT		0Ah	20h

2. The following are solutions expressed in terms of mnemonic codes and machine codes.

a) Here, we assume we have constant 0Eh available at input port OP0. The instruction NOP takes no action but adds a delay, and we can add more than one. This is a way to regulate the period of the waveform generated.

Label	Mnemonic		Address	Machine code	
START:	IN	A,OP0	00h	1Ch	
SHIFT:	SRL	A	01h	20h	
	JP	NZ, SHIFT	02h	26h	01h
	NOP		04h	00h	
	JP	START	05h	22h	00h

b) Assume that we have constant 01h available at input port OP1.

Label	Mnemonic		Address	Machine code	
	IN	A,OP0	00h	1Ch	
CHECKH:	CP	A,OP2	01h	32h	
	JP	Z,CHECKL	02h	24h	07h
INC:	ADD	A,OP1	04h	05h	
	JP	CHECKH	05h	22h	01h
CHECKL:	CP	A,OP0	07h	30h	
	JP	Z,CHECKH	08h	24h	01h
DEC:	SUB	A,OP1	0Ah	09h	
	JP	CHECKL	0Bh	22h	07h

3. The following are solutions expressed in terms of mnemonic codes and machine codes.

a) Assume we have constant 01h available at input port OP1 and constant 0Eh at input port OP0.

Label	Mnemonic		Address	Machine code	
START:	IN	A,OP0	00h	1Ch	
LOOP:	OUT	PORT0,A	01h	34h	
	SRL	A	02h	20h	
	NOT	A	03h	14h	
	ADD	A,OP1	04h	05h	
	OUT	PORT0,A	05h	34h	
	NOT	A	06h	14h	
	ADD	A,OP1	07h	05h	
	JP	NZ,LOOP	08h	26h	01h
	JP	START	0Ah	22h	00h

b) Assume we have constant 01h available at input port OP1, and that we can never have a zero in input ports OP0, OP2 or OP3.

Label	Mnemonic		Address	Machine code	
START:	IN	A,OPO	00h	1Ch	
	OUT	PORT0,A	01h	34h	
	IN	A,OP3	02h	1Fh	
PAUSE1:	SUB	A,OP1	03h	09h	
	JP	NZ,PAUSE1	04h	26h	03h
	IN	A,OP0	06h	1Ch	
DEC:	SUB	A,OP1	07h	09h	
	OUT	PORT0,A	08h	34h	
	JP	NZ,DEC	09h	26h	07h
	IN	A,OP2	0Bh	1Eh	
PAUSE2:	SUB	A,OP1	0Ch	09h	
	JP	NZ,PAUSE2	0Dh	26h	0Ch
INC:	ADD	A,OP1	0Fh	05h	
	CP	A,OP0	10h	30h	
	OUT	PORT0,A	11h	34h	
	JP	NZ,INC	12h	26h	0Fh
	JP	START	14h	22h	00h

4. The mnemonic codes are on the left and the machine codes are on the right.

Label	Mnemonic		Comment	Addr.	Machine code	
	IN	A,OP0		00h	1Ch	
DECR:	SUB	A,01h	;delay	01h	42h	01h
	JP	NZ,DECR		03h	26h	01h
LIGHTS:	LD	A,FFh	;LED on/off	05h	3Dh	FFh
	OUT	PORT0,A		07h	34h	
	OUT	PORT1,A		08h	35h	
	LD	A,00h		09h	3Dh	00h
	OUT	PORT0,A		0Bh	34h	
	OUT	PORT1,A		0Ch	35h	
	JP	LIGHTS		0Dh	22h	05h

5. Since the MP8E network's ALU cannot calculate a bitwise EXOR, it has to do it in an alternative way.

The EXOR operator can be expressed in terms of AND and OR, which can be directly calculated by the ALU in this network:

$$(A \oplus B) = \overline{A} \cdot B + A \cdot \overline{B}$$

To calculate this expression we must use an 8-bit variable to memorize the result of one of the two AND operations (let's reserve RAM location 00h for the purpose). The result of the remaining AND can be kept directly in the accumulator, which will be used to calculate the OR with the previous intermediate result. This will provide the desired result. Remember that operands A and B were previously memorized by the test program in RAM locations 01h and 02h as suggested in the exercise:

Label	Mnemonic		Comment	Address	Machine code	
	LD	A,55h	;test program	00h	3Dh	55h
	LD	(01h),A		02h	38h	01h
	LD	A,AAh		04h	3Dh	AAh
	LD	(02h),A		06h	38h	02h

The final result is memorized at location 00h.

The program begins by calculating the first intermediate result $\overline{A} \cdot B$, reading operand A from the RAM memory, negating it and calculating the AND with operand B. The result of the sequence of operations is saved at location 00h to make it possible to calculate the second intermediate result.

LD	A,(01h)	;solution	08h	3Ah	01h
NOT	A		0Ah	14h	
AND	A,(02h)	;(not A) and B	0Bh	50h	02h
LD	(00h),A	;partial result in 00h	0Dh	38h	00h

By following the same sequence of operations, we calculate the second intermediate $A \cdot \overline{B}$ result, which will then be OR-ed with the previous one.

LD	A,(02h)		0Fh	3Ah	02h
NOT	A		11h	14h	
AND	A,(01h)	;A and (not B)	12h	50h	01h
OR	A,(00h)	;OR between the two terms	14h	52h	00h

Finally, the total result of the expression is memorized in memory location 00h as ordered in the text. The program ends by putting the processor in the HALT state.

| LD | (00h),A | ;result in RAM | 16h | 38h | 00h |
| HALT | | | 18h | 01h | 01h |

6. The following is the solution expressed in terms of mnemonic codes and machine codes. This is preceded by the instructions suggested in the text.

Label	Mnemonic		Comment	Addr.	Machine code	
	LD	A,FFh	;test program	00h	3Dh	FFh
	LD	(02h),A		02h	38h	02h
	LD	A,FFh		04h	3Dh	FFh
	LD	(03h),A	;FFFFh first number	06h	38h	03h
	LD	A,01h		08h	3Dh	01h
	LD	(04h),A		0Ah	38h	04h
	LD	A,00h	;0001h second number	0Ch	3Dh	00h
	LD	(05h),A		0Eh	38h	05h
	LD	A,00h	;carry variable = 0	10h	3Dh	00h
	LD	(06h),A		12h	38h	06h
	LD	A,(02h)		14h	3Ah	02h
	ADD	A,(04h)	;add low bytes,	16h	4Ch	04h
	LD	(00h),A	;save result	18h	38h	00h
	LD	A,(03h)		1Ah	3Ah	03h
	JP	NC,NEXTB	;if carry, increment	1Ch	2Ah	2Ah
	ADD	A,01h	;the next byte	1Eh	40h	01h
	JP	NC,NEXTB	;if carry, save	20h	2Ah	2Ah
						(cont.)

	LD	(01h),A	;result in (01h)	22h	38h	01h
	LD	A,FFh	;set carry variable	24h	3Dh	FFh
	LD	(06h),A		26h	38h	06h
	LD	A,(01h)	;A= previous result	28h	3Ah	01h
NEXTB:	ADD	A,(05h)	;add high bytes	2Ah	4Ch	05h
	LD	(01h),A	;save result	2Ch	38h	01h
	JP	NC,END		2Eh	2Ah	34h
	LD	A,FFh	;if carry,	30h	3Dh	FFh
	LD	(06h),A	;carry variable = 1	32h	38h	06h
END:	HALT			34h	01h	01h

7. For this solution, we decided to use two variables: one to store the state of the counter and the other to store the inputs that we read. After initializing the state of the counter to zero, input CK is evaluated. If it is not zero, we execute the evaluation again and wait for the value to be one.

Label	Mnemonic		Comment	Addr.	Machine Code	
RESET:	LD	A,00h		00h	3Dh	00h
	LD	(01h),A	;state initialization	02h	38h	01h
	OUT	PORT0,A		04h	34h	
START:	IN	A,OP0	;read comands	05h	1Ch	
	AND	A,00001000b	;check CK	06h	44h	08h
	JP	NZ,START	;waiting for CK = 0	08h	26h	05h

Once out of the loop, the value of CK is zero. At this point, we enter a new loop, and wait for the value to be one. We leave the loop only at the rising edge of CK.

CHECK1:	IN	A,OP0		0Ah	1Ch	
	LD	(00h),A	;save in 00h	0Bh	38h	00h
	AND	A,00001000b	;check CK	0Dh	44h	08h
	JP	Z,CHECK1	;waiting for CK = 1	0Fh	24h	0Ah

Now the commands can be evaluated. First they are saved in their variable at RAM memory address 00h, then the command bits are evaluated one by one according to the priority specification, through a bitwise AND mask.

Firstly, we check if enable is activated, and if it isn't we go back to wait for the next rising edge of CK.

LD	A,(00h)	;reload commands	11h	3Ah	00h
AND	A,00000001b	;check EN value	13h	44h	01h
JP	Z,START		15h	24h	05h

Then we check the LD load input. If it is active, we jump to the LOAD label, otherwise we move ahead.

LD	A,(00h)	;check LD value	17h	3Ah	00h
AND	A,00000100b		19h	44h	04h
JP	NZ,LOAD		1Bh	26h	2Dh

We then evaluate the UP input. If it is active, the state of the counter increments; if not, it decrements.

In either case, we jump to the UPDATE label where the state and the output are updated.

	LD	A,(00h)	;check UP value	1Dh	3Ah	00h
	AND	A,00000010b		1Fh	44h	02h
	LD	A,(01h)	;get state value	21h	3Ah	01h
	JP	NZ,UP		23h	26h	29h
DN:	SUB	A,01h	;decrement it	25h	42h	01h
	JP	UPDATE		27h	22h	2Eh
UP:	ADD	A,01h	;increment it	29h	40h	01h
	JP	UPDATE		2Bh	22h	2Eh

If the counter has been ordered to load, port OP1 is acquired and we continue updating the state and the outputs so that we can begin again from the START label.

LOAD:	IN	A,OP1	;read OP1	2Dh	1Dh	
UPDATE:	LD	(01h),A	;save and copy it	2Eh	38h	01h
	OUT	PORT0,A	;to the port	30h	34h	
	JP	START		31h	22h	05h

1.8.3 Microprogramming new instructions

1. The instruction:

Mnemonic	Machine Code
LD (\<address\>),\<const\>	60h \<const\> \<address\>

Its microprogram:

Hex	LDMC	ENPC	J2	J1	J0	EFLG	EO1	EO0	F2	F1	F0	S1	S0	P1	P0	WE	Notes
60h	0	1	0	0	0	0	0	0	1	1	1	1	1	1	1	0	(a)
+1	0	1	0	0	0	0	0	0	1	1	0	0	0	0	0	1	(b)
+2	1	1	0	0	0	0	0	0	0	0	0	0	0	0	0	0	(c)

Notes:

a) The constant comes from the program memory (the PC always points to the location ahead of the one of the instruction currently being executed). This means that the paths are set to load the constant in the accumulator (P1P0 = '11' to select the program memory as input B of the ALU; F2F1F0 = '111' to copy input B of the ALU onto the output; S1S0 = '11' to store the output of the ALU in the accumulator). Also, the PC is ordered to increment.

b) The location address where the constant has to be saved comes from the program memory. Thus, the paths for loading the content of the accumulator in the RAM memory cell pointed by the program memory are defined (F2F1F0 = '110' to copy input A of the ALU onto the output; WE = '1' to order the number at the RAM input to be stored). The PC is ordered to increment to point to the next instruction to execute. On the next positive edge of the clock, the RAM memory reads the address and the input number is written in the desired location.

c) The PC is ordered to increment and the MPC to load so that the next instruction can be executed.

2. The instruction:

Mnemonic	Machine Code
CP (<address>),<const>	63h <const> <address>

Its microprogram:

Hex	LDMC	ENPC	J2	J1	J0	EFLG	EO1	EO0	F2	F1	F0	S1	S0	P1	P0	WE	Notes
63h	0	1	0	0	0	0	0	0	1	1	1	1	1	1	1	0	(a)
+1	0	1	0	0	0	0	0	0	0	0	0	0	0	0	0	0	(b)
+2	1	1	0	0	0	1	0	0	0	0	1	0	0	1	0	0	(c)

Notes:

a) The constant comes from the program memory (the PC always points to the location ahead of the one of the instruction currently being executed). This means that the paths are set to load the constant in the accumulator (P1P0 = '11' to select the program memory as input B of the ALU; F2F1F0 = '111' to copy input B of the ALU onto the output; S1S0 = '11' to store the output of the ALU in the accumulator). Also, the PC is ordered to increment.

b) Now, the address that will be acquired by the RAM (which is synchronous) on the next positive clock edge comes from the program memory. Only the PC is ordered to increment to maintain the pipeline active.

c) Now that the address has been read by the RAM, the paths to compare the RAM output value with the content of the accumulator are set (P1P0 = '10' to select the RAM memory as input B of the ALU; F2F1F0 = '001 to execute the subtraction A-B; EFLAG = '1' to store the new flag).
Notice that the result of the subtraction is not saved in the accumulator. The only purpose of instruction CP is to change the flags to compare the values and see what relationship there is between them. Also, the PC is ordered to increment to keep the pipeline active and the MPC to load to execute the next instruction.

3. The instruction:

Mnemonic	Machine Code
CP (<address1>),(<address2>)	66h <addr.1> <addr.2>

Its microprogram:

Hex	LDMC	ENPC	J2	J1	J0	EFLG	EO1	EO0	F2	F1	F0	S1	S0	P1	P0	WE	Notes
66h	0	1	0	0	0	0	0	0	0	0	0	0	0	0	0	0	(a)
+1	0	1	0	0	0	0	0	0	1	1	1	1	1	1	0	0	(b)
+2	1	1	0	0	0	1	0	0	0	0	1	0	0	1	0	0	(c)

Notes:

a) The first address comes from the microprogram (the PC always points to the location ahead of the one of the instruction currently being executed because of the pipeline mechanism), and it is read by the RAM memory on the next positive edge of the clock. Then, the PC is incremented.

b) Now, the content of the first address comes from the RAM memory, which means that the paths are set to load the constant in the accumulator (P1P0 = '10' to select the RAM memory output as ALU input B; F2F1F0 = '111' to copy input B of the ALU onto the output; S1S0 = '11' to update the content of the accumulator with the ALU output value). The PC is incremented to maintain the pipeline.

c) Now, the content of the second address, that was read on the previous positive edge of the clock, comes from the RAM memory, so the path to compare the RAM output value to the content of the accumulator is set (P1P0 = '10' to select the RAM memory output as ALU input B; F2F1F0 = '001' to subtract the content of the accumulator from the RAM output; EFLG = '1' to store the flags generated by that subtraction, the only result of interest). The PC is incremented to maintain the pipeline and the MPC is ordered to load to execute the next instruction.

4. The instruction:

Mnemonic	Machine Code
ADD (<address1>),(<address2>)	69h <addr.1> <addr.2>

Its microprogram:

Hex	LDMC	ENPC	J2	J1	J0	EFLG	EO1	EO0	F2	F1	F0	S1	S0	P1	P0	WE	Note
69h	0	1	0	0	0	0	0	0	0	0	0	0	0	0	0	0	
+1	0	1	0	0	0	0	0	0	1	1	1	1	1	1	0	0	
+2	1	1	0	0	0	1	0	0	0	0	0	1	1	1	0	0	(a)

Note:

a) The reasoning behind this exercise is similar to that of the previous one, except that the last microinstruction prepares the paths to add the value from the RAM to the content of the accumulator.

This time the result of the addition is saved in the accumulator (P1P0 = '10' to select the RAM memory output as ALU input B; F2F1F0 = '000' to add the content of the accumulator to the RAM output; EFLG = '1' to store the flags generated by the addition; S1S0 = '11' to save the result in the accumulator).

5. The instruction:

Mnemonic	Machine Code
ADD (<adr1>),(<adr2>),(<adr3>)	6Ch <adr1> <adr2> <adr3>

Its microprogram:

Hex	LDMC	ENPC	J2	J1	J0	EFLG	EO1	EO0	F2	F1	F0	S1	S0	P1	P0	WE	Notes
6Ch	0	1	0	0	0	0	0	0	0	0	0	0	0	0	0	0	(a)
+1	0	1	0	0	0	0	0	0	1	1	1	1	1	1	0	0	
+2	0	0	0	0	0	1	0	0	0	0	0	1	1	1	0	0	(b)
+3	0	1	0	0	0	0	0	0	1	1	0	0	0	0	0	1	(c)
+4	1	1	0	0	0	0	0	0	0	0	0	0	0	0	0	0	(d)

Notes:

a) The first and second microinstructions follow the same reasoning as those in the last two exercises. The first microinstruction increments the PC, waiting for <adr1> to be read by the program memory. The second one loads the value pointed by <adr1> in the accumulator and increments the PC.

b) The third address comes from the program memory, but the last address read by the RAM was the second one, so the paths to add its content to that of the accumulator are set. The result is saved in the accumulator (P1P0 = '10' to select the RAM memory output as ALU input B; F2F1F0 = '000' to add the content of the accumulator to the RAM output; S1S0 = '11' to save the result of the add in the accumulator).

c) Now that the result is ready in the accumulator, the paths are set to bring it to the memory location pointed by the third address, the last one read by the RAM (F2F1F0 = '110' to copy the value of the accumulator to the ALU output; WE = '1' to store the ALU output in the memory cell pointed by the third address). The PC is incremented to maintain the pipeline.

d) The PC is incremented to keep the pipeline active and the MPC is ordered to load to execute the next instruction.

6. The instruction:

Mnemonic	Machine Code
SUB (<adr1>),(<adr2>),(<adr3>)	71h <adr1> <adr2> <adr3>

Its microprogram:

Hex	LDMC	ENPC	J2	J1	J0	EFLG	EO1	EO0	F2	F1	F0	S1	S0	P1	P0	WE	Note
71h	0	1	0	0	0	0	0	0	0	0	0	0	0	0	0	0	
+1	0	1	0	0	0	0	0	0	1	1	1	1	1	1	0	0	
+2	0	0	0	0	0	1	0	0	0	0	1	1	1	1	0	0	(a)
+3	0	1	0	0	0	0	0	0	1	1	0	0	0	0	0	1	
+4	1	1	0	0	0	0	0	0	0	0	0	0	0	0	0	0	

The instruction:

Mnemonic	Machine Code
AND (<adr1>),(<adr2>),(<adr3>)	76h <adr1> <adr2> <adr3>

Its microprogram:

		AND (<address1>),(<address2>),(<address3>)															
Hex	LDMC	ENPC	J2	J1	J0	EFLG	EO1	EO0	F2	F1	F0	S1	S0	P1	P0	WE	Note
76h	0	1	0	0	0	0	0	0	0	0	0	0	0	0	0	0	
+1	0	1	0	0	0	0	0	0	1	1	1	1	1	1	0	0	
+2	0	0	0	0	0	1	0	0	0	1	0	1	1	1	0	0	(a)
+3	0	1	0	0	0	0	0	0	1	1	0	0	0	0	0	1	
+4	1	1	0	0	0	0	0	0	0	0	0	0	0	0	0	0	

The instruction:

Mnemonic	Machine Code
OR (<adr1>),(<adr2>),(<adr3>)	7Bh <adr1> <adr2> <adr3>

Its microprogram:

Hex	LDMC	ENPC	J2	J1	J0	EFLG	EO1	EO0	F2	F1	F0	S1	S0	P1	P0	WE	Note
7Bh	0	1	0	0	0	0	0	0	0	0	0	0	0	0	0	0	
+1	0	1	0	0	0	0	0	0	1	1	1	1	1	1	0	0	
+2	0	0	0	0	0	1	0	0	0	1	1	1	1	1	0	0	(a)
+3	0	1	0	0	0	0	0	0	1	1	0	0	0	0	0	1	
+4	1	1	0	0	0	0	0	0	0	0	0	0	0	0	0	0	

Note (for all three microprograms):

a) The reasoning behind this exercise is similar to that of the previous
   one. The three microprograms are identical, except for the third microinstruction which executes the required operation (a subtraction if
   $F2F1F0 = $ '001', a bitwise AND if $F2F1F0 = $ '010' or a bitwise OR if
   $F2F1F0 = $ '011').

7. The instruction:

Mnemonic	Machine Code
SRL (\<address\>)	80h \<address\>

Its microprogram:

Hex	LDMC	ENPC	J2	J1	J0	EFLG	EO1	EO0	F2	F1	F0	S1	S0	P1	P0	WE	Notes
80h	0	0	0	0	0	0	0	0	0	0	0	0	0	0	0	0	(a)
+1	0	0	0	0	0	0	0	0	1	1	1	1	1	1	0	0	(b)
+2	0	0	0	0	0	0	0	0	0	0	0	0	1	0	0	0	(c)
+3	0	1	0	0	0	1	0	0	1	1	0	0	0	0	0	1	(d)
+4	1	1	0	0	0	0	0	0	0	0	0	0	0	0	0	0	(e)

Notes:

a) The RAM memory reads the program memory output address on the next edge of the clock. We must allow one cycle to pass while taking no action. Since the result must be saved at the same address, which comes from the program memory, the PC is not ordered to increment so that the same address is sent to the RAM.

b) The paths are set to save the RAM output in the accumulator (P1P0 = '10' to select the RAM as ALU input B; F2F1F0 = '111' to copy input B in the ALU output; S1S0 = '11' to save the ALU output in the accumulator).

c) It commands the accumulator to right shift (S1S0 = '01').

d) The paths are set to load the content of the accumulator in the RAM location pointed by the program memory output address (F2F1F0 = '110' to copy the content of the accumulator to the ALU output; EFLG = '1' to update the flag register; WE = '1' to store the output of the ALU in the RAM memory location pointed by \<address\>).
The flag register is enabled to evaluate the accumulator and the PC is ordered to increment, to point to the next instruction.

e) The MPC is ordered to load the next instruction and the PC is incremented to keep the pipeline active.

8. The instruction:

Mnemonic	Machine Code
SLL (\<address\>)	85h \<address\>

Its microprogram:

Hex	LDMC	ENPC	J2	J1	J0	EFLG	EO1	EO0	F2	F1	F0	S1	S0	P1	P0	WE	Notes
85h	0	0	0	0	0	0	0	0	0	0	0	0	0	0	0	0	
+1	0	0	0	0	0	0	0	0	1	1	1	1	1	1	0	0	
+2	0	0	0	0	0	0	0	0	0	0	0	1	0	0	0	0	(a)
+3	0	1	0	0	0	1	0	0	1	1	0	0	0	0	0	1	
+4	1	1	0	0	0	0	0	0	0	0	0	0	0	0	0	0	

Note:

a) Here, the same reasoning is applied as in the previous exercise, except that the third microinstruction commands a left shift (S1S0 = '10').

2

A system based on the DMC8 microprocessor

Abstract In this chapter, we introduce the DMC8 microprocessor through the study of a microcomputer based on it. In particular we will present the organization of its bus, its architecture, and how it carries out instructions. Then we will start to detail the behavior over time of the processor's bus signals. Consequently we will present how memory and input/output subsystems can be organized and designed. Finally we will show how the Deeds simulator supports the development of projects based on the DMC8 microprocessor, including writing programs and functionally testing them.

2.1 The DMC8 microprocessor

In Chapter 1 we built and thoroughly studied the architecture of a precursor to a microprocessor, detailing each and every mechanism. The main goal was not to learn to program that specific device but to give a clear idea of how a general, simplified processor can work by getting into the particulars of its logical operations. The point of view was that of a microprocessor architecture designer.

From this chapter on, the goals and points of view will be different. We are no longer interested in studying the details of microprocessor architecture. Our interest is rather on the macro-functional aspect of microprocessors, that is studying the set of instructions and their influence on the components the system programmer directly interfaces with (registers, memory, input and output ports, timers...).

The explanation takes the point of view of microprocessor-based system designers, i.e. those who first have to integrate microprocessors into systems based on the tasks they must complete and then program them so that they execute those tasks correctly.

As outlined at the beginning of Chapter 1, since the dawn of the microprocessor in the early '70s, they have constantly evolved to be capable of executing faster and faster calculations. Obviously, along with this potential their architecture has grown equally complex.

From an educational point of view, this means that it might not be a good idea to use a new-generation, commercial microprocessor to introduce the subject. The first microprocessors, which were simpler, are educationally more approachable, and yet, they were designed in function of the technology and components that were available then. On the other hand, they often incorporated now outdated accessory functions and devices which makes them harder to understand for students today.

The educational approach of this book requires a microprocessor that is not very complex, similar to the first devices to be sold commercially. At the same time, it needs to be as orderly as possible and have the essential functionalities needed for teaching purposes. This is why, a few years ago, we reinvented a microprocessor to fit our educational objectives, supported by an appropriate development tool. This is how the DMC8 (Deeds Microcomputer - 8 bit) was created within the Deeds simulator environment.

So as not to distance ourselves too much from commercial devices, the DMC8 was designed with a real and historically relevant 8-bit microprocessor in mind, the Z80-CPU from Zilog. The Z80-CPU was introduced in 1976 by Federico Faggin (who founded Zilog, Inc., among other initiatives, as cited in Chapter 1), as an improvement on the previous (and popular) Intel 8080. Because it was compatible with its predecessor and offered improvements, it quickly became a market standard. Its architecture and set of instructions have survived in some microcontrollers even today.

The DMC8 has kept a large part of the internal architecture of the Z80-CPU, minus some elements to simplify the structure. Other elements have been added to make the relationship with more recent microprocessors easier. For those who are already familiar with the Z80-CPU, some of its elements like dynamic memory management and the alternative register bank have been taken out for the DMC8. Its interrupt vector management has been made simpler and more like some of the devices currently in use. The native set of instructions and the native assembly language have been largely maintained, except for instructions regarding the parts of the internal architecture that have been eliminated.

The DMC8 also has an alternative work mode called the D8080, that configures it to be much like its predecessor, the Intel 8080. It can be programmed through its original assembly language, which has remained almost totally intact.

The DMC8 is currently configured as a soft-processor, meaning it doesn't correspond to any commercially available component but its hardware can be loaded in programmable components like FPGAs[1], by using the Deeds simulation and development environment.

Therefore, we can design, simulate and then build a whole digital system that includes a DMC8 microprocessor within Deeds. The system can easily be implemented on one of the FPGA cards supported by Deeds and available commercially. The physical implementation of entire digital systems on prototype boards will be analyzed in Chapter 5.

Deeds includes the Microcomputer Emulator (Deeds-McE), which has a text editor for writing programs and an interactive debugger[2] to develop them. The Digital Circuit Simulator (Deeds-DcS) makes it possible to develop projects that include many DMC8-based microcomputer models that are resizable in terms of RAM, ROM and input/output ports.

We will introduce these development instruments toward the end of this chapter in Section 2.4.

2.1.1 The internal architecture of the DMC8 processor

Let's look again at the figure from the beginning of Chapter 1, where we introduced the Von Neumann model bus architecture. We are going to consider the internal structure of the microprocessor (the CPU, in the figure), that we implement with the DMC8.

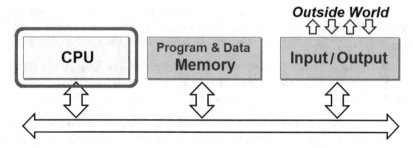

The DMC8 microprocessor is internally organized around the blocks represented by the following figure.

---

[1] FPGA (Field Programmable Gate Arrays). These are components that contain a large number of logical elements that can be internally joined and configured by a programmable connection matrix.

[2] A "debugger" is generally a software tool that identifies and eliminates "bugs", i.e. errors in programs.

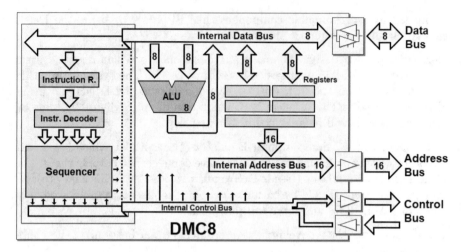

At the center of the drawing we find the 8-bit arithmetic logic unit (ALU), which can execute basic arithmetic operations (add, subtract, compare) and bitwise logical operations. The size of a microprocessor is measured by its calculation unit and since this one works with 8 bits, the DMC8 is an 8-bit microprocessor. Now let's look at the internal bus connections among the various elements that assume different names according to their functions.

To the right of the ALU we see the internal 8-bit and 16-bit registers, which can store data and, as we shall see, even memory addresses.

The 8-bit Internal Data Bus transports information (data and instructions) to and from the various internal devices. Since the data can be transferred in both directions, the connection is called bidirectional.

It is connected through bidirectional buffers toward the Data Bus, which is used to connect all the external devices (memory, input/output devices). For more about bidirectional connections and buffers see Appendix A.3.

The 16-bit Internal Address Bus allows the CPU to use an identification number to select the external device to exchange the above data with at a particular moment. The Internal Address Bus is copied to the Address Bus through unidirectional buffers so as to be available to all the devices in the system. Note that the addresses are generated by the CPU and received by all the other elements (see the arrows in the figure).

The Internal Control Bus is a heterogeneous set of control lines. Almost all of its lines come from the sequencer (at left in the figure) and control the datapath of the processor. Some of these lines are brought toward the Control Bus to manage the timing and device selection in the system. Others come from outside. Among these lines, we have the Clock and Reset, which are found in every synchronous digital systems.

On the left side of the figure, we see the circuit in charge of executing the instructions. The Instruction Register receives and stores the machine code of the instruction to execute, which was retrieved from the memory through the data bus.

The Instruction Decoder accesses the register and obtains the information necessary for the sequencer to be able to synchronously manage all the control lines of the datapath elements in rigorous sequence[3].

The sequencer is initialized by the external Reset line and synchronized by the Clock signal (the Deeds DMC8 microcomputer works with a 10 MHz frequency clock)[4].

2.1.2 Memory system structure

Let's look again at the figure of the bus-organized Von Neumann model, but this time, let's focus on the memory system, which contains both data and programs.

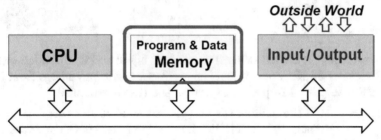

We have seen that the DMC8 has a 16-line address bus so it can address up to 2^{16} locations. Thus the overall "memory space" is $2^{16} = 64$ kB[5]. Also, the data bus has 8 bits so the memory locations will be of that size.

For the moment, we won't make a distinction among the different technologies that the memory components (RAM or ROM) could be made with. Here we will only consider the logical-functional aspects (for more information on the subject see Appendix A).

---

[3] The DMC8 sequencer does not support the pipeline, since it must keep compatible with the functioning of the Z80-CPU processor. As we will see further on (Section 2.2.3), different clock cycles (at least four) will be required to execute the instructions. Here, we will not go into greater detail on the internal structure of the sequencer.

[4] The DMC8 has been tested on FPGA boards up to 50 MHz clock.

[5] 'kbytes'. The prefix 'k' (kilo) is defined by the International System of Units (SI) as 1000. However, in some areas of information technology and in this book the prefix 'k' indicates 1024 (2^{10}). The International Electrotechnical Commission (IEC) addressed this ambiguity defining the prefix 'ki' as 1024, which should be used instead, but this hasn't really entered in the current practice.

As we saw in Chapter 1, the program should always start at location 0000h. This requirement goes for the DMC8 as well, so the ROM should be allocated as of that address.

To concisely describe the memory system, the system designer normally uses something called a Memory Map, as in the examples in the figure below. The map shows the type of memory that the designer made available to the microprocessor, by address intervals, i.e. which RAM or ROM areas are physically there and at what addresses.

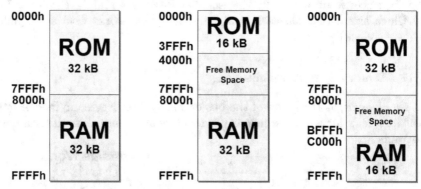

In the example on the left, the memory space is full and it is divided into two equal areas of 32 KB each. When the processor sends an address from 0000h to 7FFFh, the ROM responds, otherwise the RAM responds.

In the example in the middle, the designer chose to reduce costs and provide our system with only 16 KB of ROM. Evidently, the size of the program in this case is relatively small so we simply need to use a physically smaller ROM.

In the example on the right, the program needs more ROM space while its calculation requirements are small so it will suffice to have less RAM.

In the second and third examples, we see "Free Memory Space". This simply means that no physical memory devices correspond to those addresses and the programmer cannot use those areas because for all intents and purposes they don't exist. Reading those addresses would provide random values depending on the electrical state of the connections.

2.1.3 Bus parts and RAM and ROM memory management

Now let's add detail to the microcomputer structure. In our system, the CPU is the only element that pilots the address bus, which is connected to the elements present. The identifiers transmitted on the address bus allow the CPU to fulfill its role as controller and arbiter in the system.

We have just looked at the division of memory space in ROM and RAM. Now let's do a careful study of the parts of the bus (addresses, data and controls), just as we did for the internal structure of the CPU.

In the following figure, the bus parts (addresses, data and controls) have been separated out and the memory has been divided into the RAM and ROM subsystems. Notice that both receive the addresses generated by the CPU and are connected to the data bus. Nevertheless, as seen in Chapter 1, the ROM can only provide it's contents (i.e. it can only be read by the CPU), whereas the RAM can do a bidirectional data exchange (read/write). Remember, in our system the programs we write reside in the ROM while we use the RAM as a work space.

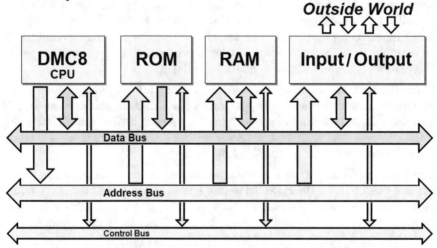

The control bus makes it possible to manage selection and timing of all the elements in the microcomputer. For example, when the CPU needs to connect to the RAM, has to specify if the current operation is write or read, and indicate the execution times for that task. To do this, the CPU controls two lines of the control bus: READ and WRITE (not shown in the figure). Further on, we will examine the timing of these signals and other control lines in greater detail.

2.1.4 Input/output ports

The previous figure also shows the connection block to the "outside world", the Input/Output system. Let's take another look at the Von Neumann model, and this time focus on the Input/Output system.

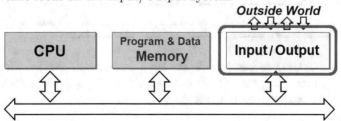

The Input/Output system can be designed in very different ways according to the needs of each different application. It is composed of various ports, each dedicated to every connection to the outside world that we need.

In Chapter 1 we were introduced to ports from a design perspective. We also wrote simple programs using dedicated IN/OUT instructions. In the DMC8 we find these instructions as well with some difference in how they are used.

We use the term "input port" to refer to a circuit that receive data from peripheral units outside our system, like a keyboard or a mouse but also devices like position, temperature, pressure or speed sensors. The term "output port" refers to a circuit that can transfer data from the microcomputer to an external unit like a graphics card or a printer but also different types of "actuators" like small displays, motors or relays.

In smaller systems we often find dedicated input ports to simply read push-buttons and switches, while output ports pilot LEDs. In this section we will introduce these last examples and focus on reading switches and activating LEDs. We will deal with design details further on, but for now we will study these devices only from a functional point of view.

The following figure provides a symbolic representation of the switches at the left and the LEDs at the right. In reality some analog circuit elements would be required to connect the switches and the lights to the ports. Here, we have omitted them so as to concentrate on the logical aspect of the network.

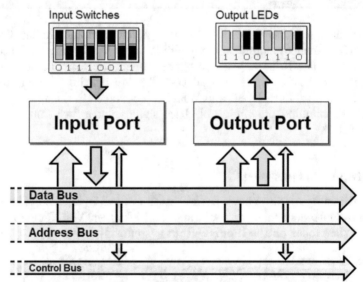

As we can see, the processor reads the eight switches in parallel through the input port (further on, we will use the instruction IN, which transfers what is read from the port to accumulator A).

To choose whether to turn on/off each of the eight LEDs, the processor writes a byte on the output port in parallel, through an OUT instruction (which copies the value in the accumulator to the port).

The DMC8 has 8-bit ports because its architecture and data bus have 8 bits. Also, the DMC8 can address up to 256 output ports and 256 input ports.

Finally, note that even more complex input/output ports can be built upon the structures we are studying here. These structures are always part of any project since they are considered to be their fundamental building blocks.

2.1.5 DMC8 connection lines

The figure below gives a detailed view of the connection lines ("pins") used to join the DMC8 with the rest of the system. It also briefly describes the functions of the individual pins. Their specific use will be explained in greater detail further on in the section on bus timing.

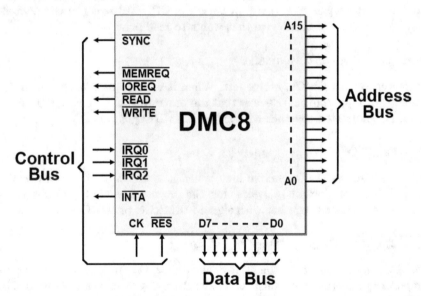

A15 .. A0   (Address bus)

The address bus with 16 output lines transports the addresses for memory and for Input/Output devices.

D7 .. D0   (Data bus)

The data bus, with 8 bidirectional lines (that use tri-state buffers, see Appendix A.3) allows for the exchange of information to and from the memory (data and instructions) and the Input/Output circuits.

$\overline{\text{SYNC}}$ (Synchronization)

Generated by the DMC8, active low. It signals to the outside that the processor is retrieving an instruction from the memory at that moment. All microprocessors have one and it is generally used for diagnostic purposes. This signal allows laboratory diagnostic instruments to synchronize with it and acquire the sequence of all the instructions the microprocessor is executing.

$\overline{\text{MEMREQ}}$ (Memory request)

Generated by the DMC8, active low. When activated, it signals to the memory devices in the system to prepare because the CPU has set (on the address bus) the identifier of the memory location that wants to read or write.

$\overline{\text{IOREQ}}$ (Input/output request)

Generated by the DMC8, active low. When it is activated, it signals to the input/output devices to activate because the CPU has sent (on the address bus) the identifier of the device that wants to read or write.

$\overline{\text{READ}}$ (Read data command)

Generated by the DMC8, active low. When it is activated, it signals to the memory (or input/output) devices that the processor wants to read data from them. It is activated together with signal $\overline{\text{MEMREQ}}$ or $\overline{\text{IOREQ}}$.

$\overline{\text{WRITE}}$ (Write data command)

Generated by the DMC8, active low. When it is activated, it signals to the memory (or input/output) devices that the processor wants to write data to them. It is activated together with signal $\overline{\text{MEMREQ}}$ or $\overline{\text{IOREQ}}$.

$\overline{\text{IRQ2}}$, $\overline{\text{IRQ1}}$, $\overline{\text{IRQ0}}$ (Interrupt requests)

These are lines received by the DMC8, active low, that order the processor to temporarily interrupt the execution of the current sequence of instructions in favor of a new sequence that handles the device that requested the interrupt.

Next we will deal broadly with this topic and the programming techniques that derive from it. For now, we limit the scope to a brief introduction and an everyday life example to clarify the subject.

Let's imagine you are studying for a difficult test. Someone calls you but your phone is on silent mode so you don't pick up and you continue studying. If the ringer is on, however, you put a bookmark on the page you're studying, close the book, answer the phone and focus on the conversation. When the call is over, you open the book to the page you were on and start studying where you had left off.

Let's go through that again, but from the perspective of the processor, which is executing a program. If it does not want to be interrupted, it ignores any interrupt request and continues. If it accepts interrupts, when a request comes, it first saves all the information about what it was working on, interrupts the execution of the current program.

Then, it launches another program whose purpose is to handle the device that requested the interrupt. When the tasks that the device requested have been carried out, the handling program is closed. The interrupt ends and the processor goes back to the information it had saved and restarts the program where it had left off, as if there had been no interrupt.

The program can control whether or not to enable the interrupt mechanism. If the interrupt mechanism is enabled and one or more $\overline{IRQ2}$, $\overline{IRQ1}$ or $\overline{IRQ0}$ lines are activated from the outside, the processor interrupts the sequence of the program and starts to execute a specific program called the "interrupt handler". When an interrupt is requested there are seven different handlers available depending on which lines are activated: $\overline{IRQ2}$, $\overline{IRQ1}$ or $\overline{IRQ0}$. See the table below.

$\overline{IRQ2}$	$\overline{IRQ1}$	$\overline{IRQ0}$	Interrupt Handler	Address
1	1	1	No request	-
1	1	0	Interrupt 1	0008h
1	0	1	Interrupt 2	0010h
1	0	0	Interrupt 3	0018h
0	1	1	Interrupt 4	0020h
0	1	0	Interrupt 5	0028h
0	0	1	Interrupt 6	0030h
0	0	0	Interrupt 7	0038h

If none of the lines is activated, (row 1 in the table) no interrupt is requested. Otherwise, if even only one of the three lines is activated, the request is sent and (if interrupts are enabled) the corresponding interrupt handler is executed. The table shows the address where the handling program must start, according to the combination of the values presented to $\overline{IRQ2}$, $\overline{IRQ1}$ and $\overline{IRQ0}$. After the handler has been executed, the processor goes back to executing the interrupted program.

The interrupt mechanism is very commonly used in microprocessor systems. It allows for rapid responses to the devices that request interventions and so it makes it possible to create the so-called "real-time systems", i.e. systems designed to respond to the needs of a plant in a bounded amount of time.

INTA   (Interrupt Acknowledge)

Generated by the DMC8, active low. When it activates, it signals to the outside that the interrupt request (made by activating at least one of lines $\overline{IRQ2}$, $\overline{IRQ1}$ or $\overline{IRQ0}$) has been accepted and its interrupt handler is about to be launched. It deactivates upon that launch.

$\overline{RESET}$

This is a signal received by DMC8, active low. Activating this line brings all the processor's internal registers to zero, disables the interrupt mechanism and initializes the sequencer.

CK

This is the clock input that synchronizes all the internal operations.

2.1.6 DMC8 processor programming model

The set of instructions the DMC8 can execute is rather large and that is a subset of that of the Z80-CPU. The instructions will be catalogued systematically in Chapter 3, from page 211 on.

Furthermore, numerous powerful addressing modes are available. The addressing mode is the ability to use various ways to find data to work on, as we have seen in Sections 1.6.2 and 1.6.7. We will systematically introduce the DMC8's addressing modes in Chapter 3 from page 203 on.

For now, we will only point out that the DMC8 belongs, in a very general sense, to the category of Complex Instruction Set Computers (CISC). As the name makes clear, a CISC processor has a highly specialized set of instructions. In contrast to CISCs we have Reduced Instruction Set Computers (RISC)[6].

It would be impractical and unhelpful to go into detail on the individual basic operations carried out within the processor while it executes instructions.

---

[6] The first computers belonged to the CISC category, but initially there was no such distinction. RISC were introduced later with the goal to make processors with more orderly and efficient architecture, that were optimized on a limited number of quickly executable instructions. It is interesting to note that the commercial battle between RISC and CISC has never produced a real winner. The elegant architectural improvements of the RISC, were met with "brute force" on the part of CISC producers. That is, they raised clock frequencies to very high values. Personal computers are generally based on CISC processors, while tablets and smartphones have RISC. The industrial world uses both. The processor we designed in Chapter 1 is not developed enough to belong to either category, although the pipeline structure of the sequencer brings it toward the RISC category.

Instead, it would be very advantageous to learn to recognize and use the instructions available, and understand what the effects of executing them are. The goal is to be able to write programs that can carry out precise tasks based on functional specifications, and in so doing develop useful programming techniques.

To reach this objective it is not necessary to have the complete circuit schematic, so producers usually offer a high-level description of what the processor can do and how these operations affect its main components. This type of description is normally called a "programming model" and usually it's represented by a block diagram.

A programming model shows only the main working parts of the architecture as well as their most important interactions. All the details of the hardware have been purposely hidden to make it easier to understand the fundamental mechanisms that govern the microprocessor operations.

The following figure shows the programming model of the calculation elements and the internal registers of the DMC8. Here we can see a number of registers, some specialized, others for general use. They will be described one by one in the following section.

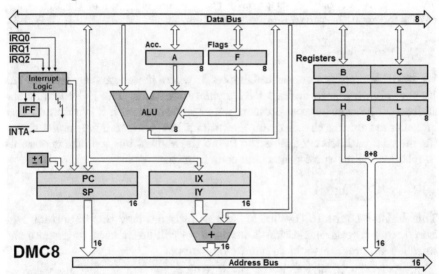

When we use the DMC8 in the D8080 mode, which makes it compatible with the Intel 8080 microprocessor, the model looks a bit simplified.

It is shown in the following figure. Notice that the only visible difference in the architectural elements of the D8080 modality is the absence of IX and IY index registers.

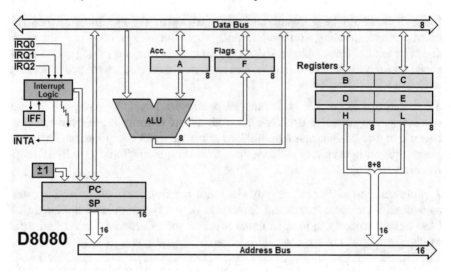

2.1.7 The internal elements of DMC8 processor architecture

What follows is a brief description of the internal elements of the processor, with a focus on the programming model figures.

PC   Program counter

We saw this register/counter in Chapter 1, where it was smaller. Here, it has 16 bits, allowing it to address 64K memory locations. The PC contains the address of the next instruction to execute and, as we know, it is automatically incremented during the program execution. The programming model shows the internal path that connects the PC to the address bus to make it possible to retrieve the instruction machine code from the memory.

SP   Stack pointer

This register points to the Stack[7]. We will discuss how this important register/counter works in Section 3.4, where we will see it used to execute the so-called subprograms and to handle interrupts.

Suffice it to say for now that the 16-bit SP register seen in the figures can be connected to the address bus and that the location at this address is the top of a RAM memory area called the Stack.

The Stack area is useful to temporarily store information (data or addresses) that we need to retrieve soon after. As we will see, its functioning is a little

---

[7] The Stack is a data structure where pieces of information are stored one on top of the other, like a stack of books on a desk. Once stacked, the information can be retrieved in reverse order from the way it was inserted. This method to store and retrieve data is called "Last In, First Out" (LIFO).

more complex from the point of view of the hardware involved, however from the programmer's perspective, it is relatively easy to use and more importantly, it brings a lot of advantages.

A   Accumulator register

We already worked with this type of register in Chapter 1. In the DMC8, it functions in the same way and it has 8 bits. The accumulator is a privileged register because it is involved by default in many of the CPU operations.

We've seen that its name has historical roots because, for some types of calculation, it makes it possible to gradually accumulate intermediate results without having to store them in the memory each time. In our processor, this is the most important register for data processing.

B   register
C   register
D   register
E   register
H   register
L   register

These are general purpose 8-bit registers for supporting and speeding up data processing. They store data without having to continually access external memory, which would take much longer.

BC   registers pair
DE   registers pair
HL   registers pair

Registers B, C, D, E, H and L can be used in pairs as if they were 16-bit registers. We call the three available pairs BC, DE and HL. Since these coupled registers are connected to the address bus, they are used in many instructions to address memory, as we will see in Section 3.2.

Remember that these are still the same registers as before, except that they are used in pairs. They are not other registers. Here is an example of how they work together: if we load number B4h in register B and then F3h in register C, register BC will hold number B4F3h.

F   Flags register

We worked with flags in Section 1.4.3. In the DMC8 they are formally grouped in one single 8-bit register, which actually only uses 6 bits. This register is often called the "status register" because the flags indicate the status of the calculation system after an arithmetic or logical operation.

We introduced only two types of flag in the processor studied in Chapter 1, one to signal if the last add or subtract in the ALU has generated a carry (or a borrow) and the other to signal that the result of the ALU is zero. The DMC8 has four more. Below, we have the flag positions in register F and an explanation of each follows.

Bit 7	Bit 6	Bit 5	Bit 4	Bit 3	Bit 2	Bit 1	Bit 0
S	Z	–	H	–	P/V	N	C

C   Carry flag

The carry flag stores the carry from the most significant bit generated by the ALU (for an add), or the borrow (for a subtract) (C=1 signals that there is a carry or a borrow, C=0 if there isn't). This is also used in shift operations.

Z   Zero flag

The zero flag signals if the ALU result is zero after a logical or arithmetic operation (Z=1 signals that all the bits of the result are at zero, Z=0 signals that at least one of them isn't). Further on, we will see that whole classes of instructions, like those that transfer data have no effect on this flag.

S   Sign flag

The sign flag has the same value as the most significant bit generated by the ALU (in fact, in 2's complement arithmetic, S=0 signals if the result of an operation is positive and S=1 if it is negative).

P/V   Parity/overflow flag

This flag, referred to as either (P or V) carries out two different functionalities depending on what operation has been performed. For an arithmetic two's complement operation it signals if there has been overflow; while for logical operations it signals if the number of '1s' in the result is even. In D8080 mode, the overflow isn't evaluated and this flag is referred to only as P (parity).

H   Half-carry flag

H is significant only for BCD arithmetic instructions. It signals if there is a carry from the 4 least significant to the 4 most significant bits in the accumulator. In D8080 mode this flag is referred to as AC.

N   Subtract flag

The subtract flag signals if the last operation executed was a subtract (N=1) or not (N=0). In D8080 mode, this flag does not exist.

IX   Index register 'X'

IY   Index register 'Y'

Index registers IX and IY are 16-bit registers made for specialized use as indexes to address memory and to facilitate access to some data structures. They will be examined in Section 3.2, in relation to addressing modes. These registers are not included in the D8080 mode.

IFF   Interrupt flip-flop

Finally, the DMC8 also has another flip-flop, which is not included in the flags register F. Flip-flop IFF enables or disables the interrupts mechanism outlined before. In IFF = 0, interrupts are disabled. Flip-flop IFF is associated to the interrupt logic that receives the three request lines $\overline{IRQ2}$, $\overline{IRQ1}$ and $\overline{IRQ0}$. The logic evaluates the requests and if IFF='1', the processor is interrupted and the interrupt handler is launched, as seen in Section 2.1.5.

2.1.8 The sequencer and instruction execution

As we saw in Chapter 1, in terms of the internal operation, a microprocessor is a digital system controlled by a sequencer that finds and executes instructions. After the Reset is activated, the sequencer goes into an infinite loop of states that includes the instructions Fetch, Decode and Execute.

As outlined before, unlike the processor designed in Chapter 1, here we have no instruction pipeline, i.e. the three operations do not overlap but are executed one after the other, as shown in the following figure.

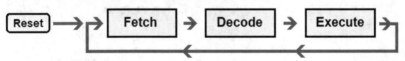

When the system is turned on, there is nothing significant in the RAM since it is a volatile memory, which keeps its contents only when it is powered as it is composed of flip-flops. For these reasons, when a microprocessor system is turned on it has to look to the ROM where it finds the program that will be executed.

Reset brings the PC of the DMC8 to zero and this means the first instruction is found at address 0000h. We'll call this the "reset address"; it is where all our programs begin from[8].

---

[8] Remember that other types of microprocessors can require Reset addresses other than zero.

2.1.9 An initial example of programming

In Chapter 1 we saw that programs from a hardware perspective are nothing other than sequences of binary codes stored in memory that the microprocessor can fetch, recognize and execute one by one. A program, as a sequence of executable binary code, is also called machine code or "object code".

As demonstrated in Chapter 1, programming in machine code is particularly inconvenient. We know that we can avoid treating each operation in numeric form and instead, use the easier and more memorable mnemonic code to write programs. In fact, ever since the first computers were produced, Assembly language has been provided. It allows programmers to write in mnemonic code and has a 1:1 ratio between lines of code the programmer writes textually (each with one instruction) and the corresponding machine code.

Each type of processor has its own machine code and Assembly language. The DMC8 has its own, which is a subset of that of the Z8-CPU. In D8080 mode, the DMC8 can be programmed by using the Intel 8080's Assembly language since the Z80-CPU is compatible with the 8080 at the machine code level. So it is possible to write the same exact program using the two different languages interchangeably[9].

In Chapter 1, the mnemonic codes used to program our processor were chosen because they were as similar as possible to DMC8 codes. The small example of a DMC8 program below uses codes that are very similar to those we have seen before. The following figure shows it written in the text editor of the Deeds-McE microcomputer emulator. We will discuss this tool in detail in Section 2.4 at the end of this chapter.

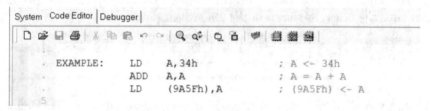

```
    EXAMPLE:     LD     A,34h          ; A <- 34h
                 ADD    A,A            ; A = A + A
                 LD     (9A5Fh),A      ; (9A5Fh) <- A
```

In the first line, as described in the comment after the semicolon[10], we load constant 34h in register A. In the second line, we add register A to itself, thus doubling its content. In the next line, the content of A is saved in the memory, at the 16-bit address location specified in parentheses. The program continues, but here, we will only consider the first three instructions.

After assembling the program (see Section 2.4), we have the machine code loaded into the emulated memory, ready to execute.

---

[9] However, in this book, we do not use D8080 mode.

[10] A semicolon indicates that a comment has been inserted from that point to the end of the line. We can also use semicolons to draw separators and frames, which are useful in making the program more readable.

The figure below is a sample from the emulator.

Addr	Op Code	Label	Instruction	Comment
0000	3E34	EXAMPLE	LD A,34H	; A <- 34h
0002	87		ADD A,A	; A = A + A
0003	325F9A		LD (9A5FH),A	; (9A5Fh) <- A

As we see in the figure, the emulator carries both the mnemonic codes, as written in the editor (on the right), and the program machine codes translated by the assembler (on the left). For practice, use what you've learned in Chapter 1 and try to interpret intuitively the machine codes allocated in the memory and made visible by the emulator.

The first column on the left shows the location addresses where the microprocessor finds the instruction machine code, while the machine code itself is carried over to the second column[11]. From column three on, the instructions written by the programmer are reported, allowing us to see the correspondence with the machine codes.

Concentrating on the first two columns, we see that the assembler has allocated the first instruction to address 0000h by default. This address is the ROM location the program will start executing from at Reset.

At address 0000h in column two, we find machine code 3Eh, which is the translation of instruction "LD A,<const>", and next to it is constant 34h, the operand of the instruction. They are represented without spaces because the emulator compacts them. Obviously we understand that constant 34h is at address location 0001h, after the instruction machine code.

At location 0002h we find number 87h, the translation of "ADD A,A".

Byte 32h is allocated at address 0003h, in row three. It is the machine code of the instruction "LD (<address>),A", followed by operand bytes 5Fh and 9Ah, at addresses 0004h and 0005h respectively. They represent the least significant byte ("the low part") and the most significant byte ("the high part") of the 16-bit address in parentheses.

To our eyes, the two bytes appear inverted but in reality, their order was dictated by a precise arrangement, the "little endian"[12].

---

[11] The addresses and machine codes are in hexadecimal format even though this is not expressly marked with an 'h'.

[12] When a 16-bit number is stored in the DMC8, the low part is always allocated first (at the lowest address) followed by the high part (the next address). In this way, the system respects "little endian" order. In other systems that save the most significant byte first, the order is called "big endian".

2.1.10 An example of instruction execution

Imagine you are starting to execute the program considered here through a reset. The processor fetches the machine code of the first instruction (3Eh) at ROM location 0000h and copies it to the instruction register (see Section 2.1.1), as shown in the figure below.

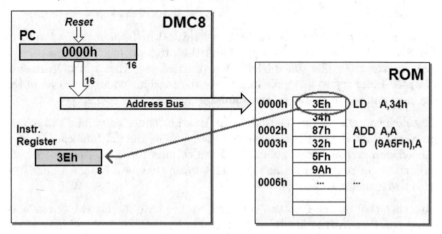

Now, the number in the instruction register is decoded and the sequencer starts with the corresponding flow of internal operations.

The machine code is followed by constant 34h. Meanwhile the PC has been incremented and now points to that constant, which is then loaded in register A (see the figure below).

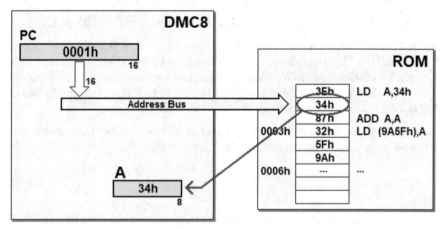

The execution of the instruction took 7 clock cycles (4 to fetch and decode the machine code, and 3 to read the constant)[13].

---

[13] In Section 2.2 we will study memory access timing in detail.

Meanwhile, the PC has been incremented again and is now pointing to 0002h, where the second machine code (87h) is fetched.

After executing the ADD A,A that was requested, the processor saves the result by overwriting register A, which will be 68h. See the figure below. This instruction was executed in 4 clock cycles.

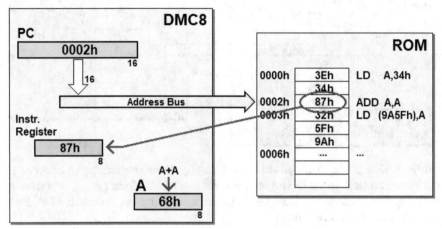

The third machine code is 32h, allocated at address 0003h. Note that the address specified as the instruction operand was placed by the assembler right after the machine code. This 16-bit number is divided into two bytes, with the least significant part (5Fh) preceding the most significant part (9Ah).

Executing this instruction takes 13 clock cycles and is quite complex. The first 4 cycles are spent fetching and decoding the instruction machine code (see the figure below).

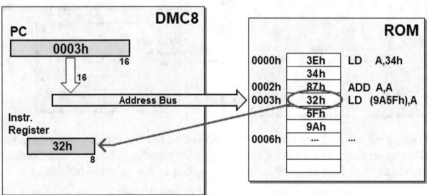

The sequencer increments the PC to access the byte that follows the machine code. Three more clock cycles are used to read the least significant bit of the address (5Fh). The following three clock cycles are used to read the most significant bit (9Ah), after raising the PC by one increment again.

As we can see in the following figure, the two bytes are recombined in a internal temporary register (not mentioned in the programming model).

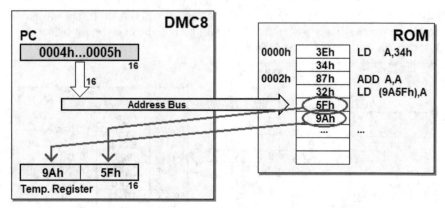

Only at this point does the processor finally have the complete address to do the requested write in the memory. In the last 3 clock cycles, it writes in memory the value taken from the accumulator, using the recombined address 9A5Fh. See the figure below.

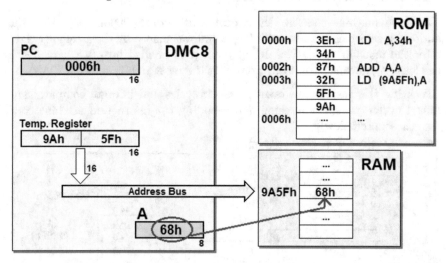

Notice that the address bus is not connected to the PC for this event as in the other operations, but rather to the temporary register. The processor accesses the RAM with this address to write the value in the selected location.

Meanwhile the PC has been incremented to 0006h, and will be used to fetch the next instruction machine code (not shown here), continuing with the program execution.

2.2 Bus signals and timing

In Section 2.1 we analyzed the functional relationship among the various elements of the system that are organized around the bus and we looked at an example of instruction execution. In this section we examine the signals involved and their timing in greater detail. Then we will design the memory sub-system and the input/output ports.

2.2.1 The clock, synchronization and initialization of the system

Every part of the system is synchronous so before we discuss signal timing we should review some basic concepts. We have seen that the microprocessor needs a clock signal to synchronize all its internal operations so the system must have a "Clock Generator" with the right frequency. See the figure.

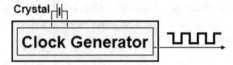

The symbol at the upper left represents a quartz crystal, which is used to stabilize the frequency of the signal[14]. Let's consider that in our system, the generator's frequency[15] is set at 10MHz, with a 50% duty cycle. Obviously, the clock is connected to the CPU but it is also available to the control bus so that it can be used by all the devices in the system that may need it.

The following figure shows another generator, which is necessary for initializing the system, the "Reset Generator".

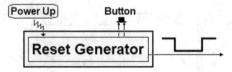

---

[14] Quartz crystals are often used in clock generators because of their piezo-electric properties. If this type of crystal is cut into the right size and covered in two layers of conductive plate, it can vibrate at a precise frequency. The vibration is harmonic, similar to that of a tuning fork but its frequency is in the tens of megahertz. Due to its piezo-electric properties and the layers of conductive plate, it produces an electric signal with a time behavior analog to that of the mechanical vibration. A suitable electric circuit amplifies this signal and feeds the mechanical vibration in order to sustain it. Then, the sinusoidal signal is reduced to two levels (high/low) and produced at the output of the generator.

[15] Note that it is not necessary for the clock frequency to always be as high as possible in any system. Here, the chosen frequency is suitable for a medium-performance control system. The frequency might also be quite low, for example, when we need to save energy as with a home thermostat, which needs a clock at a low frequency to guarantee a long life of the battery.

When the system is turned on, it sends the Reset signal to the CPU and, through the control bus, to the whole system. Its job is to correctly initialize all the sequential elements in the system when it is turned on. The Reset line is activated for the duration of a tenth of a millisecond[16].

This generator can have a push button that makes it possible to reset the system manually. This may be accessible to the user like on a desktop computer, or accessible to specialized maintenance service only, as on smartphones.

2.2.2 The physical behavior of the bus

The bus transfers information from one module to another in the system. Normally when we think of a signal generated by one logic gate and received by another we don't think about the physical actions that propel the signal along the physical connection between the two gates. This is because we take for granted that the distance is short; think of the connection between two physically adjacent logic gates on the surface of a silicon chip, at a distance of a fraction of a μm.

Still, in a digital system with a bus and connections between separate cards we have longer connections from a few to tens of centimeters. The real physical behavior of a bus and of long connections can be less straightforward than we may think. The following figure shows a generic element 'A' that sends an address on the bus, which is received by 'B'.

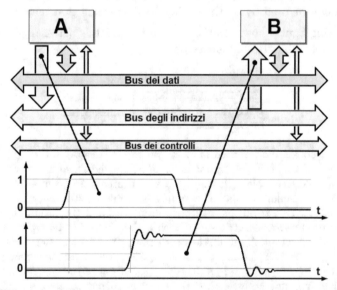

---

[16] The circuit is designed to activate the Reset signal as soon as the power is on in order to initially freeze all of the devices present during the system's power-up stage. As soon as the supply voltages have reached the nominal value, the generator releases the Reset line, allowing the system to activate.

In a vacuum the propagation time of an electromagnetic wave is at the speed of light: about 3.3 nS per meter. In a propagation medium, the speed is reduced as a function of the physical and geometric properties of the medium itself.

Signals that travel between logical devices can reach propagation times of the order of 20nS per meter, which is comparable to the times of the devices themselves. These non-negligible propagation times highlight other issues like signal reflections in a long connection[17].

Other physical parameters are in play such as parasitic inductance and capacitance, or the mutual electromagnetic coupling between wires, etc. In the previous figure we show a signal that is delayed in time and distorted in shape when it travels from 'A' to 'B'.

Basically, the immediate effect of this is that the particular electrical and geometric characteristics of a bus will limit how fast it can transfer data. Designers must be aware of this.

2.2.3 Clock cycles, machine cycles and instruction cycles

Aside from the physical limitations of the bus, we must keep in mind the time constraints imposed by the devices involved. The processor, memory and input devices have to communicate with each other, which means that the time characteristics of all the elements have to be compatible. Bus signal timing is normally defined by processor timing. The designer must adapt the other devices' behavior to the processor timing.

The DMC8's bus timing is simple to determine since it can be broken down into a very small number of standard sequences called "machine cycles". Multiple machine cycles combined together determine the processor's bus timing during the execution of instructions (the "instruction cycles"). Among the standard sequences we can see at the processor pins, we find the following machine cycles:

— Instruction fetch cycle
— Read-write memory access cycle
— Input/output access cycle (read in input, write in output)
— Interrupt acknowledge cycle

---

[17] If we vary the tension at the output of a logic gate, this variation is propagated as a wave edge with a finite propagation time along the connection toward the logic gate on the other end. When the variation gets to the destination, we have a bounce, due to the receiving device not completely absorbing the energy of the signal. Some of the signal goes backward toward the gate that had generated it. At that point, there will be another bounce forward and so on until the energy of the signal is exhausted. All these unwanted signals overlap. This can bring about serious distortions in the signals themselves and slow down the system. There are technical solutions that can mitigate these effects (which we will not examine here).

In Section 2.1.10, we examined the execution of instruction LD (9A5Fh),A in detail. When we look at the bus, we can see that this instruction can be broken down into a succession of four standard machine cycles:

— A fetch cycle (reading its opcode)
— Two memory access cycles (reading the two bytes of the operand)
— Another memory access cycle (writing the number in the memory)

In the following we examine the timing of the main machine cycles and leave that of the interrupt to Chapter 4.

2.2.4 The fetch cycle

The memory retrieval sequence of the first instruction byte, the opcode, is standard for all instructions. See the figure below.

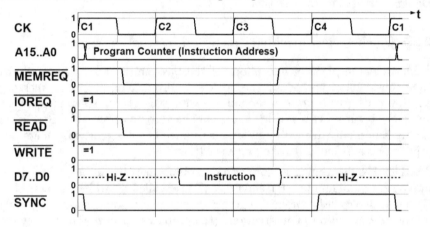

The fetch cycle takes a total of four clock cycles: in the first three cycles, the microprocessor reads the opcode while the fourth cycle is for decoding and starting execution.

With the rising edge of clock cycle C1, the processor sends the content of the Program Counter on the (A15..A0) address bus. On the falling edge of the same cycle, the memory request line ($\overline{\text{MEMREQ}}$) and read line ($\overline{\text{READ}}$) are activated. In response to these two signals, the memory is active and once the access time is over, puts the content of the addressed location (the instruction code) on the D7..D0 data bus.

Notice that the data bus is normally in the high impedance (Hi-Z) state, except when one of the devices activates (as in this case with the memory). The memory and the data bus deactivate after lines $\overline{\text{MEMREQ}}$ and $\overline{\text{READ}}$ go back to a resting state.

The microprocessor captures the instruction code and loads it in the Instruction Register on the falling edge of clock cycle C3. As of this edge, the instruction opcode starts being decoded in the processor and as of cycle C4 the sequencer starts executing it. Whatever the type of instruction, on the rising edge of the next cycle C1, the processor increments the Program Counter to set it up for the next retrieval.

For most instructions that only request access to internal registers, the execution of the instruction ends with the rising edge at the end of C4 and the beginning of C1. Let's consider the example of instruction LD A,B. Executing this instruction doesn't require memory access cycles because it involves only resources that are internal to the microprocessor[18]. Therefore, on the edge between C4 and C1, the processor directly begins fetching a new instruction.

If the instruction requires external access, as with instruction LD (9A5Fh),A (the previous example), one or more memory access machine cycles will follow the end of clock cycle C4.

During clock cycles C1, C2 and C3 the processor activates line $\overline{\text{SYNC}}$, to indicate that the processor is executing a fetch cycle. $\overline{\text{SYNC}}$ is a signal of synchronization that shows what the processor is doing from the outside[19], and is active only during the fetch sequence.

2.2.5 Read/write memory access cycles

These memory access cycles consist of three clock cycles each. In both the read and the write cases, the address location the processor wants to read or write is placed on the (A15..A0) address bus at the beginning of cycle C1. That address remains valid until the end of cycle C3.

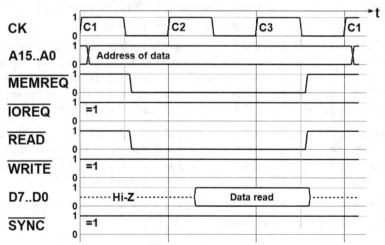

---

[18] As a matter of interest, for internal architectural reasons, the content of B is not copied into A by the end of clock cycle C4, as one would expect. Rather, it is executed soon after, during the fetch cycle of the next instruction.

[19] $\overline{\text{SYNC}}$ is normally used only by diagnostic instruments, as for example the "Logic State Analysers". When they are connected to the microprocessor, they allow us to track programs in execution in real time.

The time sequence of the read memory cycle (see the figure above) is very similar to that of the fetch cycle we have just examined. In this case, however, the data address can come from any of the 16-bit registers (HL, BC, DE, IX, IY), not only from the Program Counter. The byte that is read is copied in a CPU register.

In the write memory cycle (see the figure below) line $\overline{\text{WRITE}}$ is activated on the falling edge of C2, one cycle after $\overline{\text{MEMREQ}}$ is activated.

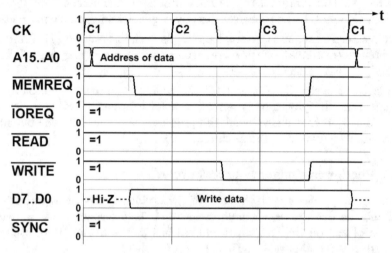

Line $\overline{\text{WRITE}}$ is then released together with $\overline{\text{MEMREQ}}$ on the falling edge of C3. This is the edge that stores data into the memory.

The reason to start the $\overline{\text{WRITE}}$ signal later is because we need to allow time for older types of RAM to complete the operation of selecting the addressed location before moving on to the actual writing. More recent memory devices don't have this problem. Notice that the processor sets the byte to write on the data bus rather early and keeps it there until the end of clock cycle C3.

Referring back to the example of instruction LD (9A5Fh),A, the complete time sequence is shown in the following figure. The sequence has been divided into two parts for reasons of space. It is made up of one fetch cycle (M1), two read memory cycles (M2, M3) and one write memory cycle (M4).

We suggest that you compare this time sequence with the description of the execution of the instruction in Section 2.1.10.

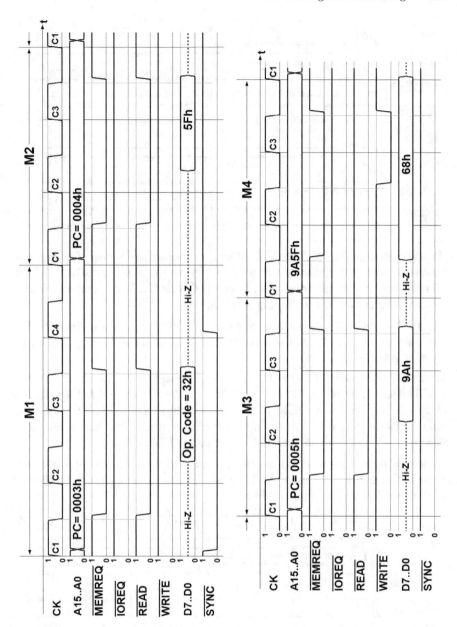

2.2.6 Input/output access cycles

Input and output access cycles last for four clock cycles each. They are generated by the processor during the execution of IN and OUT instructions. The following figure shows the input cycle.

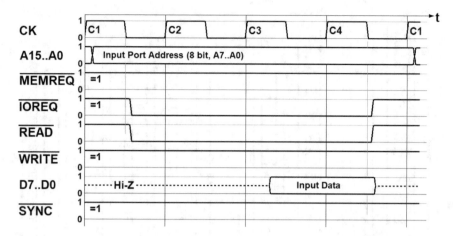

At the beginning of clock cycle C1 the identifier of the input port that the processor wants to communicate with is put on the low part A7..A0 of the address bus. It remains there until the end of cycle C4. With 8 address wires, we can differentiate up to 256 input ports.

The port's address is normally specified as an operand of the instruction IN, but there is also a variant of this instruction that retrieves it from register C, as we will see.

With the falling edge of cycle C1, line $\overline{\text{IOREQ}}$ (request to activate input/output devices) is activated. It will then be deactivated on the falling edge of cycle C4. While an instruction IN is being executed, line $\overline{\text{READ}}$ requests the selected input device to activate and provide the input data to the bus. The processor acquires the data on the falling edge of C4.

As for an access sequence to an output port, when instruction OUT is executed, the dynamic of lines A7..A0 and $\overline{\text{IOREQ}}$ is very similar to that of instruction IN. See the figure below.

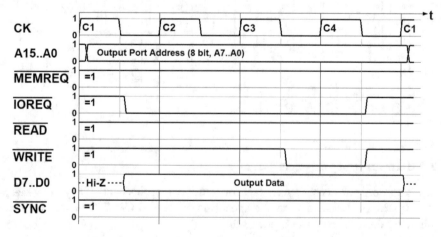

In this case, line $\overline{\text{WRITE}}$ is activated. The processor uses it to order the output device to acquire the data the processor has in the meantime put on the bus. This occurs as of the falling edge of C1 and is maintained until the end of C4. The device typically acquires the number on the rising edge of the clock C4 or on the rising edge of line $\overline{\text{WRITE}}$.

Now let's consider a timing diagram example that shows an instruction OUT that copies register A to an output port.

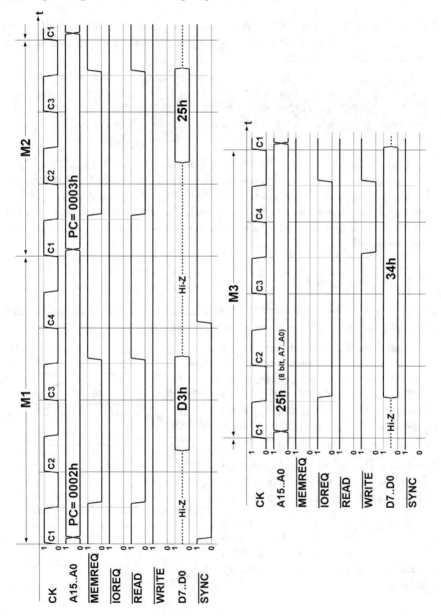

Below, an excerpt from the program that generates this timing sequence during the execution of the OUT instruction.

It loads constant 34h in register A and then copies its content to the port at address 25h. After assembling the program, the machine code related to these two source lines appears like this in the emulator:

Addr	Op Code	Label	Instruction	Comment
0000	3E34	EXAMPLE	LD A,34H	
0002	D325		OUT (25H),A	

The timing diagram shows a sequence of 3 memory cycles (M1, M2 and M3). Let's compare what we see there with the row highlighted in the emulator. In M1, the processor fetches its machine code (D3h). In read memory cycle M2, operand 25h is acquired (the port address). Finally, in output cycle M3 the processor sends the data (34h) onto the data bus, writing it to output port at address 25h.

2.2.7 Inactive cycles

In the case of some instructions besides the machine cycles we have examined so far, we can also observe some seemingly inactive clock cycles.

In actual fact, it is only the bus that is inactive during these cycles, while the processor continues working internally. In this inactive cycle, the value on the address bus is irrelevant, all the control signals are inactive and the data bus is in Hi-Z (see the figure on the right).

Let's take the example of instruction INC HL, which increments the content of the 16-bit HL register by one.

After the four clock cycles of the fetch, it takes two more clock cycles for the instruction to be completely executed internally, while the bus is externally inactive.

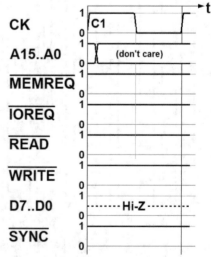

2.3 Input/output and memory subsystems

We have seen that the processor addresses the memory with 16 wires (A15..A0), so it can select up to 2^{16} locations (one byte each). We have also seen that this space can be completely or partially utilized and that we have both ROM and RAM memory.

To describe the memory in a microprocessor system from the perspective of a programmer, we have used the so-called memory map. The input/output ports are addressed with only 8 lines (A7..A0) and those that are available in the system are included in the port map.

In this section, we will take the perspective of a hardware designer and examine how to design input/output and memory subsystems for our system based on what we have seen so far.

2.3.1 Memory subsystems

As we saw with time sequences, the processor controls access to the memory through the $\overline{\text{MEMREQ}}$, $\overline{\text{READ}}$ and $\overline{\text{WRITE}}$ connections. Line $\overline{\text{MEMREQ}}$ is activated by the microprocessor to signal the beginning of a memory access cycle (and line $\overline{\text{IOREQ}}$ is activated in I/O cycles).

So to design the memory system we need to remember that it has to activate in response to line $\overline{\text{MEMREQ}}$ and obviously, we need to take $\overline{\text{READ}}$ and $\overline{\text{WRITE}}$ into account to distinguish between read and write. We must also connect the address bus to select the individual cell addressed by the processor inside the memory devices.

A memory system can generally be composed of multiple memory components (both RAM and ROM). We use the address provided by the processor for two combined purposes:

— to activate one single memory chip (that contains the addressed location) at a time
— to select the desired location inside the activated chip.

In our system, there is a total of 64 kB of address space. Within this limit, we can decide the size of the ROM and RAM areas according to need. We must also allocate these areas among the individual memory devices that we want to use. Let's imagine we only have 4-kbyte memory devices and we want to allocate a total of 8 kB of ROM and 8 kB of RAM.

Our system's memory map will look like the one below, where the ROM is allocated as of location zero and the RAM at the bottom of the available space. The remaining space for memory is available for further expansions.

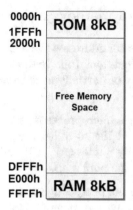

Remember that when the system is reset, the first instruction has to be at address zero in ROM. If it were otherwise, the processor would begin reading memory locations that haven't been initialized yet, trying to interpret them as instructions, which would certainly make the system malfunction.

In the figure on the right we see two 4-kbyte memory components (ROM and RAM) that are available in the Deeds simulator library.

They both have 12 address wires (A11..A0) and 8 data lines (A7..A0). Input line CS (Chip Select) activates their functions.

In the ROM, when line OE (Output Enable) is activated, it enables the tri-state buffers of the output data lines. The RAM component we have chosen uses bidirectional data lines (see Appendix A.3).

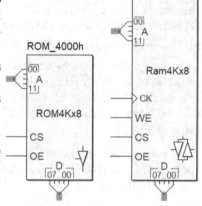

This is a synchronous RAM in the sense that read/write operations correspond with the rising edge of the clock CK.

Writing is enabled by input WE (Write Enable). To write a number in the memory, we must bring it to the data lines, activate WE = 1 and wait for the rising edge of the clock.

The read function is enabled only when write is inactive (WE = 0). When this condition is met, line OE (just as in the ROM) enables the output tri-state buffers, making it possible to read the addressed number.

Now let's go back to the memory map and look in greater detail at binary addresses. See the figure below. As we can see, the configuration of the four most significant bits of the addresses (A15..A12) decidedly determines which of the 4-kbyte memory devices should be selected.

Hex	A_{15}	A_{14}	A_{13}	A_{12}	A_{11}	A_{10}	A_9	A_8	A_7	A_6	A_5	A_4	A_3	A_2	A_1	A_0	Comment
0000h	0	0	0	0	0	0	0	0	0	0	0	0	0	0	0	0	
0001h	0	0	0	0	0	0	0	0	0	0	0	0	0	0	0	1	
.	.	.	.	.	.	.	.	.	.	.	.	.	.	.	.	.	4-kbyte
↓	.	.	.	.	↓	↓	↓	↓	↓	↓	↓	↓	↓	↓	↓	↓	ROM
.	.	.	.	.	.	.	.	.	.	.	.	.	.	.	.	.	
0FFEh	0	0	0	0	1	1	1	1	1	1	1	1	1	1	1	0	
0FFFh	0	0	0	0	1	1	1	1	1	1	1	1	1	1	1	1	
1000h	0	0	0	1	0	0	0	0	0	0	0	0	0	0	0	0	
1001h	0	0	0	1	0	0	0	0	0	0	0	0	0	0	0	1	
.	.	.	.	.	.	.	.	.	.	.	.	.	.	.	.	.	4-kbyte
↓	.	.	.	.	↓	↓	↓	↓	↓	↓	↓	↓	↓	↓	↓	↓	ROM
.	.	.	.	.	.	.	.	.	.	.	.	.	.	.	.	.	
1FFEh	0	0	0	1	1	1	1	1	1	1	1	1	1	1	1	0	
1FFFh	0	0	0	1	1	1	1	1	1	1	1	1	1	1	1	1	
2000h	0	0	1	0	0	0	0	0	0	0	0	0	0	0	0	0	
.	.	.	.	.	.	.	.	.	.	.	.	.	.	.	.	.	Free
.	.	.	.	.	.	.	.	.	.	.	.	.	.	.	.	.	Space
DFFFh	1	1	0	1	1	1	1	1	1	1	1	1	1	1	1	1	
E000h	1	1	1	0	0	0	0	0	0	0	0	0	0	0	0	0	
E001h	1	1	1	0	0	0	0	0	0	0	0	0	0	0	0	1	
.	.	.	.	.	.	.	.	.	.	.	.	.	.	.	.	.	4-kbyte
↓	.	.	.	.	↓	↓	↓	↓	↓	↓	↓	↓	↓	↓	↓	↓	RAM
.	.	.	.	.	.	.	.	.	.	.	.	.	.	.	.	.	
EFFEh	1	1	1	0	1	1	1	1	1	1	1	1	1	1	1	0	
EFFFh	1	1	1	0	1	1	1	1	1	1	1	1	1	1	1	1	
F000h	1	1	1	1	0	0	0	0	0	0	0	0	0	0	0	0	
F001h	1	1	1	1	0	0	0	0	0	0	0	0	0	0	0	1	
.	.	.	.	.	.	.	.	.	.	.	.	.	.	.	.	.	4-kbyte
↓	.	.	.	.	↓	↓	↓	↓	↓	↓	↓	↓	↓	↓	↓	↓	RAM
.	.	.	.	.	.	.	.	.	.	.	.	.	.	.	.	.	
FFFEh	1	1	1	1	1	1	1	1	1	1	1	1	1	1	1	0	
FFFFh	1	1	1	1	1	1	1	1	1	1	1	1	1	1	1	1	

When we analyze the map we see that if A15..A12 = '0000', the first of the two ROMs, the one that will contain the first four kbytes, should be selected from address 0000h to 0FFFh. If A15..A12 = '0001', the second of the two ROMs, the one that will respond to addresses 1000h and 1FFFh, should be activated.

The remaining 12 lines (A11..A0) are connected directly to the device wires of the same name so as to address locations on the inside. The following figure shows the ROM memory sub-system, which fulfills our considerations about addresses.

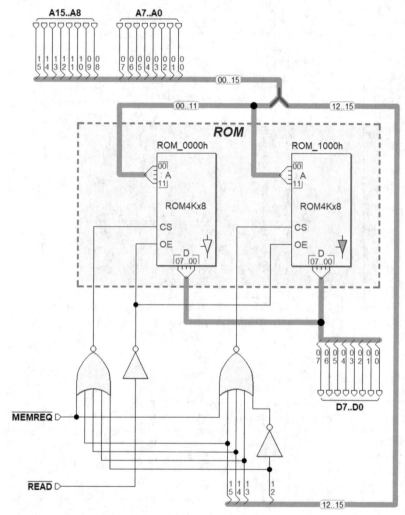

As we can see, the address wires (A15..A0) are split into two sections. Lines A11..A0 are directly connected to the corresponding address lines for the two memory components, while lines A15..A12 are used by a small decode network that activates the CS inputs as described in the table. Namely, A15..A12 = '0000' for the ROM at the left of the figure and A15..A12 = '0001' for the one on the right.

The decode network also takes line $\overline{\text{MEMREQ}}$ into account, which has to be at '0' so that the memory activates.

Therefore the CS time trends, except for propagation delays, correspond to that of the processor's $\overline{\text{MEMREQ}}$ signal. The processor's $\overline{\text{READ}}$ command is used to drive both OE enable lines of the tri-states on the output lines.

The RAM sub-system network shown below is very similar to the previous one, especially in the way the addresses are connected.

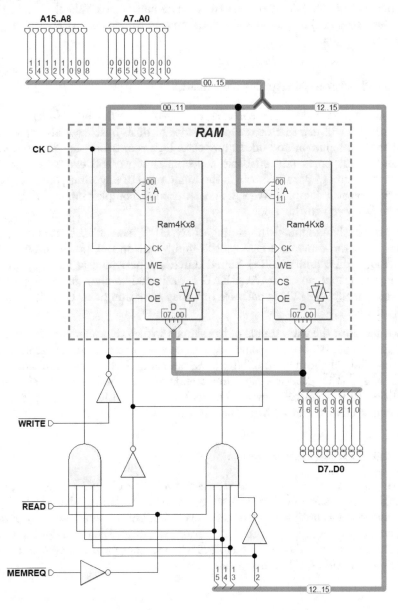

Clearly, the decode network of the address is different; it is set to activate the component on the left for A15..A12 = '1111', and the one on the right for A15..A12 = '1110' (consistent with the memory map).

The $\overline{\text{READ}}$ command for the RAMs is generated by the processor and used to activate the tri-states of the output lines by means of the OE enable lines. Similarly, line $\overline{\text{WRITE}}$ is connected to write commands WE of both RAMS. Needless to say, these commands work only on the device selected at a given moment.

2.3.2 The Input/Output subsystem

For simplicity's sake, let's consider only 8-bit parallel input/output ports. Initially we will use them to read the state of push-buttons and/or switches, and to turn lights on and off. In any case, let's remember that more complex input/output ports rely on basic concepts that we will examine now. The basic structure of parallel ports is the foundation of any input/output system design, however complex. We will encounter more complex input/output ports in the follow-up to this book.

We previously examined the timing of IN/OUT instructions (Section 2.2.6). The processor controls access to the input and output ports through lines $\overline{\text{IOREQ}}$, $\overline{\text{READ}}$ and $\overline{\text{WRITE}}$. Signal $\overline{\text{IOREQ}}$ is activated at the beginning of an input/output access cycle. $\overline{\text{READ}}$ is activated so that an input port can be read, while $\overline{\text{WRITE}}$ is activated so that something can be written on an output port.

If we compare the time trends of the signals we see that there are no substantial differences between the memory access cycles and the input/output access cycles. The only important difference is that signal $\overline{\text{IOREQ}}$ is activated, rather than $\overline{\text{MEMREQ}}$; otherwise the time trends of the signals are very similar. In addition to the control signals, the bus lines involved in the project are data bus wires D7..D0 and address bus wires A7..A0.

2.3.2.1 Parallel output ports

If we go back to the output port access cycle from Section 2.2.6, we notice that the processor produces the data to write on the falling edge of clock cycle C1, and keeps it until the end of clock cycle C4. Let's connect an 8-bit parallel register to the data bus in order to capture the data. See the following figure. This involves activating write enable E when the data are available on the bus while instruction OUT is being executed.

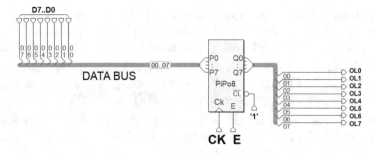

From the timing diagram we see that signal $\overline{\text{WRITE}}$ performs well as an enable signal, given its time trend. When $\overline{\text{WRITE}}$ is active, the data to write is on the bus and can be captured by the register on the rising edge of clock C4. Since this is a register, once the number is written, it remains available on the outputs until the next time it is rewritten.

Line $\overline{\text{WRITE}}$ is not enough on its own; we need to add a condition on $\overline{\text{IOREQ}}$, (which must remain active) and on wires A7..A0, so that the write only happens on the port identified by the address the processor generated.

This means we need to add some logic gates and an address recognition circuit, which compares the address sent by the processor to a locally set constant. The following figure shows a possible version of the parallel output port where we use a magnitude comparator to recognize the address.

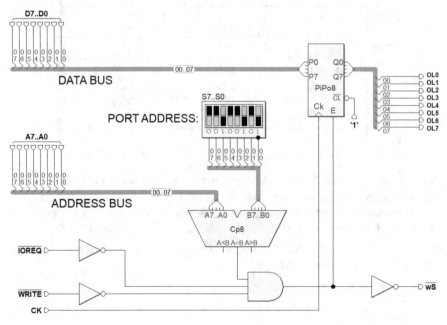

The figure shows eight dip-switches that allow us to define the address assigned on the port. They are shown for the general purposes of explanation but they actually can be substituted by simple constants established at design time[20].

In the schematic, address bus lines A7..A0 are compared to the port address, which is set at 25h ('00100101') in this example. So, the comparator output is placed in AND with processor signals $\overline{\text{IOREQ}}$ and $\overline{\text{WRITE}}$.

The enable is then brought to the output under the name $\overline{\text{wS}}$ ("Write Strobe", active low). $\overline{\text{wS}}$ signals to the outside that a byte has been written on the port and is available on output lines OL7..OL0.

So, enable E copies the time trend of $\overline{\text{WRITE}}$ but only if $\overline{\text{IOREQ}} = 0$ and the address bus has the correct address, as shown in the timing simulation reported in the figure below.

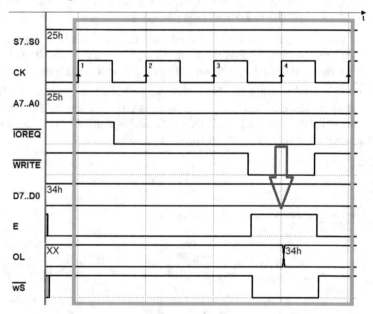

When instruction OUT (25h),A is executed, the processor puts number 25h on the address bus so our hardware recognizes that address and generates E with the time trend seen in the figure. In the simulation, the number provided to the bus (D7..D0) is set at 34h, as in the example in Section 2.2.6.

On the rising edge of clock 4, the number on the data bus is loaded in the register, so it exits lines OL7..OL0.

---

[20] There are "plug-and-play" systems that make it possible to reprogram addresses. To achieve this, the address of the port is set by another register, which is itself writable by the processor, but this is outside the scope of this book.

2.3.2.2 Parallel input ports

If we look at the input access cycle (Section 2.2.6), we see that the processor expects to capture the data on the bus (that comes from the port) on the rising edge of clock CK. This occurs after activating lines $\overline{\text{IOREQ}}$ and $\overline{\text{READ}}$ in the time interval between the falling edge of C1 and that of C4.

To make sure the input port's data can be delivered to the data bus, we must use tri-state buffers (see Appendix A.3), which can connect the data lines to read to the bus during the right time interval.

On the left side of the figure below, we see the eight tri-state buffers that we need to connect each data bus line (D7..D0) to the corresponding input line (IN7..IN0). The eight tri-state enable lines are all connected together to one line E.

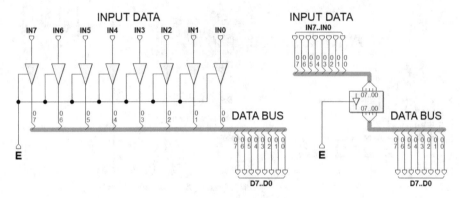

On the right side of the figure we see the network that we will use in the following. It is functionally identical in that the eight buffers, for convenience's sake, are enclosed in one component controlled by enable E.

Note that the simplest input ports do not even need to use a register (we will introduce registers further on in more complex examples). Here it is enough to assume that any element that provides data to the input port on lines IN7..IN0 keeps that data stable for the time it takes for them to be acquired by the processor.

Enable E needs to be given by a proper decoding circuit and address recognition mechanism like those used for the output port.

On the machine cycle's timing diagram (Section 2.2.6), the timing of signals $\overline{\text{IOREQ}}$ and $\overline{\text{READ}}$ suggests using these lines to condition the enable of the buffer. This way, the data provided to the bus from the outside will be available on the rising edge of the clock C4, as required by the processor.

Clearly, we should not forget to condition E to the address the processor brought to wires A7..A0, as in the output port.

The schematic in the figure below reflects this.

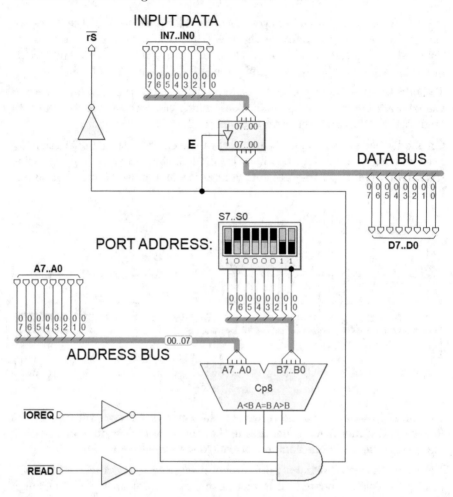

Address bus wires A7..A0 are compared to the address of the port. They are programmable with eight switches. In our example, the port address is set to '10000011' (83h). As we saw with the output port, in a real project they could be replaced by constants unless we wish to keep manually changing the port address.

An AND conditions the comparator output by the activation of lines $\overline{\text{IOREQ}}$ and $\overline{\text{READ}}$. Enable E is also carried to the output as $\overline{\text{rS}}$ (Read Strobe, active low), so that the outside is informed about the instant when the input data on IN7..IN0 is read.

Let's look at an example, simulating the execution of instruction IN (83h),A.

The processor has placed number 83h on the address bus and now expects to acquire the number from the input port on edge 4 of the clock. See the figure.

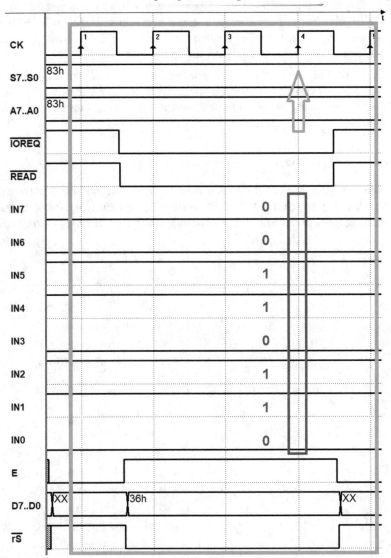

Number '00110110' = 36h is on input lines IN7..IN0 and we imagine it is kept stable by the outside element that provided it. As we can see, the tri-state buffers enabled by E force this number on the data bus. The number remains available on the bus until the falling edge of cycle 4.

The diagram does not show this explicitly but we can trust that the processor acquires the number on the bus (36h) and stores it in accumulator A on rising edge 4 (indicated by the arrow).

2.3.2.3 Specificities of output ports

An output port can be written but if its network is like the one examined here, it won't be able to be read simply because the hardware to do it is lacking. To read an output port, unlike a memory location, it must be supported by an input port and connected so that its lines can be read.

Usually we do not use hardware for reading for reasons of economy. The most economically advantageous solution would be to keep a software copy in the RAM of the software needed to remember the set of zeroes and ones produced in the output port.

2.3.2.4 The I/O Map

It is common practice to represent a map of the input/output devices in the system, similarly to what we did for memory. As an example, a simple system with only parallel ports would fit the description given in the I/O map below.

HEX	A_7	A_6	A_5	A_4	A_3	A_2	A_1	A_0	Ports (description example)
00h	0	0	0	0	0	0	0	0	Input - 8 Light Sensors
01h	0	0	0	0	0	0	0	1	Input - 8 Switches
02h	0	0	0	0	0	0	1	0	Input - 8 Switches
03h	0	0	0	0	0	0	1	1	Input - 8 push-buttons
04h	0	0	0	0	0	1	0	0	Input - Generic 8 lines (A)
05h	0	0	0	0	0	1	0	1	Input - Generic 8 lines (B)
.			.			.			...
.			.			.			...
63h	0	1	1	0	0	0	1	1	Input - 8 Switches
63h	0	1	1	0	0	0	1	1	Output - 8 LEDs
.			.			.			...
.			.			.			...
FAh	1	1	1	1	1	0	1	0	Output - Generic 8 lines (A)
FBh	1	1	1	1	1	0	1	1	Output - Generic 8 lines (B)
FCh	1	1	1	1	1	1	0	0	Output - 7-segment display (low)
FDh	1	1	1	1	1	1	0	1	Output - 7-segment display (high)
FEh	1	1	1	1	1	1	1	0	Output - 8 LEDs
FFh	1	1	1	1	1	1	1	0	Output - 8 LEDs

Note that at address 63h there are both an input port and an output port. The two devices can never be in conflict since they are activated in a mutually exclusive way: one in write and the other in read (signals $\overline{\text{READ}}$ and $\overline{\text{WRITE}}$ never activate together).

2.3.2.5 Port address spaces

There are two different schools of thought about the way to connect the processor to input and output devices. The difference is in addressing ports, which could be done by a separate port address space, as we saw in the DMC8. Alternatively, it could be done by sharing a single address space where both memory and input/output devices are allocated.

The first type is traditionally called "I/O standard". It has two separate control signals $\overline{\text{MEMREQ}}$ for the memory and $\overline{\text{IOREQ}}$ for the ports. The two address spaces are completely separate and the memory access instructions (ex. LD (2035h),A) and port access instructions (IN and OUT) are different, as we have seen. This method is also used in Intel processors for example.

The second type is called "I/O Memory Mapped" and is used in Motorola's 68000 series and their descendants, as well as in ARM (Advanced Risc Machine) processors, for example.

The microprocessors that use this method don't have specialized input/output instructions because in terms of hardware, the processor doesn't distinguish between memory devices and input/output devices. The one address space is used both for memory chips and for ports and the distinction between the two comes only with the address. The same instructions used for memory are used to write and read ports.

In terms of programming, this is considered an advantage in certain respects since it makes it possible to apply the power of the instructions used for memory to the ports. There is, however, also a disadvantage; the lack of specialized input/output instructions makes the assembly code much less readable.

2.4 Introduction to Deeds-McE

As explained in the foreword, the Deeds environment includes three software modules. The most important among them is the Deeds-DcS (Digital Circuit Simulator), which has been used in this book to build and simulate all the digital networks presented here, for example those in Chapter 1. The environment also has the Deeds-FsM (Finite State Machine Simulator), which was also used at the beginning of Chapter 1, on page 1.2.1.3.

An example of the third module is the Deeds-McE (Microcomputer Emulator), that we looked at briefly in Section 2.1.9 of this chapter. In this section we will begin to deal with the Deeds-McE, Microcomputer Emulator's main features, which will be treated in greater detail throughout the rest of this book.

2.4.1 The microcomputer components of the Deeds-DcS

The Deeds-DcS library has two components: the "DMC8 Microcomputer" and the "DMC8 Enhanced Microcomputer".

2.4.1.1 The "DMC8 Microcomputer"

This is the basic microcomputer. It has a CPU, configurable ROM and RAM from 1 to 32 kB, as well as 4 parallel input ports and 4 parallel output ports available. See the figure below.

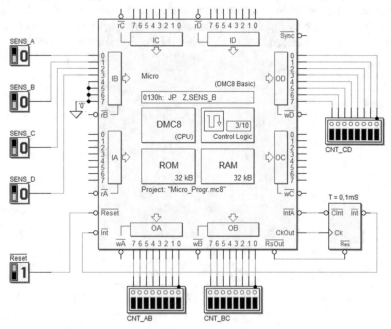

It has only one interrupt request input ($\overline{\text{Int}}$). Inside it includes one clock (10 MHz) and a reset generator for rebooting.

The example in the figure has a component working on a generic application (that is beside the scope of this introduction). The schematic shows the connections among some of the input and output lines and a "timer" component connected to line $\overline{\text{Int}}$.

Further on, we will analyze how this component works, in a discussion on interrupt techniques. For now, it is important only to know that the timer makes it possible to send an interrupt request to the CPU at regular time intervals with a period that we set in the design phase.

In the figure, the component has 4 input ports (IA, IB, IC and ID) at the upper left and 4 output ports (OA, OB, OC, OD) at the lower right, for a total of 32 input lines and as many output lines. Each port makes a signal line available ($\overline{\text{rA}}$, $\overline{\text{rB}}$, $\overline{\text{rC}}$, $\overline{\text{rD}}$ for read and $\overline{\text{wA}}$, $\overline{\text{wB}}$, $\overline{\text{wC}}$, $\overline{\text{wD}}$ for write). For a description of how they work, see Sections 2.3.2.1 and 2.3.2.2.

On the left, we see external input $\overline{\text{Reset}}$ and interrupt request input $\overline{\text{Int}}$. At the upper right, we have output $\overline{\text{Sync}}$, retrieved directly from the processor, while at the lower right we have interrupt acknowledge output $\overline{\text{IntA}}$, the clock output CkOut and the reset generator output $\overline{\text{RsOut}}$.

At the center of the figure, from higher to lower, we have the instruction being executed (during the simulation), the selected CPU (DMC8 or D8080), the ROM and RAM with their respective capacities and finally the name of the project loaded in the component. The small area labeled "Control Logic", which is animated during the simulation, shows the progress of the clock cycles.

The internal bus isn't accessible but the ports make it possible to connect anything necessary to the microcomputer. This is limited, of course, by the number of ports available. For testing purposes, we will soon see how to take advantage of all the available options during the simulation.

The following figure shows the block schematic of the inside of the microcomputer. At the center, we see the ROM and RAM memory subsystems. Next to the CPU are the reset generator and the clock generator. If you notice the Reset connections, the CPU can be reset by the internal generator and also by the external $\overline{\text{Reset}}$ input. Output $\overline{\text{RsOut}}$ is the external copy of the reset sent to the CPU.

The $\overline{\text{Int}}$ line is connected to all the processor's interrupt request lines ($\overline{\text{IRQ2}}$, $\overline{\text{IRQ1}}$ and $\overline{\text{IRQ0}}$). By activating $\overline{\text{Int}}$, we are asking the CPU to launch interrupt handler 7, which starts at address 0038h. See the table in Section 2.1.5.

The figure shows the 8-bit input and output ports connections (IA, IB, IC, ID and OA, OB, OC, OD), each with its own signalling line ($\overline{\text{rA}}$, $\overline{\text{rB}}$, $\overline{\text{rC}}$, $\overline{\text{rD}}$ and $\overline{\text{wA}}$, $\overline{\text{wB}}$, $\overline{\text{wC}}$, $\overline{\text{wD}}$, respectively).

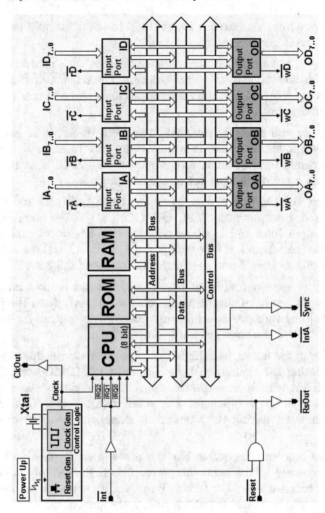

2.4.1.2 The "DMC8 Enhanced Microcomputer"

This version is very similar to the basic DMC8 Microcomputer seen above but it offers double the input and output lines. They are divided into 8 parallel input ports and 8 parallel output ports. This version also offers 7 interrupt lines. For everything else, the description of the basic DMC8 Microcomputer also applies to this version.

In the figure below, the schematic shows 64 LEDs connected to the output ports, an ON/OFF switch connected to input port IA and a timer connected to interrupt line 7 at the bottom right.

The component symbol shows the 8 input ports (IA, IB, IC, ID, IE, IF, IG and IH) at the top left, and 8 output ports (OA, OB, OC, OD, OE, IF, IG and IH) at the bottom right.

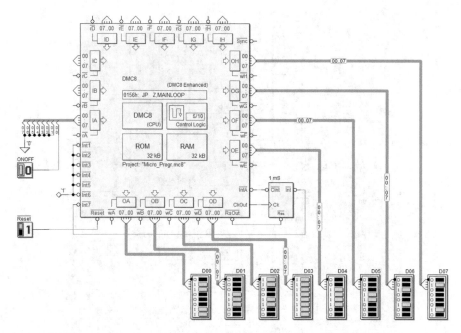

As before, each port makes a synchronization signal available: $(\overline{rA}, \overline{rB}, \overline{rC}, \overline{rD}, \overline{rE}, \overline{rF}, \overline{rG}, \overline{rH}$ for read and $\overline{wA}, \overline{wB}, \overline{wC}, \overline{wD}, \overline{wE}, \overline{wF}, \overline{wG}, \overline{wH}$ for write). Everything else has the same names and functionalities. As before, the internal bus is not available.

The next page shows a block schematic of the "enhanced" version. Its internal architecture is almost identical to the one we saw for the basic component with two visible differences. The number of available ports has doubled and the interrupt logic is different.

The interrupt logic makes 7 interrupt lines available: $\overline{Int7}$, $\overline{Int6}$, $\overline{Int5}$, $\overline{Int4}$, $\overline{Int3}$, $\overline{Int2}$ and $\overline{Int1}$. They are handled by a priority encoder that pilots the processor's 3 interrupt request lines: $\overline{IRQ2}$, $\overline{IRQ1}$ and $\overline{IRQ0}$. The encoding allows us to have up to 7 devices to handle with the interrupt technique. This guarantees that the intervention priority will be managed. See the table below.

Int7	Int6	Int5	Int4	Int3	Int2	Int1	IRQ2	IRQ1	IRQ0	Interrupt	Address
1	1	1	1	1	1	1	1	1	1	No request	-
1	1	1	1	1	1	0	1	1	0	Int. 1	0008h
1	1	1	1	1	0	-	1	0	1	Int. 2	0010h
1	1	1	1	0	-	-	1	0	0	Int. 3	0018h
1	1	1	0	-	-	-	0	1	1	Int. 4	0020h
1	1	0	-	-	-	-	0	1	0	Int. 5	0028h
1	0	-	-	-	-	-	0	0	1	Int. 6	0030h
0	-	-	-	-	-	-	0	0	0	Int. 7	0038h

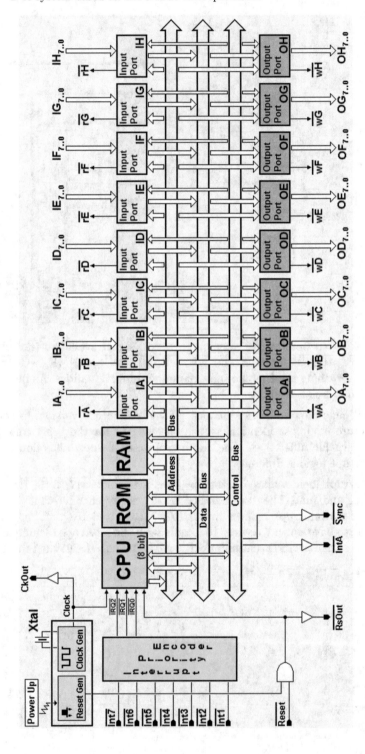

When the encoder has multiple simultaneous requests on lines $\overline{\text{Int7}}$, $\overline{\text{Int6}}$, $\overline{\text{Int5}}$, $\overline{\text{Int4}}$, $\overline{\text{Int3}}$, $\overline{\text{Int2}}$ and/or $\overline{\text{Int1}}$, will produce the code for the highest line in the output, as seen in the previous table.

Let's take an example: if two devices request an interrupt at the same time, one on line $\overline{\text{Int6}}$ and the other on line $\overline{\text{Int3}}$, the one connected to line $\overline{\text{Int6}}$ has the higher priority. So, the encoder generates handler 6's request (at address 0030h), and the processor is interrupted and acts on the request.

Now let's imagine that the device whose request was fulfilled deactivates line $\overline{\text{Int6}}$ and that the other device keeps requesting the interrupt on line $\overline{\text{Int3}}$. Now the encoder interrupts the processor and generates handler 3's request (starting at address 0018h).

When the second request has been fulfilled, line $\overline{\text{Int3}}$ deactivates and the processor is finally free from interrupts. We will study this more in depth in Section 4.4.

2.4.2 Developing a program

Developing a program starts with analyzing the project specifications, identifying the modules it will be made up of, and finally writing the source code (in our case, in assembly language).

The code-writing process is helped by a text editor. The program writing phase is followed by translation into machine code.

Its functionality can be checked by simulation or actually executing the code on a prototype card. Each programming error is corrected by coming back to the text editor. After that we repeat checking the program's functionality.

2.4.2.1 Writing source code

The Deeds-McE microcomputer emulator can be used as a text editor. Among its main features are syntax highlighting and unlimited Undo and Redo, aside from all the classic text editor commands.

The following figure shows the emulator with a file open inside the code editor (a program written in assembly language). The text appears in columns because the tabulator is pre-defined for assembly sources (columns of labels, mnemonic codes, operands and comments). At the upper part under the menu, we can see that the code editor is contained in one of the three pages that can be selected by choosing the appropriate heading.

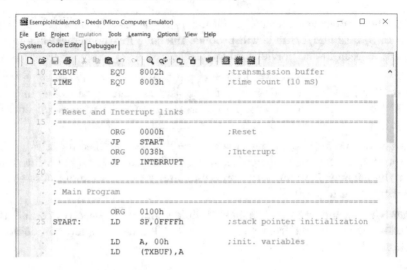

If we click on the "System" heading, a window like the one below will open. This page allows us to define the system's hardware configuration according to the needs of our project.

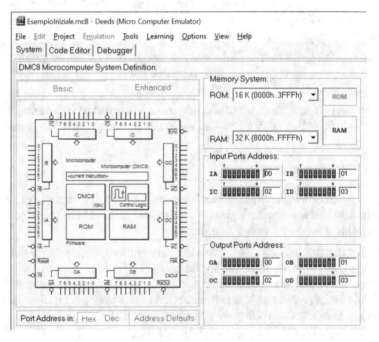

In the example in the figure above, we have chosen to use the basic version of the microcomputer (shown on the left), and to set a configuration with 16 kB of ROM and 32 kB of RAM (at the upper right), and default port addresses: 00h, 01h, 02h and 03h (in this order both for input and for output).

It is important to define the hardware used by the system at the beginning of the design phase to avoid inconsistencies between the written code and the system configuration.

2.4.2.2 Translating source code into machine code

After defining the system and writing the program in the editor, we need to translate that program into machine code. We click the icon indicated by the orange arrow (see below) to activate the assembler.

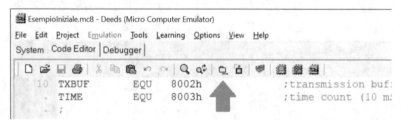

When the program is translated and there are no syntactical errors in the source code, the message below will appear at the bottom of the window to indicate that we can now check its functionality.

```
Assembling file "EsempioIniziale.mc8"
First pass.....
Second pass.....
-> Code assembled and loaded with success!   [DMC8 Instruction Set]
```

2.4.2.3 Emulation and program verification

Now let's click on the button indicated by the blue arrow in the figure below.

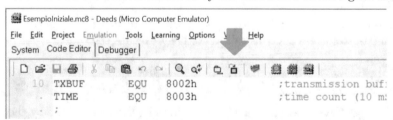

This will bring us to the "debugger" page. The debugger is a tool that allows us to verify code functionality and it's the third heading at the top of the page. See the figure on the next page. The debugger has a decidedly complex interface but it is actually well organized as we will see.

The page shows everything related to the state of our system, such as the content of the processor's registers, the memory locations, the state of the input/output ports and the program being executed.

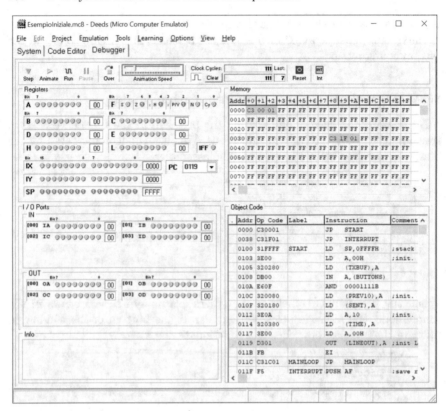

Now let's look at the sections that the debugger interface is divided into. We will enlarge them and comment on them one by one.

The control bar

At the upper part of the figure, it is shown enlarged below. The control bar has the buttons and controls to interactively execute the code.

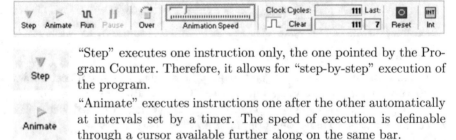

"Step" executes one instruction only, the one pointed by the Program Counter. Therefore, it allows for "step-by-step" execution of the program.

"Animate" executes instructions one after the other automatically at intervals set by a timer. The speed of execution is definable through a cursor available further along on the same bar.

"Run" executes the program at the highest speed allowed by the personal computer in use (for example with a 2 GHz CPU, we get an execution like that of a real 2 MHz clock).

"Pause" (unsurprisingly) pauses the emulation if it had previously been started by "Animate" or "Run". One can start execution again by clicking on "Step", for example.

 "Over" has two fixed positions (pressed or released). If it is active in the "Step" and "Animate" modes, the emulator executes the subprograms as if they were one instruction.

The subprograms will be examined next in Section 3.4 Chapter 3. For now, it is enough to say that they are sequences of instructions that can be called with the "CALL" instruction.

 This cursor helps the user regulate the speed of execution in the animation mode. The minimum is one instruction per second.

 The upper field in the clock section shows the number of clock cycles that have run as of system reset.

The "Clear" button makes it possible to set the number in the lower field to zero. This number represents the number of clock cycles run as of the most recent action on the button (or as of system reset). This partial counter is useful, for example, when we want to assess a program's execution times.

 "Reset" controls the microcomputer's $\overline{\text{Reset}}$ input. When it is pressed, the emulated system is kept in its initial state. When it is released, it goes back to responding to emulation commands.

 In the basic version of the microcomputer, this button controls the microcomputer's $\overline{\text{Int}}$ inputs and makes it possible to manually request an interrupt of the processor.

 In the "enhanced" version, rather than just one button, there are seven, one for each interrupt line.

The register section

The "Registers" section shows the internal state of the processor, namely what is contained in all the registers and flags (see the following figure). Note, for example, accumulator A at the upper left and the flags at the right represented one by one but grouped as register F. The content of the registers is updated during the course of execution of the machine code.

This section also gives us the option of manually changing the value of the registers. We can do that in binary mode by pressing the small, round, colored buttons of the individual bits or in text mode by inserting the number in the field at the right of each register (the context-sensitive menu in this section allows us to set the format to decimal or hexadecimal).

It is useful to change the content of the registers in the program test phase to shorten working times.

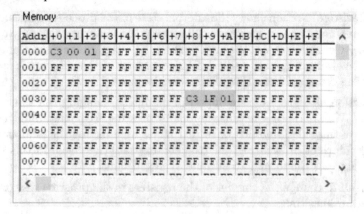

For example, if there is a loop in the code, under some circumstances it might be useful to force an early exit from that loop by manually changing the value of the register used as a counter.

Flags are only changeable in binary mode since they are individual bits. Manually changing the value of a flag can also be useful for example to cause or prevent a conditional jump for testing purposes.

The content of the Program Counter cannot be freely changed. We can only assign it the instruction addresses of the program currently loaded in the memory. Changing the value of the Program Counter makes it possible to bypass parts of the code when we need to, to go and test individual parts of the program.

The memory section

The memory section is shown below. The address of the first memory location in a row is shown on the left in hexadecimals. The addresses of the other locations on the same row are given by that address plus the number carried over at the top of the columns themselves.

Addr	+0	+1	+2	+3	+4	+5	+6	+7	+8	+9	+A	+B	+C	+D	+E	+F
0000	C3	00	01	FF	FF	FF	FF	FF	FF	FF	FF	FF	FF	FF	FF	FF
0010	FF	FF	FF	FF	FF	FF	FF	FF	FF	FF	FF	FF	FF	FF	FF	FF
0020	FF	FF	FF	FF	FF	FF	FF	FF	FF	FF	FF	FF	FF	FF	FF	FF
0030	FF	FF	FF	FF	FF	FF	FF	FF	C3	1F	01	FF	FF	FF	FF	FF
0040	FF	FF	FF	FF	FF	FF	FF	FF	FF	FF	FF	FF	FF	FF	FF	FF
0050	FF	FF	FF	FF	FF	FF	FF	FF	FF	FF	FF	FF	FF	FF	FF	FF
0060	FF	FF	FF	FF	FF	FF	FF	FF	FF	FF	FF	FF	FF	FF	FF	FF
0070	FF	FF	FF	FF	FF	FF	FF	FF	FF	FF	FF	FF	FF	FF	FF	FF

For example, in the row identified by address 0030h the location containing 1Fh is found at address 0039h (0030h + 9h).

For our convenience, the emulator color codes the background of the different memory areas according to their destination: ROM, RAM, non-initialized area, area containing code, etc. In the example above, we can distinguish two 3-byte ROM areas that contain machine code (two jumps) from the areas not explicitly initialized.

The grid also gives us the option of manually changing the content of the memory locations, but only in the RAM area. To edit a location, we simply need to click on it, delete the previous value and write the new number (in hexadecimals).

Remember that only the RAM can be edited because the emulator places the user on the same level as the CPU (which can't write in the ROM). The ROM was programmed in the previous phase, following the compilation of the program. When the user starts using the debugger, the ROM is ready to use and no longer modifiable at "run-time", as in a real system.

The context-sensitive menu lets us move to another address quickly (see the figure below) whereas the scroll bar on the right is impractical for large movements.

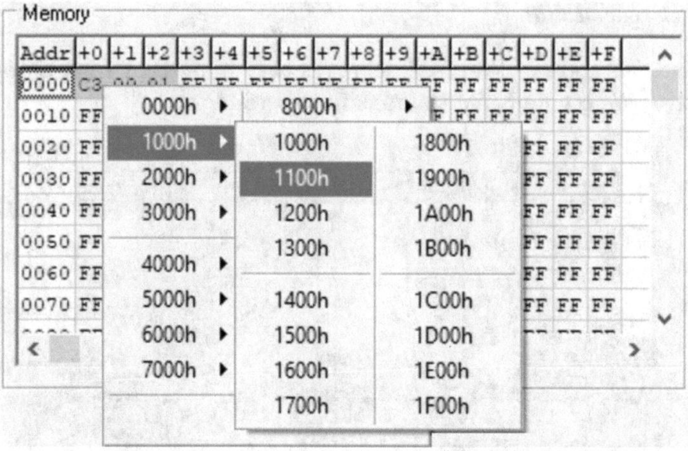

The input/output port section

The following figure shows the section on the states of the input/output ports for the "basic" microcomputer. It allows us to change the state of input ports IA, IB, IC and ID, similarly to what we saw for the register section.

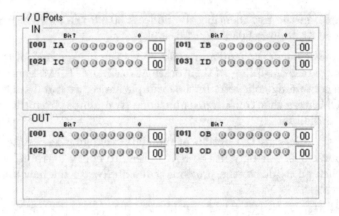

The state of output ports OA, OB, OC and OD, however, cannot be changed by the user, as they are generated by the microcomputer.

We have seen that port addresses are defined on the system definition page. This example shows the default addresses. In the "enhanced" version, the empty spaces we see in this section are filled by the other ports (we have 8 input ports and 8 output ports in total).

The object code section

The executable code section (see below) is very important from an operative perspective. We see multiple columns in this window.

Addr	Op Code	Label	Instruction		Comment
0000	C30001		JP	START	
0038	C31F01		JP	INTERRUPT	
0100	31FFFF	START	LD	SP,0FFFFH	;stack
0103	3E00		LD	A,00H	;init.
0105	320280		LD	(TXBUF),A	
0108	DB00		IN	A,(BUTTONS)	
010A	E60F		AND	00001111B	
010C	320080		LD	(PREV10),A	;init.
010F	320180		LD	(SENT),A	
0112	3E0A		LD	A,10	;init.
0114	320380		LD	(TIME),A	
0117	3E00		LD	A,00H	
0119	D301		OUT	(LINEOUT),A	;init L
011B	FB		EI		
011C	C31C01	MAINLOOP	JP	MAINLOOP	
011F	F5	INTERRUPT	PUSH AF		;save r

Columns two and three show the instruction address and its machine code, respectively (both in hexadecimals).

The following columns show the corresponding source code (label, mnemonic code, operands and comment), which has generated the machine code shown on the left.

To learn to read this grid, let's look at the first row. We find opcode C3h at address 0000h, followed by bytes 00h and 01h (we read: C30001 in the box). As we can see, this instruction takes up 3 bytes. In column 5, we read "JP START", the instruction's source code. Further down, the START label corresponds to address 0100h, which is split into two bytes (in little-endian convention) and placed immediately after opcode C3h.

When we test our program, the instructions being executed are highlighted row by row in this window. When execution is paused, the instruction highlighted is the one that still needs to be executed.

It is possible to insert "breakpoints" in the code. They let the system know when to stop code execution. During execution, if a breakpoint is found, execution is immediately stopped before the instruction indicated. This gives the user the chance to proceed in step mode, for example.

Breakpoints are very useful because they allow us to very quickly execute parts of code that have already been tested and also to stop where we need to more closely examine the execution of instructions, possibly one by one. To insert or eliminate a breakpoint, we go to the grid and select the row of the instruction we want the emulator to stop at, then we use the context-sensitive menu (see the following figure).

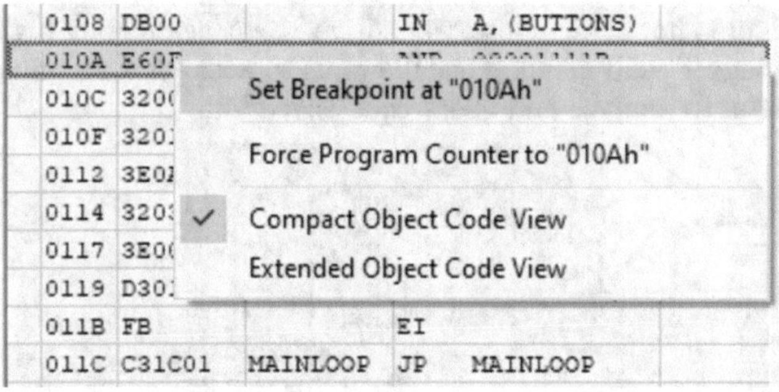

After we insert a breakpoint, a little red square will appear to highlight it, as shown in the following example.

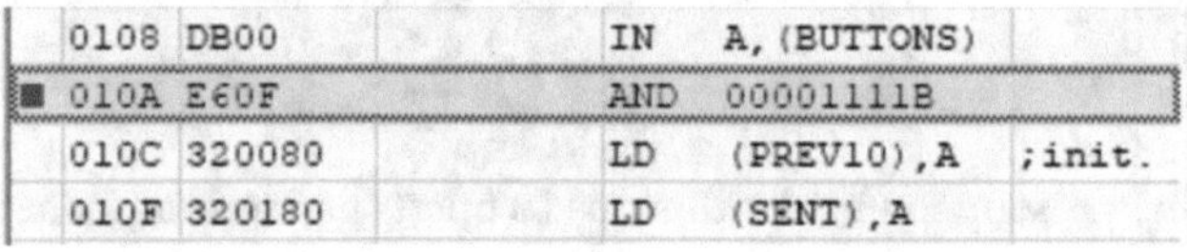

2.4.3 Configuring the microcomputer component

After checking the program's logical functionality in the Deed-McE, we can load the developed project in the DMC8 microcomputer component inserted into the Deeds-DcS schematic. The figure below shows the component's context-sensitive menu, where the "Load DMC8 Project" menu item is highlighted. It accepts project files with an ".mc8" extension. The project includes the source assembly code as well as the memory and port configurations.

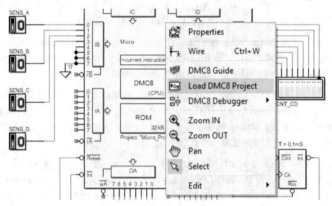

The system compiles the source program contained in the project and loads the microcomputer's ROM with the resulting machine code ready to execute when the whole circuit is simulated.

The next figure shows the context-sensitive menu again, but highlights the "DMC8 Debugger" sub-menu. This allows us to activate a certain number of windows useful to analyze and check code execution.

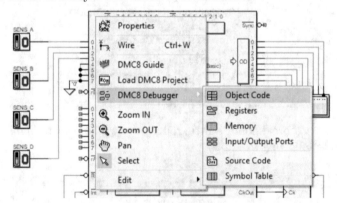

As we can see, these menu items allow us to open the sections from the Deeds-McE that we saw previously. Here, they can be put on the screen separate from one another if we like, each in a stand-alone window. In the next few chapters, we will examine how to use these windows and conduct a timing simulation of a network that includes a microcomputer.

2.5 Exercises

The digital content pages of the book on the Deeds simulator website have outlines of the schematics, diagrams and/or programs to complete for each exercise. Those same web pages also have the files for the solutions, so that students can check their work.

2.5.1 Memory systems

1. Define the memory map of a system that includes a 1-kbyte ROM allocated as of location 0000h, and two 1-kbyte RAMs connected from location 8000h on.

2. Take the memory system described in the following schematic, and obtain its memory map by analyzing the logic highlighted in section S.

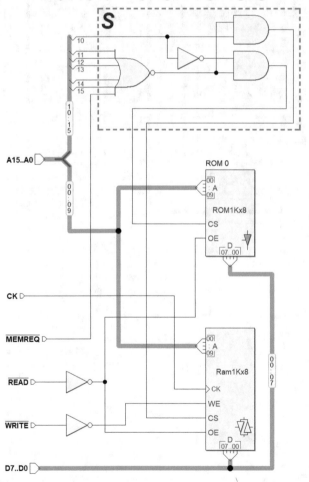

3. In the schematic below, a memory system is composed of four 1-kbyte ROMs and four 1-kbyte RAMs. Draw its memory map (we suggest beginning your analysis at the logic highlighted in section S).

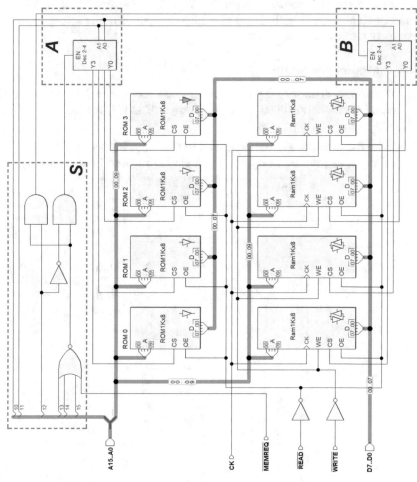

4. Complete the schematic of the partially defined memory system below so that the 8 kB of ROM are allocated as of address 0000h, and the area for the 32 KB of RAM ends at address FFFFh.

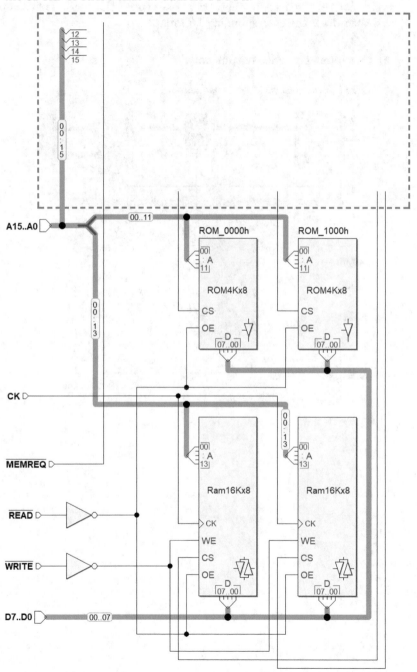

2.5.2 Parallel input/output ports

1. Taking the parallel port systems in the following figures, write the Boolean expressions of the decoding networks that control the enables of each port and then draw the corresponding I/O map.

a) One input port, one output port.

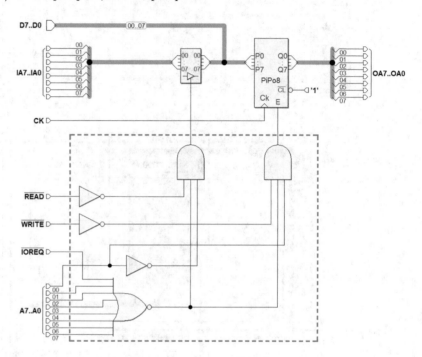

b) Four input ports.

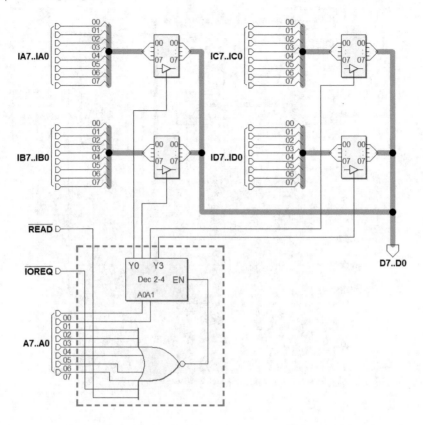

c) Two input ports, two output ports.

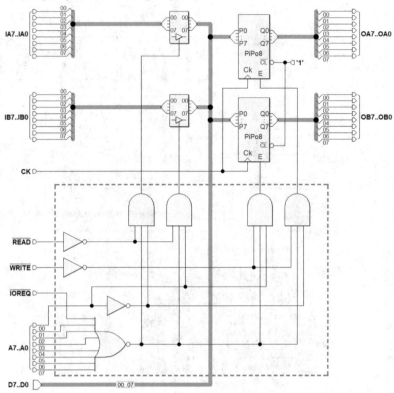

2. Consistently with the I/O map below, design the decoder for a system
   with eight parallel output ports. Write the Boolean expressions for the
   enables of each port. The schematic to fill in can be found on the next
   page.

Hex	A_7	A_6	A_5	A_4	A_3	A_2	A_1	A_0	Ports	
F8h	1	1	1	1	1	0	0	0	OA	Output
F9h	1	1	1	1	1	0	0	1	OB	Output
FAh	1	1	1	1	1	0	1	0	OC	Output
FBh	1	1	1	1	1	0	1	1	OD	Output
FCh	1	1	1	1	1	1	0	0	OE	Output
FDh	1	1	1	1	1	1	0	1	OF	Output
FEh	1	1	1	1	1	1	1	0	OG	Output
FFh	1	1	1	1	1	1	1	1	OH	Output

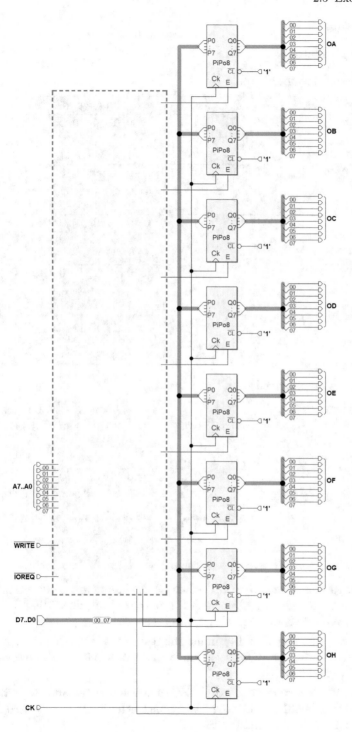

2.6 Solutions

2.6.1 Memory systems

1. The requested memory map.

Hex	A_{15}	A_{14}	A_{13}	A_{12}	A_{11}	A_{10}	A_9	A_8	A_7	A_6	A_5	A_4	A_3	A_2	A_1	A_0	Comment
0000h	0	0	0	0	0	0	0	0	0	0	0	0	0	0	0	0	1 kB
↓	.	.	.	.	.	.	.	.	.	.	.	.	.	.	.	.	ROM
03FFh	0	0	0	0	0	0	1	1	1	1	1	1	1	1	1	1	
0400h	0	0	0	0	0	1	0	0	0	0	0	0	0	0	0	0	Free
↓	.	.	.	.	.	.	.	.	.	.	.	.	.	.	.	.	Space
7FFFh	0	1	1	1	1	1	1	1	1	1	1	1	1	1	1	1	
8000h	1	0	0	0	0	0	0	0	0	0	0	0	0	0	0	0	1 kB
↓	.	.	.	.	.	.	.	.	.	.	.	.	.	.	.	.	RAM
83FFh	1	0	0	0	0	0	1	1	1	1	1	1	1	1	1	1	
8400h	1	0	0	0	0	1	0	0	0	0	0	0	0	0	0	0	1 kB
↓	.	.	.	.	.	.	.	.	.	.	.	.	.	.	.	.	RAM
87FFh	1	0	0	0	0	1	1	1	1	1	1	1	1	1	1	1	
8800h	1	0	0	0	1	0	0	0	0	0	0	0	0	0	0	0	Free
.	.	.	.	.	.	.	.	.	.	.	.	.	.	.	.	.	Space
FFFFh	1	1	1	1	1	1	1	1	1	1	1	1	1	1	1	1	

2. The memory map derived from the network.

Hex	A_{15}	A_{14}	A_{13}	A_{12}	A_{11}	A_{10}	A_9	A_8	A_7	A_6	A_5	A_4	A_3	A_2	A_1	A_0	Comment
0000h	0	0	0	0	0	0	0	0	0	0	0	0	0	0	0	0	1 kB
↓	.	.	.	.	.	.	.	.	.	.	.	.	.	.	.	.	ROM
03FFh	0	0	0	0	0	0	1	1	1	1	1	1	1	1	1	1	
0400h	0	0	0	0	0	1	0	0	0	0	0	0	0	0	0	0	1 kB
↓	.	.	.	.	.	.	.	.	.	.	.	.	.	.	.	.	RAM
07FFh	0	0	0	0	0	1	1	1	1	1	1	1	1	1	1	1	
0800h	0	0	0	0	1	0	0	0	0	0	0	0	0	0	0	0	Free
.	.	.	.	.	.	.	.	.	.	.	.	.	.	.	.	.	Space
FFFFh	1	1	1	1	1	1	1	1	1	1	1	1	1	1	1	1	

The NOR gate only generates a '1' when all the inputs are at '0'. So that the following AND gates can generate a '1' in the output, lines $\overline{\text{MEMREQ}}$, A15, A14, A13, A12 and A11 all have to be at '0'.

If they are, we go ahead and examine the rest of the inputs at the AND gates. If A10 = '0', the ROM selection is enabled. If A10 = '1', however, the RAM is selected.

Therefore, if memory request $\overline{\text{MEMREQ}}$ is active and the address is between 0000h and 03FFh the ROM is enabled. The RAM is activated in the area between 0400h and 07FFh.

3. As suggested in the text, we begin the analysis by the NOR gate. We know that it only generates a '1' when all the inputs are at '0', so A15, A14 and A13 must be '0' and, similarly, line $\overline{\text{MEMREQ}}$ must be active.

Now, supposing we have a '1' in the output of the NOR gate, let's examine the second logical level (the AND gates). If A12 = '0', the ROM bank selection circuit is enabled (shown in section A). If A12 = '1', the RAM bank decoder (section B) is enabled.

So we conclude that if the memory request $\overline{\text{MEMREQ}}$ is active and the address is between 0000h and 0FFFh the ROM bank is enabled. If the address is between 1000h and 1FFFh, the RAM bank is enabled. The resulting memory map is shown below.

Hex	A_{15}	A_{14}	A_{13}	A_{12}	A_{11}	A_{10}	A_9	A_8	A_7	A_6	A_5	A_4	A_3	A_2	A_1	A_0	Comment
0000h	0	0	0	0	0	0	0	0	0	0	0	0	0	0	0	0	4 x 1 kB
↓	.	.	.	.	.	.	.	.	.	.	.	.	.	.	.	.	ROM
0FFFh	0	0	0	0	1	1	1	1	1	1	1	1	1	1	1	1	
1000h	0	0	0	1	0	0	0	0	0	0	0	0	0	0	0	0	4 x 1 kB
↓	.	.	.	.	.	.	.	.	.	.	.	.	.	.	.	.	RAM
1FFFh	0	0	0	1	1	1	1	1	1	1	1	1	1	1	1	1	
2000h	0	0	1	0	0	0	0	0	0	0	0	0	0	0	0	0	Free
.	.	.	.	.	.	.	.	.	.	.	.	.	.	.	.	.	Space
FFFFh	1	1	1	1	1	1	1	1	1	1	1	1	1	1	1	1	

4. To complete the schematic, we need to define the address decoders. With the description in the text in mind, let's allocate the two ROM devices so that one of them activates in the interval between addresses 0000h..0FFFh, and the other between 1000h and 1FFFh.

Given that the overline of $\overline{\text{MEMREQ}}$ indicates that this input is "active low" and not a true negation, the expressions for the ROM decoders are the following.

$$\text{CS}_{\text{ROM\_0000h}} = \overline{(\overline{\text{MEMREQ}}) + A_{15} + A_{14} + A_{13} + A_{12}}$$
$$\text{CS}_{\text{ROM\_1000h}} = \overline{(\overline{\text{MEMREQ}}) + A_{15} + A_{14} + A_{13} + \overline{A_{12}}}$$

Similarly, to define the RAM decoder, one of the two components activates in the interval 8000h..BFFFh, and the other between C000h and FFFFh. The two expressions are shown below.

$$\text{CS}_{\text{RAM\_8000h}} = \overline{(\overline{\text{MEMREQ}})} \cdot A_{15} \cdot \overline{A_{14}}$$
$$\text{CS}_{\text{RAM\_C000h}} = \overline{(\overline{\text{MEMREQ}})} \cdot A_{15} \cdot A_{14}$$

The complete schematic is shown on the next page.

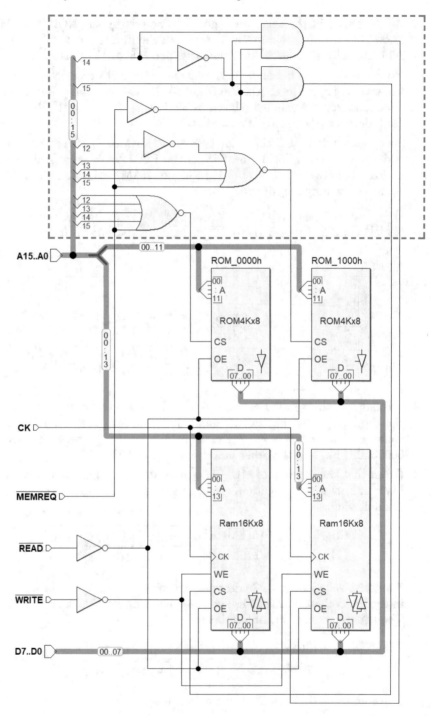

2.6.2 Parallel input/output ports

1. Assuming that the overline of $\overline{\text{IOREQ}}$, $\overline{\text{READ}}$ and $\overline{\text{WRITE}}$ does not indicate that there are explicit negations, but that these inputs are "active low", the expressions for decoding the ports and the input/output maps are as follows.

a) To make it easier to read, let's define the following intermediate expressions:

$$EN = \overline{(\overline{\text{IOREQ}}) + A_7 + A_6 + A_5 + A_4 + A_3 + A_2 + A_1}$$

Thus, the decoding expressions are:

$$E_{IA} = \overline{A_0} \cdot \overline{(\overline{\text{READ}})} \cdot EN$$
$$E_{OA} = A_0 \cdot \overline{(\overline{\text{WRITE}})} \cdot EN$$

The map of the ports follows.

Hex	A_7	A_6	A_5	A_4	A_3	A_2	A_1	A_0	Ports	
00h	0	0	0	0	0	0	0	0	IA	Input
01h	0	0	0	0	0	0	0	1	OA	Output

b) Here, as above, it makes sense to define an intermediate expression:

$$EN = \overline{(\overline{\text{IOREQ}}) + (\overline{\text{READ}}) + A_7 + A_6 + A_5 + A_4 + A_3 + A_2}$$

The resulting expressions:

$$Y0_{IA} = (\overline{A_1} \cdot \overline{A_0}) \cdot EN$$
$$Y1_{IB} = (\overline{A_1} \cdot A_0) \cdot EN$$
$$Y2_{IC} = (A_1 \cdot \overline{A_0}) \cdot EN$$
$$Y3_{ID} = (A_1 \cdot A_0) \cdot EN$$

The I/O map.

Hex	A_7	A_6	A_5	A_4	A_3	A_2	A_1	A_0	Ports	
00h	0	0	0	0	0	0	0	0	IA	Input
01h	0	0	0	0	0	0	0	1	IB	Input
02h	0	0	0	0	0	0	1	0	IC	Input
03h	0	0	0	0	0	0	1	1	ID	Input

c) When we have defined the following:

$$EN = \overline{(\overline{IOREQ}) + A_7 + A_6 + A_5 + A_4 + A_3 + A_2 + A_1}$$

We get the expressions:

$$E_{IA} = \overline{A_0} \cdot \overline{(READ)} \cdot EN$$

$$E_{IB} = A_0 \cdot \overline{(READ)} \cdot EN$$

$$E_{OA} = \overline{A_0} \cdot \overline{(WRITE)} \cdot EN$$

$$E_{OB} = A_0 \cdot \overline{(WRITE)} \cdot EN$$

The map (notice that an input port and an output port are both located at the same address).

Hex	A_7	A_6	A_5	A_4	A_3	A_2	A_1	A_0	Ports	
00h	0	0	0	0	0	0	0	0	IA	Input
00h	0	0	0	0	0	0	0	0	OA	Output
01h	0	0	0	0	0	0	0	1	IB	Input
01h	0	0	0	0	0	0	0	1	OB	Output

2. The requested network including eight output ports. Let's define the following intermediate expression:

$$EN = \overline{(WRITE)} \cdot \overline{(IOREQ)} \cdot A_7 \cdot A_6 \cdot A_5 \cdot A_4 \cdot A_3$$

We get the following expressions for the decoder:

$$E_{OA} = \overline{A_2} \cdot \overline{A_1} \cdot \overline{A_0} \cdot EN$$

$$E_{OB} = \overline{A_2} \cdot \overline{A_1} \cdot A_0 \cdot EN$$

$$E_{OC} = \overline{A_2} \cdot A_1 \cdot \overline{A_0} \cdot EN$$

$$E_{OD} = \overline{A_2} \cdot A_1 \cdot A_0 \cdot EN$$

$$E_{OE} = A_2 \cdot \overline{A_1} \cdot \overline{A_0} \cdot EN$$

$$E_{OF} = A_2 \cdot \overline{A_1} \cdot A_0 \cdot EN$$

$$E_{OG} = A_2 \cdot A_1 \cdot \overline{A_0} \cdot EN$$

$$E_{OH} = A_2 \cdot A_1 \cdot A_0 \cdot EN$$

The next page shows the complete schematic of the port system.

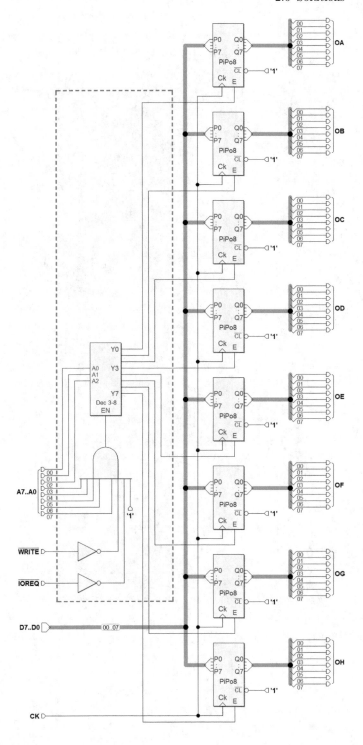

3

Programming the DMC8

Abstract In this chapter, after a brief and general introduction to the programming languages and their compilers, the DMC8 assembly language is presented. The different sections will describe the processor's addressing modes, categorize the instructions available, and provide examples to illustrate their functionality. Particular attention will be given to jumps, delay loops, call and return from subprograms, and to the usage of the stack. The end of the chapter will provide several programming examples. We will deal with the emulation of combinational and sequential networks, the calculation of simple mathematical expressions, and the emulation of finite state machines.

3.1 Introduction to assembly language programming

If we were to design a control system for an industrial plant, for example, we would face both hardware and software problems. Once the hardware system architecture is defined, we would need to work on the management and control software for the plant. The program we write will gear our system toward the specific control application.

The tasks our program has to do can be described by an algorithm. An algorithm can be expressed in any form, such as in words (like a recipe or directions) or through images (like a map of the emergency exits in a building). We can use graphical languages (such as flow charts) or programming languages.

In order for our microprocessor-based system to carry out the requested tasks, we need to express the algorithm in a language the processor can understand, machine language.

3.1.1 Programming languages

As we have seen since Section 1.2, translating a textually expressed algorithm understandable by people into machine code is a long, difficult process prone

© The Author(s), under exclusive license to Springer Nature Switzerland AG 2022 189
G. Donzellini et al., *Introduction to Microprocessor-Based Systems Design*,
https://doi.org/10.1007/978-3-030-87344-8_3

to errors. This is why we need software that can automate the process. We were introduced to this in Section 1.2.6 and developed the idea by seeing the Deeds-McE emulator in Section 2.4.2.2.

Generally speaking, there are different translators associated to different languages. A program written in any programming language is called a "source program".

3.1.1.1 Languages at a "high level" of abstraction

The more understandable a language is to humans, the higher the "level of abstraction" with respect to the machine. Languages with a high level of abstraction have very expressive syntactic constructs that make it possible to write algorithms that are short and easy for humans to understand. When these constructs are translated, they generally give rise to a large number of machine instructions.

Some languages with a high level of abstraction are Basic, Fortran, Cobol, Pascal, Object Pascal, Modula-2, C, C++, Java and C#. Each of these offers a different syntactic construct in order to facilitate writing algorithms according to the different programming paradigms. Microprocessor manufacturers normally offer support for the most common high level languages such as C and C++.

Translators used for high level languages are commonly called "compilers". When a source program is compiled, the result is a program in machine code that the processor can directly execute. Once the machine code is available and memorized, the compiler will only need to be used again if the source is changed.

Another type of translator called an "interpreter" is used to translate the source code directly at the moment of execution. That is outside the scope of this book.

3.1.1.2 Languages at a "low level" of abstraction

Low level languages like Assembly (specific to each processor) have mnemonic-symbolic instructions that correspond one-to-one with the machine instructions that the microprocessor can directly execute.

For a given processor, there is a specific "assembler" that programmers use. An assembler is a software application that translates a program written in mnemonic assembly code into the corresponding machine code.

Assembly language allows the programmer to directly manipulate registers and bits inside the machine. It is typically used in situations that call for particularly efficient execution. Programming in assembly code is arduous work and especially time consuming. Every minute detail must be programmed and even apparently tiny, mundane particulars must be specified. Using assembly also requires the programmer to have a solid understanding of the hardware that executes the program.

For ordinary, (more or less complex) programs, these disadvantages lead one to favor using high level, more concise and understandable languages. Often a mixed technique is used: a high level language to write most of the code, and assembly code for parts that need special optimization (time and/or memory). Once the different modules are translated into machine language, they are "connected" together to make a single, executable machine code thanks to another software called a "linker".

A program written in assembly code is not "portable", that is it is specific for that microprocessor and the hardware around it. Reusing a certain program on a different machine may require a complete rewriting of the code. Programs written in standard high level language, however, are more interchangeable among different machines and microprocessors as long as the right compilers are used.

3.1.1.3 From the source code to machine code that is executable in the system

To summarize, compiling source code produces object code, which is directly executable by the machine (except for the case of interpreters). Object code can be obtained by translating high level or low level languages. The figure below shows the consecutive phases of programming a system.

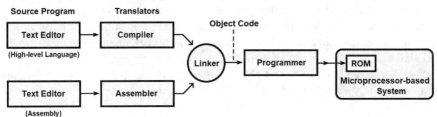

If the program is made up of multiple, interdependent modules of code due to different translations, the linker intervenes, giving us one object code (dividing a program into modules is very useful if the project is complex).

The machine code then needs to be loaded in the memory of the system where it will run. This can be done by a a purely software module called a "loader", when possible. In many cases, however, it is still necessary to use a hardware "programmer". The programmer lets us permanently load the machine code into the system's ROM memory so it is ready to execute when the computer is turned on.

In the rest of the book, we will use assembly language. This choice is meant to keep the reader in contact with microprocessor system architecture, allow for total control over the system hardware and promote a better understanding of how it interacts with the programming.

3.1.2 DMC8 assembly language

In assembly language, each line of source code is divided into a few fields that must be separated by spaces or tabs. Each line contains only one instruction. When we write a program in assembly, we need to conceive of our page as virtually divided into four columns (or fields), as shown in this example:

Label	Mnemonic	Operand	Comment
	. . .	. . .	. . .
	LD	A,(34F0h)	; read a byte from memory
	LD	D,2Ah	; the initial count value in register 'D'
LOOP:	DEC	D	; decrement the counter
	JP	NZ,LOOP	; until register 'D' is zero-ed
	. . .	. . .	. . .
	. . .	. . .	. . .

The meaning of the four fields:

Label   It is used as a reference for the program's jumps (the programmer only inserts it when necessary).

Mnemonic   It contains the mnemonic code of the instruction.

Operand   It includes the operand (or operands) of the mnemonic code of the instruction.

Comment   As an optional text, it helps to understand the code.

Generally, comments should begin with a semi-colon but some assemblers accept any symbol: they assume that everything following an instruction, which is recognized as complete, is a comment. A comment usually does not repeat the task carried out by the instruction since it is already clear in the mnemonic code. Rather it should explain the programmer's intent in words so that the user better understands the algorithm implemented.[1] Comments are essential as they help the programmer make his/her own code readable for others as well as for themselves.

3.1.2.1 A sample program written in assembly DMC8

A sample program is shown here below. It adds two numbers represented in 8 bits. The data are retrieved from the system's ROM memory and the result is saved in the RAM.

```
Label        Mnem./Operand        Comment
SUM:         LD    A,(0500h)       ;load the first operand
             LD    B,A
             LD    A,(0501h)       ;load the second operand
             ADD   A,B             ;add the two operands
             LD    (8000h),A       ;save the sum in RAM
             HALT
```

---

[1] In this first phase, the comments in the code do not follow this best practice since we want to better support the understanding of the mnemonic codes that we encounter for the first time.

By looking at the code and reading the comments, we can distinguish four consecutive phases:

1. Acquisition of (reading) the first operand, from the ROM to the CPU
2. Acquisition of (reading) the second operand, from the ROM to the CPU
3. Adding the two operands inside the CPU
4. The CPU stores (writes) the result to the RAM.

Phase one was coded using two instructions. The DMC8 does not have an instruction like 'LD B,(0500h)' and transferal to register B has to be done in two steps, passing through register A. Often an instruction does not allow all the operands that would seem admissible from a syntactical perspective. When in doubt, it is always a good idea to consult the summary tables showing DMC8 instructions, in order to avoid errors (See Appendix C).

Once the first three instructions are executed, the operands are transferred from the ROM memory to registers A and B. The ADD instruction adds the two operands in A and B and puts the result in A. Then, the second to last instruction transfers the result of the addition to the RAM memory. The last instruction, HALT, stops the processor. Hereafter, this instruction will be used very rarely. Except in special cases, it makes little sense to stop the processor.

Remember that the assembler does not distinguish between capital and lower case letters (technically speaking, it is not "case-sensitive"). This means writing "sum: ld a,(0500h)" would have achieved the same thing.

3.1.3 Constants and variables

High level languages offer the option of assigning a "type" to constants and variables. By assigning this attribute to constants and variables when they are declared, we let the compiler help us write our code. For example, the compiler can warn us if we are making an error in our code by transferring a 32-bit constant to a 16-bit variable.

High level languages also free us from worrying about where our constants and variables will be allocated, how they will be implemented or whether they will overlap, etc. This is because the compiler itself takes care of those important specifics for us.

There are also assemblers that let us assign a type to constants and variables. Nevertheless, a programmer using assembly has to get used to manipulating data without using types, partly because not all assembly languages support assigning types to variables, especially those with small, simple microprocessors. So programmers should decide directly which memory locations to use for a certain constant or variable and how many bytes to reserve for them.

In the example we are considering, the programmer chose to take the operands from two ROM locations at addresses 0500h and 0501h. Since this is previously programmed read only memory, the two locations contain numbers that are constants, from the perspective of the processor. Further ahead, we will see how we can set up a constant in ROM.

The programmer also decided where to allocate the variable that will assume the result: in the RAM at address 8000h. These addresses are normally chosen through a consideration of the available space in the ROM and RAM, and a comparison with the requirements of other software modules there may be.

Notice that, in our example, a label has been placed before the first instruction. Here, the programmer wanted to label the sequence of instructions to make it more readable, which is why they used the 'SUM' string, to make its function manifest.

3.1.4 The EQU directive

To meet the needs of the programmer, all assemblers allow for assigning an identifying symbol to variables similarly to high level languages. If we take advantage of this, we can rewrite the program as it appears below. For the first time, we encounter the EQU directive, EQU being an abbreviation of 'equal'.

```
RESULT    EQU    8000h       ;RESULT variable (the container of the result)
;
SUM:      LD     A,(0500h)   ;load the first operand
          LD     B,A
          LD     A,(0501h)   ;load the second operand
          ADD    A,B         ;add the two operands
          LD     (RESULT),A  ;save the sum in the variable RESULT
          HALT
```

Note that a directive is not an instruction, but rather a command for the assembler, inserted into the source code. The assembler uses that command to correctly translate the source code, but it doesn't insert any instruction corresponding to the directive in the machine code. This is why directives are also called pseudo-instructions.

In the example, by means of the EQU directive, we have declared that RESULT is a variable at memory address 8000h. However, this is an abstraction on our part; we are actually only declaring the corresponding value of a symbol. In fact, the EQU directive simply declares that RESULT is a symbol and that it is equal to the value 8000h. It is only by extension that we take the liberty of saying that RESULT is a variable, simply because we use the symbol RESULT in the code rather than the explicit address of a memory location, writing LD (RESULT),A. In other words, the variable is found at address RESULT = 8000h and we can abbreviate this by saying that the variable itself is called RESULT, identifying the variable with its address.

For easy readability, the programmer chooses to define the variables by declaring their respective symbols and corresponding addresses at the beginning of the program. Note that when we define a variable, we need to choose an address for it and take care that the space assigned to it doesn't overlap with other variables. We can define variables made up of multiple bytes and reserve the necessary space for them. In this case, the identifying symbol of the variable is usually the address of the first byte in the memory.

Aside from the readability of the code, there is another advantage related to using a symbol instead of a number. If a programmer at a given moment modifies a program by changing the allocation of the variable RESULT, they can do it by editing the row where RESULT is defined, without being forced to look through all the rows of source code that have address 8000h (in this example, that's only one row but in a full program there could be many scattered everywhere).

An important syntactic rule requires us to give symbols a name beginning with a letter of the alphabet so that the numbers are easy to distinguish from the symbols. For hexadecimal numbers, which can begin with a letter, we must add a leading zero before the number. This helps the assembler to deal with situations that would have been ambiguous. For example, if we write LD A,(FEBAH), the assembler interprets the subject in parentheses as a symbol called FEBAH. If we had wanted to write a hexadecimal number in parentheses, we would have had to write 0FEBAH (with a leading zero), and it would have been interpreted as a number (equal to 65210_{10}).

The assembler translates the source code by scanning it twice. When it scans for the first time ("Pass 1"), it doesn't produce machine codes, but identifies all the addresses they will be allocated to, instruction by instruction. It does this by assigning precise addresses to the labels it finds to the left of the instructions. These addresses are memorized in the "symbol table". This table also holds all the symbols defined through the EQUs.

When it scans for the second time ("Pass 2"), the table has values corresponding to all the labels and symbols used. Thus, every time a symbol or label appears in the source code, the assembler uses the table to substitute it with the corresponding value (in our example, in place of RESULT, the value 8000h is inserted in the machine code).

3.1.5 The ORG directive

Every DMC8 instruction translated into machine code takes one to four bytes of memory. If nothing is specified, the assembler translates the code bearing in mind that the machine code is allocated in the memory as of address 0000h.

The programmer can still specify a different address to allocate the various parts of the code. They do this by using the ORG directive, abbreviation of 'origin'. The ORG <address> directive orders the assembler to allocate the code that follows the directive as of the address specified as an operand.

Below, the previous example is shown again with some added rows of code which use the ORG directive multiple times.

```
RESULT    EQU   8000h        ;RESULT variable (the container of the result)
;
          ORG   0000h        ;the following instruction is allocated at 0000h,
          JP    0100h        ;where we place a jump to the program SUM
;
          ORG   0100h        ;the following instruction is allocated at 0100h
;
SUM:      LD    A,(0500h)    ;load the first operand
          LD    B,A
          LD    A,(0501h)    ;load the second operand
          ADD   A,B          ;add the two operands
          LD    (RESULT),A   ;save the sum in the variable RESULT
          HALT
```

The first ORG commands the assembler to allocate instruction JP 0100h, which follows the directive, at address 0000h. Due to the above mentioned default, this actually wouldn't be necessary but it is a good idea to insert it anyway to make the code clear and easy to read. We know that after the system is reset, the first instruction is taken from the processor at address 0000h. In our example, that instruction is a jump to location 0100h.

The second ORG requires the assembler to place the following instruction, LD A,(0500h) labeled as SUM, at address 0100h. So, the second ORG places our program as of address 0100h, and the first ORG connects the hardware reset to the program, thanks to the jump to 0100h.

One might ask why we didn't allocate the program directly at address 0000h since the processor starts here, instead of using a jump and making it start at 0100h. The reason is that several locations after 0000h are reserved. In fact, the locations from 0008h to 0038h are reserved for handling interrupts. See Section 4.4. A classic tactic is to circumvent the first 256 locations and jump to address 0100h, found after all the reserved locations.

3.1.6 The DB and DW directives

Programmers often need to define constants. In our example, we know that the two operands of the addition were taken from the ROM. To define one or more 8-bit constants we use the DB (Define Byte) directive and to define 16-bit constants (subdivided into two bytes) we use the DW (Define Word) directive.

Let's apply DB to our example. See the following listing. Here, we see there is an added ORG (over the previous version of the example) after the sequence of instructions. ORG defines the address (0500h) where the assembler will allocate what follows.

The first DB orders the assembler to insert the byte with value 34h at address 0500h. Note that, like the other directives, DB is not translated into executable machine code; it is not an instruction.

```
;--- Definition of variables -----------------------------------------------
RESULT        EQU    8000h          ;RESULT variable (the container of the result)
;
;--- Link to system Reset --------------------------------------------------
              ORG    0000h          ;the following instruction is allocated at 0000h,
              JP     0100h          ;where we place a jump to the program SUM
              ORG    0100h          ;the following instruction is allocated at 0100h
;
;--- The program -----------------------------------------------------------
SUM:          LD     A,(OPE_1)      ;load the constant OPE_1 in B (from ROM)
              LD     B,A
              LD     A,(OPE_2)      ;load the constant OPE_2 in A (from ROM)
              ADD    A,B            ;add the two operands
              LD     (RESULT),A     ;save the sum in the variable RESULT
              HALT
;
;--- Definition of constants -----------------------------------------------
              ORG    0500h          ;we define constants starting from 0500h
OPE_1:        DB     34h            ;OPE_1 is used as first operand
OPE_2:        DB     12h            ;OPE_2 as second operand
```

It does, however, allow us to insert the constant defined by its argument in the ROM. The second DB asks to insert byte 12h in the next location (at address 0501h).

Before each of the DBs we have inserted a label: OPE_1 and OPE_2, respectively. Thus, we make sure the assembler inserts the two labels in the symbol table so as to associate the symbol OPE_1 to the address of the first of the two bytes (34h), and OPE_2 to the address of the second (12h) (0500h and 0501h, respectively). Instruction LD A,(OPE_1) will be translated in the second pass as if we had written LD A,(0500h). The same goes for the other instruction.

Finally, note that we would have been able to omit ORG 0500h. If we had, the assembler would have allocated the two DBs immediately after the program code (rather than at the address we chose) without jeopardizing the program's functionality.

The following figure shows the result of the compilation as seen in the emulator in the object code section. The program is ready to execute.

Object Code

Addr	Op Code	Label	Instruction	Comment
0000	C30001		JP 0100H	;where we place a jump to the program SUM
0100	3A0005	SUM	LD A,(OPE_1)	;load the constant OPE_1 in B (from ROM)
0103	47		LD B,A	
0104	3A0105		LD A,(OPE_2)	;load the constant OPE_2 in A (from ROM)
0107	80		ADD A,B	;add the two operands
0108	320080		LD (RESULT),A	;save the sum in the variable RESULT
010B	76		HALT	

We have seen that the second column shows the memory addresses while the third shows the bytes that make up the machine code. At address 0000h we find the code of the jump instruction (C3h), followed by the address to jump to (0100h, divided into two bytes: high and low, 00h and 01h).

Now let's look at the memory section to see the result of the compilation from this point of view. See the figure below.

Memory

Addr	+0	+1	+2	+3	+4	+5	+6	+7	+8	+9	+A	+B	+C	+D	+E	+F
0000	C3	00	01	FF	FF	FF	FF	FF	FF	FF	FF	FF	FF	FF	FF	FF
0010	FF	FF	FF	FF	FF	FF	FF	FF	FF	FF	FF	FF	FF	FF	FF	FF
0020	FF	FF	FF	FF	FF	FF	FF	FF	FF	FF	FF	FF	FF	FF	FF	FF

The colored background indicates which locations the result of the compilation was inserted in. The others show a default value[2]. At address 0000h we find the machine code for the jump to address 0100h, as was shown in the section on object code.

Let's move a little further ahead in the memory. See the figure below.

Memory

Addr	+0	+1	+2	+3	+4	+5	+6	+7	+8	+9	+A	+B	+C	+D	+E	+F
00F0	FF	FF	FF	FF	FF	FF	FF	FF	FF	FF	FF	FF	FF	FF	FF	FF
0100	3A	00	05	47	3A	01	05	80	32	00	80	76	FF	FF	FF	FF
0110	FF	FF	FF	FF	FF	FF	FF	FF	FF	FF	FF	FF	FF	FF	FF	FF

At 0100h we find that part of our program's code that was allocated right here, as we know, due to the 'ORG 0100h' directive. We can see that it takes up 12 bytes and we also see the machine codes that were visible in the object code window from 3Ah to 76h.

Let's move even further ahead (see the figure below) and look at locations 0500h and 0501h, which contain constants 34h and 12h, those defined by the two DB directives.

Memory

Addr	+0	+1	+2	+3	+4	+5	+6	+7	+8	+9	+A	+B	+C	+D	+E	+F
04F0	FF	FF	FF	FF	FF	FF	FF	FF	FF	FF	FF	FF	FF	FF	FF	FF
0500	34	12	FF	FF	FF	FF	FF	FF	FF	FF	FF	FF	FF	FF	FF	FF
0510	FF	FF	FF	FF	FF	FF	FF	FF	FF	FF	FF	FF	FF	FF	FF	FF

What would have happened if we had omitted the last ORG directive, as we alluded to previously? This is how the result of the compilation would appear, with the two constants 34h and 12h allocated right after the program code.

Memory

Addr	+0	+1	+2	+3	+4	+5	+6	+7	+8	+9	+A	+B	+C	+D	+E	+F
00F0	FF	FF	FF	FF	FF	FF	FF	FF	FF	FF	FF	FF	FF	FF	FF	FF
0100	3A	0C	01	47	3A	0D	01	80	32	00	80	76	34	12	FF	FF
0110	FF	FF	FF	FF	FF	FF	FF	FF	FF	FF	FF	FF	FF	FF	FF	FF

---

[2] In a real system, the non-programmed locations will show a value dependant on the physical characteristics of the memory. Here the emulator puts them at FFh. We can ignore these values with no consequences.

The syntax of DB and DW makes it possible to define many types and formats of constants. They can be numbers, ASCII characters, or whole tables of numbers and characters. Below are some examples of how they are used.

```
AVALUE      EQU    3Fh                  ;define symbol AVALUE as 3Fh
;
;--- DB examples ------------------------------------------------------------
            ORG    0800h                ;These addresses are in the ROM space,
                                        ;all the values are stored in ROM
CONST1:     DB     AVALUE               ;store 3Fh in the ROM location 0800h
BYTES:      DB     0,1,2,3,4,5,6,7      ;store 8 bytes, as of 0801h:
                                        ;       [00h]
                                        ;       [01h]
                                        ;       [02h]
                                        ;       [03h]
                                        ;       [04h]
                                        ;       [05h]
                                        ;       [06h]
                                        ;       [07h]
;
ASCII:      DB     "ABCDEF"             ;store 6 ASCII codes, in this order
DIGITS:     DB     "0123456789"         ;store other 10 ASCII characters
            DB     "A""B""C"            ;store 5 ASCII codes: [A],["],[B],["],[C]
;
NUMBERS:    DB     8,-8,-58, AVALUE     ;store 4 constants in memory, in order:
                                        ;       [08h]  = +8
                                        ;       [F8h]  = -8  (in two's compl. code)
                                        ;       [C6h]  = -58 (in two's compl. code)
                                        ;       [3Fh]  = AVALUE
;
MIX:        DB     "XY","ZW",0,0FFh     ;(mixed definitions are allowed)
;--- DW examples ------------------------------------------------------------
WORDS:      DW     8,-8,1033,AVALUE     ;store in ROM four 16-bit constants,
                                        ;each one broken in two bytes:
                                        ;       [08h]
                                        ;       [00h]  = 0008h = +8
                                        ;       [F8h]
                                        ;       [FFh]  = FFF8h = -8
                                        ;       [09h]
                                        ;       [04h]  = 0409h = 1033
                                        ;       [3Fh]
                                        ;       [00h]  = 003Fh = AVALUE (16-bit extended)
;
            DW     0,0,0,0,0,0,0,0      ;store 16 bytes of zeroes
;----------------------------------------------------------------------------
```

Now let's analyze the rows of this last example, from the beginning (shown below for easy consultation). The first row defines the symbol AVALUE = 3Fh through the EQU directive.

```
AVALUE      EQU    3Fh                  ;define symbol AVALUE as 3Fh
```

This definition makes it so that the assembler adds this symbol and its corresponding value to the symbol table. Through EQU, the programmer has only defined a symbol for use in compiling the program. Note that EQU does not load this value in the memory; the only directives that can load constants in memory are DB and DW.

A few rows ahead, in fact, we find a DB directive with the symbol AVALUE
as its argument.

```
;--- DB examples -------------------------------------------------------
          ORG    0800h              ;These addresses are in the ROM space,
                                    ;all the values are stored in ROM
CONST1:   DB     AVALUE             ;store 3Fh in the ROM location 0800h
```

With these rows of code, the assembler inserts value 3Fh, which was just
assigned at symbol AVALUE, in the ROM location of address 0800h. The
following figure shows the content of the ROM memory. The byte at location
0800h is highlighted.

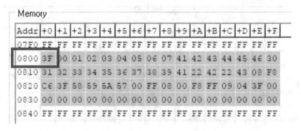

At the BYTES label (see below) DB has multiple arguments separated by
commas. Each argument starts with a byte; they are inserted in the memory
in the order in which they are written. In this case, the numbers are written
in decimal form and will be translated into binary code by the assembler.

```
BYTES:     DB    0,1,2,3,4,5,6,7    ;store 8 bytes, as of 0801h:
                                    ;        [00h]
                                    ;        [01h]
                                    ;        [02h]
                                    ;        [03h]
                                    ;        [04h]
                                    ;        [05h]
                                    ;        [06h]
                                    ;        [07h]
```

The following figure shows the 8 bytes allocated in the ROM.

Memory																	
Addr	+0	+1	+2	+3	+4	+5	+6	+7	+8	+9	+A	+B	+C	+D	+E	+F
07F0	FF	FF	FF	FF	FF	FF	FF	FF	FF	FF	FF	FF	FF	FF	FF	FF
0800	3F	00	01	02	03	04	05	06	07	41	42	43	44	45	46	30
0810	31	32	33	34	35	36	37	38	39	41	22	42	22	43	08	F8
0820	C6	3F	58	59	5A	57	00	FF	08	00	F8	FF	09	04	3F	00
0830	00	00	00	00	00	00	00	00	00	00	00	00	00	00	00	00
0840	FF	FF	FF	FF	FF	FF	FF	FF	FF	FF	FF	FF	FF	FF	FF	FF

The three DBs following the ASCII label define the string of characters. The
syntactic rule is that individual ASCII characters or strings of them must be
enclosed in quotation marks (ex. "A", "BCD").

```
ASCII:    DB    "ABCDEF"        ;store 6 ASCII codes, in this order
DIGITS:   DB    "0123456789"    ;store other 10 ASCII characters
          DB    "A""B""C"       ;store 5 ASCII codes: [A],["],[B],["],[C]
```

At the DIGITS label, the argument of DB is a string of 10 ASCII characters corresponding to symbols of decimal numbers which are translated into the respective binary codes 30h, 31h, ..., 39h, in sequence.

If we include quotation marks in the string so that the assembler considers them as characters and not as delimiters, we need to write them twice (""), as shown in the third row of the example. When this source row is translated, we find the codes 41h, 22h, 42h, 22h and 43h in the memory. The figure below shows the bytes allocated by these rows of code.

Memory

Addr	+0	+1	+2	+3	+4	+5	+6	+7	+8	+9	+A	+B	+C	+D	+E	+F
07F0	FF	FF	FF	FF	FF	FF	FF	FF	FF	FF	FF	FF	FF	FF	FF	FF
0800	3F	00	01	02	03	04	05	06	07	41	42	43	44	45	46	30
0810	31	32	33	34	35	36	37	38	39	41	22	42	22	43	08	F8
0820	C6	3F	58	59	5A	57	00	FF	08	00	F8	FF	09	04	3F	00
0830	00	00	00	00	00	00	00	00	00	00	00	00	00	00	00	00
0840	FF	FF	FF	FF	FF	FF	FF	FF	FF	FF	FF	FF	FF	FF	FF	FF

The first line here below, at the NUMBERS label, defines 4 bytes.

```
NUMBERS:   DB    8,-8,-58, AVALUE   ;store 4 constants in memory, in order:
                                    ;    [08h]  = +8
                                    ;    [F8h]  = -8  (in two's compl. code)
                                    ;    [C6h]  = -58 (in two's compl. code)
                                    ;    [3Fh]  = AVALUE
;
MIX:       DB    "XY","ZW",0,0FFh   ;(mixed definitions are allowed)
```

Negative numbers are translated into two's complement code. Since the assembler codes them on one byte, we can only represent negative numbers from -128 to -1. A '+', or no sign at all forces the assembler to code the number as an unsigned integer with a value from 0 to 255. This means the programmer must pay close attention because writing -1, 255 or FFh is the same thing from the perspective of the result produced in the assembler.

The fourth number is defined by using the symbol AVALUE again, which for the assembler, has the value 3Fh. At the MIX label, we have an example of mixed parameters.

Here below, the bytes allocated by these last directives are highlighted.

Memory

Addr	+0	+1	+2	+3	+4	+5	+6	+7	+8	+9	+A	+B	+C	+D	+E	+F
07F0	FF	FF	FF	FF	FF	FF	FF	FF	FF	FF	FF	FF	FF	FF	FF	FF
0800	3F	00	01	02	03	04	05	06	07	41	42	43	44	45	46	30
0810	31	32	33	34	35	36	37	38	39	41	22	42	22	43	08	F8
0820	C6	3F	58	59	5A	57	00	FF	08	00	F8	FF	09	04	3F	00
0830	00	00	00	00	00	00	00	00	00	00	00	00	00	00	00	00
0840	FF	FF	FF	FF	FF	FF	FF	FF	FF	FF	FF	FF	FF	FF	FF	FF

Below, we have some examples of how the DW directive is used.

```
WORDS:      DW    8,-8,1033,AVALUE  ;store in ROM four 16-bit constants,
                                    ;each one broken in two bytes:
                                    ;      [08h]
                                    ;      [00h]  = 0008h = +8
                                    ;      [F8h]
                                    ;      [FFh]  = FFF8h = -8
                                    ;      [09h]
                                    ;      [04h]  = 0409h = 1033
                                    ;      [3Fh]
                                    ;      [00h]  = 003Fh = AVALUE (16-bit extended)
            ;
            DW    0,0,0,0,0,0,0,0   ;store 16 bytes of zeroes
```

At the WORDS label, DW defines the 16-bit constants divided into pairs of bytes, in the memory. Consider the first argument, which is number 8. Since it is coded on 16 bits, it will appear as 0008h, but it is divided into two bytes: 08h (low part) and, then 00h (high part).

The number -8 is converted by the assembler into the number FFF8h in two's complement code with the lower part, F8h, first.

The number 1033, converted into binary with no sign, reads as 0409h in Hex with the lower part, 09h, inserted first.

For the fourth operand, we recall the symbol AVALUE. Extended to 16 bits and expressed in Hex, it is worth 003Fh, divided into bytes 3Fh and 00h.

In the last row, we define 8 zeroes but, since this is a DW directive, we will get 16 memory locations at zero (8 numbers, two bytes each).

Finally, the figure below shows the result of the assembler having allocated the various arguments of the two DWs, as they appear in the ROM of the emulator. Note the 16 locations at zero as of address 0830h.

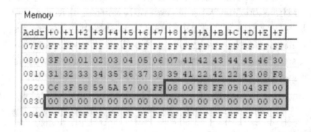

3.2 Addressing modes

In Section 3.1 we looked at some of the instructions the DMC8 microprocessor can execute. An instruction can operate on one or more operands. The data the instructions operate on can:

— be contained in the internal registers of the CPU.
— be contained in the memory locations.
— come from peripheral Input/Output circuits.

As mentioned in Section 1.6.2, the term "addressing mode" indicates the way in which the CPU retrieves the data to work on for a given instruction.

We have already seen other ways to transfer data (LD instructions) that can be distinguished based on the different locations of the source operands and destination operands. For example:

    LD   A,(OPE_1)   ; Copy the content of memory location OPE_1 in A.
    LD   B,A   ; Copy the content of A in B.

Note that even though we are dealing with data transfer operations with the same mnemonic code LD, the source operands are in two different places in the two cases. The first instruction copies the byte from memory to A. The second instruction refers only to the internal registers.

The options of addressing modes in the DMC8 are listed here:

— IMMEDIATE addressing mode
— EXTENDED IMMEDIATE addressing mode
— DIRECT addressing mode
— REGISTER INDIRECT addressing mode
— INDEXED INDIRECT addressing mode
— REGISTER addressing mode
— IMPLIED addressing mode
— BIT addressing mode
— MODIFIED addressing mode

An instruction can have zero, one or two operands. The instruction adopts a specific addressing mode for each operand. Because of the limitation on the number of bits used in machine codes, not all of the combinations of addressing modes are permitted.

For example, one cannot directly transfer a byte from one memory location to another with just one instruction. The DMC8 doesn't have an instruction of the type "LD (3500h),(1F00h)", so it must be replaced with the following pair of instructions:

    LD   A,(1F00h)   ; Copy the content of memory location 1F00h in A.
    LD   (3500h),A   ; Copy the content of A in memory location 3500h.

Generally, when a data transfer cannot be made with one instruction, we need to use at least two, and a register for transit, often accumulator A. Now let's look at the specifics about the main addressing methods used in the DMC8 microprocessor.

3.2.1 IMMEDIATE addressing mode (8-bit data)

As seen in Chapter 1, with immediate addressing, the 8-bit operand is put immediately after the instruction opcode.

Examples:

Mnemonic	Machine code	
LD A, 00h	3Eh	[opcode]
	00h	[immediate data to set in register A]
LD B, 5Fh	06h	[opcode]
	5Fh	[immediate data to set in register B]

Note that the data to load in the destination register is found in the byte following the opcode. After fetching the opcode, the microprocessor reads the source operand from the next memory location using the PC to address it.

3.2.2 EXTENDED IMMEDIATE addressing mode (16-bit data)

With extended immediate addressing, the operand is made up of 16 bits (divided into 2 bytes) and is found immediately after the instruction opcode, as in the following example[3]:

Mnemonic	Machine code	
LD HL, 0C035h	21h	[opcode]
	35h	[immediate data, low part, to set in register L]
	C0h	[immediate data, high part, to set in register H]

The processor reads the two bytes following the opcode and transfers them to parts L and H respectively, of paired register HL. As before, the PC is used to address the two bytes one after the other.

Other examples:

```
LD   SP, 0FF00h   ; Inizialize the Stack Pointer register to FF00h.
LD   BC, 0FAFFh   ; Set register BC = FAFFh.
```

---

[3] Note that we need to place a zero before the hexadecimal number so that the assembler can correctly distinguish the symbols (which have to begin with a letter) from numbers (which have to begin with a numeral).

3.2.3 DIRECT addressing mode

We have direct addressing when we find the address of the memory location (that has the number to read or write) immediately after the opcode. There are two types of direct addressing modes that differ according to the size of the operand (8 bits or 16).

3.2.3.1 DIRECT addressing mode (8-bit data)

Examples:

Mnemonic	Machine code	
LD (0E000h),A	32h	[opcode]
	00h	[memory address, low part]
	E0h	[memory address, high part]
...	...	
IN A,(20h)	DBh	[opcode]
	20h	[port address (8-bit)]

In the first example we find address E000h after the opcode. It is divided into bytes 00h and E0h. After the processor recombines the two bytes together, it uses that number to address the memory and write the A content on it.

In the second example, after the opcode of instruction IN, the processor finds the (8-bit) address of the input port that we want to read. The port of address 20h is copied to register A.

Notice the brackets. They let the assembler know that the number is to be read (or written) in the memory cell pointed by the address in brackets.

Other examples:

```
LD      A,(0E000h)    ; Copy the content of the memory
                      ; location E000h into register A
LD      (845Fh),A     ; Copy the content of register A
                      ; into the memory location 845Fh.
OUT     (0FFh),A      ; Copy the content of register A
                      ; to the output port at address FFh.
```

3.2.3.2 Direct addressing mode (16-bit data)

Example:

Mnemonic	Machine code	
LD HL,(820Fh)	2Ah	[opcode]
	0Fh	[memory address, low part]
	82h	[memory address, high part]

As in the previous example, the processor finds the address of the data to transfer immediately after the opcode, but this time its size is 16 bits, since its destination is a 16-bit register.

The processor uses the specified address to retrieve the first byte from the memory. It will be loaded in register L. Right after, the address is incremented and used to read the next byte in the memory, which will be copied in register H. The overall result is that the 16-bit number is transferred, even though it was done in two separate 8-bit steps.

Other examples:

```
LD    (8F00h),HL    ; Copy the content of register HL into the
                    ; memory locations 8F00h and 8F01h.
LD    BC,(1A00h)    ; Copy the two memory locations 1A00h and 1A01h
                    ; into the C and B parts of register BC, respectively
```

3.2.4 REGISTER INDIRECT addressing mode

With register indirect addressing, the content of a 16-bit register (BC, DE or HL), that was specified in the instruction, is used to address the memory location to read or write. To find the address of the data, the processor executes an intermediate step: consulting an internal register, hence the name.

In the next example, the memory byte whose address is found in register HL is transferred to register C. Naturally, register HL needs to be initialized with the desired address beforehand.

Mnemonic	Machine code
LD C, (HL)	4Eh [opcode]

Notice how compact the machine code is: just one byte. Everything required is contained in the opcode with no need for further specifications. This addressing mode is convenient when we have to go to the same location multiple times or when the location address is "calculated" (incremented or decremented, for example) as in data table management.

Here is an example of how it is used:

```
LD    HL, 80F0h   ; Initialize HL register at 80F0h, this 16-bit number
                  ; will be used as memory address.
LD    (HL), 3Fh   ; Write the value 3Fh into the memory location
                  ; 80F0h, pointed by the register HL.
INC   HL          ; Increment the HL content (it becomes 80F1h).
LD    (HL), 12h   ; Write the value 12h into the next memory location,
                  ; that is pointed by HL = 80F1h.
INC   HL          ; Increment the HL content (it becomes 80F2h).
...               ; ... (and so on)
```

In this example, the memory address is specified only once at the beginning and it is progressively incremented. We are initializing a table of values one by one and specifying only the first address. Therefore, this address can identify the whole table.

In the following example, 50 memory locations are initialized to zero as of the location at address A700h. We specify only the address of the first one and every time the loop repeats, we write a location and address it with register HL, then we increment its content in order to address the next location. The loop repeats 50 times using register B as a counter.

```
          LD    HL,0A700h  ; Initialize HL with the address A700h.
          LD    B,50       ; Initialize B with the value 50
LOOP:     LD    (HL),00h   ; Write the number 00h into the location
                           ; pointed by the content of register HL.
          INC   HL         ; Increment the address contained in HL
          DEC   B          ; Decrement register B (used as a 'counter')
          JP    NZ,LOOP    ; Repeat the loop until B is zeroed
          ...
```

As seen in this last example, indirect addressing makes it possible to point to memory locations by managing the addresses directly from the program at the moment of execution.

In high level languages, "pointers"[4] are based on the options offered by indirect addressing. Therefore, in assembly, the tasks the pointers can do can be reproduced by the programmer using this addressing mode.

3.2.5 INDEXED INDIRECT addressing mode

This addressing mode is similar to the previous register indirect one but it has an added feature. In this case, the address of the memory read/write location is calculated by adding the value of the byte that follows the instruction opcode to the content of one of the index registers (IX or IY). That byte is called "displacement", a constant that can be either positive or negative (between -128 and +127).

Example:

Mnemonic	Machine code	
LD B, (IY+2Fh)	FDh	[opcode, first byte]
	46h	[opcode, second byte]
	2Fh	[displacement $= +47_{10}$]

---

[4] In high level languages, "pointers" are variables that let us "point to" other variables. In other words, a "pointer" variable contains the address of another variable. A pointer can be incremented or decremented and, generally, calculated.

Here, the opcode is made up of two bytes: FDh and 46h, followed by the displacement, which has value 2Fh = $+47_{10}$ in this example. This instruction reads the content of the memory location whose address is obtained by adding constant 2Fh to the content of register IY, and copies it to register B.

Other examples:

```
LD   IX, 9000h      ; Initialize IX = 9000h
LD   A, (IX+33h)    ; Copy the content of the memory location
                    ; pointed by register IX + 33h into register A.
CP   (IX+01h)       ; Compare the content of register A (implied)
                    ; with the content of the memory location
                    ; pointed by the address in register IX + 01h
```

Displacement constants are for addressing individual variables inside a "data structure". In high level languages, a data structure[5] is a kind of container that groups multiple homogeneous or heterogeneous variables together, allowing us to reference the group of variables in a single, cumulative way.

For example, a structure can collect information about sending a package in the mail, like its weight, length, depth, width, shipping date, etc., all in order. Imagine for simplicity's sake that the variables of weight and size take up one memory location (one byte) each. If this structure is allocated to address IX = B000h, the weight variable is found at address B000h, while the length, depth and width are dislocated further by one constant (+1, +2 and +3, the displacement), and so on with all the other variables depending on their size.

Let's remember that IY and IX are 16-bit registers that cannot be divided into upper and lower parts. Also, pay attention to the fact that we can't add a calculated displacement to the address specified by these registers at the moment of execution. This is because we are dealing with a constant displacement, defined when the programmer wrote the program; it can't be changed by the processor during the execution of the program.

3.2.6 REGISTER addressing mode

With this addressing mode, the information that specifies the register(s) to be worked on is already contained in the opcode.

Example:

Mnemonic	Machine code
LD A,D	7Ah [opcode]

---

[5] In high level languages such as C and C++, data structures are called "struct"; in Pascal and Delphi, they are called "record", etc.

The instruction asks to copy the content of register D in A. The operands are both registers and we don't need to specify the addresses.

The register addressing mode is often used next to other modes, for example in data transfer instructions where we can specify one or more source or destination register, as in this instruction: "LD B,(HL)". In this case, we are dealing with "register indirect" addressing mode with respect to the source operand (HL), whereas for destination B we have simple "register" addressing mode.

Other examples:

```
CP    B   ; Compare the content of register A (implied) with the
          ; content of register B (the flags are affected by the result)

AND   L   ; Calculate the bitwise AND between the content of A
          ; (implied) and the content of L (the result is stored in A)
```

3.2.7 IMPLIED addressing mode

In some instructions, the requested operation requires the use of a predefined register. In arithmetic and logical operations, for example, register A is the destination of the results by default (see some of the previous examples).

Examples:

Mnemonic	Machine code
ADD A,L	85h [opcode]
SCF	37h [opcode]
NOP	00h [opcode]

The instruction ADD A,L explicitly specifies the two addends but register A, which memorizes the results, is implied.

The instruction SCF (Set Carry Flag) acts exclusively on the Carry bit in the flag register, forcing it to 1. We will see that it is often used in conjunction with shift and rotate instructions.

We also find implied addressing in the NOP. We have already seen this instruction, which takes no action but simply makes 4 clock cycles go by. It is one of the instructions with implied addressing mode, in that there are no operands to specify.

3.2.8 BIT addressing mode

Bit addressing is used only in the instructions RES, SET and BIT, which we will describe more broadly in Section 3.3. This makes it possible to specify a single bit among the eight, which constitute the byte contained in the register or in the memory location specified by the instruction argument.

Example:

Mnemonic	Machine code
RES 2,B	CBh [opcode, first byte]
	9Eh [opcode, second byte]

In this example, the instruction RES brings the bit in position 2 of register B to zero. The bits are numbered from bit 0 (LSB) to bit 7 (MSB) and the bit selection index is included in the opcode. The bit index can only be a constant between 0 and 7, and not a variable (the number of the bit can't be specified with a register, for example).

Other examples:

```
SET   3,A      ; The bit in position 3 of register A is set to '1'
SET   5,(HL)   ; The bit in position 5 of the memory location pointed
               ; by register HL is set to '1'.
BIT   0,E      ; Check if bit in position 0 of register E is '1' or '0',
               ; store the result into the Zero flag (Z).
```

3.2.9 MODIFIED addressing mode

Modified addressing is only used by RST ("Restart") instructions, which force the processor to jump to one of eight predefined memory addresses (0000h, 0008h..., 0038h) by using only one instruction byte[6]. In the following example, all we need to know is that the processor is forced to jump to location 0038h:

Mnemonic	Machine code
RST 38h	FFh [opcode]

To understand this instruction more completely, first we'll need to go into more detail on some concepts in Section 3.4 where we will learn about "subprogram calls".

---

[6] In other processors, this type of instruction is named INT, or TRAP.

3.3 Types of instructions

Let's place the instructions supported by the DMC8 into the following functional categories.

— Data transfer instructions
— Arithmetic/logic instructions
— Shift/rotate instructions
— Bit manipulation instructions
— Jump instructions
— Instructions to call and return from subprograms
— Input/output instructions
— CPU control instructions

3.3.1 Data transfer instructions

These are also called "Load" instructions. They allow us to copy data from register to register in the CPU or from a CPU register to the memory and vice versa. There are instructions to work on 8-bit data and 16-bit data (in pairs of bytes). They come in the form shown below:

$$LD \quad \text{<destination>, <source>}$$

As we have seen before, this is read as "copy the content of the <source> to the <destination>".

3.3.1.1 Data transfer instructions (8-bit)

Depending on the addressing mode, the source element can be:

An immediate value (8-bit):

LD	A,3Bh	; Load the value 3Bh (0011.1011b) in A
LD	B,72	; Load the decimal value 72 (0100.1000b) in B
LD	L,01001000b	; Load the binary value 0100.1000 in L

The content of an 8-bit register:

LD	A,B	; Load the content of B in A
LD	D,A	; Load the content of A in D
LD	H,E	; Load the content of E in H

The content of a memory location:

LD	A,(0F001h)	; copy the memory byte found at address F001h ; into register A
LD	C,(HL)	; copy the memory byte located at the address ; specified by the register HL into register C

The destination element can be:

A register (8-bit):

| LD | B,E | ; copy the content of register E into register B |
| LD | D,00h | ; copy the costant zero in D |

A memory location, if the source element is a register:

LD   (0F305h),A   ; copy the contents in register A in the memory
                    ; location located at address F305h
LD   (HL),A   ; copy the contents in register A in the memory
                    ; location pointed by register HL

As mentioned in Section 3.2, this microprocessor does not support all the combinations of addressing modes and we need to check the instruction tables from time to time to see if a certain combination is supported. See Appendix C. The list of 8-bit data transfer instructions is available on page 600.

3.3.1.2 Data transfer instructions (16-bit)

The processor actually executes 16-bit data transfer instructions as if they were sets of two consecutive 8-bit data transfer instructions. The syntax is the same as the 8-bit version but the operands are 16-bit data:

$$LD \quad <\text{destination}>, <\text{source}>$$

The registers that can be classified as operands are the "paired" registers BC, DE and HL or the (proper 16-bit) registers SP, IX and IY. Depending on the addressing mode, the source element can be:

An immediate value (16-bit):

LD   BC,5F3Dh   ; copy the value 5F3Dh (0101.1111.0011.1101b)
                    ; into the 16-bit register BC

The content of a 16-bit register:

LD   SP,HL   ; copy the content of HL into the SP register

The contents of two consecutive memory locations (16 bits in total):

LD   HL,(0F140h)   ; copy in HL the two bytes found in the memory
                    ; locations at addresses F140h (low part → L)
                    ; and F140h+1 (high part → H)

The pair of the locations is indicated only by the address of the first one. Following little-endian convention, the 16-bit operands allocated in the memory are split into two bytes.

The least significant one is placed at the address of the pair, while the most significant one is placed at the next address. The destination element can be:

A 16-bit register:

LD   IX,8001h   ; copy the value 8001h into register IX

Two consecutive memory locations (16 bits in total):

LD   (0F140h),HL   ; copy the two bytes found in the lower and
                      ; upper parts of the 16-bit register HL,
                      ; respectively in the memory locations
                      ; located at addresses F140h and F140h+1

Sixteen-bit data transfer operations are another case where not all addressing modes are supported. For an overall framework of this, consult the summary table on page 601, in Appendix C.

This group includes transfer instructions for byte pairs related to the "Stack" memory area, which we have yet to study. For the sake of completeness, two examples are available below:

PUSH HL   ; saves the contents of HL on the top of the stack
POP   IX   ; retrieves the contents of IX from the stack

We will discuss these instructions in detail further on in Section 3.4 on subprograms and the stack area. This class of instructions is very important; we will use them frequently.

3.3.2 Arithmetic and logic instructions

Arithmetic and logic instructions allow us to work with both 8- and 16-bit numbers. This group of instructions, which relies on the ALU (Arithmetic-Logic Unit) is especially interesting to use because it is supported by various addressing modes, making it possible to write sufficiently versatile programs. The list of all the arithmetic logic operations supported here is available on pages 603..606, in Appendix C.

As we will see, the operations supported by the DMC8's ALU are basic; for example, we have no instructions to directly execute multiplications or division. In any case, any non-implemented function can be carried out through a sequence of basic instructions.

Clearly, the speed of execution would be much higher if the instructions were supported by bespoke hardware inside the processor (think of the "math co-processors" in the most powerful processors).

3.3.2.1 Arithmetic instructions (8-bit data)

The DMC8's eight-bit arithmetic instructions usually use accumulator A as the first operand and destination for the result. If there is a second operand, it can be:

— An 8-bit immediate value.
— The content of an 8-bit register.
— The content of a memory location, specified by the indirect addressing mode using register HL or index registers IX or IY.

The ADD instruction

Let's look at the ADD instruction, which adds two operands. The following examples show all the supported addressing modes for the source operand:

```
ADD   A,01h       ; sum A with an immediate value (01h)
ADD   A,B         ; with a register (here: B)
ADD   A,(HL)      ; with the memory location pointed by HL
ADD   A,(IX+16)   ; with the memory location pointed by IX +16
ADD   A,(IY+32)   ; with the memory location pointed by IY +32
```

This instruction adds the source in question to accumulator A and saves the result in A. The previous content of A is obviously lost. After the operation, the Carry flag is set to one if there has been a carry in the operation. If not, it is set to zero.

The ADC instruction

ADC is an important variation on the ADD instruction. It takes into account the Carry flag value during the sum. When it makes the sum, it also adds the Carry flag value, which comes from a previous operation. In other words, it adds the specified source to accumulator A plus 1 if the Carry flag is already at one, and puts the result in A.

As with ADD, the previous content of A is lost and the Carry flag is set to 1 in the output if there is a carry in this operation. The previous value of the Carry, which was used to make the calculation, is overwritten.

The use of ADD and ADC with indirect addressing, an example

Let's imagine we have extracted a sample of code from a broader program. This code orders a 16-bit operation where the add is broken down into two 8-bit operations. We need to define two variables OPE_A and OPE_B. They contain the 16-bit operands to add. The variable RESULT will hold the result of the add. See the figure below.

	high	low	
OPE_A	2F	03	+
OPE_B	15	A0	=
RESULT	44	A3	

OPE_A → 8000	03	low
8001	2F	high
OPE_B → 8002	A0	low
8003	15	high
RESULT → 8004	A3	low
8005	44	high

The left hand figure shows the operation to carry out as we might write it with pen and paper. First, we must add the low bytes of the operands together. This gives us the low part of the result. Then we must add the high bytes along with any carry from adding the low bytes, to calculate the upper part of the result.

The two bytes that constitute the variables are placed in memory according to little-endian convention. So if we allocate them in consecutive locations as of address 8000h, the high and low bytes will appear as shown in the right hand figure. In the code, they will be defined thusly:

```
OPE_A     EQU    8000h       ; first operand (16-bit)
OPE_B     EQU    8002h       ; second operand (16-bit)
RESULT    EQU    8004h       ; result (16-bit)
```

Notice that attention has been paid to the space each variable occupies. Their addresses are all at a distance of two locations from each other.

We will use register indirect addressing to access the variables. First we point to the low part found at the addresses defined above, and then we point to the high part by incrementing the "pointers" (the registers containing the variable address).

So firstly, we copy the addresses of the three variables OPE_A, OPE_B and RESULT in registers BC, HL and DE respectively.

```
ADD16:    LD     BC,OPE_A    ; BC addresses the low part of OPE_A
          LD     HL,OPE_B    ; HL addresses the low part of OPE_B
          LD     DE,RESULT   ; DE addresses the low part of RESULT
```

Now we use these registers as pointers to the variables. Let's move the low part of OPE_A in the accumulator, then we add the low part of OPE_B. The result is then saved in the low part of the variable RESULT.

```
          LD     A,(BC)      ; get the low part of variable OPE_A
          ADD    A,(HL)      ; add it to the low part of variable OPE_B
          LD     (DE),A      ; save the partial result (low part)
```

Note that any carry generated by the add is now in the Carry flag. We will use it soon to execute the sum of the high parts. To do this, we must increment all our pointers by one, that is registers BC, HL and DE, which will address the high part.

```
        INC   BC              ; update the pointers to the high part
        INC   HL
        INC   DE
```

Finally, we repeat an adding sequence similar to that of the previous one except that the registers now target the high part of the variables and the ADC substitutes the ADD so that it also sums any carry from the low part.

```
        LD    A,(BC)          ; get the high part of variable OPE_A
        ADC   A,(HL)          ; add it to the high part of variable OPE_B
        LD    (DE),A          ; save the high part of the final result
```

Remember that this example was meant as an introduction. In fact, the processor has 16-bit arithmetic instructions that would make it possible to solve the task more easily, as we will see next.

The SUB instruction

Let's look at the SUB instruction, which subtracts two operands. The first is always the accumulator and the second is defined by the programmer and can be of the same types as those meant for the ADD instruction.

SUB subtracts the specified source from register A and overwrites it with the result (the previous content of A is overwritten). If the subtract has generated a borrow, it is put in the Carry flag[7].

The following examples are similar to those shown for ADD and they list the acceptable addressing modes.

```
SUB   39h          ; subtract (from register A) an immediate value (39h)
SUB   E            ; the content of a register (here: E)
SUB   (HL)         ; the memory location pointed by HL
SUB   (IX+22)      ; the memory location pointed by IX +22
SUB   (IY+44)      ; the memory location pointed by IY +44
```

We skip the identification of the first operand because this is implied by the mnemonic code SUB. The first operand of a SUB instruction is always implicitly accumulator A. As we shall soon see, the first operand can also be register HL for the ADD and ADC instructions.

The SBC instruction

In this case as well, we have a variant that keeps track of the borrow "in input", the SBC instruction. This instruction subtracts the specified source from accumulator A (minus 1 if the Carry flag indicates that there was a borrow) and saves the result in A.

---

[7] The processor does not have a Borrow flag so the normal Carry flag will be used in its place.

As with SUB, the previous content of A is overwritten with the result. The output borrow is handled as in the previous SUB and the new value of the Carry flag overwrites the previous one.

The CP instruction

CP ('Compare') is an important instruction since allows us to perform arithmetic comparisons. It compares accumulator A to the source indicated by the operand by actually doing an algebraic subtraction without memorizing the result. The operands, including the one contained in accumulator A remain unchanged.

The table below shows the supported addressing modes. They are the same as those of the SUB with the sole difference being that the result is not saved.

```
CP    75h          ; compare A to the constant 75h
CP    B            ; to the content of a register (here: B)
CP    (HL)         ; to the memory location pointed by HL
CP    (IX+56)      ; to the memory location pointed by IX + 56
CP    (IY+12)      ; to the memory location pointed by IY + 12
```

The useful information, (the result of the compare instruction) is memorized in the Carry, Zero and Sign flags as shown in the table below ('s' is the value of the operand).

Result	Carry	Zero	Sign
A > s	0	0	0
A = s	0	1	0
A < s	1	0	1

Sign is important in compares between signed numbers represented in two's complement code. For comparisons between 8-bit unsigned numbers only Carry and Zero flags matter.

We saw conditional jumps in Chapter 1 (Section 1.4). They are used in the DMC8 in practically the same way, apart from a few small technical differences. The bit of program shown below compares the content of a variable with constant values and decides where to jump according to the result of the compare. We will examine jumps in detail further ahead in Section 3.3.5.

Examples of how the CP instruction is used

In the sequence of instructions here, the CP instruction is used along with conditional jumps.

```
CHECK:    LD    A,(VAR)      ; get the value to be tested
          CP    3Ah          ; compare register A to the constant 3Ah
          JP    Z,EQUAL1     ; jump if they are equal
```

In the first row, we take a value from the memory (from variable VAR). In the second, we compare the value, which is now in the accumulator, to a constant. In the third row, we find a conditional jump at label EQUAL1 (the part of the code labelled this way is not shown).

The jump is executed only if A = 3Ah, that is if the result of the subtract between the accumulator and the constant is zero. If it is, the processor activates the Zero flag. The first operand of instruction Z asks to jump if the flag is active. If the result is other than zero, the processor does not jump but goes on to execute the instruction after the jump. In our case that is another CP, which compares register A with another constant. See below.

```
CP      20h              ; compare register A to the constant 20h
JP      Z,EQUAL2         ; jump if A = 20h
                         ; continue forward if not equal
```

Remember that the result of the subtraction executed by the CP is not saved in A. So, the original value to compare is still there, unchanged, as it was taken from variable VAR. The conditional jump to label EQUAL2 is executed only if A = 20h, that is if A - 20h gives a result of zero.

The parts of the code labelled EQUAL2 and EQUAL1 are not shown in the code (it is assumed that they carry out tasks that correspond to the value identified, but they are irrelevant for the purposes of our discussion).

Now let's assume that the value in A proves to be different from 20h, so the processor goes ahead and executes the instruction immediately under the jump. After we exclude that the the result is equal, we evaluate if the value is greater or lesser than 20h. Based on the table above, we can use the Carry flag, which is set to zero if the value in A is greater than the constant, so we write:

```
          JP    NC,MAJOR   ; jump to MAJOR if the value is > 20h

MINOR:    ...              ; go ahead to MINOR if it is < 20h, executing
          ...              ; the corresponding instruction sequence,
          JP    CONTINUE   ; and then skip the code labeled as MAJOR.

MAJOR:    ...              ; code to execute only if value > 20h,
          ...              ; then, go on with...

CONTINUE:                  ; the rest of the program
```

The label MINOR could be omitted because it is not mentioned anywhere. However, it makes the code more readable because it identifies the part that manages situations where the value is < 20h. The jump to CONTINUE was inserted to prevent executing the part of the code related to < 20h, after the code sequence related to greater than 20h.

The CPL and NEG instructions

CPL (Complement) and NEG (Negative) are among the 8-bit arithmetic instructions. Neither has operands since they work implicitly and only on the accumulator.

CPL inverts all the bits of A, giving us the "one's complement" of its content. This instruction can also be considered from a logical perspective, as a NOT operator in that it "negates" all the bits in the accumulator. Note that the CPL instruction does not change the Zero or Carry flag, nor any other flag.

After it inverts all the bits in the accumulator, NEG adds a '1' to the result, giving us the "two's complement" of the number in the register. NEG regularly changes all the flags in the processor.

A simple example showing the functionality of the two instructions is shown here below. First we calculate two's complement of a number through a pair of instructions (it calculates one's complement for the accumulator, and adds one), then it takes the partial result and calculates it again but only using the NEG instruction. By definition, we get the initial number.

```
LD    A,(08000h)    ; get a byte from the RAM
CPL                 ; calculate one's complement of register A
ADD   A,1           ; add '1' to the number in A
NEG                 ; calculate two's complement of register A
                    ; obtaining the initial number again
```

3.3.2.2 16-bit arithmetic instructions

16-bit arithmetic instructions of the DMC8 operate with the HL paired register, the IX register or the IY register as an accumulator. The second operand can only be a 16-bit internal register and not all the combinations are possible. The DMC8 can add (with or without a carry in input) and subtract (with a borrow in input). Here are some examples:

```
ADD   HL,BC         ; 16-bit addition, HL = HL + BC
ADD   HL,HL         ; double the content of HL (this is a little trick)
ADD   IX,DE         ; 16-bit addition, IX = IX + DE
```

These are not real 16-bit operations because the ALU is 8 bits. The operation is broken down internally into two consecutive phases, first on the "low" byte and then on the "high" byte. From the outside, however, this can't be seen (other than from the time it takes to execute the operation).

Let's have another look at the example of an add of 16-bit numbers. For convenience's sake, let's restate the definition of the three variables OPE_A, OPE_B and RESULT, each of which is two bytes. As in the previous case, RESULT accepts the sum of the contents of OPE_A and OPE_B.

```
OPE_A      EQU    8000h        ; first operand (16-bit)
OPE_B      EQU    8002h        ; second operand (16-bit)
RESULT     EQU    8004h        ; result (16-bit)
```

We cannot use indirect addressing mode with 16-bit data transfer instructions, so we use direct addressing mode to acquire data in registers BC and HL.

```
ADD16:     LD     BC,(OPE_A)   ; copy OPE_A into register BC
           LD     HL,(OPE_B)   ; copy OPE_B into register HL
```

One single instruction makes it possible to calculate the 16-bit sum of the contents of BC and HL, and overwrite the result in HL. In the end, a 16-bit data transfer operation copies the resulting sum in RESULT.

```
           ADD    HL,BC        ; add the two 16-bit operands
           LD     (RESULT),HL  ; save the result in memory
```

As we can see, 16-bit arithmetic instructions easily carry out their task. They do not, however, support indirect addressing mode, making them less versatile to use in a program that executes calculations with a higher number of bits (32 or 64, for example). This is why is is preferable in this case to work with 8-bit instructions, which are better supported and more complete in terms of available addressing modes.

Example of a 64-bit adding algorithm

This example is an addition of two integer numbers coded in 64 bits that are represented by 8 bytes each. The left-hand figure below shows the operation as we might write it with pen and paper.

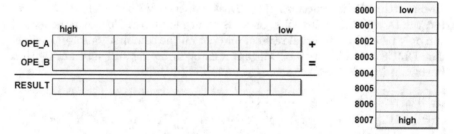

First we should add the least significant bytes together and get the least significant byte of the result. Then we add the bytes of the second column on the right together, along with the carry from the first column. We reserve the carry produced here for column 3. We continue moving to the next column and repeating this operation until we get to the last column on the left, which has the most significant bytes.

Let's take the example of variable OPE_A, which we can place at address 8000h. Let's assume that the eight bytes it is made of follow little-endian convention. The bytes will be ordered as shown in the right hand figure, with the least significant byte at address 8000h and the others following and occupying up through location 8007h.

Let's define the space reserved for the variables in the code, keeping in mind that their size in the memory is 8 bytes. Let's also define constant NBYTE, which we use to count the bytes the operation is broken down into[8].

```
OPE_A       EQU    8000h        ; first operand (8-byte)
OPE_B       EQU    8008h        ; second operand (8-byte)
RESULT      EQU    8010h        ; result (8-byte)
NBYTE       EQU    8            ; number of partial operations
```

We use indirect addressing mode to address, one by one, the bytes the numbers are broken up into in the memory. Let's dedicate registers IX and IY to address the bytes of the operands and HL for the bytes of the result.

```
ADD64:      LD     IX,OPE_A     ; initialize addresses into IX, IY and HL
            LD     IY,OPE_B
            LD     HL,RESULT
```

Let's add the least significant bytes with an ADD. The carry resulting from the operation will be saved in the Carry flag.

```
            LD     A,(IX)       ; execute the first partial sum
            ADD    A,(IY)       ; of the least significant bytes
            LD     (HL),A       ; save the partial result
```

The next seven bytes that the sum is broken down into are handled by a loop that repeats seven times by counting on register B. Before starting the loop, register B is initialized with constant NBYTE and immediately decremented by one since the least significant byte has already been handled.

```
            LD     B,NBYTE      ; initialize the loop counter B
            DEC    B            ; B = (NBYTE - 1)
```

The loop begins with label ADDBYTE. The first three instructions in the loop increment registers IX, IY and HL so that they address the next byte (of the operands and result) to handle.

```
ADDBYTE:    INC    IX           ; update the addresses
            INC    IY
            INC    HL
```

Let's add the bytes of the operands and the carry resulting from the previous sum, thanks to the ADC instruction.

```
            LD     A,(IX)       ; execute the partial sum, taking into
            ADC    A,(IY)       ; account the carry resulting from the
            LD     (HL),A       ; previous sum, and save the partial result
```

---

[8] This code can easily be modified to support numbers of different dimensions. We simply need to change constant NBYTE and the way the variables are assigned.

Finally, let's go back to the label ADDBYTE and repeat the loop until the count, done by decrementing register B, goes to zero.

```
DEC    B                     ; decrement the loop counter and repeat
JP     NZ,ADDBYTE   ; the loop until finished the partial sums
...
```

At the end, the program leaves the loop and goes on to execute other instructions that follow it (but this is beyond the scope of this example and these instructions are not shown).

3.3.2.3 Logic instructions

The logic instructions for the DMC8 are as follows: AND, OR and XOR[9]. They only support 8-bit operations; one of the operands is implicit and is always accumulator A. The other operand is supported by the same addressing modes as 8-bit arithmetic instructions. Below is an example for each addressing mode that we can use with the AND instruction:

```
AND    01h          ; A in AND with the immediate value 0000.0001₂ (01h)
AND    B            ; with register B
AND    (HL)         ; with the memory location pointed by HL
AND    (IX+16)      ; with the memory location pointed by IX +16
AND    (IY+32)      ; with the memory location pointed by IY +32
```

The result of the logical operation always overwrites the operand in accumulator A and is executed "bitwise". Therefore, it executes 8 identical logical operations in parallel, one for each bit, and each bit independently of the others. For example, instruction AND B executes operations thusly:

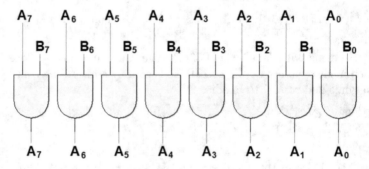

Analytically:

$A_7 \leftarrow A_7$ and B_7, $A_6 \leftarrow A_6$ and B_6, $A_5 \leftarrow A_5$ and B_5, $A_4 \leftarrow A_4$ and B_4,
$A_3 \leftarrow A_3$ and B_3, $A_2 \leftarrow A_2$ and B_2, $A_1 \leftarrow A_1$ and B_1, $A_0 \leftarrow A_0$ and B_0

---

[9] We have already dealt with instruction CPL, which can be used like NOT (remember though that CPL does not change flags).

Examples of the use of the AND instruction

Logic operations change the Zero flag, which is activated if all the bits of the result are zero. This is why these instructions are very useful to test groups of bits by "bitmasking".

Below, we see the first example using the AND instruction:

```
LD      A,(0F000h)       ; get a byte from RAM and load it into A
AND     00000111b        ; A = (A and 0000.0111₂). Bits 7, 6, 5, 4 and 3
                         ; are zeroed thanks to the "bitmask" 0000.0111₂,
                         ; while bit 2, 1 and 0 does not change.
JP      Z,ZERO           ; test if bits 2, 1 and 0 are all zero
                         ; and then jump to ZERO if they are
```

In this code, the number loaded in register A is masked in the sense that some bits are forced to zero (because they are put in AND with the corresponding 0 of the mask), while the others are left unchanged (since they are in AND with a 1). The following table shows a summary of the operation.

Initial data	a_7	a_6	a_5	a_4	a_3	a_2	a_1	a_0	(the bits contained in A)
Bit mask	0	0	0	0	0	1	1	1	(in AND with A - bitwise)
Result	0	0	0	0	0	a_2	a_1	a_0	(all zero if '$a_2a_1a_0$' = '000')

Since the masking obtained by the AND has set bits 7, 6, 5, 4, and 3 to zero, a "zero" result depends on bits 2, 1 and 0 being zero. The conditional jump instruction then checks if the result of the operation is zero.

In the second example (below) we test the same group of bits (in position 2, 1 and 0), but this time it is to check if they are all at one.

```
LD      A,(0F000h)       ; get a byte from RAM and load it into A
CPL                      ; invert all data bits contained in A
AND     00000111b        ; A = (A and 0000.0111₂). Bits 7, 6, 5, 4, and 3
                         ; are zeroed thanks to the "bitmask" 0000.0111₂,
                         ; while bit 2, 1, and 0 do not change.
JP      Z,ALL1           ; test if bits 2, 1, and 0 are all zero, this
                         ; happens if originally they were all equal to one.
                         ; jump to ALL1 if it is so
```

As we can read in the comments, the goal is reached by masking in exactly the same way as the previous case, but after negating all the bits of the original number as specified in the following table.

Initial data	a_7	a_6	a_5	a_4	a_3	a_2	a_1	a_0	(the bits contained in A)
Inverted data	$\overline{a_7}$	$\overline{a_6}$	$\overline{a_5}$	$\overline{a_4}$	$\overline{a_3}$	$\overline{a_2}$	$\overline{a_1}$	$\overline{a_0}$	(inverted by the CPL instr.)
Bit mask	0	0	0	0	0	1	1	1	(in AND with A - bitwise)
Risultato	0	0	0	0	0	$\overline{a_2}$	$\overline{a_1}$	$\overline{a_0}$	(all zero if '$a_2a_1a_0$' = '111')

This allows us to find zeroes in the accumulator corresponding to the bits that do not interest us, whereas those we are testing have their original value, but negated. Overall, the result in A will be zero only if the bits we are interested in were all originally at one.

Example of the use of the XOR instruction

The XOR instruction is very versatile and can be used to invert the value of some of the bits in the accumulator, as in the following example. After loading a constant in the register, we enter an infinite loop where we make the content of accumulator A exit on a parallel output port at each repetition, but each time we change the value.

```
        LD    A,11110000b   ; load a value in A
LOOP:   OUT   (00h),A       ; output the value on the port at address 00h
        XOR   00001111b     ; A = (A exor 0000.1111₂). Thanks to the bit
                            ; mask, bits 7, 6, 5, and 4 remain unchanged,
                            ; while bits 3, 2, 1, and 0 are inverted
        JP    LOOP          ; repeat the loop indefinitely
```

As suggested by the comments inserted in the code, the XOR instruction inverts the bits corresponding to the '1s' of the mask each time the loop repeats, as specified in the following table.

Register A	a_7	a_6	a_5	a_4	a_3	a_2	a_1	a_0	(the bits contained in A)
Bit mask	0	0	0	0	1	1	1	1	(in XOR with A - bitwise)
Result	a_7	a_6	a_5	a_4	$\overline{a_3}$	$\overline{a_2}$	$\overline{a_1}$	$\overline{a_0}$	

The properties of the EXOR operation make it so that the accumulator goes back to the initial value at the next repetition of the loop as shown below.

Register A	a_7	a_6	a_5	a_4	$\overline{a_3}$	$\overline{a_2}$	$\overline{a_1}$	$\overline{a_0}$	(the bits inverted before)
Bit mask	0	0	0	0	1	1	1	1	(in XOR with A - bitwise)
Result	a_7	a_6	a_5	a_4	a_3	a_2	a_1	a_0	(the initial values again)

Thus, the values 1111.0000_2 and 1111.1111_2 alternate on the output port.

The examples of the use of logic instructions show some of the particularities of assembly programming and they merit a brief remark. Clearly, in assembly, we sometimes use rather involved techniques, which require a bit of attention to apply.

Yet, they are often necessary to get the best advantage of what the hardware we are using has to offer. If we are programming in assembly, our goal is to get the most compact and efficient programs possible. This means that our programs may become difficult to read or understand. To overcome this, we try to make the best use of the comments we can insert in our code.

3.3.2.4 Increment and decrement instructions (8-bit)

The DMC8's 8-bit increment and decrement instructions work on a register or a memory location reached by the indirect addressing mode (through HL, IX or IY). Here are some examples:

```
INC    A            ; increment register A
INC    B            ; increment register B
INC    (HL)         ; increment the memory location pointed by HL

DEC    H            ; decrement register H
DEC    (IX)         ; decrement the memory location pointed by IX
DEC    (IY+3)       ; decrement the memory location pointed by IY+3
```

Increment and decrement instructions always raise or lower by one unit. They affect the Zero and Sign flags but not the Carry flag. Increment and decrement operations work as cyclic counters, as shown in the following example.

```
LD     C,11111110b  ; set register C at the maximum value less one (FEh)
INC    C            ; increment register C, now it contains $11111111_2$ (FFh)
INC    C            ; increment register C again, now it contains $00000000_2$
DEC    C            ; decrement register C, that returns to $11111111_2$ (FFh)
```

These instructions make it possible, for example, to use the processor's registers as loop counters, as shown in the previous examples.

3.3.2.5 Increment and decrement instructions (16-bit)

16-bit increment and decrement instructions can only work on registers BC, DE, HL, IX and IY, and not on memory locations. Here as well, the increment or decrement is only by one unit and the operation is cyclical.

```
INC    HL           ; increment the paired register HL
DEC    IX           ; decrement index register IX
```

16-bit INC and DEC instructions do not affect the flags and we should keep this in mind when we write programs[10], as we will see in the following.

In the example below, we initialize 32 memory locations with a value of 00h as of location 8000h. Below are the definitions of the address of the beginning of the RAM area and of the number of locations involved.

```
MEM       EQU  8000h     ; memory area (address of the first location)
NLOC      EQU  32        ; number of locations
```

---

[10] This limitation has been kept for the sake of compatibility with the processor the DMC8 derives from (and the Z80 itself inherited it from the older I8080).

We use HL indirect addressing mode to write memory locations. So let's copy the address of the beginning of the area (MEM) in HL and the number of locations to count (NLOC) in B.

```
INIT_RAM:  LD    HL,MEM     ; set HL to the memory area address
           LD    B,NLOC     ; set B to the number of locations to initialize
```

The writing loop in the memory repeats 32 times. We write the constant value 00h into the memory cell pointed by HL. We increment HL immediately after so that it points to the next location every time the loop repeats.

```
LOOP:      LD    (HL),00h   ; write the value in memory
           INC   HL         ; increment the address to use
```

Every time it repeats, we evaluate the number of cells left to write by decrementing register B. The processor will leave the loop only when this number reaches zero and the Zero flag is activated.

```
           DEC   B          ; decrement the count of cells left
           JP    NZ,LOOP    ; repeat loop until the count will reach zero
```

In the example shown here below, we execute the same task except that there are many more locations to initialize (2048), so one 8-bit register won't be enough to count them.

Rather than register B alone, we will pair it with register C (as BC, 16-bit register). Here are the definitions, similar to the previous example.

```
MEM        EQU   8000h      ; address of the memory area
NLOC       EQU   2048       ; number of locations (greater than 255!)
```

Let's limit ourselves to discussing the differences from the previous case. The number of locations to count (NLOC) is loaded in BC:

```
INIT_RAM:  LD    HL,MEM     ; set HL to the memory area address
           LD    BC,NLOC    ; set BC to the number of locations
```

This loop is almost identical to the previous case except for the fact that the number of locations is counted on the 16-bit BC paired register. Since the 16-bit decrement doesn't change the flags we have had to do a little trick to force the Zero flag to change so that the conditional jump can behave as we expect. The trick can be seen in the two added instructions after the decrement of BC.

```
LOOP:      LD    (HL),00h   ; write the constant 00h in memory
           INC   HL         ; increment the address to use
           DEC   BC         ; decrement the count of the locations
           LD    A,B        ; check if BC has been zeroed, executing a
           OR    C          ; bitwise OR between its low and its high part,
           JP    NZ,LOOP    ; and repeat the loop until BC goes to zero
```

The solution is relatively simple even though it makes the code more complicated to read. It consists in calculating the OR between the low part C and the high part B of the paired register BC. Due to the characteristics of the bitwise OR function the result in the accumulator is only zero overall if neither of the two halves of the register contains a 1, thus if BC is completely at zero. The Zero flag is therefore activated by the OR only in this case.

3.3.3 Rotate and shift instructions

Rotate and shift instructions work on 8-bit numbers. The operand can be any internal 8-bit register or a memory location that is reachable by the indirect addressing mode (through HL, IX or IY). The list of rotate and shift instructions can be found in Appendix C, from page 607 to 609.

RLC instructions

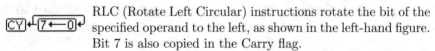 RLC (Rotate Left Circular) instructions rotate the bit of the specified operand to the left, as shown in the left-hand figure. Bit 7 is also copied in the Carry flag.

This is obviously not an arithmetic carry, but this property is interesting because it makes it possible to test the flag (with a conditional jump instruction, we see the examples in the following) and therefore to know the value of the bit that moves out of bit 7 at each rotate. When eight rotates have been executed, the bits of the operand return to their original position. Below are some examples of supported addressing modes for instruction RLC.

RLC   A   ; rotate to the left (8 bits) register A
RLC   B   ; register B
RLC   (HL)   ; the memory location pointed by HL
RLC   (IX+8)   ; the memory location pointed by IX+8
RLC   (IY+15)   ; the memory location pointed by IY+15

These addressing mode examples are also valid for all the other rotate and shift instructions examined here so they will not be repeated further on.

The RRC instruction

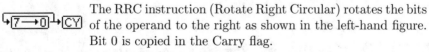 The RRC instruction (Rotate Right Circular) rotates the bits of the operand to the right as shown in the left-hand figure. Bit 0 is copied in the Carry flag.

This instruction also makes it possible to test the value of the bits by rotating them to the right rather than to the left.

The RL instruction

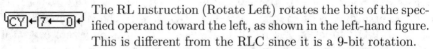 The RL instruction (Rotate Left) rotates the bits of the specified operand toward the left, as shown in the left-hand figure. This is different from the RLC since it is a 9-bit rotation.

During the rotation, the Carry flag is copied in bit 0 and, after the rotation, it is overwritten with the value of bit 7. The Carry flag is used as if it were the ninth bit of the operand. The complete rotation is made up of nine consecutive instructions of this sort. As we will see in the examples, this instruction is useful to rotate multiple byte numbers.

The RR instruction

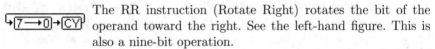 The RR instruction (Rotate Right) rotates the bit of the operand toward the right. See the left-hand figure. This is also a nine-bit operation.

The Carry flag preceding the rotation is copied in bit 7 and the flag is overwritten by bit 0 afterward. This instruction is also useful for rotating large numbers.

The SLA instruction

The SLA instruction (Shift Left Arithmetic) shifts the bit of the operand to the left and inserts a '0' at the right. Bit 7 is copied in the Carry flag and is not carried to the other side.

This left shift is defined as "arithmetic" because the operation is equal to multiplying the initial number by 2.

In fact, after the left shift, all the '1s' of the number are in a position where their "weight" is twice what it was initially. Any overflow from the calculation is saved in the Carry flag. Let's look at the SLA B instruction for example: if we initially have the number 0001.0100_2 ($= 20_{10}$) in register B, all the bits will be moved to the left after execution, giving us 0010.1000_2 ($= 40_{10}$).

The SRL instruction

The SRL instruction (Shift Right Logic) shifts the bits of the operand to the right and inserts a '0' at the left. Bit 0 is copied in the Carry and, similar to the previous case, is not carried to the other side.

The SRA instruction

The SRA instruction (Shift Right Arithmetic) shifts the bit of the operand to the right but keeps bit 7 unchanged, understood as the sign bit. Bit 0 is copied in the Carry flag.

This right shift is defined as "arithmetic" because the operation is equal to dividing the initial number by 2 and it supports negative numbers represented in two's complement.

This is a division because after the instruction is executed, all the digits of the number are in a position where their weight is cut by half of the initial number, as they have been right shifted. After execution the Carry flag contains the rest of the division.

If the initial number is negative, bit 7 is '1'. After the shift, it won't change so we can say we have "kept the sign". If we take number $1111.0000_2 = -16_{10}$ into consideration, after the right shift it is 1111.1000_2, which is -8_{10}.

Actually, the SRL instruction also can divide by two, as it is a logic instruction, but it doesn't support negative numbers.

The use of the RLC instruction: An example

In the example below, let's imagine that an output port is connected to eight LEDs, one for each line, that light up if they are driven by '1'. After initializing A to 0000.0001_2, let's make its content go out onto the port. We'll see only one light on: the one connected to bit 0.

```
        LD     A,00000001b    ; load the bit pattern 00000001₂ in A
LOOP:   OUT    (00h),A        ; output A onto the port at address 00h
        RLC    A              ; rotate left A of one position
        JP     LOOP           ; repeat the loop indefinitely
```

After executing instruction RLC, the configuration of the bits in A will be 0000.0010_2 and if we execute instruction OUT again, the LED lit will be the one in position 1. After 8 rotations, the LED lit will be the one in position 0 again. The rotation on the port will repeat infinitely.

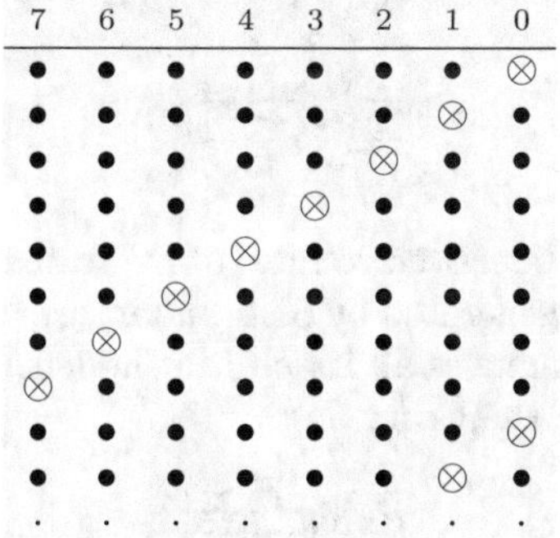

The use of instructions SLA and RL: An example

This example has two output ports connected to a total of 16 LED lights. Here, we cyclically left rotate a lit LED among the 16 positions in sequence, starting from the configuration shown below.

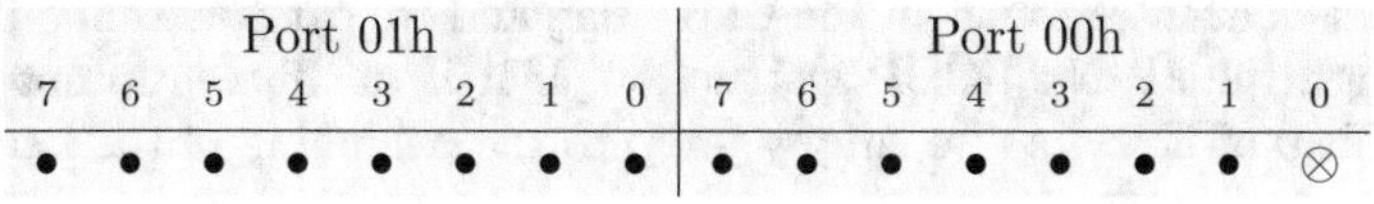

Since we have to rotate a '1' to 16 different positions, we've chosen to use the DE register to memorize the state of the LEDs. Let's start by initializing the register with a single '1' in position 0, before entering into the loop:

```
LD    DE,01h        ; set DE to the initial 16-bit pattern
```

The first four instructions in the loop copy DE content to the ports, specifically the high part D on port 01h and the low part E on port 00h.

```
LOOP:  LD    A,D        ; copy register D onto output port 01h
       OUT   (01h),A
       LD    A,E        ; copy register E onto output port 00h
       OUT   (00h),A
```

The table below shows the last part of the loop.

```
       SLA   E          ; shift left register E
       RL    D          ; rotate left register D
       JP    NC, LOOP   ; jump if Carry flag is zero, because E is okay
       SET   0,E        ; else, set to '1' bit 0 of E
       JP    LOOP       ; repeat the loop indefinitely
```

Before we infinitely repeat the loop with JP LOOP, we need to update the configuration of the bits in DE by rotating them left.

The issue is that we don't have a specialized instruction to rotate register DE and all those available work on 8 bits.

Let's think about the solution. We have seen that the SLA E instruction moves all the bits of E to the left, copies bit 7, which goes out to the Carry flag and places a '0' at the right.

$$\boxed{CY} \leftarrow \boxed{7 \leftarrow 0} \leftarrow 0$$
$$E$$

Bit 7 of E still needs to be moved to bit 0 of D. To that end, instruction RL D retrieves the Carry flag generated by SLA and inserts it at the right of register D. As the same time, it moves all its bits to the left. The bit that exits from bit 7 is in turn saved in the Carry flag.

$$\boxed{CY} \leftarrow \boxed{7 \leftarrow 0} \leftarrow$$
$$D$$

In summary, we have created a left shift of all of register DE, thanks to two separate 8-bit operations, by means of the Carry flag. Yet, to complete our task, which requires a complete rotation of DE, the bit that has exited from bit 7 of register D should be moved in bit 0 of E.

That bit is currently found in the Carry flag and we can determine if it is at '0' through the JP NC,LOOP instruction. If it is at '0', we do not need to correct bit 0 of E and so we simply jump to the beginning of the loop.

If it is not at '0', we must bring bit 0 of E to '1'. This task is done here by the SET 0,E instruction (which we will look at in detail further on). The resulting LED rotation will look like this.

	Port 01h								Port 00h							
	7	6	5	4	3	2	1	0	7	6	5	4	3	2	1	0
	●	●	●	●	●	●	●	●	●	●	●	●	●	●	●	⊗
	●	●	●	●	●	●	●	●	●	●	●	●	●	●	⊗	●
	●	●	●	●	●	●	●	●	●	●	●	●	●	⊗	●	●
	●	●	●	●	●	●	●	●	●	●	●	●	⊗	●	●	●
	●	●	●	●	●	●	●	●	●	●	●	⊗	●	●	●	●
	●	●	●	●	●	●	●	●	●	●	⊗	●	●	●	●	●
	●	●	●	●	●	●	●	●	●	⊗	●	●	●	●	●	●
	●	●	●	●	●	●	●	●	⊗	●	●	●	●	●	●	●
	●	●	●	●	●	●	●	⊗	●	●	●	●	●	●	●	●
	●	●	●	●	●	●	⊗	●	●	●	●	●	●	●	●	●
	●	●	●	●	●	⊗	●	●	●	●	●	●	●	●	●	●
	●	●	●	●	⊗	●	●	●	●	●	●	●	●	●	●	●
	●	●	●	⊗	●	●	●	●	●	●	●	●	●	●	●	●
	●	●	⊗	●	●	●	●	●	●	●	●	●	●	●	●	●
	●	⊗	●	●	●	●	●	●	●	●	●	●	●	●	●	●
	⊗	●	●	●	●	●	●	●	●	●	●	●	●	●	●	●
	●	●	●	●	●	●	●	●	●	●	●	●	●	●	●	⊗

The RLCA, RRCA, RLA and RRA instructions

We mention these instructions for the sake of completeness since they were kept in the set of DMC8 instructions (and Z80 as well) to be compatible with the I8080 processor. They perform the same functions as the RLC, RRC, RL and RR instructions, respectively when we set register A as operand.

The RLCA, RRCA, RLA and RRA instructions have the advantage of being faster because operand A is implicit. At the same time, they are limited by the fact that they work only on that register. They also only change the Carry flag while instructions RLC, RRC, RL and RR also affect the Zero flag and Sign flag.

3.3.4 Bit manipulation instructions

Bit manipulation instructions (BIT, SET and RES) work on a single 8-bit operand, which can be a register (A, B, C, D, E, H or L), or an indirectly addressed memory location (through HL, IX or IY). The complete list of bit manipulation instructions and their addressing modes is on page 610 in Appendix C.

The BIT instruction

This instruction allows us to test a specific bit of the operand. A constant indicates the bit's position (from 7 to 0). The position of the bit is set by the programmer and cannot be changed while the program is being executed. Here are a few examples below:

```
BIT    0,A              ; test bit 0 (LSB) of register A
BIT    3,D              ; test bit 3 (counting from right) of register D
BIT    7,(HL)           ; test bit 7 (MSB) of the memory location
                        ; pointed by HL
```

The BIT instruction does not change the content of the register or the memory location indicated by the operand, but only changes the Zero flag. If the bit selected turns out to be zero, the flag is set, otherwise it is cleared.

The SET and RES instructions

The other two instructions are SET and RES (Reset). Here as well, the bit is indicated with a position anywhere from 7 to 0. SET forces the specified bit to '1'. RES, on the other hand, zeroes it. Here are some examples:

```
RES    0,A              ; clear bit 0 (LSB) of register A
SET    7,A              ; set to '1' bit 7 (MSB) of register A
RES    5,E              ; clear bit 5 (counting from right) of register E
SET    7,(HL)           ; set to '1' bit 7 (MSB) of the memory location
                        ; pointed by HL
```

The use of the BIT and RES instructions: An example

The system has an input port and an output port, both of which are allocated at address 00h. We acquire a command signal on the line 0 of the input port. Following the signal's transition from zero to one, we need to increment the binary number generated on the output port that is (we imagine) connected to eight LEDs. The number must be set to zero at system reset and the count must be cyclical from 0 to 127.

Let's maintain the state of the count in register B. The state is set to zero at the start of the program and, since the count must be visible from the output port, let's copy it on the port (going through A).

```
START:  LD   B,00h        ; clear register B, used to store the count state
        LD   A,B
        OUT  (00h),A      ; output the state on port at address 00h
```

The specifications require us to reveal the input command's transition from 0 to 1. To do this, first we need to verify that the input signal is zero and if it isn't, wait for it to go to zero.

```
TEST0:  IN   A,(00h)      ; read the input port at address 00h
        BIT  0,A          ; test bit 0
        JP   NZ,TEST0     ; if it is at '1', repeat from TEST0
```

After the input port is read, we use the BIT instruction to test the value of the line in position 0. The conditional jump will keep us in the loop on TEST0 until the line goes low. If the line is already low, we go on.

```
TEST1:   IN     A,(00h)        ; read the input port again
         BIT    0,A            ; test bit 0
         JP     Z,TEST1        ; if it is at '0', repeat from TEST1
```

This second loop on TEST1 seems identical to TEST0 but with one difference related to the condition of the jump. This time we stay in the loop if the input signal is still at zero, while we exit if it has gone to one.

Exiting the second loop means that we have identified the transition from zero to one so we increment the count while respecting the cyclicity between 0 (0000.0000_2) and 127_{10} (0111.1111_2). A simple solution is to set bit 7 or B to zero after the increment by using the RES instruction.

```
         INC    B              ; increment the counter (B), but clear always
         RES    7,B            ; the bit 7 to grant the cyclicity 0 - 127
```

The number is then copied onto the port so it is visible and we return to wait for a new transition by jumping to TEST0.

```
         LD     A,B
         OUT    (00h),A        ; display the state (B) on the output port
         JP     TEST0          ; repeat the loop indefinitely
```

3.3.5 Jump instructions

We have seen jumps in Section 1.4 and in many examples of programming. In this section, we will examine them more systematically and considering some of their applications. The available jump instructions are listed on page 611 in Appendix C.

Jump instructions are very important since they allow the programmer to explicitly change the order of execution of the instructions that make up the program. Jump instructions act on the content of the Program Counter (PC), the register responsible for the sequence of execution of the instructions, as we have seen in Chapter 1.

3.3.5.1 Unconditional jumps

If we examine the tables of data transfer instructions, the PC would seem inaccessible to the programmer. We find no explicit instruction in the tables that allows us to read or write its content. Yet, the execution of a jump consists precisely in transcribing the address specified by the operand field inside the PC as if it were a data transfer instruction with the PC as destination. We have seen that the unconditional jump instruction has the following structure:

```
         JP     <address>      ; jump to the specified 16-bit address
```

The address is often defined through a symbol (JP LOOP for example) to make the code more readable.

As we have seen, the execution of a jump instruction loads the specified address in the PC. The next instruction will be fetched from the location specified by the <address> field, not from the address right after that of the current instruction.

The forward jump: An example

For greater comprehensibility, the left hand side of the example has the addresses of the instructions, as they are allocated in the memory. They are all merely indicative, irrelevant to our scope, except for the jump instruction. When the processor reaches address 1100h, it executes an unconditional jump. The next instruction to be executed is at location 2000h. Everything that we see in between (in this example, from location 1103h through location 1999h) is simply ignored by the processor.

Address	Instruction		; Comment
10FFh	LD	A,0	; (a generic previous instruction)
1100h	JP	2000h	; jump to location 2000h
1103h	LD	B,3Fh	; this instruction is not executed, because
	...		; the processor jumped to address 2000h
2000h	LD	B,A	; from here on it will execute another sequence
	...		

The backward jump: An example

The example shows an infinite loop created by a backward jump. Every time the loop is repeated, we make the content of A go out on the port and then increment it. We use a label to identify the address at the beginning of the loop, which is the address specified by the operand of the jump instruction.

	LD	A,0	; clear A, used as a counter
LOOP:	OUT	(00h),A	; display the A content on the output port
	INC	A	; increment A and
	JP	LOOP	; repeat the loop indefinitely

The instructions between the LOOP label and the JP LOOP jump are repeated infinitely. Infinite loops are rare unless we want a program to repeat cyclically and indefinitely until the system is reset.

Further ahead, we will see that "interruptions" allow us to temporarily suspend the execution of an infinite loop to start working on another program, as mentioned in Section 2.1.5

3.3.5.2 Conditional jumps

The most important type of jump from a computational perspective is the conditional jump, which we were introduced to in Section 1.4.

These jumps are executed by the processor only when the flags have a specific value. As we have seen, they allow the processor to make decisions. The syntax of conditional jump instructions is as follows:

$$\text{JP} \quad <\text{condition}>,<\text{address}>$$

As with unconditional jumps, the <address> field indicates the location to jump to. The <condition> syntactically precedes the address and determines whether the jump will be executed or not at "run-time". Possible scenarios:

Condition	Operation	Flag
Z	Jump if zero result	$Z = 1$
NZ	Jump if not zero result	$Z = 0$
C	Jump if Carry	$C = 1$
NC	Jump if not Carry	$C = 0$
P	Jump if positive	$S = 0$
M	Jump if negative	$S = 1$
PE	Jump if parity even	$P = 1$
PO	Jump if parity odd	$P = 0$

"Parity" indicates if the number of '1s' in the result is odd or even.

Here are some examples of how the most common conditional jumps are used (based on the Zero and Carry flags):

```
JP   C, 11A0h      ; jump to the location 11A0h if Carry Flag is '1'
JP   NC, 3F00h     ; jump to the location 3F00h if Carry Flag is '0'
JP   Z, 1F00h      ; jump to the location 1F00h if Zero Flag is '1'
JP   NZ, 8000h     ; jump to the location 8000h if Zero Flag is '0'
```

A jump is executed only given the right condition, otherwise the processor goes on to the instruction right after that of the jump.

3.3.5.3 Indirect jumps

Among the instructions the processor can execute are "indirect jumps" (unconditional). These instructions make it possible to jump to an address determined at run-time, that has to be loaded in a 16-bit register ahead of time. All the possible variations of this instruction are as follows:

```
JP   (HL)      ; jump to the address specified in HL
JP   (IX)      ; jump to the address specified in IX
JP   (IY)      ; jump to the address specified in IY
```

In the types of jumps we have seen before, the address was defined at "design-time" (when the program was written). The example below shows the basic operations required to execute an indirect jump, including loading the address to jump to in a 16-bit register.

```
        LD      HL,ADDR  ; load the jump address in HL
        JP      (HL)     ; jump to the address just loaded in HL
        ...
ADDR:   NOP              ; the sequence of instructions where we jump
```

Replacing a normal jump with an indirect jump for no particular reason would be useless and would make writing the program clumsy and convoluted. Instead, indirect jumps are advantageous when the next task must be chosen among a specific number of alternatives based on the results of the operations done by the program itself. With only a few lines of code, we can efficiently choose a jump address among the many choices available. It makes sense to collect them in dedicated "jump tables" as explained below.

Jump tables

Let's assume that we have written a program that executes a specific algorithm that returns a number (0, 1, 2...) into register A. We will use this number to choose the next sequence of instructions to execute.

```
        <instruction>      ; the program executes an algorithm that
        <instruction>      ; returns a number (0,1,2...) in A
        LD   A,(0F000h)    ; the number identifies the sequence to execute
```

For simplicity's sake, this example has only four different sequences, but a real system would clearly have many more. Let's label the sequences FIRST, SECOND, THIRD and FOURTH. They can have the following basic structure:

```
FIRST:     <instruction>    ; code sequence handling the FIRST task
           <instruction>    ; ...
           JP CONTINUE

SECOND:    <instruction>    ; code sequence handling the SECOND task
           <instruction>    ; ...
           JP CONTINUE

THIRD:     <instruction>    ; code sequence handling the THIRD task
           <instruction>    ; ...
           JP CONTINUE

FOURTH:    <instruction>    ; code sequence handling the FOURTH task
           <instruction>    ; ...

CONTINUE: ...               ; the program will continue from here on
```

From the perspective of the assembler, FIRST, SECOND, THIRD and FOURTH are just four 16-bit constants (four addresses).

Let's collect these four constants in the following table and define them by using the DW directive (introduced in Section 3.1.6):

```
JTABLE:  DW   FIRST      ; address of the FIRST task     (index = 0)
         DW   SECOND     ; address of the SECOND task    (index = 1)
         DW   THIRD      ; address of the THIRD task     (index = 2)
         DW   FOURTH     ; address of the FOURTH task    (index = 3)
```

The four DWs insert in the ROM four consecutive 16-bit constants that have the value of the corresponding sequences' addresses. Let's use this "jump table" to retrieve the address to jump to based on the number provided by the start of the program. We will use this number as an index in the table as shown in the comments of the code[11].

We can see that the first row of the table is placed at the address identified by the label JTABLE, while the following rows are at addresses that increase two by two. Therefore, if we keep in mind the index received in A, the address of every row in the table can be expressed as JTABLE + (2 · <index>).

So, let's multiply the index by two and transfer it[12] to register DE.

```
        SLA   A          ; multiply the index by 2
        LD    E,A        ; transfer it in register E and clear register D
        LD    D,00h      ; now, the doubled index is in DE
```

Let's copy the address of the table in HL and add DE (in a 16-bit operation).

```
        LD    HL,JTABLE  ; load the jump table address in HL
        ADD   HL,DE      ; add the doubled index to HL
```

At this point HL has the address of the desired constant to retrieve from the table. This constant, in turn, must be copied in HL to be able to execute the desired indirect jump. First, however, we must retrieve the constant by transferring it to another register (DE for example, as we no longer need its content) given that we are still using HL.

```
        LD    E,(HL)     ; copy in E the low part of the constant from
        INC   HL         ; the table and, after incrementing HL,
        LD    D,(HL)     ; copy in D also the high part of the constant
```

Now, let's move the content of DE in HL, by using two 8-bit transfers since there is no suitable 16-bit data transfer instruction to do this.

---

[11] These tables could also be constructed in the RAM so they are replaceable at run-time. In some high-level languages, (C++, C#, Java, Object Pascal, etc.) we have the concept of "polymorphism". Briefly, this is the possibility of a program or part of a program to assume a different appearance or behavior according to the circumstances (these subjects are, however, beyond the scope of this book).

[12] We cannot transfer the content of an 8-bit register directly into a 16-bit register, so we copy A in the low part of DE and then we set the high part to zero.

Finally, we execute the indirect jump through HL.

```
LD      L,E          ; copy the address retrieved from the table in
                       HL
LD      H,D
JP      (HL)         ; Finally, jump to the desired code sequence
```

The last six lines of code could also be replaced by the five rows here below. They are a bit more efficient but far less readable.

```
LD      A,(HL)       ; copy in A the low part of the constant taken
INC     HL           ; from the table and, after incrementing HL,
LD      H,(HL)       ; copy in H the high part of the constant.
LD      L,A          ; Now we have in HL the complete address and,
JP      (HL)         ; finally, jump to the desired code sequence
```

3.3.5.4 Delay loops

Now let's look at an important conditional jump application: "delay loops". Microprocessor systems are often required to generate sequences of control signals with precise delays between one action and the next. Think of a burglar alarm: before the alarm goes off, the legitimate owner needs a few seconds to deactivate it.

To get the processor to let some time pass inert, we must execute a delay cycle as in this example:

```
        LD      C,255        ; initialize C as "loop counter"
LOOP:   DEC     C            ; decrement the counter at every loop
        JP      NZ,LOOP      ; jump backward if the result of decrement
        ...                  ; is still not zero, else exit the loop
```

Before the loop is executed, register C is assigned a value specifically calculated to make a specific amount of time pass. Afterward, register C, used as a counter, is cyclically decremented until it reaches zero. At that point, the time interval will be up and it can go on to execute the next code.

The only target of this bit of code is to make an amount of time go by, to generate a delay. Let's try to calculate it.

The instruction tables available in Appendix C show that our instructions take the following number of clock cycles to be executed:

```
        LD      C,255        ; 7 clock cycles
LOOP:   DEC     C            ; 4 cycles
        JP      NZ,LOOP      ; 10 cycles (they don't depend on the flag value)
```

Since C is initialized at FFh in the beginning, the loop is executed 255 times. The total number of clock cycles N is:

$$N = 7 + 255 \times (4 + 10) = 3577$$

Let's assume that our microprocessor works with a clock frequency of $F_{ck} = 10$ MHz. The duration of a clock cycle is then $T_{ck} = 1/F_{ck} = 100$ nS, so the amount of time it takes T_d is:

$$T_d = N \times 100nS = 3577 \times 100nS = 357700nS = 0,3577mS$$

Calculating the number of repetitions

The delay of a loop can be easily and precisely calibrated both by changing the initial value of C and by adding other instructions in the loop to increase the amount of time it takes to finish. For example, we can insert two NOPs and leave the initial value of the counter as X (to be determined).

```
          LD    C,<X>       ; 7 clock cycles
LOOP:     NOP               ; 4 cycles
          NOP               ; 4 cycles
          DEC   C           ; 4 cycles
          JP    NZ,LOOP     ; 10 cycles (they don't depend on the flag value)
```

The total number of clock cycles we get (in function of X) is:

$$N = 7 + X \times (4 + 4 + 4 + 10) = 7 + X \times 22$$

So we retrieve the number X to load in the register in function of N:

$$X = \frac{N - 7}{22}$$

If we wanted to get a delay of $T_d = 0,5mS$, keeping in mind that $T_{ck} = 100nS$, we should set the number of cycles as:

$$N = \frac{T_d}{T_{ck}} = \frac{0,5mS}{100nS} = 5000$$

Let's calculate X to get 5000 clock cycles:

$$X = \frac{N - 7}{22} = \frac{5000 - 7}{22} \approx 226,95$$

rounding up the the nearest integer number, X = 227. If we use this number to initialize register C, we get:

$$T_d = (7 + 227 \times (4 + 4 + 4 + 10)) \times 100nS = 5001 \times 100nS = 0,5001mS$$

which is a very good approximation of the amount of time, just $0,1\mu S$ off.

Delays using nested loops

If we need more consistent delays, we can rely on techniques that are a bit more complex, such as nested loops.

```
           LD      D,255        ; initialize D as the outer loop counter
                                ;
LEXT:      LD      C,255        ; initialize C as the inner loop counter
                                ;
LINT:      DEC     C            ; decrement the inner loop counter
           JP      NZ,LINT      ; jump backward to LINT if the counter is not
                                ; zero yet, otherwise exit the inner loop and
           DEC     D            ; decrement the outer loop counter
           JP      NZ,LEXT      ; jump backward to LEXT if it is not zero yet,
           ...                  ; otherwise exit the outer loop too and go on
```

Here above, we can see the loop from the previous example, in the lines in the center, nested one inside the other with the same format but using a different register as a counter. In this example, the outer loop executes 255 repetitions. For each one of these, the inner loop executes its own 255 repetitions. Each counter is initialized just before the beginning of the corresponding loop.

Intuitively, the approximate overall execution time is given by the product of the time the internal loop takes multiplied by the number of outer loop repetitions. The time will ultimately be about proportional to the product of the values we load in the two counters. In any case, when we do the calculations in detail, we get:

$$T_d = (7 + (7 + 255 \times (4 + 10) + 4 + 10) \times 255) \times 100nS = 91,5712mS$$

that is, less than a tenth of a second.

Delay loops with 16-bit counter

Another way to get consistent delays is to use 16-bit registers for loop counts, as in the following example:

```
           LD      BC,65535     ; initialize the 16-bit loop counter (BC)
                                ;
LOOP:      DEC     BC           ; decrement the loop counter
           LD      A,B          ; check if register BC has been zeroed
           OR      C            ; executing an OR operation between B e C
           JP      NZ,LOOP      ; jump backward if BC is not zero,
           ...                  ; otherwise go on
```

In this example, we get the count by using the DEC BC instruction, which carries out a 16-bit decrement operation. Unfortunately, this type of 16-bit instruction does not check the flags, as we know. This means that it is we who must check that register BC has gone to zero after the decrement. As we did in a previous example (see Section 3.3.2.5), we need to insert one or more instructions that change the flags.

We have added a pair of instructions that execute an OR between the high part B and the low part C of register BC. This OR produces a null result and activates the Zero flag only when all the bits of register BC are at zero. Otherwise, the processor continues to repeat the loop. When we do the delay calculation, this example gives us:

$$N = (10 + 65.535 \times (6 + 4 + 4 + 10)) = 1.572.850$$

$$T_d = 1.572.850 \times 100nS = 157,285mS$$

which is about 1/6 of a second.

Generating very long delays

We can always combine the methods we've seen to obtain very consistent delays. We must note, however, that it is not very efficient to spend a long time on delay loops for a microprocessor system. This would entail wasting the potential of the system because we require it to be inert when it could be doing useful tasks.

As we will see further on in Section 4.6, we can measure even very long delays by using the technique of interrupts and timers without sacrificing the potential of the processor.

Checking delay times in the emulator

In Section 2.4.2.3 we introduced the Deeds-McE emulator as a means to check programs. Let's open a program in the emulator (see figure below) that includes a delay loop with a 16-bit counter.

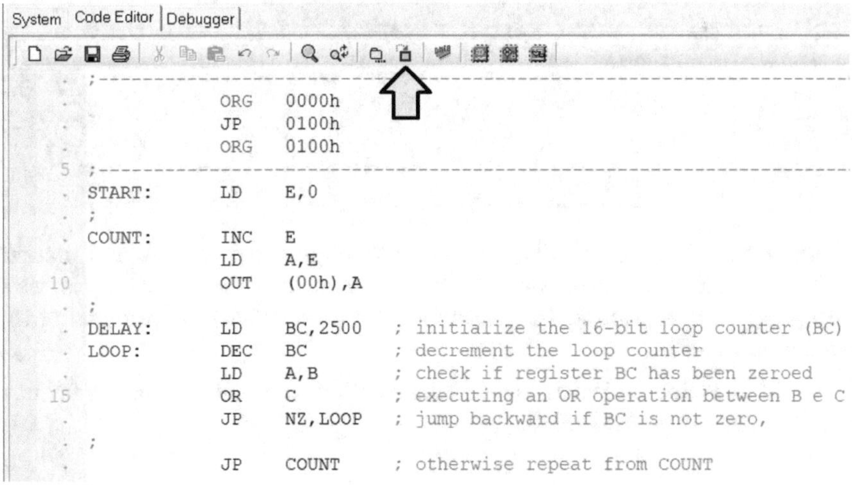

```
                ORG    0000h
                JP     0100h
                ORG    0100h
;-------------------------------------------------------------
START:          LD     E,0
;
COUNT:          INC    E
                LD     A,E
                OUT    (00h),A
;
DELAY:          LD     BC,2500   ; initialize the 16-bit loop counter (BC)
LOOP:           DEC    BC        ; decrement the loop counter
                LD     A,B       ; check if register BC has been zeroed
                OR     C         ; executing an OR operation between B e C
                JP     NZ,LOOP   ; jump backward if BC is not zero,
;
                JP     COUNT     ; otherwise repeat from COUNT
```

On the output port, the program visualizes a binary number that increments at every repetition of the infinite loop beginning at the label COUNT.

Each time the number is written on the port a delay loop is executed. The loop is repeated 2,500 times. This is almost identical to the previous example; the only change is the number loaded at the beginning into the register BC. Therefore, we apply the expression previously used to calculate the duration (in clock cycles):

$$N = (10 + 2.500 \times (6 + 4 + 4 + 10)) = 60.010$$

We want to use the emulator's debugger to verify that the delay actually has this value. We use the commands "Run" and "Clear" on the command bar (see the figure below, shown by the yellow and green arrows, respectively) and read the clock cycle count in the field marked by the blue arrow.

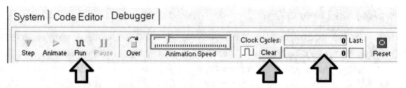

Before launching the program, let's define two "breakpoints" that will allow us to interrupt execution at pre-established points. As we can see in the following figure, after selecting the line with the instruction we want to stop at, we use the context menu indicated by the yellow arrow to assign it a breakpoint.

Addr	Op Code	Label	Instruction		Comment
0000	C30001		JP	0100H	
0100	1E00	START	LD	E,0	
0102	1C	COUNT	INC	E	
0103	7B		LD	A,E	
0104	D300		OUT	(00H),A	
0106	01C409	DELAY	LD	BC,2500	the 16-bit loop counter
0109	0B		*Set Breakpoint at "0106h"*		the loop counter
010A	78				register BC has been zeroe
010B	B1		*Force Program Counter to "0106h"*		g an OR operation between B
010C	C20901	✓	*Compact Object Code View*		kward if BC is not zero,
010F	C30201		*Extended Object Code View*		e repeat from COUNT

The figure below shows the two breakpoints highlighted in pink and marked with a red square. Since we need to measure how long the delay loop is, we'll stop program execution at the beginning (location 0106h) and the end of the same loop (010Fh).

The light blue line indicates the next instruction to execute, which is for now still the first (at address 0000h).

Object Code

. Addr	Op Code	Label		Instruction	Comment
0000	C30001		JP	0100H	
0100	1E00	START	LD	E,0	
0102	1C	COUNT	INC	E	
0103	7B		LD	A,E	
0104	D300		OUT	(00H),A	
■ 0106	01C409	DELAY	LD	BC,2500	; initialize the 16-bit loop counter
0109	0B	LOOP	DEC	BC	; decrement the loop counter
010A	78		LD	A,B	; check if register BC has been zero
010B	B1		OR	C	; executing an OR operation between
010C	C20901		JP	NZ,LOOP	; jump backward if BC is not zero,
■ 010F	C30201		JP	COUNT	; otherwise repeat from COUNT

We execute the program by clicking on the "Run" command. Emulation stops when the instruction corresponding to the first breakpoint is fetched (address 0106h at the white arrow in the figure below) but hasn't yet been executed.

0103	7B		LD	A,E	
0104	D300		OUT	(00H),A	
■ 0106	01C409	DELAY	LD	BC,2500	⇐ alize the 16-bit loop counter
0109	0B	LOOP	DEC	BC	ment the loop counter
010A	78		LD	A,B	; check if register BC has been zero
010B	B1		OR	C	; executing an OR operation between
010C	C20901		JP	NZ,LOOP	; jump backward if BC is not zero,
■ 010F	C30201		JP	COUNT	; otherwise repeat from COUNT

At this point, let's go to the command bar and click on the "Clear" command (green arrow in the figure below), to set the field indicated by the light blue arrow to zero.

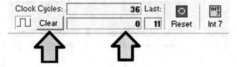

The upper field shows the absolute number of clock cycles as of reset, while the lower field shows the number of clock cycles since it was zeroed.

Let's press the "Run" command again.

0103	7B		LD	A,E	
0104	D300		OUT	(00H),A	
■ 0106	01C409	DELAY	LD	BC,2500	; initialize the 16-bit loop counter
0109	0B	LOOP	DEC	BC	; decrement the loop counter
010A	78		LD	A,B	; check if register BC has been zero
010B	B1		OR	C	; executing an OR operation between
010C	C20901		JP	NZ,LOOP	backward if BC is not zero,
■ 010F	C30201		JP	COUNT	⇐ rwise repeat from COUNT

Program execution restarts and then stops again at the second breakpoint at the end of the delay loop (see the white arrow). Now, we can see the partial count of the clock cycles as they appear in the previously zeroed field and can verify that the number is what we had thought: 60.010.

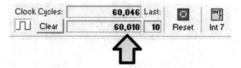

3.3.6 CPU control instructions

This small group of instructions lets us, for example, enable or disable the interrupt mechanism and check the state of the processor. The summary table can be found in Appendix C on page 614.

The EI and DI instructions

The Enable interrupt and Disable interrupt instructions will be examined within the overall context of the interrupt "mechanism" in Section 4.4. These are the two instructions:

   EI   ; enable interrupt mechanism

   DI   ; disable interrupt mechanism

The NOP instruction

We will include the instruction that carries out no operation at all:

   NOP   ; no operation

As we saw in Chapter 1, this instruction's only effect is to increment the Program Counter (PC) to execute the next instruction. It can also be used as a 4-clock-cycle delay.

The HALT instruction

This instruction stops the processor:

   HALT   ; stop program execution and enter the HALT state

The execution of the program is stopped because the internal sequencer is "frozen" in stand-by. This situation is made evident because the external activity stops. The CPU can only exit the HALT state through a reset or an interrupt request.

If the system is reset, the program recommences execution at address 0000h. If there is an interrupt request, the processor reactivates to execute the handling routine (as we will see in Section 4.4) and when it finishes that, program execution recommences as of the instruction after the HALT.

SCF and CCF instructions

Finally, the following two Carry flag manipulation instructions belong to the group of CPU control instructions:

SCF   ; set Carry flag to '1'
CCF   ; complement (invert) the Carry flag value

Notice the absence of an instruction that explicitly sets the Carry flag to zero. In order to do that, we could obviously execute an SCF and a CCF in sequence or any of the logic instructions (AND, OR or XOR), in that they set the Carry flag to zero. For example:

OR   A   ; execute the operation A ← (A or A)

sets the Carry flag to zero without changing the content of the accumulator.

3.3.7 Input/output instructions

Input/output instructions allow the DMC8 to communicate with "outside devices" by means of the ports we examined in Section 2.3.2. As we have seen, the processor only uses the 8 least significant wires of the address bus ($A_7...A_0$), so up to 256 input devices and 256 output devices can be handled.

The input/output instructions that we have seen in the examples are shown here below. We are required to use accumulator A to read or write a port and the addressing mode is direct. Furthermore, the flags are not changed.

IN   A,(<address>)   ; read the input port at the specified <address>
OUT   (<address>),A   ; write the output port at the specified <address>

Here is an example of how IN and OUT instructions are used in this format. This system has an input port and an output port which are both allocated at address 00h.

Register B is continually changed according to the value assumed by bit 0 of the input port: if '1', it is incremented, if '0', it is decremented. The program visualizes the content of register B on the output port.

```
START:   LD    B,0        ; initialize B (the counter state)
LOOP:    IN    A,(00h)     ; read the input port at address 00h
         BIT   0,A         ; check the requested count direction
         JP    NZ,INCR     ; jump if bit0 = 1, we need to increment the count
DECR:    DEC   B           ; otherwise, decrement the count
         JP    OUTPUT      ; jump and update the output port
INCR:    INC   B           ; increment the count
OUTPUT:  LD    A,B         ; copy the counter state on
         OUT   (00h),A     ; the output port at address 00h
         JP    LOOP        ; repeat the loop indefinitely
```

Indirect addressing mode is also available for input and output instructions where the address of the port is specified by register C (see the table in Appendix C on page 613).

| IN | r, (C) | ; read the port at the address specified by register C |
| OUT | (C), r | ; write the port at the address specified by register C |

Using this format, we are no longer required to use the accumulator to read or write a port since we can use any 8-bit register (r could be A, B, C, D, E, H or L). Also, the IN instruction affects the flags and this makes it possible to directly execute tests on the ports.

Here below is an example of the use of the OUT instruction with indirect addressing mode. The system has 8 output ports allocated at contiguous addresses from low to high (from 00h to 07h). The program is tasked with copying the 8 memory locations as of address 8000h on the ports.

We load the address of the first memory cell in HL and that of the first output port in C. Register B counts the number of repetitions of the loop:

	LD	HL,8000h	; the address of the table in memory
	LD	C,00h	; the address of the first port
	LD	B,8	; number of ports involved (and memory cells)

In the first part of the loop, we use registers HL and C to address the memory location we take the number from and the output port where we transcribe it, respectively.

| LOOP: | LD | A,(HL) | ; read the memory location pointed by HL in A |
| | OUT | (C),A | ; write the value in A to the port pointed by C |

So, in the second part of the loop, we update the registers at the addresses of the next memory and port, and we decrement the loop counter.

	INC	HL	; point to the next memory location
	INC	C	; point to the next output port
	DEC	B	; decrement the loop counter
	JP	NZ,LOOP	; repeat the loop 8 times, then exit the loop

After the eight memory locations are copied onto the ports, register B goes to zero, and we exit the loop.

3.3.8 Subprogram call and return instructions

Before discussing subprogram call and return instructions, we need to examine the function of a special area of RAM memory, the "stack". To this end, all of Section 3.4 is dedicated to this category of instructions. The subprogram call and return instruction table is available in Appendix C on page 612.

3.4 Subprograms and the Stack area

A subprogram is a program that is executed (or "called") by another program. In turn, a subprogram can call yet other subprograms. The main purpose of grouping one or more specific functions in a subprogram is to make those functions "reusable", able to be called multiple times within a program.

A collection of subprograms is called a "library". Often, programmers create their own subprogram libraries so that they can avoid rewriting the same things again and again, and often they reuse them in different projects and programs.

So that a subprogram can be called by another program (or subprogram), a few simple rules need to be followed, which we will see further on. These rules depend on the way in which the subprograms are implemented in the hardware of the specific processor being used. We will examine from the perspective of the processor how a program can "call" a subprogram and how this can consequently "return" to the caller.

Before we proceed, we need to examine how the processor handles a special area of RAM memory called the "Stack" (as in stack of books).

3.4.1 The Stack and the Stack Pointer

The Stack is a privileged area of the memory where the microprocessor can temporarily save and retrieve various types of data. To be clear, this area is special only because of the use we make of it; from a hardware point of view it has nothing new, an area in RAM memory that the programmer chooses.

The basic difference in using normal memory and using the stack area is in the order used to read and write data. We normally access memory by indicating the operation we want to execute: read or write, and the location address is specifically chosen by the programmer as we have seen. For example:

LD   (8000h),A   ; write A into a specific memory location

LD   A,(8000h)   ; retrieve data from the same memory location

In some cases, however, it is convenient not to have to know the specific address of the data to transfer. We don't have to if the processor has an automatic mechanism that decides the memory addresses for us while we only deal with the data to transfer.

A stack allows us to organize objects one on top of the other and worry about nothing more than the sequence for storing and recovering them. The apparent disadvantage is the fact that we cannot immediately access objects under the top of the stack.

To keep the stack from toppling over, we would always have to put and retrieve the object on top[13].

In some cases, however, this apparent limitation can be an advantage. We will soon see that it is very convenient to temporarily store data at the top of the stack and then retrieve them when necessary.

The processor provides us with internal hardware devices that automatically take care of the reading and writing of data on the Stack. This support frees us from having to deal with the specific memory location addresses where the data are written. We only need to pay attention to the order we put the data in so we can take them out in the opposite order.

Inside the processor there is a dedicated 16-bit register called the Stack Pointer (SP), and it is appropriately handled by the sequencer, which can increment it and/or decrement it. When the processor accesses the Stack memory area, it sends the content of register SP on the address bus to select the memory cell to be read or written. The processor contains the Stack Pointer and its handling logic, while the RAM connected to the processor contains the Stack area (see the figure below)[14].

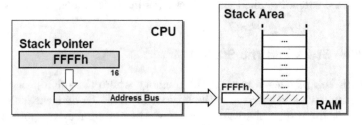

Since the memory area used for the Stack is not predefined, programmers are free to choose any available part of the system RAM to be used as the Stack area. To do this, they simply need to insert the initialization of register SP among the first program startup instructions.

LD   SP,<nn>   ; initialize the Stack Pointer at the address <nn>

When programmers assign an initial value to register SP, they are defining the bottom of the memory area dedicated to the Stack. The figure shows that SP has been initialized at address FFFFh, the last location possible in the RAM. Please note that we have actually only defined the location we will start filling it at (backward), not the size of the area. We will discuss this further on.

The processor uses the Stack area automatically for calling subprograms and for returning from those calls. The Stack area is also automatically used by the interrupt mechanism. However, programmers can use some specialized data transfer instructions such as PUSH and POP, which make it possible to work directly with the Stack area.

---

[13] A memory structure handled thus is called LIFO (Last In, First Out).

[14] Make sure not to confuse the terms "Stack" and "Stack Pointer".

The PUSH and POP instructions

When programmers want to expressly save data on the Stack and then re-trieve them, they use the PUSH and POP instructions. The example below illustrates how these instructions are used:

```
LD      SP,0FFFFh   ; initialize the Stack Pointer at FFFFh
...
...                 ; first part (omitted): the program uses BC and HL
...
PUSH BC             ; save the content of register BC on the Stack
PUSH HL             ; save the content of register HL on the Stack
...
...                 ; second part (omitted): BC and HL are overwritten
...
POP  HL             ; retrieve the previous content of HL from the Stack
POP  BC             ; retrieve the previous content of BC from the Stack
...
...                 ; third part (omitted): we return to use the previous
                    ; contents of BC and HL again, just recovered
```

In this example, as before, we chose to initialize register SP at the address of the last available RAM location.

Let's assume that tasks involving registers BC and HL are carried out in the first part of the program (the example does not show the code of this part). Let's further assume that the values obtained in BC and HL are important because they need to be used again later and they should be preserved.

However, we will need to use BC and HL again in the second part of the program, but for different purposes. So, their original contents will be over-written. To use registers BC and HL here, we need to temporarily save their contents and retrieve them later.

PUSH instructions can help here because they make it possible to save the values in BC and HL at the top of the Stack without using expressly defined variables in the memory, thus avoiding having to choose and assign their addresses in the program.

The operand of PUSH and POP instructions can only be one of the following: AF, BC, DE, HL, IX or IY[15]. These names identify the well known 16-bit registers, plus AF. This acronym defines the set of accumulator A and of the F flag register[16].

---

[15] The processor only allows us to save 16-bit data in the Stack. Since the memory locations are 8 bits, whatever is saved is divided into a low byte and a high byte (following little-endian convention).

[16] Writing PUSH AF makes the processor save accumulator A first, followed by the F flag register at the top of the Stack.

The figure below deals with the execution of the first three instructions in our example (the stack is represented in the figure as a sort of glass). On the left hand side we see the situation downstream of the initialization of register SP, which targets memory location FFFFh at the bottom of the Stack (still empty here):

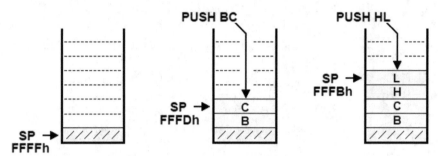

In the middle, the processor has finished executing the PUSH BC instruction. The processor has used register SP to address the memory over two successive writes, each time decrementing it by one. The result being that the content of register BC is copied on the Stack and divided into two parts: B and C. The top has been raised by two positions and register SP now targets the highest filled location, FFFDh.

Note that the Stack grows toward high in the drawing but as it does, the top address goes down. This is a design choice found in almost all existing microprocessors.

The right hand side of the figure shows the content of the Stack after PUSH HL is executed. The content of register HL has been copied on the Stack and register SP now targets the new top of the Stack (SP = FFFBh).

Saving the current content of the two registers now allows us to freely reuse them for other computing purposes (the second part of the code has been left out in the example, but we can assume that the content of the two registers has been overwritten because the code was executed).

In the third part of the program, we need to get the previous contents of the registers back, so we simply retrieve them from the Stack by executing two POP instructions and pay attention to the order we take them in.

On the left hand side of the following figure, we see the Stack is in the situation we left it in before.

In the center, we see that the Stack has been emptied of its two uppermost bytes by the POP HL instruction, which refreshed the previous content in HL. After the two reads targeting the memory with register SP are executed, SP is incremented twice and now targets (SP = FFFDh), the new (lower) top of the Stack.

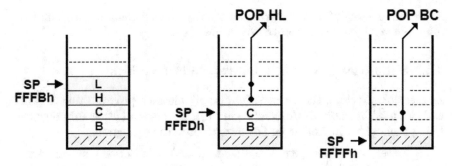

Finally, after POP BC is executed (right hand side of the figure), the previous values in register BC will also be refreshed and the Stack will be empty again (SP = FFFFh)[17].

Extending this example more generally, if it were necessary to save all the registers we would need to use all the available PUSH instructions.

PUSH	AF	; save the content of all registers in the Stack
PUSH	BC	; A and F, BC, DE, HL, IX, IY
PUSH	DE	;
PUSH	HL	
PUSH	IX	
PUSH	IY	

After executing any calculation using these registers, we will have to refresh all of them. Downstream of the execution of all the POP instructions, the previous condition will be completely restored.

POP	IY	; recover the previous contents of
POP	IX	; all the registers from the Stack
POP	HL	
POP	DE	
POP	BC	
POP	AF	

Lastly, let's go over some important concepts:

— Using PUSH and POP instructions, programmers do not have to explicitly address the memory locations that are used, that is they don't have to define their addresses, except for initializing register SP one time.

— The number of PUSHes executed must be followed by the same number of POPs. If not, the Stack fills up but doesn't empty, or vice versa.

— The sequence for retrieval has to be flipped vis à vis the sequence for saving, given the nature of the Stack itself.

---

[17] Note that POP has not deleted the values that were in the Stack, but they are not shown in the figure since formally they no longer have meaning (later Stack operations will overwrite them).

After the analysis of how it works, we see the Stack's main application: handling the "call" and the "return" of subprograms.

3.4.2 Subprograms and call and return instructions

As mentioned before, the term "subprogram" means a program that is called and executed by another program; it, in turn, can also call other subprograms. Using them gives the following advantages in writing programs:

— a better use of memory space since a task is described by one single segment of code that can be executed as many times as we like
— a neat improvement in the readability of the source code
— better reusability of the code.

A subprogram can be called by means of the CALL instruction:

CALL   <address>   ; call the subprogram at the given <address>

To go back to the calling program, the subprogram must be terminated by the following instruction:

RET   ; return to the calling program

How the CALL and RET instructions are used: An example

To explain how these two instructions are used, let's look at a simple example that we can imagine was extracted from a larger program. The initialization of register SP must always be placed at the beginning of the program.

START: LD   SP,0FFFFh ; initialize the Stack Pointer
          ...

Later in the program, the subprogram AVERAGE is called according to the needs of the calculation.

```
LD    B,34       ; set up the operands in the registers B and C
LD    C,15
CALL  AVERAGE  ; call the subprogram AVERAGE
LD    E,A        ; the result, returned in A, is copied to register E
...
```

Afterwards, it is called again with different operands, perhaps multiple times.

```
...
LD    B,56       ; set up new operands in the registers B and C
LD    C,22
CALL  AVERAGE  ; call the subprogram AVERAGE
LD    D,A        ; the result, returned in A, is copied to register D
...
```

The subprogram, shown below, calculates the average of two operands, which are expected to be found in registers B and C.

```
AVERAGE:  LD    A,B      ; copy the content of register B in register A,
          ADD   A,C      ; to add it to the content of register C
          SRA   A        ; the result is divided by 2 (with a right shift)
          RET            ; return to the caller, with the result in register A
```

As we read in the comments, the result is returned to the caller in the accumulator. Note that the subprogram ends with the RET instruction.

Before it executes the subprogram with the CALL instruction, the calling program provides the operands to work on by loading their values in the established registers. When the average is calculated, the RET instruction makes the control of execution return to the calling program and the instruction immediately after the CALL is executed. In this example, we assume we are using the result by copying it in another register of the processor.

In programming jargon, the program "passed the parameters" to the subprogram, and the result was "returned" to the calling program. In our example, the parameters and the result were passed and returned "by registers".

The CALL instruction and the Stack

At first glance, the CALL instruction looks like the JP instruction. One similarity is that they both accept an address to jump to as an operand. The CALL instruction, however, is a special jump; before executing the jump to the subprogram, it saves the "return address" on the Stack.

The figure below gives an example of what the instruction CALL inserts in the Stack. On the left hand side of the figure, we see the Stack before CALL is executed, while on the right, we see the stack containing a copy of the PC (broken into two bytes) after CALL is executed.

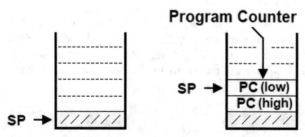

By the time the processor saves this information, the PC has already been updated by the internal sequencer so it contains the address of the first instruction after the CALL, which is exactly what needs to be executed after the return of the subprogram.

The RET instruction and the Stack

This may be less obvious but the RET instruction is also a special jump. It has no operands but it is able to jump from the bottom of the subprogram and return to the calling program. In fact, it retrieves the address of the jump from the top element on the Stack.

That element is the previously mentioned "return address" that was saved by the CALL instruction. On the left hand side of the figure below, we see the Stack, which contains the return address saved before, during the CALL instruction execution.

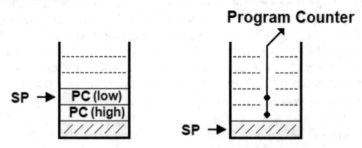

The right hand side of the figure shows the result after RET was executed. The address that was at the top of the Stack has been retrieved and loaded into the Program Counter. In other words, the processor has jumped to the instruction following the CALL that called the subprogram.

Can the JP instruction be used to call a subprogram?

Note that the return address that is saved in the Stack is different for every CALL instruction that calls the subprogram even if it remains the same. This prevents us from using ordinary JP instructions because they have no mechanism for identifying the caller.

To demonstrate this, let's pretend for a moment that we have no CALL or RET instructions but only ordinary jumps. Let's modify the previous example by jumping to subprogram AVERAGE with a JP instruction. Let's add a RETURN1 label, which allows us to return and execute the rest of the program.

```
          ...
          LD    B,34
          LD    C,15
          JP    AVERAGE  ; jump to the subprogram AVERAGE
RETURN1:  LD    E,A         ; we add this label to be able to return here
          ...
```

Since we have no RET instruction, we need to put a jump at the bottom of the subprogram that will allow us to go back.

```
AVERAGE:  LD    A,B      ; the subprogram is the same as before
          ADD   A,C      ; but the last instruction is different
          SRA   A
          JP    RETURN1  ; jump to return to the caller
```

So far, this code seems to work perfectly. Let's try to intervene in the second call and see what happens.

```
          ...
          LD    B,56
          LD    C,22
          JP    AVERAGE  ; jump to the subprogram AVERAGE
RETURN2:  LD    D,A      ; we add a different label to return here
          ...
```

Since we have to return to the instruction after the jump, we need to add a new label (RETURN2) in front of it, which is obviously different from the previous one. Unfortunately, we already have a jump at the bottom of the subprogram, but it's pointing to RETURN1, the return address of the previous call.

If we only use jumps, we won't be able to call a subprogram more than once, undermining one of their main advantages. If we use CALL and RET, however, we don't need to worry about who is calling the subprogram or where it has to return from because the Stack works automatically.

Compliance with common rules, and errors

We have seen that the return address's storage and recovery mechanism is completely automatic and transparent for the programmer using the CALL and RET instructions. Our job is only to allocate the Stack area by initializing the Stack Pointer.

The programmer's responsibility begins when it is necessary to also use the PUSH and POP instructions. There is often the need for a subprogram to preserve the content of the registers intact.

Consider the following example :

```
SUBPROG:  PUSH  BC      ; save registers B, C, A and F
          PUSH  AF
          ...           ; here the subprogram tasks
          ...           ; registers B, C, A and F are used and modified
          ...
          POP   AF      ; retrieve the previous contents of the registers
          POP   BC
          RET
```

In subprogram SUBPROG, the content of B, C, A and the flags is saved in the Stack before the assigned tasks are executed. The tasks have been omitted here for the sake of simplicity.

After using these registers to do the calculations, and before returning to the caller thanks to the RET instruction, the subprogram retrieves the contents saved in the beginning from the Stack. Thus, the caller finds the content of the registers intact after the subprogram is executed.

To get a close look at how this program behaves and highlight the dynamics of the Stack, we need to introduce a generic calling program. We only need to see how it initializes SP and calls our subprogram.

```
START:  LD      SP,0FFFFh   ; initialize the Stack Pointer
        ...
        CALL  SUBPROG    ; call the subprogram SUBPROG
        NOP              ; note that this NOP represent any
        ...              ; instruction following the CALL
```

Let's look at the figure below. On the left-hand side, once register SP is initialized, the Stack is empty. After the subprogram is called, the Stack contains the return address i.e., the address of the instruction after CALL. At this point, the processor has made the jump to SUBPROG and executes its first instruction.

After the PUSH BC instruction is executed, the contents of registers B and C are found in the Stack in the order in which they are inserted. The following PUSH AF adds the contents of the accumulator and the flags on the Stack.

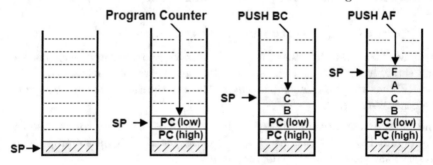

At this point, the subprogram performs its task, free to use the registers, whose contents have been saved on the Stack.

Now let's look at the conclusion of the subprogram, where the processor executes POP AF. Consider the following figure. From the situation drawn on the left we move to the second-to-the-left image where the previous contents of registers A and F are retrieved and restored in the processor.

A similar situation happens for POP BC. After POP BC is executed, (second-to-the-right image) only the return address that was saved by the CALL instruction remains in the Stack.

As we have seen, RET simply retrieves it and replaces it in the PC, producing a jump to the first instruction after the CALL instruction. At this point, the Stack is empty again.

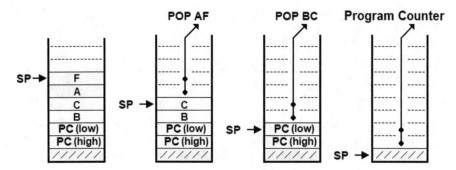

To insure that the whole mechanism keeps working, it seems clear that the programmer needs to respect some general rules:

— A RET (in the subprogram) must correspond to every CALL.
— A POP must follow every PUSH in a mirror pattern.
— The order in which the contents of the registers is retrieved must be flipped vis à vis the order in which they are inserted.

The example below shows a trivial but serious error:

```
SUBPROG:  PUSH  BC      ; save registers B, C, A and the flags F
          PUSH  AF

          ...

          ...
          POP   BC      ; is something missing here?
          RET
```

Here, the programmer has forgotten POP AF, provoking some interesting (if lethal) consequences:

— The saved values of A and the flags are not restored.
— POP BC retrieves what were the contents of A and the flags from the top of the Stack, not from BC.
— RET retrieves the previous content of register BC from the top of the Stack instead of the return address from the calling program.

Consequently, program execution continues incorrectly and unpredictably as of the address in BC since the processor "returns" to the completely wrong address. This is the most serious consequence of all since it will inevitably cause the system to crash.

In the following example of a subprogram, the programmer has only made a mistake in the order the data is retrieved from the registers.

POP BC retrieves the previous contents of A and F from the top of the Stack and loads them in B and C. At the same time, the previous contents of B and C are loaded in A and F.

```
SUBPROG:  PUSH  BC        ; save registers B, C, A and the flags F
          PUSH  AF
          ...
          POP   BC        ; the order of the two POPs is reversed!
          POP   AF
          RET
```

This inversion will clearly generate errors in the calling program, which will find the contents of the two registers switched. However, RET will find the previously saved address at the top of the Stack, as expected.

The following example also has a significant oversight.

```
SUBPROG:  LD    A,B
          ...
          ...
          CALL  SUBPROG
          ...
          ...
          RET
```

The error is that the programmer called the subprogram from inside itself, triggering an uncontrolled mechanism of "recursion". The processor can never execute RET, because first, it meets the CALL instruction which calls the subprogram again. The real problem is caused by the Stack: each time CALL is executed, a new return address is saved at the top of the Stack, so it continually grows without going back down.

This processor and others like it are not equipped with an automatic check on the value assumed by the Stack Pointer. This means that the Stack will have grown so much that it will invade other memory areas. The program could even overwrite other data areas and make everything crash.

Special uses of the Stack

There are still cases where it could be useful to break the rules, as long as its done conscientiously. Let's look at the following example where we have apparently exchanged the order of the POP instructions. What happens to registers BC and DE if we call this subprogram?

```
EXBCDE:   PUSH  BC
          PUSH  DE
          POP   BC
          POP   DE
          RET
```

Among all the instructions available, there is no instruction to switch the contents of two 16-bit registers. A judicious use of the PUSH and POP instructions can resolve this problem.

Here, we invert the POPs on purpose to insert the content of DE in BC and vice versa. At any time we need to switch the content of the two registers, we execute the following call:

CALL EXBCDE

What follows is another particularly interesting and possibly useful infringement. Among the jump instructions (see the table on page 611 in Appendix C) we find the following indirect jump.

JP   (HL)   ; jump to the location specified by register HL

The only usable registers, however, are HL, IX and IY. BC and DE can't be used here. Here is how creativity, along with a masterful use of the Stack, solves the problem:

```
PUSH  BC       ; save the content of register BC on the Stack
RET            ; retrieve it as an address and jump there
```

This pair of instructions inserts two bytes (B and C) in the Stack but then removes them and so this actually doesn't change the contents. Instead of retrieving a return address from the Stack, RET takes the content of BC. This is intentional and the processor jumps to the address specified by BC (which was clearly pre-assigned).

As we can see from these examples, it would be easy to mistake some unorthodox programming techniques for actual errors. A comment written beside the code can help clarify our intentions.

Nested calls and the size of the Stack

Since it is possible to call subprograms from inside a subprogram, we want to know the limits of this, specifically in relation to the Stack area. See this example of a bit of code that has a small number of nested calls:

```
MAIN:       ...              ; this is an infinite loop
            CALL  SUB_1      ; in which we call the subprogram SUB_1
            ...
            JP    MAIN
```

Subprogram SUB_1, in turn, contains two more calls:

```
SUB_1:      ...              ; (subprogram SUB_1)
            CALL  SUB_2      ; call SUB_2
            ...
            CALL  SUB_4      ; call SUB_4
            ...
            RET
```

Subprogram SUB_2 also has a call:

```
SUB_2:      ...                    ; (subprogram SUB_2)
            CALL   SUB_3           ; call SUB_3
            ...
            RET
```

Finally, subprograms SUB_3 and SUB_4 do not call any others:

```
SUB_3:      ...                    ; (subprogram SUB_3)
            ...
            RET

SUB_4:      ...                    ; (subprogram SUB_4 )
            ...
            RET
```

For every call, a return address is added at the top of the Stack so two more bytes occupy the area each time. We need to consider that we can do any conspicuous use of PUSH and POP inside the subprograms. The sequence of calls in this example is graphically highlighted in the figure below:

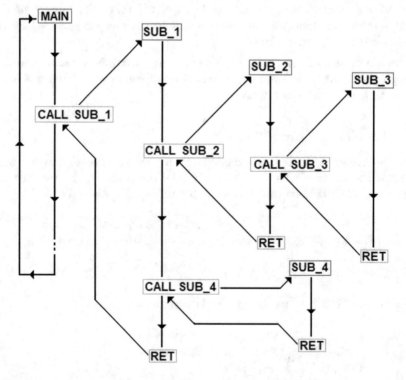

If we follow the sequence of calls, we can easily see that as of the MAIN cycle, the Stack keeps growing and growing from the SUB_1 call to the moment when the SUB_3 subprogram enters.

Then, as the program is executed in the order indicated by the arrows, the subprograms end one by one until they reach the MAIN cycle and we have an empty Stack again.

In a real program, we might have hundreds of nested calls or more depending on the type of application. When we also take PUSH and POP instructions into consideration, the situation can become critical.

Excessive nesting can cause the Stack to overflow from the space the programmer had established for it. To be safe, it is always a good idea to provide more than enough space in the memory for the Stack[18], and try to estimate our needs in terms of nested calls and the use of PUSH and POP instructions.

Conditional CALL and RET instructions

CALL and RET instructions also have conditional form, like conditional jumps (see the aforementioned table on page 611 in Appendix C). The following shows an example of a conditional CALL at the result of an operation, and underneath, the same thing obtained by a normal conditional jump.

Note that the program is easier to read with the conditional CALL.

```
        ...
        DEC   B              ; decrement register B
        CALL  Z,SUBPROG ; if B has been zeroed, call the subprogram
        LD    D,E            ; anyway, continue to execute the program
        ...
```

Although the code from the example below is totally equivalent from a logical point of view, it is written in a contorted way.

```
        ...
        DEC   B              ; decrement register B
        JP    NZ,NOEX        ; if B has not been zeroed, skip the CALL
        CALL SUBPROG         ; otherwise, call the subprogram SUBPROG
NOEX:   LD    D,E            ; anyway, continue to execute the program
        ...
```

---

[18] Note that the most complex microprocessors have protection mechanisms for this.

3.5 Programming examples

The examples in this section allow us to approach assembly programming through simple exercises[19]. The first set of examples is inspired by the emulation of the typical logical functions of digital components.

The interesting point of these first examples is reproducing the behavior of systems with a procedural approach (executing operations one by one in succession), while their real logical processes are normally parallel (executed by logic gates and flip-flops).

This requires us to do a sort of "logical breakdown" of their behavior, which we believe serves a very useful educational purpose. The later examples are inspired by real systems or parts of them.

3.5.1 Emulation of combinational logic

In a combinational network, the outputs are a direct function of the inputs. A procedural system that emulates this behavior has to continually check the inputs and produce corresponding outputs. We will obtain much longer propagation delays compared to real logical networks, but our intent is only to reproduce the logical functioning of the network in order to gain confidence in programming in assembly and its relationship with the system hardware. See Chapter 2).

3.5.1.1 The NOT gate

This is the network we will emulate:

An input X and an output $\overline{X}$ are connected to the NOT component. Similarly, we'll connect an input X and an output $\overline{X}$ to our system (see Section 2.4) through the two ports IA and OA, respectively (see the figure at the right). We'll connect the unused inputs to '0'.

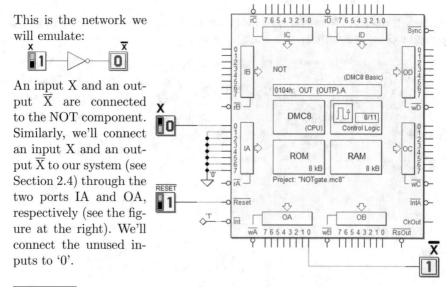

---

[19] Remember that all the programs and networks cited in the text are available on the Deeds website and ready to analyze, simulate and modify.

Cyclically, we should (a) read the input, (b) invert the bits we are interested in, and (c) produce the output. The following is the resulting code:

```
INP         EQU   00h        ; define symbolic names to identify
OUTP        EQU   00h        ; input and output ports (IA and OA)
;
            ORG   0000h      ; at the reset location:
            JP    START      ; define a jump to the program start
            ORG   0100h      ; define the program start address
;
START:      IN    A,(INP)    ; read the input port value (0000.000X)
            XOR   00000001b  ; invert only the bit of interest (bit 0)
            OUT   (OUTP),A   ; send the result to the output port
            JP    START      ; repeat all indefinitely
```

In the first couple of lines, the symbols INP and OUTP define the addresses of ports IA and OA. Remember that the RESET hardware makes the processor execute the instruction found at address 0000h. Here, we have inserted a jump to our program allocated as of 0100h (for the motivations for this intermediate jump, see Section 3.1.5).

For information on programming the microcomputer component, refer to Section 2.4, specifically page 174.

3.5.1.2 The two-input AND gate

The schematic below shows an AND gate with two inputs X, Y and an output that we'll call XY (since it is equal to X · Y).

Like in the last case, we'll connect the two inputs X and Y to our system at input port IA and we'll retrieve output XY from port OA, as seen in the figure on the right. In this network as well, we'll set the unused inputs to '0'.

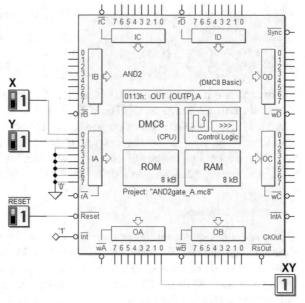

For the NOT, we only read the input, negated it and brought the result to the output. It wasn't necessary to execute a test on the input. Here, we have to evaluate the combination of input X and Y and we can use many different techniques for that.

The first technique

The first technique consists of testing one bit after the other through the BIT instruction. By evaluating the two values, we can choose between two different sequences. The code is as follows (the initial definitions are identical to those of the previous example):

```
INP           EQU   00h          ; define symbolic names to identify
OUTP          EQU   00h          ; input and output ports (IA and OA)
              ORG   0000h
              JP    START
              ORG   0100h
;
START:        IN    A,(INP)      ; read X and Y from the input port
              BIT   0,A          ; check X (on bit 0)
              JP    Z,OUT0       ; if X = 0 then XY = 0, jump to OUT0
              BIT   1,A          ; otherwise, if X = 1, check Y (on bit 1)
              JP    Z,OUT0       ; jump to OUT0 if Y = 0
;
OUT1:         LD    A,00000001b  ; otherwise is X = Y = 1 and hence XY = 1
              JP    OUTPUT
;
OUT0:         LD    A,00000000b  ; set XY = 0
OUTPUT:       OUT   (OUTP),A     ; copy the result to the output port
              JP    START        ; repeat all indefinitely
```

We test to see if input X (bit 0) is at '0'. If so, we don't need to check the other bits and we jump to OUT0. If X = '1', we also have to check input Y (bit 1). If this bit is at '0', we go to OUT0, otherwise we have identified the condition X = Y = '1' so the output of AND should be brought high. The two different configurations of bits are loaded in OUT0 and OUT1, then the OUT instruction makes them exit on port OA.

The second technique

In the second version of the program, let's look at both inputs at the same time through the use of a bitmask (see Section 3.3.2.3). By bitmask, we mean a configuration of zeroes and ones that we use as an operand in a logic operation (AND, OR and XOR). This method is usually faster than testing individual bits because bitmasking generally allows us to evaluate the value of a group of bits in one operation. Below, we see the second version (we have omitted the definitions of the first lines):

```
START:       IN    A,(INP)       ; read X and Y from the input port
             XOR   0FFh          ; invert all the bits we have read from input
             AND   03h           ; thanks to the bitmask, bit 7..2 are zeroed
                                 ; Zero flag is set to '0' if !X = !Y = '0',
             JP    Z,OUT1        ; that is, jump if X = Y = '1'
;
OUT0:        LD    A,00000000b   ; otherwise XY is set to '0' in A
             JP    OUTPUT
;
OUT1:        LD    A,00000001b   ; set XY = 1 in A
OUTPUT:      OUT   (OUTP),A      ; copy the result to the output port
             JP    START         ; repeat all indefinitely
```

In sum, to verify that bits 0 and 1 are simultaneously at '1', we invert all the bits read by the port and force the bits we are not interested in to zero. If bits 0 and 1 were both high before these operations, now all the bits of the result are at zero and the Zero flag allows us to evaluate this condition, so we'll only bring the output high if X = Y = '1'.

The third technique

The third technique is more compact and uses the shift instruction SRL to move one of the two bits in the other's position so that we can directly execute the AND instruction that is native to the processor. Let's remember that it works on 8 bits in parallel but here, we are only interested in the bit in position 0. The code is shown below[20]:

```
START:       IN    A,(INP)       ; copy the value of the port INP to A
             LD    B,A           ; copy the bit pattern to B
             SRL   B             ; shift right that bit pattern
             AND   B             ; execute an AND between the two patterns
             OUT   (OUTP),A      ; copy the resulting bit pattern to OUTP
             JP    START         ; repeat all indefinitely
```

3.5.1.3 The two-input multiplexer

The specifications of the two-channel multiplexer (shown in the figure at the right) are the following:

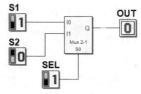

- OUT = S1 if SEL= 0
- OUT = S2 if SEL = 1

---

[20] Here as well, we have omitted the first lines of code for brevity. We will do the same in the future if they are not necessary.

As in the previous cases, we connect SEL, S1 and S2 in order to bits 0, 1 and 2 of input port IA, and OUT to bit 0 of the output port OA.

The first technique

The first technique is to follow the given specifications to analyze the bits one by one and generate the output in line with the component's logical function. We test the value of selection input SEL and then proceed with the value of the selected input in mind. Based on this, we jump to OUT0 or OUT1, depending on the value to generate.

```
START:      IN     A,(INP)       ; read the input lines
            BIT    0,A           ; check SEL
            JP     NZ,S2         ; if SEL = '1' jump to S2
S1:         BIT    1,A           ; else check the line S1
            JP     Z,OUT0        ; if S1 = '0' then output = '0'
            JP     OUT1          ; if S1 = '1' then output = '1'
S2:         BIT    2,A           ; check the line S2 (only if SEL = '1')
            JP     Z,OUT0        ; if S2 = '0' then output = '0'
            JP     OUT1          ; if S2 = '1' then output = '1'
OUT0:       LD     A,00000000b
            JP     OUTPUT
OUT1:       LD     A,00000001b
OUTPUT:     OUT    (OUTP),A
            JP     START
```

The second technique

The second technique is much more compact, quicker and takes advantage of the options the shift instructions offer.

By right shifting the bits read from the input port, we get two simultaneous intermediate results: (a) the value of SEL moves to the Carry flag and (b) the value of input S1 is shifted to position 0 (the same as output OUT).

So if the flag (or SEL) is 0, in position 0 we already have the correct value, which is equal to S1. Otherwise, we right shift A another step to be able to present the value of S2, this time at the output.

```
START:      IN     A,(INP)       ; read port, now S2, S1, SEL are in A
            SRL    A             ; SEL value is moved into Carry flag
            JP     NC,OUTPUT     ; if SEL = '0', S1 is already in position 0
            SRL    A             ; otherwise shift S2 in position 0
OUTPUT:     OUT    (OUTP),A
            JP     START
```

3.5.1.4 The 3-8 decoder

We connect a 3-8 decoder to three inputs (C, B and A) and eight output lines (U7..U0), as shown in the figure at the right. The outside sets the binary number CBA, whose value is between '000_2' and '111_2'.

The device activates the output that corresponds to the CBA number and leaves the other seven inactive, following the truth table on the right.

To emulate the component, we connect lines C, B and A to the microcomputer's input port IA, and we retrieve the other eight outputs from port OA.

See the figure on the right. In this case, we assume that the bits of input port IA that the decoder didn't use (7..3) are connected to the lines used for other undocumented tasks, marked here with hatched lines.

Since we don't know what happens on these lines, we need to make sure that their value (of no interest here) doesn't interfere with evaluating C, B and A in our code.

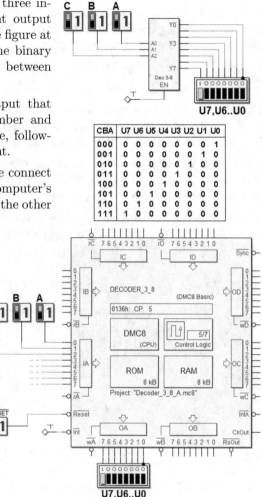

U7,U6..U0

CBA	U7	U6	U5	U4	U3	U2	U1	U0
000	0	0	0	0	0	0	0	1
001	0	0	0	0	0	0	1	0
010	0	0	0	0	0	1	0	0
011	0	0	0	0	1	0	0	0
100	0	0	0	1	0	0	0	0
101	0	0	1	0	0	0	0	0
110	0	1	0	0	0	0	0	0
111	1	0	0	0	0	0	0	0

In the following, we offer different forms the decoder can take if we use a different decoding technique for each.

Linear decoding

Linear decoding consists of comparing the given value with all the possible values, one after the other, in order to identify the corresponding action to take. This has the advantage of simple, repetitive code but the great disadvantage of producing the result in a time linearly dependent on the value assumed by the input. If the value we are looking for is the first one, the result comes immediately; if it is the last, the result is produced after testing all the ones before it. When the number of possible values is very large, this is rather inefficient.

At the beginning of the loop, we read the port and mask the bits that don't interest us, setting them to zero with an AND instruction. What is left in the accumulator is the binary number CBA (at a value from 0 to 7).

```
START:      IN     A,(INP)
            AND    00000111b    ; mask the bits that don't interest us
```

Then the compares are done one by one. The advantage of using the CP instruction is that the accumulator's value doesn't change after the comparison. If the value in A is not '000$_2$', we jump to the next compare and so on.

If the value is '000$_2$' the configuration of the corresponding output is loaded in A. Then we jump to the OUTPUT label where the accumulator is copied on the output port and the loop is repeated.

```
TEST0:      CP     0              ; CBA = '000'?
            JP     NZ,TEST1       ; jump if it is not
            LD     A,00000001b    ; U7..U0 output pattern, for CBA = '000'
            JP     OUTPUT
;
TEST1:      CP     1              ; CBA = '001'?
            JP     NZ,TEST2       ; jump if it is not
            LD     A,00000010b    ; U7..U0 output pattern, for CBA = '001'
            JP     OUTPUT
;
TEST2:      CP     2              ; CBA = '010'?
            JP     NZ,TEST3       ; jump if it is not
            LD     A,00000100b    ; U7..U0 output pattern, for CBA = '010'
            JP     OUTPUT
;
TEST3:      CP     3              ; CBA = '011'?
            ...
```

Clearly, the code is very repetitive, which is why part of has been omitted. The last compare is with number 6 because number 7 is evaluated by exclusion (i.e. if we get a negative result on all the other compares, the value must necessarily be 7).

```
TEST6:      CP     6              ; CBA = '110'?
            JP     NZ,LAST7       ; jump if it is not
            LD     A,01000000b    ; U7..U0 output pattern, for CBA = '110'
            JP     OUTPUT
;
LAST7:      LD     A,10000000b    ; U7..U0 output pattern, for CBA = '111'
;
OUTPUT:     OUT    (OUTP),A
            JP     START
```

Decoding through decision trees

A decision tree allows us to compare bits C, B and A, one by one and produce the corresponding output. This gives us eight possibilities, as shown below:

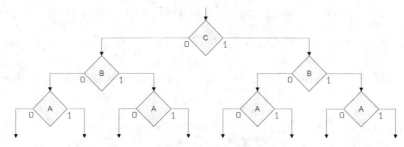

Even though the figure representing the decision tree is simple from the logical point of view, we will have to introduce a large number of jumps when we translate it into assembly code. This means that this method has the disadvantage of requiring a rather involved code.

However, it reaches the goal quickly by doing the same number of tests as there are bits used to code the number in the input (all of the possibilities shown here require 3). With linear decoding, we might use $(N - 1)$ tests to decode a number of inputs no larger than N, whereas here, we need $\lceil \log_2(N) \rceil$ (i.e. the number of bits used to code it).

Clearly, we begin the loop by reading the port. We only need to set the bits that interest us to zero since the tests are executed on individual bits. In the following code, we execute the first series of tests with BIT instructions. As shown in the code itself, a jump occurs if the bit being tested is at '1'.

In other words, we only proceed straight ahead without jumping if C, B and A are all at zero (the furthest left part of the figure above). We get to label A_low, where we set the corresponding output configuration in A ('00000001_2').

```
START:      IN      A,(INP)        ; get C, B and A lines from the input port
            BIT     2,A            ; test bit C (bit 2)
            JP      NZ,C_high      ; jump if C = '1', otherwise go on
C_low:      BIT     1,A            ; test bit B (bit 1)
            JP      NZ,B_high      ; jump if B = '1', otherwise go on
B_low:      BIT     0,A            ; test bit A (bit 0)
            JP      NZ,A_high      ; jump if A = '1', otherwise go on
;
A_low:      LD      A,00000001b    ; U7..U0 output pattern, for CBA = '000'
            JP      OUTPUT
```

With a little patience, we can analyze the flow of the code and distinguish all the possible paths described in the decision tree.

```
A_high:      LD      A,00000010b    ; U7..U0 output pattern, for CBA = '001'
             JP      OUTPUT
;
B_high:      BIT     0,A            ; test bit A (bit 0)
             JP      NZ,A_high_B    ; jump if A = '1', otherwise go on
;
A_low_B:     LD      A,00000100b    ; U7..U0 output pattern, for CBA = '010'
             JP      OUTPUT
A_high_B:    LD      A,00001000b    ; U7..U0 output pattern, for CBA = '011'
             JP      OUTPUT
;
C_high:      BIT     1,A            ; test bit B (bit 1)
             JP      NZ,B_high_C    ; jump if A = '1', otherwise go on
B_low_C:     BIT     0,A            ; test bit A (bit 0)
             JP      NZ,A_high_C    ; jump if A = '1', otherwise go on
;
A_low_C:     LD      A,00010000b    ; U7..U0 output pattern, for CBA = '100'
             JP      OUTPUT
A_high_C:    LD      A,00100000b    ; U7..U0 output pattern, for CBA = '101'
             JP      OUTPUT
;
B_high_C:    BIT     0,A            ; test bit A (bit 0)
             JP      NZ,A_high_BC   ; jump if A = '1', otherwise go on
;
A_low_BC:    LD      A,01000000b    ; U7..U0 output pattern, for CBA = '110'
             JP      OUTPUT
;
A_high_BC:   LD      A,10000000b    ; U7..U0 output pattern, for CBA = '111'
OUTPUT:      OUT     (OUTP),A
             JP      START
```

Every path ends with the corresponding bit pattern to produce in the output and with a jump to the OUTPUT label. Here, where all the paths are together, the accumulator is copied on the output port with the OUT instruction and we go back to the beginning of the loop with an unconditional jump.

Decoding by calculations

Decoding by calculations takes advantage of the fact that we have to activate only one bit at a time in the output. We can set the output pattern to '00000001_2' (corresponding to CBA = '000_2') and then "calculate" it for the other cases. Here, we shift this bit configuration to the left with an SLA instruction, inserted in a loop repeated for the number of times indicated by the CBA bits.

After reading the port, we mask the unwanted bits. The resulting number (CBA) is left in accumulator A (it will be used to count the following loop repetitions, as described above).

```
START:    LD    B,00000001b   ; set the default pattern in B
          IN    A,(INP)       ; read the number CBA to decode
          AND   00000111b     ; mask the bits that don't interest us and,
                              ; at the same time, modify the Zero flag
LOOP:     JP    Z,OUTPUT      ; jump to OUTPUT if CBA is zero
          SLA   B             ; otherwise, shift the content of B to the left
          DEC   A             ; count the loop repetitions modifying the
          JP    LOOP          ; Zero flag, and then repeat from LOOP
;
OUTPUT:   LD    A,B           ; copy the calculated pattern to the port
          OUT   (OUTP),A
          JP    START
```

As explained in the comments, we prepare the bit configuration in B in case CBA = '000'. The cycle that repeats CBA times (counted by the accumulator) starts at the LOOP label. If this is zero, or if it goes to zero after a cycle is repeated, we jump to the OUTPUT label.

If not, we perform a one bit left shift on register B and return to LOOP after decrementing the count. Finally at the OUTPUT label, the "calculated" bit configuration in B is copied to the output port.

Decoding by means of tables

When we decode a number by means of tables, we list all the possible outputs in a table (defined as an array of bytes, using the DB directive). The address of the table is implicitly defined by the programmer who puts the TABLE label on the row of its first location.

```
TABLE:    DB    00000001b     ; CBA = 000
          DB    00000010b     ; CBA = 001
          DB    00000100b     ; CBA = 010
          DB    00001000b     ; CBA = 011
          DB    00010000b     ; CBA = 100
          DB    00100000b     ; CBA = 101
          DB    01000000b     ; CBA = 110
          DB    10000000b     ; CBA = 111
```

After we eliminate the bits that don't interest us, the number read on the input port is used as an index for the array.

```
START:    IN    A,(INP)       ; copy the number to decode to A
          AND   00000111b     ; mask the bits that don't interest us; now,
                              ; A contains the index for reading the table
```

The address of the row of the table is loaded in the 16-bit HL register. We add index A to this address so that now HL targets the location to read (the one that corresponds to the index).

```
LD    HL,TABLE     ; HL = base address of the table
ADD   A,L          ; add index (that is in A) to register L
LD    L,A          ; and update the content of L; if the add
JP    NC,OUTPUT    ; has not generated carry, skip
INC   H            ; the increment of the high part of HL
```

Register H, the high part of HL, is only incremented if the low part has a carry when A is added.

If we use the indirect addressing mode, instruction LD A,(HL) loads the element we're looking for in A, which is then sent to the output.

```
OUTPUT:   LD    A,(HL)        ; now HL points to the desired item
          OUT   (OUTP),A      ; send it to the output (by means of A)
          JP    START
```

3.5.2 Calculating a polynomial

We need to write a program that calculates the following polynomial:

$$OA = 1.5 \cdot IA + 5.0 \cdot IB + 0.75 \cdot IC$$

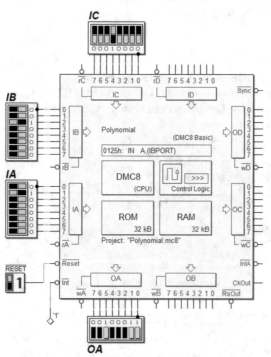

Operands IA, IB and IC are 8-bit binary numbers (with no sign) that are read on the ports with the same name (see the figure to the left). OA, also with 8 bits, is the result generated on the output port with the same name.

The peculiarity of this problem is that we multiply by real numbers, which at first might seem incalculable with a logic arithmetic unit that can only work with integer numbers.

The constants used, however, are able to be broken down so that we can do the calculation using integer number arithmetic.

In this example we are only using 8 bits for the calculations, so they will be approximate. Also, to keep the code legible, no data checks are done before or after. This means that no checks on possible overflows have been inserted into the code nor are any remainders from the integer divisions considered.

In the first lines, the code defines the addresses of the four ports used and the link to the reset location. The program starts with initializing the stack pointer SP since we use CALL, RET, PUSH and POP instructions. Then we enter the MAIN loop where subprogram POLY_ABC is called. CALL instruction is used here so that we can write the function of the calculation elsewhere and, at the same time, make the code more readable.

```
IAport      EQU    00h           ; define addresses of input ports IA, IB, IC
IBport      EQU    01h
ICport      EQU    02h
OAport      EQU    00h           ; define address of output port OA
;

            ORG    0000h
            JP     START         ; link to the reset location
            ORG    0100h
;
START:      LD     SP,0FFFFh     ; initialize the Stack Pointer
MAIN:       CALL   POLY_ABC      ; main loop
            JP     MAIN
```

In programming, a subprogram that executes a specific task, a calculation and returns the result to the calling program is called a "function".

So let's examine the "functions" that carry out the multiplications and are called inside subprogram POLY_ABC.

The following function Mult_1_5 executes the multiplication for 1.5:

```
Mult_1_5:   PUSH   BC
            LD     B,A
            SRA    B             ; B = (A / 2)
            ADD    A,B           ; A = A + B = A + (A / 2)
            POP    BC
            RET
```

As is shown in the code, the function receives the value to multiply in A and copies it in B. Here, it divides it by 2 through a right shift then adds it to the original value. The result appears in A through the following breakdown:

$$1.5 \cdot A = (1 + 0.5) \cdot A = A + A/2$$

Together, PUSH and POP save the content of register B so that the calling program finds it intact after the function is executed.

It might be interesting to think about the magnitude of the approximation of the calculations our function carries out. The SRA instruction moves the bit that exits from the right after the shift into the Carry flag. Since this is a division by 2, that bit is nothing more than the remainder. For simplicity's sake, remainders are ignored by the code, so the result is approximate. For example, if the calculation A = 53 is done by hand, it would give:

$$1.5 \cdot 53 = (1 + 0.5) \cdot 53 = 53 + 53/2 = 53 + 26.5 = 79.5$$

Whereas, our program ignores remainders and gives:

$$1.5 \cdot 53 = (1 + 0.5) \cdot 53 = 53 + \lfloor 53/2 \rfloor = 53 + 26 = 79$$

Clearly, the error in percentages is more relevant in small numbers.

```
Mult_5_0:     PUSH  BC
              LD    B,A
              SLA   B        ; B = (A * 2)
              SLA   B        ; B = (A * 4)
              ADD   A,B      ; A = A + B = A + (A * 4)
              POP   BC
              RET
```

Here above, the function Mult_5_0 was written to execute a multiplication by 5, broken down as follows:

$$5 \cdot A = (4 + 1) \cdot A = (4 \cdot A) + A$$

The function below, Mult_0_75, gives us a multiplication by 0.75 by first multiplying by 1.5 and then dividing the result by 2:

```
Mult_0_75:    CALL  Mult_1_5    ; A = A + A/2
              SRA   A           ; A = (A + A/2) /2 = A/2 + A/4
              RET
```

Finally, let's look at the subprogram that calculates the polynomial, and reads the input ports one after the other. After reading port IA, the function to multiply by 1.5 is called. The partial result is saved in B.

```
POLY_ABC:     IN    A,(IAport)
              CALL  Mult_1_5
              LD    B,A
```

A similar operation is carried out with the IB port, where another partial result is added again to B.

```
              IN    A,(IBport)
              CALL  Mult_5_0
              ADD   A,B
              LD    B,A
```

Then, after also reading IC, the final result is obtained by adding the value from the last calculation (in A) with the one in B.

```
IN     A,(ICport)
CALL   Mult_0_75
ADD    A,B
```

The subprogram finishes by sending the final result, which is in A, directly to output port OA.

```
OUT    (OAport),A
RET
```

3.5.3 Timers

We need to write a program that replicates the functionality of a timer.

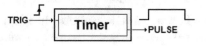

In the idle condition of the system, the PULSE output is kept at zero. The timer waits for the rising edge of input TRIG.

When the rising edge comes, it activates the PULSE output for approximately one second.

After a pulse is generated, the device goes back to the idle condition, and waits for another rising edge on input TRIG.

While PULSE is activated, TRIG is ignored.

As shown in the figure at the right, we use input port IA to receive the TRIG command on bit 0, and output OA to drive line PULSE on bit 0.

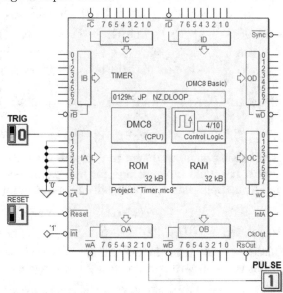

We can achieve this with a main loop and a subprogram. First we define the labels that identify the input and output ports' addresses (TRIGP and PULSEP, respectively). Then we find the usual jump from the Reset location at the start of our program.

```
TRIGP      EQU    00h              ; input port IA (TRIG line)
PULSEP     EQU    00h              ; output port OA (PULSE line)
```

```
            ORG    0000h              ; link to the RESET location
            JP     START
            ORG    0100h
```

Since we are calling a subprogram, before we enter the main cycle, we need to define where the stack area is located by assigning an appropriate RAM address to the Stack Pointer (here FFFFh).

```
START:      LD     SP,0FFFFh    ; initialize the Stack Pointer
```

At the start of the main loop, we set the output port to zero. This operation serves to initialize the PULSE output after Reset but also to set it to zero again when the processor returns to the loop's label Main.

```
MAIN:       LD     A,00h          ; set the output PULSE to zero
            OUT    (PULSEP),A
```

Right afterward, to wait for the rising edge of the clock on line TRIG, we need to make sure that it is at rest so we introduce two consecutive delay loops.

In the first loop (closed on CHECK), we make sure that TRIG goes to '0' (or that it's already at this value). In the second (UPEDGE), we enter if TRIG is at '0' and this time, we wait for the input to go from '0' to '1'.

```
CHECK:      IN     A,(TRIGP)    ; verify if the TRIG line is at '0'
            BIT    0,A          ; repeat the loop if it is not,
            JP     NZ,CHECK     ; otherwise go on
;
UPEDGE:     IN     A,(TRIGP)    ; this time wait for the rising edge
            BIT    0,A
            JP     Z,UPEDGE     ; go on if it is found
```

At the rising edge, we begin to generate the pulse by activating the PULSE output. It should be kept high for approximately one second, so we call the DELAY subprogram that will exit after that amount of time has passed.

```
            LD     A,00000001b    ; set high the output PULSE
            OUT    (PULSEP),A
```

Finally, we return to the beginning of the main loop where, as we know, the output is set to zero and we start all over again.

```
            CALL   DELAY          ; wait for about 1 second
            JP     MAIN           ; repeat backward from MAIN
```

The DELAY subprogram generates a delay by nesting two counter loops (see Section 3.3.5.4). Assuming the system clock is at 10 MHz, the inner loop (Int-Loop label) is calculated to take about 100 mS, by counting the number of repetitions on register DE.

The outer cycle (ExtLoop label) counts on register B and executes the inner loop 10 times, thus reaching the goal of a one-second delay.

```
DELAY:      LD      B,10          ; execute 10 times the inner loop
;
ExtLoop:    LD      DE,41667      ; that delays about 1 million clock cycles
IntLoop:    DEC     DE            ; decrement the counter DE and execute
            LD      A,D           ; the OR between the D and E parts of
            OR      E             ; register DE, to affect the Zero flag
            JP      NZ,IntLoop    ; repeat the inner loop
;
            DEC     B
            JP      NZ,ExtLoop    ; repeat the outer loop
;
            RET                   ; return to the caller
```

3.5.4 Finite state machines

We want to emulate the functionality of the synchronous sequential component shown in the figure at the right.

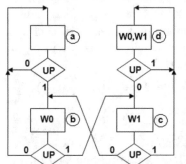

Outputs W1 and W0 can assume binary values from '00' to '11'.

The component is described in finite state machine (FSM) terms by the Algorithmic State Machine (ASM) chart in the figure on the left.

At every rising edge of the clock CK, the component increments or decrements the number W1W0 based on the value of input UP.

The operation is not cyclical in the sense that the increments stop at the top and decrements at the bottom, as shown in the ASM chart.

For example, if both W1W0 outputs are high, we are in state (d). If UP = 0, the FSM will go to the next state (c) on the next rising edge of the clock CK and activate only output W1, otherwise, it won't change states.

Let's connect CK and UP to port IA, and W1 and W0 to port OA, as shown in the schematic at the right.

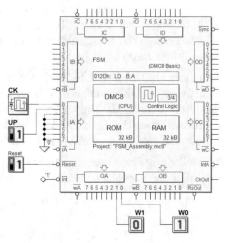

Make sure not to confuse the processor's clock and the clock CK of the component being emulated. From the perspective of the microcomputer, CK is an input like the others, connected to pin 0 of IA. Also, CK should have a far lower frequency than the processor's clock, to allow the program to manage its variations.

The program starts with the usual definitions and the link to the Reset.

```
INP         EQU    00h        ; input port IA (UP, CK lines)
OUTP        EQU    00h        ; output port OA (W1, W0 lines)

            ORG    0000h
            JP     START
            ORG    0100h
```

In the first couple lines we have the initialization of the Stack Pointer and register E. E is loaded with the current contents of the input port (i.e. the values of UP and CK) and will serve to determine the rising edges of CK.

```
START:      LD     SP,0FFFFh   ; initialize the Stack Pointer
            IN     A,(INP)     ; read the input port to set
            LD     E, A        ; the initial value of E
```

For each state of the FSM, there is corresponding piece of code labeled with the name of that state. Entering state (a), for example, means executing the following instructions identified by label STATE_A:

```
STATE_A:    LD     A,00000000b ; set W1 = 0 and W0 = 0 in state (a)
            CALL   CHECK
            JP     Z, STATE_A  ; remain in state (a) if UP = 0
                               ; otherwise go on, in state (b)
```

We will provide a detailed description of the CHECK subprogram a bit further on. As we will see, CHECK takes care of updating W1 and W0 on the output port, among other things. To do this, we have passed the state (a) output values through the accumulator to the subprogram. Based on the ASM chart, these values must both be '0'.

The main task of CHECK, however, is to return to the calling program after identifying the rising edge of CK. CHECK is also a "function", and returns the value of input UP through the Zero flag. This is why we test the flag when we return from the function and, as described in the ASM chart, we decide whether to remain in state (a) or continue to the next state (b).

The following code manages states (b), (c) and (d), in a similar way to state (a), with the necessary variations, following the logic of the ASM chart.

```
STATE_B:    LD     A,00000001b ; set W1 = 0 and W0 = 1 in state (b)
            CALL   CHECK
            JP     Z, STATE_A  ; return in state (a) if UP = 0
                               ; otherwise go on, in state (c)
```

```
STATE_C:     LD    A,00000010b  ; set W1 = 1 and W0 = 0 in state (c)
             CALL  CHECK
             JP    Z, STATE_B   ; return in state (b) if UP = 0
                                ; otherwise go on, in state (d)
STATE_D:     LD    A,00000011b  ; set W1 = 1 and W0 = 1 in state (d)
             CALL  CHECK
             JP    Z, STATE_C   ; return in state (c) if UP = 0
             JP    STATE_D      ; otherwise remain in state (d)
```

As soon as the CHECK function is called, it updates the outputs of the current state (determined by the calling program through the accumulator, as shown before). Then it waits for the next rising edge of CK. The code of the function is as follows:

```
CHECK:      OUT   (OUTP),A   ; generate the state outputs (received in A)
  ;
LOOP:       LD    A,E        ; get the previous reading of the input port
            CPL              ; invert its content and
            LD    B,A        ; transfer it into register B
  ;
            IN    A, (INP)   ; read the input port
            LD    E, A       ; save the reading as next "previous value"
  ;
            AND   B          ; find the rising edge of CK, checking bit 0
            BIT   0,A        ; if no positive edge happened,
            JP    Z,LOOP     ; jump backward, otherwise go on and test
            BIT   1,E        ; the value of UP (modifing the Zero flag)
            RET              ; return to the caller
```

The function manages how long to stay in the current state by entering a loop where it reads the value of line CK on the input port. We exit the loop when the current value of line CK is '1' and the previous value is '0', that is after detecting the rising edge.

Looking more closely at this, we see that the previous reading of the port is in register E. We retrieve it, negate all the bits with a CPL and copy this modified version in B. We are actually only interested in CK, the bit in position 0.

We get the current configuration of the port in A and we save it in E, so that at the next loop repetition, it will have what is then considered the "previous reading".

The AND between A and B and the next check of the bit in position 0 tell us if the rising edge has come. In fact, the result of the AND on that bit will only be '1' if the current CK is at '1' and the previous CK was at '0'. Before exiting and returning to the calling program, the function evaluates the current value of UP, which is on bit 1 of E, and changes the Zero flag accordingly.

3.6 Exercises

The digital content pages of the book on the Deeds simulator website have outlines of the schematics, diagrams and/or programs to complete for each exercise. Those same web pages also have the files for the solutions, so that students can check their work.

3.6.1 Emulation of digital components

Note: the components emulated via software are necessarily much slower than real ones. So, there is no reason to assume that it would be practical to emulate hardware components like logic gates and counters using microprocessors. The goal of these exercises is educational. By introducing typical, recurring problems for most real projects, these exercises teach students to reason like assembly programmers.

1. Write a program in assembly that emulates an 8-input AND.

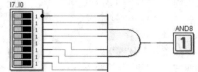

   Connect the inputs I7..I0 to the microcomputer's input port IA and the output AND8 on bit 0 of output port OA.

2. Write a program in assembly that emulates an 8-input OR.

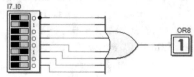

   Connect the inputs I7..I0 to the microcomputer's input port IA and the output OR8 on bit 0 of output port OA.

3. Write a program in assembly that emulates the AND-OR combinational network shown in the figure below:

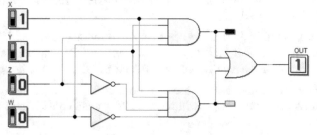

   Connect the four inputs W, Z, Y and X to bits 3, 2, 1 and 0 of the microcomputer's input IA, respectively, as well as the output OUT on bit 0 of output port OA. Remember that the unused lines of port IA could assume random values or be connected to something else and for this reason, appropriate measures need to be taken.

4. Write a program in assembly that emulates an 8-bit serial-parallel (SIPO) shift register like the one in the figure at the right.

At each rising front of CK, the content of the register shifts right by one bit (from Q7 to Q0), while input IN is loaded in Q7.

Connect the inputs of clock CK and line IN to bits 0 and 1 of the microcomputer's input port IA, respectively. Also, retrieve lines Q7..Q0 from output port OA.

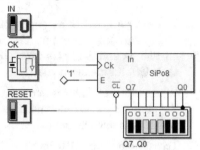

5. Write a program in assembly that emulates a 16-bit serial-parallel shift register (SIPO) like the one in the figure at the right.

At each rising front of CK, the content of the register shifts right by one bit (from Q15 to Q0), while input IN is loaded in Q15.

Connect the inputs of clock CK and the line IN to bit 0 and 1 of the microcomputer's input port IA, respectively. Also, retrieve lines Q15..Q8 from output port OA, and lines Q7..Q0 from output port OB.

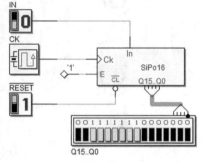

6. Write a program in assembly that emulates an 8-bit binary synchronous up counter that can be pre-loaded. The count takes place on the rising edge of the clock CK and can be seen on outputs Q7..Q0. The counter has a synchronous pre-load command, LOAD, which is active high. See the figure below.

If LOAD is active, number P7..P0 is loaded in the counter on the rising edge of CK.

If LOAD is not active, the counter advances by one unit on every rising edge of CK (the count is cyclical so when it gets to 255, it goes back to 0). Output TC (Terminal Count) is only activated when the number at the output is 255.

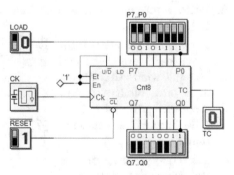

Connect the inputs of the clock CK and the LOAD command to bits 1 and 0 of the microcomputer's input port IA, respectively. Also retrieve lines Q7..Q0 of the counter from output port OA. Pre-load lines P7..P0

can be connected to the IB input port, while line TC can be generated on
bit 0 of output port OB.

7. Write a program in assembly that creates a 12-bit cyclical, synchronous,
binary up/down counter. The figure below shows a circuit version based
on two 4- and 8-bit counters. The count, which can be seen on outputs
Q11..Q0, happens on the rising edge of the clock CK.

When input CLEAR is activated, it makes it possible to synchronously set
the counter to zero on the rising edge of the clock (unlike input $\overline{\text{RESET}}$,
which acts asynchronously).

The synchronous DIR input sets the direction of the count (up if it's high;
down if it's low).

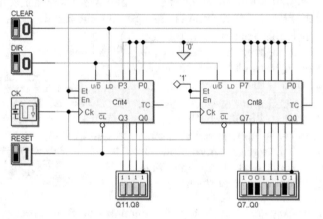

Connect lines CK, DIR and CLEAR in order, to bits 0, 1 and 2 of the
microcomputer's input port IA. Retrieve lines Q7..Q0 of the counter from
output port OB and Q11..Q8 from OA.

8. Write a program in assembly that emulates a 4-bit, synchronous Gray
code up counter.

The figure at the right
shows hardware that could
be built, where the outputs
of a pure binary counter
are converted into Gray
code through a combina-
tional network.

The count takes place on
the rising edge of the clock
CK and is generated on
outputs G3..G0.

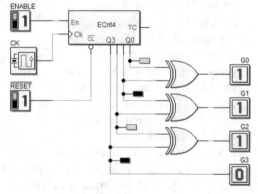

Connect inputs CK and ENABLE to bits 0 and 1 of the microcomputer's input port IA. Also, retrieve outputs G3..G0 from output port OB.

9. Write a program in assembly that emulates the functionality of a 4-digit synchronous, cyclical BCD (Binary Coded Decimal) counter. The counter has a clock CK input. The count advances by one unit at every rising edge of the clock and has an input CLEAR to initialize it. When CLEAR is at '1', the counter is set to zero on the next rising edge of CK.

Connect inputs CK and CLEAR to bits 1 and 0 of the microcomputer's input port IA, respectively. Connect the outputs of the counter to ports OA and OB in the following way:

Port OA		Port OB	
7 6 5 4	3 2 1 0	7 6 5 4	3 2 1 0
Thousands	Hundreds	Tens	Units

3.6.2 Arithmetic functions

1. Write a program in assembly that can calculate the arithmetic average of two 32-bit unsigned integer variables. The subprogram receives the address of the two variables in registers IX and IY, and has to save the result in a third variable, which is addressed by register HL. We need to preserve the content of the registers that are used.

   Note: the variables are handled in the memory according to "little endian" convention (see Section 2.1.9).

2. Write a function in assembly that calculates the arithmetic average of the contents of a table with 256 8-bit unsigned integer numbers whose address is passed by the calling program into register HL. The result is an 8-bit unsigned integer number, which has been rounded down and must be returned to register A. The function has to preserve the content of all of the processor's registers, except for A.

3. Write a subprogram in assembly that multiplies two 8-bit unsigned integer numbers.

   The calling program passes the two operands to multiply in registers C and D, while the (16-bit) result must be returned to register HL. Preserve the content of all of the registers, except HL.

   Suggestion: follow the classic algorithm for long multiplication (as if you were doing it with pen and paper), as in the example on the right (for brevity's sake, 4 bits).

```
0110×
1100=
─────
0000
0000-
0110--
0110---
───────
1001000
```

4. Write a subprogram in assembly that calculates the function:

$$Y = \lceil 127 \cdot sin(\tfrac{X \cdot 360}{256}) \rceil$$

where X is an 8-bit unsigned integer number, passed to the function in A, where the result is returned. Assume you have a table in the ROM, which is identified by the TABLE label and only contains the values of the function's positive half wave, so that it takes advantage of the symmetry of the sinusoidal wave. There is no need to preserve the content of the registers used.

3.6.3 Reusable modules and functions

1. Write a subprogram in assembly that initializes a specific RAM area, defining the start address (in register HL), the number of locations (in register C) and the ASCII code to write in the cells (in register A). There is no need to preserve the content of the registers used.

   We also need to write a test program that initializes a certain number N of RAM locations at reset, according to the table below:

Address	N	Code
C000h	32	'W'
C020h	16	'Y'
C030h	8	'Z'
C038h	8	'K'

   In the end, the test program stops the processor with a HALT.

2. Write a program in assembly that rotates a configuration of '0s' and '1s' on an output port, which were acquired from an input port. What follows are the specifications in detail:

   — The seven least significant bits of the output port are rotated cyclically, while the most significant bit remains set at '0', as shown in the following figure:

   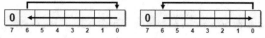

   — The bit configuration to rotate is copied directly by the 7 least significant bits of the input port.
   — Bit 7 of the input port sets the direction of the rotation; if bit $7 = 1$, it rotates right, MSB → LSB).
   — A single rotation step takes about half a second.
   — At the end of each complete 7-bit rotation, the program rereads the input port, updates the bit configuration to rotate and sets the rotation direction.
   — The program is executed at system reset.

3. Use a DMC8 microcomputer (basic version) to create a display composed by a column of 32 LEDs (a sort of "thermometer-style" linear gauge). The system receives a number between 0 and 32 on the Data input port and this number is to be represented by activating a string of LED lights from the bottom.

The following example shows that port IA has the number 22_{10} ('00010110b') so, the 22 bottom-most are lit.

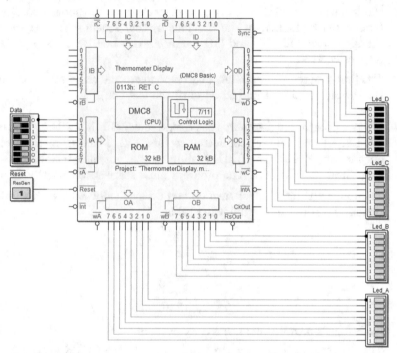

The largest number that can be represented is 32, so if a larger number is presented in the input, it would be reduced ("saturated") to 32. Assume you have a decoding table in the ROM memory that is organized as follows (allocated as of address 0200h):

```
        ORG    0200h
TABLE:  DB     00000000b,00000000b,00000000b,00000000b
        DB     10000000b,00000000b,00000000b,00000000b
        DB     11000000b,00000000b,00000000b,00000000b
        DB     11100000b,00000000b,00000000b,00000000b
        DB     11110000b,00000000b,00000000b,00000000b
        DB     11111000b,00000000b,00000000b,00000000b
        DB     11111100b,00000000b,00000000b,00000000b
        DB     11111110b,00000000b,00000000b,00000000b
        DB     11111111b,00000000b,00000000b,00000000b
        ...    ...omissis ...
```

```
      ...       ...
      DB     11111111b,11111111b,11111111b,11111000b
      DB     11111111b,11111111b,11111111b,11111100b
      DB     11111111b,11111111b,11111111b,11111110b
      DB     11111111b,11111111b,11111111b,11111111b
```

The table has 33 rows, each of which is made up of 4 bytes (32 bits). Each row contains the configuration of LED lights to activate (or leave off) on the 4 output ports. The first byte of each row corresponds to port OA (connected to Led_A, in the previous system schematic), and the other three to ports OB, OC and OD (connected to Led_B, Led_C and Led_D, respectively).

The first 4-byte row of the table corresponds to the value zero in the input; the second to value 1 and so on until the last (32).

4. Write a program in assembly that makes four LED lights flash. The lights, L3, L2, L1 and L0, are connected to bits 3, 2, 1 and 0, respectively, on output port PLED. The LEDs flash in function of the control signals read from input port PCTR. The program must start automatically at reset.

The four LED lights must turn on and off cyclically, each with a different period according to the following table:

LED	Period
L3	800 mS
L2	400 mS
L1	200 mS
L0	100 mS

Each light's on/off cycle has a duty cycle of 50%.

The program cyclically checks bits 7 and 6 of the PCTR port:

— If bit 7 is at '0', the LED lights are off; if it is at '1' they flash.

— If bit 6 is at '1', the flashing times are doubled.

5. The following program generates a square waveform (periodic two-level signal) on bit 0 of output WAVEP.

```
WAVEP     EQU    00h              ; output port OA
          ORG    0000h
          JP     START            ; jump to START at reset
          ORG    0100h
;
START:    LD     A,00000000b      ; set the initial value of A
MAIN:     OUT    (WAVEP),A        ; send the current value of A to the port
          XOR    01h              ; invert bit 0 of A
          JP     MAIN             ; repeat the loop indefinitely
```

Assuming that the processor's clock frequency is $10MHz$:

a) calculate the period of the square wave that is generated

b) change the program so that the period is $4mS$ (tolerance: ±0,5%).

6. Write a program in assembly that retransmits a parallel number received on port IA, in serial format on bit 0 of port OA (output SER in the schematic below).

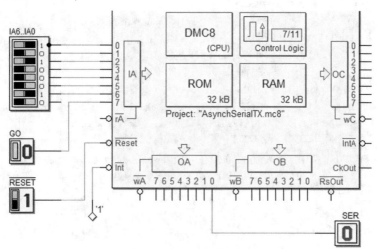

The data received on port IA is coded on 7 bits (lines IA6 .. IA0). Line IA7 is used as a validation signal for the data. A new number is considered received when IA7 moves from '0' a '1' (the GO button in the schematic). While the data is being transmitted, we can ignore line IA7, assuming (for simplicity's sake) that there will be no new data coming until the current data is transmitted.

The figure below shows the format of the serial packet:

Start	D0	D1	D2	D3	D4	D5	D6	D7	Stop

The specification for this serial transmission format requires:

— a start bit at '1'.

— 7 data bits in this order: D0 = IA0, D1 = IA1, .. D6 = IA6.

— parity $D7 = (IA0 \oplus IA1 \oplus IA2 \oplus IA3 \oplus IA4 \oplus IA5 \oplus IA6)$.

— a stop bit at '0'.

— a bit-time of $0.1mS$.

The program must execute at system reset.

Note: the processor works with a clock of 10 MHz. For the solution, the bit-times can be obtained in an approximate way.

3.7 Solutions

3.7.1 Emulation of digital components

1. The configuration of the 8 inputs '11111111' can be easily verified by a compare, using instruction CP. At the end of that operation, if the zero flag is activated, we generate a '1' at the output. The details are explained in the code comments.

```
INP       EQU   00h           ; define symbolic names for the input and
OUTP      EQU   00h           ; output ports (IA and OA, respectively)
          ORG   0000h
          JP    START
          ORG   0100h
START:    IN    A,(INP)       ; read the 8 lines from the input port
          CP    11111111b     ; check if they are all at '1'
          JP    Z,OUT1        ; jump to OUT1 if it is so
OUT0:     LD    A,00000000b   ; otherwise set A to zero and
          JP    OUTPUT        ; jump to OUTPUT to generate '0'
OUT1:     LD    A,00000001b   ; set the bit 0 of A at '1'
OUTPUT:   OUT   (OUTP),A      ; output the value of A to the port
          JP    START         ; repeat from START indefinitely
```

2. This code is almost completely identical to the one in the previous exercise. The only differences are the input test and the condition of the jump right after. Therefore it is only the different parts that are shown below:

```
START:    IN    A,(INP)       ; read the 8 lines from the input port
          OR    A             ; check if they are all at '0'
          JP    NZ,OUT1       ; jump to generate '1' if they're not
```

The input configuration '00000000', is the only one to give a '0' at the output. It is verified by means of an OR between register A and itself. This is more efficient than using CP 00000000b (4 clock cycles rather than 7).

3. Considering the 4 inputs WZYX, the configurations that should generate a '1' at the output are WZYX = '1111' and WZYX = '0011'.

The first solution: First we set the unrelated bits (7, 6, 5 and 4) to zero, then we do a comparison of the two configurations with a CP instruction. If at least one of the two corresponds, we generate a '1'; if not, we generate a '0'. Read the comments in the code.

```
INP       EQU   00h           ; define symbolic names for the input and
OUTP      EQU   00h           ; output ports (IA and OA, respectively)
          ORG   0000h
          JP    START
          ORG   0100h
```

```
START:   IN    A,(INP)      ; read W, Z, Y and X from the port
         AND   00001111b    ; mask the bits that don't interest us
         CP    00000011b    ; verify if W = Z = '0' and Y = X = '1'
         JP    Z,OUT1       ; jump if it is so
         CP    00001111b    ; verify if W = Z = Y = X = '1'
         JP    Z,OUT1       ; jump if it is so
OUT0:    LD    A,00000000b  ; otherwise set A to zero and
         JP    OUTPUT       ; jump to OUTPUT generating '0'
OUT1:    LD    A,00000001b  ; set the bit 0 of A at '1'
OUTPUT:  OUT   (OUTP),A     ; output the value of A to the port
         JP    START        ; repeat from START indefinitely
```

Second solution: The truth table is derived from the combinational network and it has been transcribed in the assembler through the DB directive. The combination of the 4 inputs WZYX is used to address the TABLE table and retrieve the output to generate. The first two lines of code have been omitted because they are identical to the previous solution. This is a very general technique that allows us to create any type of combinational function.

```
START:   IN    A,(INP)      ; read W, Z, Y and X from the port
         AND   00001111b    ; mask the bits that don't interest us
                            ; A = index to use for reading the table
         LD    HL,TABLE     ; set the table base address in HL
         ADD   A,L          ; add the index in A to L
         LD    L,A          ; and update it. If the addition
         JP    NC,OUTPUT    ; does not generate carry, jump
         INC   H            ; otherwise, increment the high part of HL
OUTPUT:  LD    A,(HL)       ; get the table item (pointed by HL) in A
         OUT   (OUTP),A     ; generate its value on the output
         JP    START
;                           ; WZYX
TABLE:   DB    00000000b    ; 0000
         DB    00000000b    ; 0001
         DB    00000000b    ; 0010
         DB    00000001b    ; 0011 → OUT = '1'
         DB    00000000b    ; 0100
         DB    00000000b    ; 0101
         DB    00000000b    ; 0110
         DB    00000000b    ; 0111
         DB    00000000b    ; 1000
         DB    00000000b    ; 1001
         DB    00000000b    ; 1010
         DB    00000000b    ; 1011
         DB    00000000b    ; 1100
         DB    00000000b    ; 1101
         DB    00000000b    ; 1110
         DB    00000001b    ; 1111 → OUT = '1'
```

4. To emulate a shift register, we need to emulate its state, i.e. the set of values assumed by its flip-flops (here, we'll use register B). Below, the input and output port definitions, the link to the reset and the initialization of registers SP and B.

INP	EQU	00h	; input port IA (lines IN and CK)
OUTP	EQU	00h	; output port OA (lines Q7..Q0)
	ORG	0000h	
	JP	0100h	
	ORG	0100h	
	LD	SP,0FFFFh	; initialize the Stack Pointer and B,
	LD	B,00h	; that contains the shift register state

In the code, we use instruction SRL to right shift the 8 bits of register B from Q7 to Q0 at each rising edge of CK. If the current value of input IN differs from zero then we set bit 7 of B (through instruction SET 7,B). Every time B changes, going back to the MAIN label, we update the output port.

MAIN:	LD	A,B	; send the contents of B to the output
	OUT	(OUTP),A	
	CALL	CLOCK	; wait for the rising edge of the clock CK
	SRL	B	; when it arrives, shift right register B and
	BIT	1,A	; check bit 1 of A (the input IN)
	JP	Z,MAIN	; if IN = 0 no action is needed, B7 = 0
	SET	7,B	; otherwise, if IN = 1, set B7 = 1
	JP	MAIN	

The rising edge of CK is identified by the CLOCK subprogram, which is called by the program in the infinite MAIN loop. When the rising edge appears, the control of execution returns to the caller.

The CLOCK subprogram is set over two loops that check the value of line CK (bit 0 of A). The first loop checks that line CK has gone back to zero or if not, waits for it to do so.

The second loop checks that CK goes to 1 and then exits the loop. Also, input bit IN remains available in A when we go back to the calling program.

CLOCK:	IN	A,(INP)	; the clock line must be at '0'
	BIT	0,A	; wait for it to go to zero if it isn't
	JP	NZ,CLOCK	
CK2:	IN	A,(INP)	; wait for the clock rising edge
	BIT	0,A	
	JP	Z,CK2	; when the rising edge arrives, return to
	RET		; the caller (with the IN value in A)

5. The solution to this exercise is very similar to that of the previous one. Here, to memorize the state of the component, we use the 16-bit register HL. Among the definitions, we have: that of the two output ports, the connection to the reset and zeroing HL.

```
INP      EQU   00h        ; input port IA (lines IN and CK)
OUTH     EQU   00h        ; output port OA (lines Q15..Q8)
OUTL     EQU   01h        ; output port OB (lines Q7..Q0)
         ORG   0000h
         JP    0100h
         ORG   0100h

         LD    SP,0FFFFh   ; initialize the Stack Pointer and HL,
         LD    HL,0000h    ; that contains the shift register state
```

At the start of the MAIN loop, we copy the state of the component (HL) to the output ports. The CLOCK subprogram, which is identical to the one in the previous exercise (see on page 290), returns the control of execution to the calling program when the rising edge of CK is detected.

```
MAIN:    LD    A,L         ; display the low part of the outputs
         OUT   (OUTL),A
         LD    A,H         ; and then the high part
         OUT   (OUTH),A

         CALL  CLOCK       ; wait for the rising edge of the clock CK
```

The right shift in the 16-bit HL register has to be broken down into two 8-bit operations, as shown in the following figure. Note that the boxes representing individual bits all contain the name of the corresponding output lines (Q15..Q0).

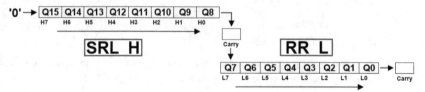

The SRL H instruction right shifts register H, inserts a '0' at the left and saves the outgoing bit in the Carry flag. The next instruction, RR L, gets that bit, inserts it at the left of register L and right shifts all the other bits (read the comments of the code).

```
         SRL   H           ; 16-bit right shift broken down into two
         RR    L           ; 8-bit shifts (of register H, and then L)
         BIT   1,A         ; check the input IN (it is already in A)
         JP    Z,MAIN      ; if IN is not at '0', it is necessary
         SET   7,H         ; to adjust to '1' the bit 7 of register H
         JP    MAIN
```

After SRL H is executed, bit 7 of H is forced to zero, as we have seen. If necessary, we bring it to '1' with the SET 7,H instruction in order to copy the value of input IN. Please refer to the solution of the previous exercise for the CLOCK subprogram code (on page 290).

6. Let's define the microcomputer ports needed for the emulation of an 8-bit synchronous, up binary counter that can be pre-loaded, according to the exercise:

```
INP      EQU   00h        ; port IA: LOAD (bit 1) and CK (bit 0)
PDATA    EQU   01h        ; port IB: Input lines P7..P0
QOUT     EQU   00h        ; port OA: Output lines Q7..Q0
TCOUT    EQU   01h        ; port OB: Output line TC (bit 0)
```

The program starts by zeroing register C, where we intend to memorize the state of the counter. Note that this operation corresponds to the zeroing of the real network's flip-flops after reset.

```
          ORG   0000h
          JP    START
          ORG   0100h
START:    LD    SP,0FFFFh
          LD    C,00h      ; initialize register C (the counter state)
```

In the MAIN loop, we first copy the state on outputs Q7..Q0 (on the QOUT port). Right after, we call the CLOCK subprogram, which waits for the rising edge of CK (we've omitted the code for this function since it is identical to that used for the exercise solved on page 290). Returning from the subprogram, everything that was read by port INP in accumulator A is available. We are interested in the value of line LOAD.

```
MAIN:     LD    A,C
          OUT   (QOUT),A    ; copy the internal state to the outputs
          CALL  CLOCK
```

Back in the calling program, at the rising edge of CK, we check the LOAD command. If it's active we jump to the DOLOAD label.

Otherwise, the count goes up 1 by incrementing C. We use the CP instruction to verify if its content is at the highest value. If (and only if) this is the case, we activate TC. We then return to the start of the loop to update outputs Q7..Q0 and wait for a new rising edge of the clock.

```
          BIT   1,A         ; if LOAD is active, jump to DOLOAD
          JP    NZ,DOLOAD   ; otherwise go on to count up
;
          INC   C           ; increment the internal state value by one
          LD    A,C
          CP    0FFh        ; if C = 255 jump to TC_ON to activate
          JP    Z,TC_ON     ; the output TC, otherwise...
```

```
TC_OFF:   LD    A,00h       ; set TC = '0'
          OUT   (TCOUT),A   ; on the port TCOUT
          JP    MAIN
TC_ON:    LD    A,01h       ; set TC = '1'
          OUT   (TCOUT),A   ; on the port TCOUT
          JP    MAIN
```

We get to the DOLOAD label if the pre-load command is active. In this case, we acquire P7..P0 from the PDATA input port and copy it to the internal state (register C). To be consistent, when we go back to the start of the loop, we update outputs Q7..Q0 as well.

```
DOLOAD:   IN    A,(PDATA)   ; LOAD is active, so read PDATA port
          LD    C,A         ; update the counter state in C
          JP    MAIN
```

7. Following the suggestions in the text of the exercise, let's define the connections to the microcomputer's ports.

```
INP       EQU   00h         ; input port IA:
                            ; CLEAR (bit 2), DIR (bit 1), CK (bit 0)
OUTH      EQU   00h         ; output port OA (lines Q11..Q8)
OUTL      EQU   01h         ; output port OB (lines Q7..Q0)
          ORG   0000h
          JP    START
          ORG   0100h
```

We memorize the 12-bit counter state in the 16-bit register HL.

```
START:    LD    SP,0FFFFh   ; initialize the Stack Pointer
CLEAR:    LD    HL,0000h    ; and the register HL (the counter state)
```

At the MAIN label, the infinite loop starts by writing the counter state on the output ports.

```
MAIN:     LD    A,L         ; copy the counter state to the ports
          OUT   (OUTL),A    ; (low part)
          LD    A,H
          OUT   (OUTH),A    ; (high part)
```

The CLOCK subprogram waits for the rising edge of CK (the subprogram code is not written here; for that, please see the solution of the exercise solved on page 290).

```
          CALL  CLOCK       ; wait for the rising edge of CK
```

The values of inputs CLEAR and DIR are available in A when the control of execution goes back to the calling program.

First, let's evaluate CLEAR[21]. If it is active, we jump to CLEAR and zero the internal state again, then go back to the main loop.

```
        BIT    2,A          ; if the CLEAR input is active,
        JP     NZ,CLEAR     ; jump to CLEAR
```

Depending on the value of DIR, we will take one of two possible routes in the code that handle the direction of the count by incrementing or decrementing the content of register HL.

```
        BIT    1,A          ; if DIR = '1', jump
        JP     NZ,GO_UP
GO_DN:  DEC    HL           ; count down (DIR = '0')
        JP     CUT
GO_UP:  INC    HL           ; count up (DIR = '1')
CUT:    LD     A,H
        AND    00001111b    ; make the count both 12-bit and cyclic
        LD     H,A
        JP     MAIN
```

The two routes come together at the CUT label where the 4 most significant bits of the count are zeroed to make it cyclical at 12 bits. We then go back and repeat the main loop.

8. As explained in the text, we connect CK to bit 0 and ENABLE to bit 1 of the microcomputer's port IA. We retrieve outputs G3..G0 from bits 3, 2, 1 and 0 of output port OB.

After we initialize the Stack Pointer, as usual, we set register C, which contains the state of the counter, to zero.

```
INP     EQU    00h          ; port IA: ENABLE (bit 1), CK (bit 0)
OUTG    EQU    00h          ; port OA: Outputs G3..G0 (Gray code)
        ORG    0000h
        JP     START
        ORG    0100h
START:  LD     SP,0FFFFh
        LD     C,00h        ; set the counter state to zero
```

In the next MAIN loop, we first call the OUTPUT subprogram, which converts the state of the counter in Gray code and copies it to the output port, as we will soon see. On the first loop execution, obviously the output is zeroed, emulating the component's behavior at reset.

---

[21] Note that in the universal counters available in the Deeds component library, the load command LD has priority over the others. Since the input CLEAR controls the LD inputs of the counters, in assembly code CLEAR must be evaluated before the DIR input to comply with the priority schedule.

The CLOCK subprogram waits for the rising edge of CK (the code for this is identical to the one proposed on page 290). When we return from the function, the value of input ENABLE is available in register A.

If ENABLE is at '0', the count is disabled so we jump back to the NOCNT label and wait for the next rising edge of CK without changing the state of the counter.

```
MAIN:    CALL  OUTPUT
NOCNT:   CALL  CLOCK
         BIT   1,A          ; if ENABLE = '0', don't change the state
         JP    Z,NOCNT      ; otherwise go on and increment its value
```

If ENABLE = '1', we go on to increment the state of the counter. The calculation is executed in the accumulator since we want to limit the count to the 4 least significant bits, as explained in the code comments.

```
         LD    A,C          ; read the state from register C
         INC   A            ; increment the state value,
         AND   00001111b    ; but limit the count to 4 bits only
         LD    C,A          ; update the state in C
         JP    MAIN
```

Now, let's look at the details of the OUTPUT subprogram. As can be seen in the circuit in the text of the exercise, a number in Gray code G3G2G1G0 is obtained by the corresponding binary number Q3Q2Q1Q0 through the following transformation:

$$G3 = Q3; \quad G2 = Q2 \text{ exor } Q3; \quad G1 = Q1 \text{ exor } Q2; \quad G0 = Q0 \text{ exor } Q1;$$

The code achieves this by right shifting the state of the counter after copying it in A and executing an XOR with the unshifted state.

```
OUTPUT:  LD    A,C          ; copy the state to A
         SRL   A            ; and shift it to the right
         XOR   C            ; convert the state in Gray code
         OUT   (OUTG),A     ; copy the state to the output port OUTG
         RET
```

9. Following the requirements of the exercise, we connect CK to bit 0 and CLEAR to bit 1 of IA, then we retrieve the 16 outputs of the counter from ports OA and OB to visualize the number on the BCD displays, as in the figure below:

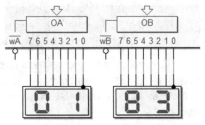

The first part of the code shows the necessary definitions of the ports, the link to the reset and the initialization of the Stack Pointer.

```
INP      EQU   00h        ; port IA: CLEAR (bit 1), CK (bit 0)
OUTH     EQU   00h        ; port OA (thousands and hundreds)
OUTL     EQU   01h        ; port OB (tens and units)
         ORG   0000h
         JP    START
         ORG   0100h
START:   LD    SP,0FFFFh   ; initialize the Stack Pointer
```

Before entering the main loop, we zero registers D, E, B and C (which are used to memorize the state of the count) at label DOCLR. Read the comments in the code.

```
DOCLR:   LD    D, 0       ; D = thousands
         LD    E, 0       ; E = hundreds
         LD    B, 0       ; B = tens
         LD    C, 0       ; C = units
```

The main loop starts with a call to the OUTPUT subprogram, which copies the state of the count from registers D, E, B and C to the output ports (we will look at the details of the subprogram later).

```
MAIN:    CALL  OUTPUT     ; update the BCD outputs
         CALL  CLOCK      ; wait for the rising edge of CK
         BIT   1,A        ; check the CLEAR command
         JP    NZ,DOCLR   ; clear the BCD outputs, if requested
```

As in an earlier exercise (see page 290), the CLOCK subprogram waits for the rising edge of CK and the value of input CLEAR remains available in register A when we go back to the calling program.

If CLEAR orders the counter to go to zero, we jump back to DOCLR and zero the state in registers D, E, B and C, and then repeat the main loop.

If not, we go on with the BCD count. Let's start with the decimal number of the units. As we can see in the following listing, we increment C, the register storing the units, making sure it does not go higher than 9.

```
ONE:     INC   C          ; increment the units
         LD    A,C
         CP    10         ; check if units go past number 9
         JP    NZ,MAIN    ; if not, go back to MAIN
         LD    C,0        ; if yes, zero the units and pass to tens
```

If it doesn't go past 9, there is no need to increment the most significant numbers of the count so this incrementing step is finished and we go back to the start of the main loop. Otherwise, we must set the units to zero and increment the tens.

The next steps on incrementing the other numbers, (tens, hundreds, thousands) are basically identical to that of units. After handling the thousands, we go back to the main loop.

```
TEN:     INC   B              ; increment the tens
         LD    A,B
         CP    10             ; check if tens go past number 9
         JP    NZ,MAIN        ; if not, go back to MAIN
         LD    B,0            ; if yes, zero the tens and pass to hundreds

HUNDR:   INC   E              ; increment the hundreds
         LD    A,E
         CP    10             ; check if hundreds go past number 9
         JP    NZ,MAIN        ; if not, go back to MAIN
         LD    E,0            ; if yes, zero E and pass to thousands

THOUS:   INC   D              ; increment the thousands
         LD    A,D
         CP    10             ; check if thousands go past number 9
         JP    NZ,MAIN        ; if not, go back to MAIN
         LD    D,0            ; if yes, zero the thousands
         JP    MAIN           ; go back to MAIN, ignore last carry
```

The subprogram OUTPUT has the task of compacting the four BCD numbers into two bytes, in order to be able to visualize them on the two output ports OA and OB. For example, the 4 bits of the tens digit are moved 4 bits over to the left so they can be inserted next to the units digit. Then, we copy the byte on the corresponding OA port.

```
OUTPUT:  LD    A,B            ; shift to the left the tens of 4 positions
         SLA   A              ; inserting 4 zeros from the right
         SLA   A
         SLA   A
         SLA   A
         OR    C              ; place the two BCD numbers side by side
         OUT   (OUTL),A       ; display tens and units
         ;
         LD    A,D            ; shift to the left the thousands of 4
         SLA   A              ; positions inserting 4 zeros from the right
         SLA   A
         SLA   A
         SLA   A
         OR    E              ; place the two BCD numbers side by side
         OUT   (OUTH),A       ; display thousands and hundreds
         RET
```

Note: the code of the CLOCK function has been omitted because it is identical to that of the previously cited example (see page 290).

3.7.2 Arithmetic functions

1. The subprogram that executes the arithmetic average of two unsigned 32-bit integer variables receives their addresses on registers IX and IY. The variables are memorized according to "little endian" convention.

0Bh
45h
FAh
56h

They are actually organized into four consecutive bytes as in the figure on the left, which shows a generic 32-bit integer variable that contains the number 56FA450Bh (broken down into the 0Bh, 45h, FAh, 56h bytes, as of the least significant one).

When we enter the subprogram, we save all the registers used except for DE and HL. As we will see, the paired register DE is not used, while the content of HL is restored by the algorithm itself.

```
MEAN32:   PUSH   AF
          PUSH   BC
          PUSH   IX
          PUSH   IY
```

The memory variables can be pointed through the indirect addressing mode, using IX and IY for the operands, and HL for the result. Clearly, we assume that IX, IY and HL have been defined by the calling program before it launches the subprogram.

Let's do the first of the partial sums, by adding the least significant bytes together.

```
          LD     A,(IX)       ; execute the first partial sum
          ADD    A,(IY)       ; of the least significant bytes
          LD     (HL),A       ; save the partial result
```

The carry is saved in the Carry flag. The other bytes are added in one loop that repeats three times. Every time the loop repeats, the addresses in IX, IY and HL are first incremented to target the next bytes. Then, the sum is executed, as with the first byte except that instruction ADC keeps the previous carry into account (read the comments in the code).

```
          LD     B,3          ; initialize the counter of the partial sums
ADDB:     INC    IX           ; update all the addresses
          INC    IY
          INC    HL
          LD     A,(IX)       ; execute the partial sum, taking in
          ADC    A,(IY)       ; count also the previous carry
          LD     (HL),A       ; save the partial result
          DEC    B            ; count the number of repetitions
          JP     NZ,ADDB      ; jump backward if not ended
```

After the 32-bit sum of the two operands is done, we must divide the result by two. We can use the shift and rotate instructions, by employing the indirect addressing mode through HL.

First, let's use instruction SRA to right shift the most significant byte, which register HL is still pointing to. The bit that exits at the right is saved in the carry flag. From here, the next rotation instruction RR retrieves the bit to insert at the left in the rotation of the second byte. The operation repeats until all the bytes are right shifted. Register HL is decremented each time and in the end, it goes back to pointing to the least significant byte of the result, as in the beginning.

```
SRA   (HL)              ; 32-bit division by 2
DEC   HL
RR    (HL)
DEC   HL
RR    (HL)
DEC   HL
RR    (HL)
```

Finally, the subprogram restores the previous content of all the registers used and then goes back to the calling program.

```
POP   IY                ; restore the registers' previous content
POP   IX
POP   BC
POP   AF
RET
```

Below, we have a simple test program to check the subprogram we've worked with. It defines VAR_RES variable that takes in the result.

```
VAR_RES  EQU   8000h          ; 32-bit result (4 bytes)
         ORG   0000h
         JP    START
         ORG   0100h
```

It also uses two 32-bit constants as operands, which are defined by the DB directive. The constants are equal so that the average can be immediately verifiable since it will be the same as the initial values.

```
TEST_A:  DB    0Bh            ; 08123F0Bh
         DB    3Fh
         DB    12h
         DB    08h
TEST_B:  DB    0Bh            ; 08123F0Bh
         DB    3Fh
         DB    12h
         DB    08h
```

After reset, we jump to the START label. The Stack Pointer is initialized and then the program loads the addresses of the two constants in IX and IY, and the address of the variable VAR_RES in HL.

```
START:   LD    SP, 0FFFFh      ; initialize the Stack Pointer
         LD    IX,TEST_A       ; address of operand A
         LD    IY,TEST_B       ; address of operand B
         LD    HL,VAR_RES      ; address of the result
         CALL  MEAN32
         HALT
```

As we can see, after calling the MEAN32 subprogram, the HALT instruction stops the processor so that we can read the result in the memory, by using the Deeds-McE debugger.

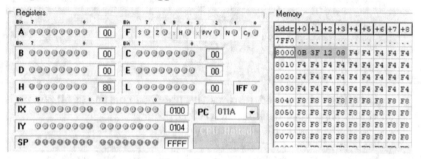

After the program is executed, we read a number equal to the constants defined by the DB directives in the four bytes of RAM as of address 8000h, as shown in the figure.

2. At the start of the subprogram, we save the registers in use on the Stack.

```
MEAN256:  PUSH  BC
          PUSH  HL
          PUSH  AF
```

To calculate the average of the 256 values in the table, we first need to calculate the sum of the entire table. Each value is at most 255 and in the worst case, the total could be as much as $255 \cdot 256 = 65280$. To express 65280 in binary code, we need $\lceil log_2(65280) \rceil = \lceil 15.9 \rceil = 16$ bits.

Since each addend is 8 bits, however, it is quicker to use a little trick: using register A to accumulate only the 8 least significant bits of the sum. When we add the elements of the table one after the other in A, we increment another register (B for example) each time we generate a carry. In the end, this other register contains the 8 most significant bits of the total 16-bit sum.

Therefore, before we get into the add loop, we set B to zero and copy the first of the values to add in A. We use the indirect addressing mode to retrieve it (remember that the specifications require the calling program to pass the address of the table through register HL).

```
         LD    B,00            ; initialize B, the high byte of the sum
         LD    A,(HL)          ; get the first value to add
```

We repeat the following loop 255 times seeing that we have already loaded the first element in A. For every repetition, we increment HL to point to the next element in the table and we add it to A. Register B is only incremented (through a conditional jump) if a carry is generated.

```
          LD    C,255        ; initialize the loop counter (255 sums)
LOOP:     INC   HL           ; point to the next item of the table
          ADD   A,(HL)       ; add that item
          JP    NC, NOCY     ; jump if there isn't carry, otherwise
          INC   B            ; increment the high part of the sum
NOCY:     DEC   C            ; decrement the counter
          JP    NZ,LOOP      ; repeat until ended
```

When the sum is completed, we then divide the result by 256. Note that dividing an integer by 256 is nothing other than right shifting the number by 8 positions, so we should right shift register B 8 times and insert each exiting bit in A.

At the end of the shifting process, this means that all the bits from B are in A. In our specific case, we only need to take the content of B, which is perfectly rounded down as required, and copy it directly to A.

Finally, our function restores the previous content of the registers and goes back to the calling program, as per specifications.

Since it is impossible to only save the flags in the Stack, we also saved A, as usual. So, to allow the result of the function to return to the accumulator, B is only copied in A after retrieving the previous content of the flags, then also refreshing the content of the other registers.

```
          POP   AF           ; restore the Flags
          LD    A,B          ; copy the result to A
          POP   HL           ; restore the previous content
          POP   BC           ; of registers HL and BC
          RET
```

The following is a possible test program for our function:

```
OUTP      EQU   00h          ; output port OA
          ORG   0000h
          JP    START
          ORG   0100h
START:    LD    SP, 0FFFFh   ; initialize the Stack Pointer
          LD    HL,TABLE
          CALL  MEAN256       ; execute the function
          OUT   (OUTP),A     ; copy the result to the output port OA
          HALT
```

TABLE, the address of the table of values, is loaded in register HL before calling function MEAN256. At the end, the result of the function can be seen on the microcomputer's output port OA.

The test table defines 256 values that were chosen to obtain an easily verifiable result. There are 16 identical lines, each defining 16 constants with the DB directive (the average of the values in a line is 8).

```
TABLE:    DB      1,2,3,4,5,6,7,8,9,10,11,12,13,14,15,8 ; 16 rows as this one
          DB      1,2,3,4,5,6,7,8,9,10,11,12,13,14,15,8
          ...
          DB      1,2,3,4,5,6,7,8,9,10,11,12,13,14,15,8
          DB      1,2,3,4,5,6,7,8,9,10,11,12,13,14,15,8
```

3. Let's write the multiplication function by following the classic long multiplication algorithm, as suggested. Here, the operation is broken down into a loop of multiple partial sums. We accumulate the sums in register HL so that we can find the result at the end of the subprogram, as required by the specifications.

At the start of the code, we save all the registers that are used, except HL, which is used to provide the result.

```
MUL8BIT:  PUSH  AF              ; save the registers in use on the Stack
          PUSH  BC
          PUSH  DE
```

We set B to zero (the high part of BC), since one of the 8-bit operands is passed to register C. If we take advantage of the processor's 16-bit operations, we can consider register BC as a multiplicand and thus, an addend for partial sums.

Register E is used as a counter for partial sums (it is set to 8 because of the 8-bit multiplication). When it reaches zero, the final result is ready in HL, which for now is set to zero.

```
          LD    B,00h           ; zero B (BC contains the multiplicand)
          LD    E,8             ; count the 8 partial sums in E
          LD    HL,0000h        ; prepare HL for the partial sums
```

In the main LOOP loop, a bit from the multiplier in D, starting from the least significant bit, is evaluated at each repetition. The bits of the multiplier are inserted in the Carry flag one by one thanks to an SRL instruction.

Then, the flag is used to decide whether to execute the current partial sum. A '0' in the i-th bit in the multiplier leads to a partial sum with zero (which is not executed as we jump to NOSUM). A '1', however, executes the sum.

```
LOOP:      SRL    D                ; move the i-th multiplier bit to the Carry
           JP     NC,NOSUM         ; if = '1' execute the 16-bit partial sum
           ADD    HL,BC
```

Remember that register BC initially contains the multiplicand, but BC is left shifted with each repetition, which prepares it for the next partial sum (whether it is executed or not).

```
NOSUM:     SLA    C                ; prepare the multiplicand for the sum,
           RL     B                ; by means of a 16-bit left shift
```

The loop is repeated until register E (the loop counter) reaches zero. At that point, the result is ready in HL.

```
           DEC    E                ; decrement the loop counter
           JP     NZ,LOOP          ; repeat until no more sums to execute
```

Finally, the registers saved in the Stack are restored and, by executing the RET instruction, the CPU control returns to the calling program.

```
END:       POP    DE               ; (if finished, the result is in HL)
           POP    BC               ; restore the registers
           POP    AF
           RET                     ; end return to the caller
```

The subprogram can be tested with the following program, which reads the two operands on the microcomputer's ports IA and IB, defined here as INP_A and INP_B, respectively. This produces the result of the multiplication on output ports OA (high part) and OB (low part). Here are the definitions of the ports and the usual link to the reset:

```
INP_A      EQU    00h              ; port IA: operand A (the multiplicand)
INP_B      EQU    01h              ; port IB: operand B (the multiplier)
RIS_L      EQU    01h              ; port OB: result (low part Q7..Q0)
RIS_H      EQU    00h              ; port OA: result (high part Q15..Q8)
           ORG    0000h
           JP     START
           ORG    0100h
```

The subprogram MUL8BIT is called inside of an infinite loop. Before the call, we retrieve the two operands from the ports of the microcomputer and we copy them in C and D, as required by the function.

```
START:     LD     SP, 0FFFFh       ; initialize the Stack Pointer
MAIN:      IN     A,(INP_A)        ; transfer the multiplicand into register C
           LD     C,A
           IN     A,(INP_B)        ; transfer the multiplier into register D
           LD     D,A
           ;
           CALL   MUL8BIT          ; call the function
```

Since the function returns the result of the multiplication in HL, we copy the content of these two registers to the output ports.

```
LD      A,L             ; copy the result to the output ports
OUT     (RIS_L),A       ; low part (L)
LD      A,H
OUT     (RIS_H),A       ; high part (H)
JP      MAIN
```

4. The given function can be pre-calculated by the programmer and memorized in a table. For each value of the input, the subprogram consults the table and produces the result.

Since trigonometric functions are symmetric, we can save on the size of the table by memorizing only the 128 positive values of the function. The table is used directly in the interval of X between 0 and 127.

For all those after, up to 255, we simply need to subtract 128 from X and read the table, as in the previous case, then invert the sign of the value.

The table will be defined as follows (only a part is shown):

```
TABLE:  DB      000         ; x = 0       (0 degrees)
        DB      003         ; x = 1
        DB      006         ; x = 2
        DB      009         : x = 3
        ...
        DB      088         ; x = 31
        DB      090         ; x = 32      (45 degrees)
        DB      092         : x = 33
        ...
        DB      127         ; x = 63
        DB      127         ; x = 64      (90 degrees)
        DB      127         : x = 65
        ...
        DB      092         ; x = 95
        DB      090         ; x = 96      (135 degrees)
        DB      088         : x = 97
        ...
        DB      006         ; x = 126
        DB      003         ; x = 127
        DB      000         ; x = 128     (180 degrees, not used)
```

First we initialize HL with the address of the table, then we save the value X in B so that we can later evaluate whether that value was between 0 and 127 (bit 7 = '0') or between 128 and 255 (bit 7 = '1'). So we mask bit 7 so that we always have an index between 0 and 127.

```
SIN127:     LD      HL,TABLE      ; HL = table's base address
            LD      B,A           ; save the bit 7 in B, before zeroing it
            AND     01111111b     ; (the table has the first half-cycle only)
```

Now we need to add the value of the index to the address in HL. We add A to the least significant byte in HL and if the operation produces a carry, we increment the most significant byte. Now that the target address is computed, we copy the value that interests us in A.

	ADD	A,L	; add A to the low part of the base
	LD	L,A	; address
	JP	NC,NOCY	; if the sum generates a carry,
	INC	H	; increment the high part H of the
NOCY:	LD	A,(HL)	; address; read the value from the table

Finally, we check the original bit 7 from X and if it is zero, we exit (note the use of the conditional RET instruction). Otherwise, we invert the sign of the value in A and we leave the function.

BIT	7,B	; check if X is in 0..127; if yes,
RET	Z	; exit the function with the value in A
NEG		; otherwise invert its sign and exit the
RET		; function with the negative value in A

Now, let's write a test program for the function that would be appropriate for a system like the one in the figure below. A counter allows us to cyclically send all the possible values of X to the microcomputer. We also use a DAC to graphically represent the function values over time.

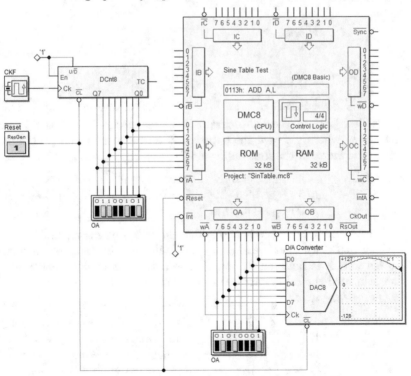

In the test program, we define the input and output ports, as well as the link to the reset. Then we read the values from the input port cyclically and produce the result of the function on the output port.

```
INP       EQU    00h          ; define symbolic names for input and
OUTP      EQU    00h          ; output ports (IA and OA, respectively)
          ORG    0000h
          JP     START
          ORG    0100h
;
START:    LD     SP,0FFFFh    ; initialize the Stack Pointer
MAIN:     IN     A,(INP)      ; read the 'X' variable from IA
          CALL   SIN127       ; call the subprogram
          OUT    (OUTP), A    ; display the function result
          JP     MAIN
```

We can change the solution strategy so that we can optimize the space used in the ROM. We'll take better advantage of the symmetry of the sinusoid by cutting the table's size by half. It now shows only a quarter of the wave. Now it will be the code that reads the values in the right order.

```
TABLE:    DB     000          ; x = 0      (0 degrees)
          DB     003          ; x = 1
          DB     006          : x = 2
          ...
          DB     088          ; x = 31
          DB     090          ; x = 32     (45 degrees)
          DB     092          : x = 33
          ...
          DB     127          ; x = 63
          DB     127          ; x = 64     (90 degrees)
```

The first part of the subprogram is the same as that of the previous strategy, where we save the value of X on register B and preliminarily take the index in the first half cycle (between 0 and 127), masking bit 7.

```
SIN127:   LD     HL,TABLE     ; HL = table's base address
          LD     B,A          ; save the bit 7 in B, before zeroing it
          AND    01111111b
```

We took the previous code and added the following instruction sequence, which checks whether the index is <63. If it is, the index is left unchanged, otherwise, it is between 64 and 127, so it is subtracted from 128 so that we can read the table mirrored (for instance, 127 becomes 1):

```
          BIT    6,A          ; the reduced index (0..127) is < 63?
          JP     Z,CONT       ; jump if it is, otherwise take advantage
          LD     C,A          ; of the symmetry of the half-wave and
          LD     A,128        ; calculate the new index to use:
          SUB    C            ; A = new index = 128 - old index
```

The remaining part of the code (where we read the table and invert the sign of the result if the index is greater than 127) is identical to that of the previous strategy.

```
CONT:     ADD   A,L         ; add A to the low part of the base
          LD    L,A         ; address
          JP    NC,NOCY     ; if the sum generates a carry,
          INC   H           ; increment the high part H of the
NOCY:     LD    A,(HL)      ; address; read the value from the table
          BIT   7,B         ; check if X is in 0..127; if yes,
          RET   Z           ; exit the function with the value in A
          NEG               ; otherwise invert its sign and exit the
          RET               ; function with the negative value in A
```

3.7.3 Reusable modules and functions

1. The WRAM subprogram initializes an area of the RAM that is defined by the starting address (passed in register HL) and by its size (in register C). It receives the ASCII code to use to write in the cells in A.

```
WRAM:     LD    (HL),A      ; write the RAM location pointed by HL
          INC   HL          ; point to the next location
          DEC   C           ; count the loop repetitions
          JP    NZ,WRAM     ; repeat loop until C goes to zero
          RET               ; return to the calling program
```

At each repetition of the loop, the content of A is copied to the location pointed by HL. We increment HL to point to the next location and we decrement C, the location (and loop) counter. We go back to the calling program when the requested number of locations has been reached.

Note that assignations apparently haven't been made before entering the loop. This is simply because this is not the job of the subprogram, but rather the calling program, which has to define the contents of HL, C and A, as required by the specifications.

The following shows what the test program looks like:

```
          ORG   0000h
          JP    START
          ORG   0100h
START:    LD    SP,0FFFFh   ; initialize the Stack Pointer
          LD    HL,0C000h   ; define the RAM area start address
          LD    C,32        ; number of location to write
          LD    A,"W"       ; with this ASCII code
          CALL  WRAM        ; call the subprogram
```

As we can see, before we call WRAM, we load HL, C and A with the values established in the specifications. In other words, we "pass the parameters" from the caller to the subprogram by means of the registers.

The program is very repetitive since only the values of the parameters change. Finally, everything finishes with a HALT, as required.

```
         LD     HL,0C020h
         LD     C,16
         LD     A,"Y"
         CALL   WRAM
;
         LD     HL,0C030h
         LD     C,8
         LD     A,"Z"
         CALL   WRAM
;
         LD     HL,0C038h
         LD     C,8
         LD     A,"K"
         CALL   WRAM
;
         HALT
```

To see the write operations in the RAM inside the Deeds-McE debugger, we suggest setting the memory visualization frame as shown in the figure below. The example shows the content of the RAM memory after the program is executed.

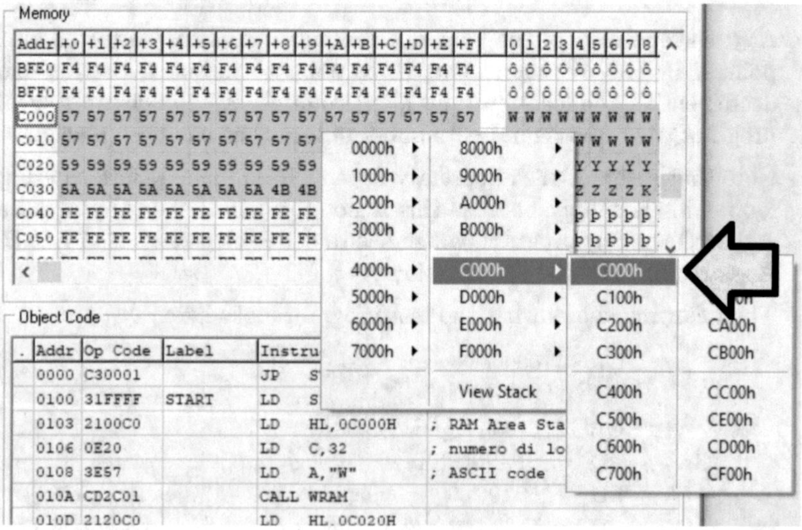

To see other locations, use the context menu to choose the address, as shown in the figure.

2. At reset, the program jumps to START, where it initializes the Stack Pointer and then enters an infinite loop where it executes the rotation.

```
INP       EQU    00h         ; define symbolic names for input and
OUTP      EQU    00h         ; output ports (IA and OA, respectively)
          ORG    0000h
          JP     START
          ORG    0100h
START:    LD     SP,0FFFFh    ; initialize the Stack Pointer
```

At the MAIN label, the start of the loop, we assign register B the number of shifts to execute to get a complete rotation. Then we acquire the bit configuration we will rotate from the input port. We will rotate this configuration to the left or right depending on the value of bit 7.

```
MAIN:     LD     B,7         ; initialize the rotation counter
          IN     A,(INP)     ; read the input port
          BIT    7,A         ; check the bit of the rotation verse
          JP     NZ,RIGHT    ; '1' = right, '0' = left
```

For a left rotation, we zero bit 7 in A (it should always be displayed at '0') and then we copy the register on the output port.

```
LEFT:     AND    01111111b   ; leave on only the 7 bits of interest, and
          OUT    (OUTP),A    ; copy the bit pattern to the output port
```

Let's call the DELAY subprogram (that we will explain further ahead), so that there is a half a second pause between two writes.

```
          CALL   DELAY       ; execute a delay loop of about 0.5S
```

The figure at the right describes the content of A. The bits are indicated in letters from 'a' to 'g'.

The shift instruction SLA makes it possible to left shift all the bits in A by inserting a '0' at the right. The figure on the left shows the content of A before execution and after.

```
          SLA    A           ; shift left the byte
```

However, to execute a rotation on 7 the bits required by the specifications, the bit in position 0 has to assume the value of the previous bit in position 6 ('a'). Let's look at the remaining part of the code:

```
          BIT    7,A         ; check if bit 7 is at '0' and jump if
          JP     Z,NOSETL    ; it is so, because bit 0 is already at '0'
          SET    0,A         ; otherwise set bit 0 at '1'
NOSETL:   DEC    B           ; count the repetitions
          JP     NZ,LEFT     ; repeat from LEFT seven times,
          JP     MAIN        ; then return to the label MAIN
```

After the execution of instruction SLA, we find bit 'a' shifted into position 7 and we check it with the BIT instruction.

If bit 'a' is at '1', we force bit 0 to that value with SET, otherwise we leave it at '0'.

The figure at the right shows register A before the execution of SET and after.

We can ignore the fact that 'a' is in position 7, since it will be zeroed when it goes back to the LEFT label, before the OUTP port is written. If seven rotations have been executed, we go back to the beginning at MAIN.

The right rotation is executed similarly, but with the necessary changes due to the different rotation direction.

```
RIGHT:    AND    01111111b      ; leave on only the 7 bits of interest, and
          OUT    (OUTP),A       ; copy the bit pattern to the output port
          CALL   DELAY          ; execute a delay loop of about 0.5S
          SRL    A              ; shift right the byte, and jump if bit 0,
          JP     NC,NOSETR      ; now in the Carry flag, was at '0'
          SET    6,A            ; otherwise set bit 6 at '1'
NOSETR:   DEC    B              ; count the repetitions
          JP     NZ,RIGHT       ; repeat from RIGHT seven times,
          JP     MAIN           ; then return to the label MAIN
```

The figure on the right shows the content of register A before the execution of SRL and after. The bits have been right shifted and a '0' has been inserted at the left.

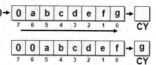

Our focus is on bit 'g', now moved from position 0 into the Carry flag.

We need to rotate seven bits, as we can see in the figure on the right.

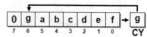

The bit in the Carry flag must be copied in A, in position 6, where SRL has moved the '0', which had been in bit 7.

The DELAY subprogram deserves a special explanation; its only effect is to make time go by.

```
DELAY:    PUSH   AF             ; save A and the Flags
          LD     HL, 0B4D9H     ; define the number of repetitions
LOOP:     DEC    HL             ;  6 clock cycles
          PUSH   AF             ; 11
          POP    AF             ; 10
          PUSH   AF             ; 11
          POP    AF             ; 10
          PUSH   AF             ; 11
          POP    AF             ; 10
          PUSH   AF             ; 11
          POP    AF             ; 10
```

```
        LD    A,H        ; 4
        OR    L          ; 4
        JP    NZ,LOOP    ; 10
        POP   AF         ; restore the contents of A and Flags
        RET
```

The PUSH and POP pairs, seen in the code don't do anything special, except to make time go by. Assuming a $10MHz$ clock, to get an approximately $0.5S$ delay, we need to go through 5 million of the processor's clock cycles. A loop repetition lasts 108 clock cycles so $(5000000/108) \approx 46297_{10} = $ B4D9h, which we have loaded in HL, the loop counter.

Finally, a consideration about checking programs in the emulator when there are loops that produce long delays. A step-by-step execution of the program can be quite impractical unless the delay subprograms are expressly (and temporarily) excluded, so that it will be quicker to check the function of the code.

In the following example, an RET has been inserted in the first line of the subprogram, while the rest of the line has been excluded through the addition of a semicolon after the RET[22].

```
DELAY:   RET   ; PUSH AF    ; save A and the Flags
```

In this manner, the calling program remains intact, calling the subprogram as usual, while the subprogram has, in fact, been excluded[23].

3. We will use the given table to solve the problem, even though it is not the only method to get the bit configuration as of the given number to copy on the ports to turn on the LED lights. See Section 3.5.1.4.

The solution that uses the table, however, is certainly the most efficient from the point of view of execution times (it is less so from the perspective of memory usage). Let's define the addresses of the ports and the MaxNum constant at 32, then let's link the program to the system reset.

```
DATA      EQU   00h        ; input port IA
MaxNum    EQU   32         ; maximum value of the input
Led_A     EQU   00h        ; output port OA
Led_B     EQU   01h        ; output port OB
Led_C     EQU   02h        ; output port OC
Led_D     EQU   03h        ; output port OD (a total of 32 LEDs)
          ORG   0000h
          JP    START
          ORG   0100h
```

---

[22] The added semicolon turns all the following into a comment.
[23] In programming jargon, we say that a temporary "patch" has been executed.

After the initialization of the Stack Pointer, three subprogram calls make the code more legible and better organized.

```
START:    LD     SP,0FFFFh    ; initialize the Stack Pointer
MAIN:     IN     A,(DATA)     ; read data from the input port
          CALL   LIMIT        ; ensure that the value rests in the limits
          CALL   DECODE       ; decode the value
          CALL   OUTPUT       ; update the column of LEDs
          JP     MAIN
```

The names chosen for the subprograms help us understand their function. After reading the Data port, we limit (saturate) the value to the highest number. Then, we decode it by getting the bit configuration to copy on the output ports.

The value limiting function, LIMIT, compares the value in the accumulator with the MaxNum constant. If A is greater than or equal to MaxNum, the Carry flag is zeroed, otherwise it is set at '1'. If it is set at '1', A is lesser than MaxNum and we exit (RET C), because the number is in the desired interval. If the Carry flag is set at '0', the number is overwritten with the maximum value, giving us the required limited value.

```
LIMIT:    CP     MaxNum       ; compare the value with 32
          RET    C            ; return if the value is less than 32
          LD     A,MaxNum     ; otherwise substitute it with 32
          RET
```

Through the index received in the accumulator, the DECODE subprogram takes the address of the first byte in the line corresponding to that index. The index has to be multiplied by 4, since a group of 4 bytes corresponds to every LED configuration in the table.

The table's address is TABLE, since this label is placed ahead of its first byte. If we copy that address in register HL and add that to the index multiplied by 4, we get the address of the first byte in the line corresponding to the given number.

Note that the add is simplified (we only add the low part), because it will never produce a carry in the high part, since the table is allocated at address 0200h and the index will always be less than 255.

```
DECODE:   LD     HL,TABLE     ; base address of the table
          SLA    A            ; multiply the index by 4,
          SLA    A            ; because every row contains 4 bytes
          ADD    A,L          ; add the new index to the table address
          LD     L,A          ; now, HL points to the row of interest
          RET
```

Now, we have the address of the first byte in the requested line in HL. The OUTPUT subprogram goes to take this byte by means of the indirect addressing mode through HL and copies it to port IA (Led_A).

The operation is repeated three more times and the INC HL instruction increments the address in HL each time. The LED configuration corresponding to the given number appears on the 4 output ports.

```
OUTPUT:   LD    A,(HL)          ; read the less significant byte from table
          OUT   (Led_A),A       ; and copy it to the port Led_A
          INC   HL              ; increment HL to point to the next byte
          LD    A,(HL)          ; ...and so on, for all the other bytes
          OUT   (Led_B),A       ; to the port Led_B
          INC   HL
          LD    A,(HL)
          OUT   (Led_C),A       ; to the port Led_C
          INC   HL
          LD    A,(HL)
          OUT   (Led_D),A       ; to the port Led_D
          RET
```

4. This solution defines a subprogram OUTPUT that handles the flashing times and the conditioning by the control inputs.

If we carefully read the table of flashing times defined in the text, we see that the LED flashing times halve consecutively. This suggests that we should conform their time trend to that of a binary 4-bit counter where each output bit starting from the least significant bit flashes twice as frequently as the one before it. We call the OUTPUT subprogram for each of the combinations produced by the count.

We define the control port address (PCTR) and the output port address (PLED), as well as the link to the system reset.

```
PCTR      EQU   00h           ; port IA: control inputs
PLED      EQU   00h           ; port OA: output LEDs
          ORG   0000h
          JP    START
          ORG   0100h
```

We initialize the Stack Pointer, then zero register A, which we will use to memorize the state of the LED lights. In the main loop, we call the OUTPUT subprogram at each repetition to force the current LED configuration to exit onto the port PLED.

```
START:    LD    SP,0FFFFh      ; initialize the Stack Pointer
          LD    A,00000000b    ; set the initial LEDs state in A
MAIN:     CALL  OUTPUT         ; call the subprogram OUTPUT
          INC   A              ; generate the new light configuration
          AND   00001111b      ; maintain the count cyclical at 4 bits
          JP    MAIN
```

Before repeating the loop, we increment the content of A, which gives us the next configuration of LEDs lit. The bitmasking done by the AND instruction allows us to maintain the count cyclical at 4 bits.

As mentioned before, OUTPUT also handles the control inputs giving us the correct visualization timing of each combination. After saving the state of the LED lights in B, it reads the control port, and copies that in H. If bit 7 = '0', we zero the state of the LED lights (turning them off), otherwise, we take up the current state that was saved in B, thus moving the sequence forward. When flashing is enabled again, it will start with all the lights off.

After visualizing the state of the LED lights on the output port we call the delay subprogram DEL50mS. If bit 6 is at '1', we call DEL50mS again to double the delay time.

```
OUTPUT:   LD     B,A           ; save A in B
          IN     A,(PCTR)      ; read the control bits 6 and 7 from the
          LD     H,A           ; port and save them in register H
          BIT    7,H           ; check bit 7
          JP     NZ,LAMP       ; if it is at '0', switch off all the LEDs
          LD     A,00h
          JP     LAMP1
LAMP:     LD     A,B           ; if it is at '1' get the LEDs state
LAMP1:    OUT    (PLED),A      ; switch on/off the LEDs
          CALL   DEL50mS       ; wait for 50 mS
          BIT    6,H           ; check bit 6
          JP     Z,NORMAL      ; if it is at '0', the half period is 50ms
          CALL   DEL50mS       ; otherwise it is doubled
NORMAL:   RET
```

Finally, we examine the delay generation done by the DEL50mS subprogram. It must be set at 50 mS, since the lights must be on for half the time and off for half the time in one period (note that 50 mS is the least common multiple of all the half periods).

```
DEL50mS:  PUSH   AF            ; save A and the Flags
          LD     BC,7575       ; initialize the delay loop counter
LOOP:     PUSH   HL            ; four instructions useful only
          PUSH   BC            ; to add delay in the loop
          POP    BC
          POP    HL
          DEC    BC            ; count the loop repetitions
          LD     A,B           ; check if BC goes to zero through a
          OR     C             ; bitwise (B OR C) that affects the flags
          JP     NZ,LOOP       ; repeat until BC goes to zero
          POP    AF            ; restore the contents of A and Flags
          RET
```

If the processor's clock frequency is $10MHz$, each clock cycle lasts $100nS$. We can easily check that the operations inside the LOOP loop make 66 clock cycles go by (consult the instruction tables in Appendix C). The loop counter, BC, is set at 7575. when we do the calculations, we see that $(7575 \cdot 66) \cdot 100nS$ gives us approximately the $50mS$ required.

5. a) We calculate the period of the square wave generated. Here, we show the code defined in the text in relation to the main loop.

```
START:    LD    A,00000000b    ; set initial value in A
MAIN:     OUT   (WAVEP),A      ; send the current value to the port
          XOR   01h            ; invert the bit 0 of A
          JP    MAIN           ; repeat indefinitely the loop
```

Before entering in MAIN, the accumulator is zeroed. Then its value is cyclically copied to the port (with the OUT instruction), and right after, bit 0 is inverted (with an XOR instruction). Going back to MAIN (with the JP instruction), we repeat the sequence, resulting with bit 0 alternating cyclically between the values of '0' and '1'.

If we consult the instruction tables (see Appendix C), we see that OUT is executed in 11 clock cycles, XOR in 7 and JP in 10. So a loop repetition lasts 28 cycles and this is the time that passes between two transitions of the square wave in the output. The period of the signal is $28 \cdot 2 = 56$ clock cycles (see the figure below):

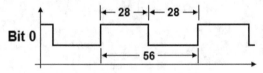

The clock frequency is $10MHz$, therefore the length of the cycle, computed as the inverse of the frequency, is $100\,nS$ ($10^{-7}S$). Hence, the duration of the period of the square wave is $56 \cdot 100\,nS = 5600\,nS = 5.6\,\mu S$ (corresponding to a frequency of approximately $179KHz$).

b) To change the period of the square wave generated by bringing it to $4\,mS$, we need to add a delay (the DELAY subprogram shown below) inside the main loop. Between one variation of the output and the next, $2\,mS$ must go by (half of the period required in the text):

```
START:    LD    SP,0FFFFh      ; initialize the Stack Pointer
          LD    A,00000000b    ; set the initial value in A
MAIN:     OUT   (WAVEP),A      ; send the current value to the port
          XOR   01h            ; invert the bit 0 of A
          CALL  DELAY          ; generate a delay
          JP    MAIN           ; repeat indefinitely the loop
```

Since now we use the Stack mechanism (to call a subprogram and return from it), we have also added the initialization of the Stack Pointer in the first couple lines of the program.

Now we must write the code of DELAY, taking care to preserve the content of A used in the main cycle. We take the simplest structure of a delay loop that we know (see Section 3.3.5.4), making sure to add a RET after the loop. We try to set the counter at the highest value possible.

```
DELAY:    LD    C, 255        ; 7 clock cycles
LOOP:     DEC   C             ; 4
          JP    NZ,LOOP       ; 10
          RET                 ; 10
```

We get $7 + (4 + 10) \cdot 255 + 10 = 3587$ cycles, corresponding to

$$3587 \cdot 100\, nS = 358700\, nS \approx 0.36\, mS,$$

which is a bit less than the $2\, mS$ required. So we go on to a nested loop structure like the following example where we leave indicated the values to load in the outer loop counters (m) and inner loop counters (q), but we indicate the clock cycles of each instruction.

```
DELAY:     LD    C, <m>          ; 7 clock cycles
LOOP:      LD    B, <q>          ; 7
NESTED:    DEC   B               ; 4
           JP    NZ,NESTED       ; 10
           DEC   C               ; 4
           JP    NZ,LOOP         ; 10
           RET                   ; 10
```

We have seen that the clock cycles of the repetition were 28 in the version of the program that didn't call a DELAY.

Let's evaluate how many clock cycles we need to make $2\, mS$ go by. We divide this time by the length of one cycle ($100\, nS$):

$$2\, mS \,/\, 100\, nS = 2000000\, nS \,/\, 100\, nS = 20000 \text{ cycles}.$$

Also, considering that CALL is executed in 17 clock cycles, the subprogram should produce a delay of:

$$20000 - 28 - 17 = 19955 \text{ clock cycles}.$$

The specifications define a tolerance of 0.5%. Let's apply this percentage to the number of cycles of the half period.

$$20000 \cdot (\pm 0.005) = \pm 100 \text{ tolerance cycles}.$$

To make sure we keep within the required tolerance, we can define the delay of the internal loop (which is repeated m times) at a value of less than 100 clock cycles.

Let's look at the part that will be repeated m times:

```
LOOP:       LD    B, <q>       ; 7 clock cycles
NESTED:     DEC   B            ; 4
            JP    NZ,NESTED    ; 10
            DEC   C            ; 4
            JP    NZ,LOOP      ; 10
```

Let's express the delay generated by this sequence of instructions in function of q. The delay must be less than 100 clock cycles.

$$(7 + (4 + 10) \cdot q + 4 + 10) < 100$$

From this equation, we derive:

$$q < \frac{(100 - 7 - 4 - 10)}{(4 + 10)} \approx 5.6 \ .$$

So we choose to assign this number ($q = 5$, rounded down) to register B. The delay is:

$$(7 + (4 + 10) \cdot 5 + 4 + 10) = 91 \text{ clock cycles}$$

Earlier, we saw that the subprogram execution time has to be 19955 clock cycles. To calculate m we must take the duration of RET and the instruction that initializes register C from this value, since they are outside the LOOP loop. Then we divide by the number of clock cycles calculated for the internal delay:

$$m = \frac{19955 - 10 - 7}{91} \approx 219.1$$

which we round to the nearest whole number ($m = 219$). The subprogram code in the final version, is as follows:

```
DELAY:      LD    C, 219       ; 7 clock cycles
LOOP:       LD    B, 5         ; 7
NESTED:     DEC   B            ; 4
            JP    NZ,NESTED    ; 10
            DEC   C            ; 4
            JP    NZ,LOOP      ; 10
            RET                ; 10
```

Based on the values obtained from m and q, taking the main loop into account, the length of an half period is:

$$28 + 17 + (7 + (7 + (4 + 10) \cdot 5 + 4 + 10) \cdot 219) + 10 = 19991$$

This value is well within the required tolerance. We can check this by calculating the percent in relation to the expected value.

$$\frac{20000 - 19991}{20000} \cdot 100 \approx 0.045\% \ \ll \ 0.5\% \quad \text{(c.v.d.)}.$$

6. Let's define the port addresses and the link to the reset. On start, we initialize the Stack Pointer and the output port, which zeroes line SER.

```
IA          EQU    00h          ; define name and address of the
OA          EQU    00h          ; input and output ports (IA and OA)
            ORG    0000h
            JP     START
            ORG    0100h
START:      LD     SP,0FFFFh     ; initialize the Stack Pointer
            LD     A,00000000b   ; and the output port OA (SER = '0')
            OUT    (OA),A
```

The program's main loop is divided into two subprogram calls. READ is a function that waits for a new number on the input port, while SEND transmits it in serial form on line SER. As indicated in the text, we know it is impossible to receive new data while we are still transmitting the current one, so we can safely separate these two tasks.

```
MAIN:       CALL READ            ; wait for the data
            CALL SEND            ; send it on the serial line
            JP   MAIN
```

On the rising edge of bit 7, the READ function copies input port IA to the accumulator. This function is made up of two consecutive wait loops. The first loop checks that bit 7 is at '0' and if it isn't, it waits until it is. The second loop waits for the rising edge and then goes back to the calling program when it arrives.

```
READ:       IN     A,(IA)        ; wait for bit 7 = '0'
            BIT    7,A
            JP     NZ,READ
EDGE:       IN     A,(IA)        ; wait for the transition
            BIT    7,A           ; from '0' to '1' of bit 7
            JP     Z,EDGE
            RET
```

The SEND subprogram is divided into three parts. The first part zeroes bit 7 read from the port (it should not be transmitted), then computes the parity bit to insert at the end of the packet.

The operation the text requires ($IA0 \oplus IA1 \oplus IA2 \oplus IA3 \oplus IA4 \oplus IA5 \oplus IA6$) is actually executed by the processor automatically, at the time the AND instruction is executed, and the result is memorized in the parity flag.

```
SEND:       AND    01111111b     ; mask bit 7 and calculate the parity
            JP     PE,PAR0       ; jump if parity is even
PAR1:       OR     10000000b     ; otherwise set at '1' the parity bit
PAR0:       LD     C,A           ; save into register C the packet to send
```

The flag is tested by a conditional jump (PE = "Parity Even") and the result tells us whether (or not) to insert a '1' in position 7 of the byte to transmit. Then, the byte is copied in register C.

We start serial transmission by writing a '1' (the "Start" bit) on line 0 of the output port. This value has to be kept active on line SER for the duration of a "bit time", which is why we call the DELAY subprogram, which will make 0.1 mS go by:

```
LD      A,00000001b   ; set the "Start bit" at '1'
OUT     (OA),A        ; send it to the line SER
CALL    DELAY         ; wait for 0.1 mS
```

The third part of the subprogram consists of a loop that sends all the data bits and the Stop bit in sequence. The loop counter is set at 9 (7 data bits + 1 parity bit + 1 stop bit).

At each repetition, we collect the next bit to transmit from register C and we make the bit in position 0 exit onto the port. This bit corresponds to line SER (see the figure below left, which shows the content of register C at the first sending).

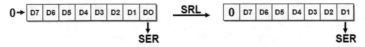

The value is kept on the output for a bit-time (because of the call to DELAY). Then, the register is right shifted through an SRL instruction so that the next bit to transmit is prepared in position 0 (the right hand side of the figure shows the content of C ready for the second sending).

Note that the SRL instruction also inserts a '0' from the left. We can use this to have the ninth bit to transmit (the Stop bit) available. The loop is actually repeated 9 times (until register B is zeroed).

```
        LD    B,9           ; initialize the counter of the bits to send
LOOP:   LD    A,C           ; get the (remaining) data bits to send
        AND   00000001b     ; zero the bits not of interest for us
        OUT   (OA),A        ; transmit the current bit on line SER
        CALL  DELAY         ; wait for 0.1 mS
        SRL   C             ; right shift the remaining bits to send
        DEC   B             ; count the bits remained to transmit
        JP    NZ,LOOP       ; repeat from LOOP until ended
        RET
```

All we need for a delay is an 8-bit loop counter. If we set D = 70, we get a delay of $(7 + (4 + 10) \cdot 70 + 10) = 997$ clock $\approx 0.1 mS$.

```
DELAY:  LD    D,70          ; 7 clock cycles
WAIT:   DEC   D             ; 4
        JP    NZ,WAIT       ; 10
        RET                 ; 10
```

4

Interfacing with external devices

Abstract In this chapter, we will learn how to interface the microcomputer with external devices and how to transfer data between different systems. First, we will introduce the concept of "handshake". Then we will present techniques such as "polling", "interrupts" and the use of "timers", which allow convenient management of the communication. Each concept will be presented theoretically and then deepened with one or more examples developed using both software and specialized hardware. The chapter ends with the presentation of several systems developed exploiting the above mentioned concepts (e.g. pulse generators, sinusoidal waveform generators, object counters, sensor reading, and asynchronous communication).

4.1 Managing communication with external devices

In the previous chapters, we have used parallel ports mostly to acquire the states of switches and drive LED lights. Switches and LED lights are "peripheral devices" (or "peripherals", or "external devices") from the point of view of microcomputers. In this simple case, we have not posed the problem of synchronizing the peripherals and the system. In fact, LED lights are always accessible to the processor (that can turn them on and off when it wants) as are switches (that can be checked at any moment).

In general, a peripheral can be something more complex like a printer, keyboard, mouse or network card... For example, at the moment that the processor tries to use a printer, it might be turned off, out of paper or blocked due to a previous error (like the paper getting stuck).

Furthermore, many types of peripherals may remain occupied for a certain amount of time in which they cannot accept (or send) information. It would, therefore, be unwise to expect to interact with these devices at any moment without first determining their current state. So, for many types of peripherals, the processor will need to check whether the device is available to exchange information.

The data transfer is handled by a dedicated synchronization protocol called "handshake". This is generally implemented by adding lines and hardware devices to the data connection, and dedicated software-handling procedures from the connection. In this book, we will not deal with standard connection specifications, which are generally very complex and vary broadly. Rather, we will examine the underlying introductory concepts.

In this chapter, we will deal with the topic of "interrupts", which are often necessary to handle exchanges of data between the processor and peripherals to the best advantage. As alluded to previously, (see Section 2.1.5), the processor has lines ($\overline{IRQ2}$, $\overline{IRQ1}$ and $\overline{IRQ0}$) that allow it to interrupt the execution of a currently running program in order to start another program specifically designed to handle the situation that caused the interrupt. As we will see further on in this discussion, the interrupt "mechanism" allows us to efficiently resolve communication with devices.

The figure below shows a case in which a microcomputer has to transmit data to a standard peripheral. We introduce data lines and other lines (shown in the figure as "control signals"), which will be used to synchronize the transmission.

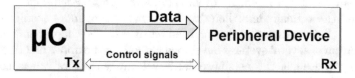

Even when the roles are reversed (the microcomputer receives data from the peripheral), it is generally necessary to use control signals to synchronize data reception, as shown in the following figure.

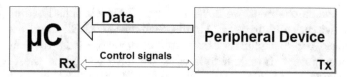

These signals are indispensable for preventing data loss or duplication in the communication between devices. The control signals are handled by the software and dedicated hardware. When we create these connections and the hardware to support them, we say we are "interfacing" peripheral devices to the microcomputer.

An interface[1] makes it possible to connect two systems and have them adapt to each other to permit an efficient exchange of data. From the hardware perspective, an interface is a set of input/output ports and a specific hardware and software logic that allows us to connect a certain peripheral to the computer system.

---

[1] The term "interface" derives from the Latin words "inter" (between) and "facies" (faces).

To make the situation clearer, let's bring back the example of the printer, which we imagine is connected to a personal computer. Assume we have to print a very long document. When the microprocessor sends the document to the printer, that last uses its own internal memory. Once it accepts a certain percentage of the document, the printer will be unable to receive the rest until it has printed out that first percentage, thus freeing up its memory.

Printing this takes up a much longer time than the processor's normal run times. As the printing is taking place, we need to temporarily stop data transmission and start it again when the printer's memory is free.

We have dealt briefly with the need to temporarily slow down or stop data transmission. Now, let's take a look at another example, which is diametrically opposed to that case.

Imagine we receive data from an alphanumeric keyboard. The typing speed is not constant (it depends on a human), so a long time can go by between two successive keystrokes. Here, the processor does not need to handle a large amount of data, but rather it must pay attention to the keyboard only when there is a keystroke. In other words, the challenge is to prevent the processor from uselessly waiting for data for an excessive amount of time, which could be costly in terms of system efficiency.

In general, this also applies when two microprocessor systems communicate with each other, as in the following figure.

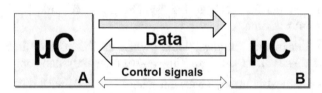

Considering the examples reported above, a printer is in fact a system constructed around a microcomputer, as is an alphanumeric keyboard. So, the concept of "peripheral" depends on the point of view. From the perspective of system A (left hand side in the figure above), we can think of system (B) as its peripheral, but the opposite is also true.

For simplicity's sake, we will refer in this section to parallel interfaces, that is where data are exchanged through (in our case, 8-bit) parallel connections.

We mentioned that in creating a specific interface, we must design a dedicated handshaking network between the processor and the peripheral device. This network can be implemented in many ways. As seen in the previous figures, we must add wires to the connection beyond those for data.

The connected parts use these additional wires to generate a series of call and response signals in both directions. It should be noted that in general, there is no "universal" solution, and each device requires specific choices, case by case. Obviously, there are pre-made input/output devices on the market for every type of processor. They may be for general use or specialized for specific applications, able to satisfy the broadest range of connection requirements.

Rather than analyzing commercial devices, in the following, we take a design approach progressively introducing examples of device connections, starting with some basic points.

4.1.1 The unidirectional handshake

For the discussion below, let Tx (data transmitter) and Rx (data receiver) be the units that communicate with each other. We have seen that these can be a computer and a peripheral or vice versa, or even two computers connected together. The figure below shows a device Tx that sends data to a receiver Rx using parallel data transmission.

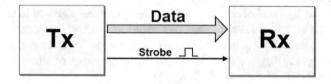

The device that is transmitting data signals this through a dedicated validation pulse, associated with the information being sent. Here, this is called a Strobe pulse. When Rx receives this "synchronization" pulse, it can acquire and use the information.

We call the handshake "unidirectional" when all the control lines go in one direction (Tx → Rx). In the figure above, we have only one control line.

This method is usable in any situation where Rx deals with data reasonably quickly and so is always ready to receive the next data from the transmitter. If this is not the case, we need to choose a more complex type of handshake (we will analyze this case further on).

The figure below shows a data transfer sequence where Tx first loads number 3Fh on the data lines and then generates a pulse on the Strobe line. Rx acquires the number (for example on the rising edge of Strobe) and then waits for the next one.

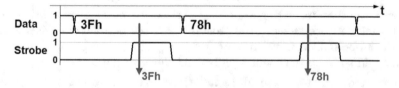

Number 78h is sent, thus repeating the sequence. It is also acquired by Rx at the new Strobe pulse (the red arrows symbolize the acquisition of the data).

Note: The Strobe signal is necessary because the receiver must know when a new piece of information is available. The two communicating systems are often asynchronous, that is they cannot share the same clock. The Strobe signal is always necessary, regardless, because the receiver must know when a new piece of information is available and it is not enough to transmit the data to "notify" the receiver. For example, think of sending some text in ASCII code, character by character, which contains a word with a double letter (like the word "cool"). Between the first and second 'o', the data lines do not change. In the absence of a synchronization signal, the receiver doesn't know that we are sending two consecutive letter 'o's' since it can only observe the data line values.

Example of a parallel interface with a unidirectional handshake

Using what we've learned about parallel ports in past chapters, let's try to design a unidirectional handshake interface. Let's imagine Tx is our computer system and we need to send data to the external device, Rx, which is fast enough to acquire all the data we send it.

As we can see in the following figure, to transmit data, all we need is an 8-bit parallel output port, which we will call a "Data Port". For the Strobe signal, we can use a line retrieved from a second output port (generally, a port dedicated to managing handshake signals is called a "Control Port").

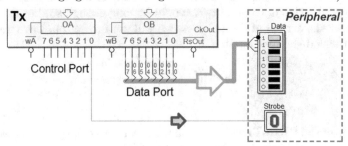

We complete the design choices with appropriate handling software, as suggested in the example below. The program begins with the usual definitions, link to the reset and initialization of the Stack Pointer. Before entering the main cycle, we also initialize the output ports and register B.

```
CTRLP     EQU     00h          ; OA and OB output ports
DATAP     EQU     01h
          ORG     0000h        ; link to the reset
          JP      START        ; jump to the program start
          ORG     0100h
START:    LD      SP,0FFFFh    ; initialize the Stack Pointer
          LD      A,00000000b  ; and the output ports
          OUT     (DATAP),A
          OUT     (CTRLP),A
          LD      B,0          ; (simulated) data to trasmit
```

We use register B to simulate the data to transmit. In a real application, the data can be taken from any type of calculation, or come from reading a file or another source. For our example, every number is generated as a test, incrementing B before sending each time. A delay (CREATE) has been inserted to allow us to also simulate the time that a real system would take to get the number.

```
MAIN:        CALL   CREATE     ; simulate the time that is necessary
             INC    B          ; to 'create' a new number to transmit
```

After the new value is retrieved from B, it is sent to the Data Port (DATAP). So we bring the Strobe line to '1' by writing this value on bit 0 of the Control Port (CTRLP). For the moment the other bits don't interest us, so we write them all at '0'.

```
             LD     A,B         ; copy the number to register A
             OUT    (DATAP),A   ; and transmit it to the peripheral
             LD     A,00000001b ; set the Strobe line to '1'
             OUT    (CTRLP),A   ; on the Control Port
```

Suppose that a duration of about $0.5mS$ is required for the Strobe pulse. The line must be brought to '0' after this time, so a call to a delay subprogram (PTIME) has been inserted.

The duration of the pulse is chosen on the basis of the timing specifications of the overall system. Generally, it should be long enough so that the receiver can detect the arrival of the pulse, as we will see further on. Let's go back to MAIN to send the new number.

```
             CALL   PTIME       ; keep Strobe high for about 0.5 mS
             LD     A,00000000b ; set the Strobe line to '0'
             OUT    (CTRLP),A
             JP     MAIN
```

The code of the CREATE subprogram is shown here below. It includes a delay loop with a duration that in our case is purely symbolic.

```
CREATE:      LD     C,50        ; wait for the 'creation' of new data
DCR:         DEC    C           ; 4 +
             JP     NZ,DCR      ; 10 = 14 cycles; 14 x 50 = 700 cycles +
             RET                ; 17 (call) +7 (ld C) +10 (ret) = 734 cycles
```

The PTIME subprogram is executed in approximately $0.5mS$. Read the comments in the code (suppose that the clock cycle takes $100nS$).

```
PTIME:       LD     C,142       ; calculate the pulse duration time
PLOOP:       PUSH   BC          ; 11 +   (PUSH and POP added to delay)
             POP    BC          ; 10 +
             DEC    C           ; 4 +
             JP     NZ,PLOOP    ; 10 = 35 cycles; 35 x 142 = 4970 cycles +
             RET                ; 17 (call) + 7 (ld) + 10 (ret) =
                                ; 5004 cycles = about 0.5 mS
```

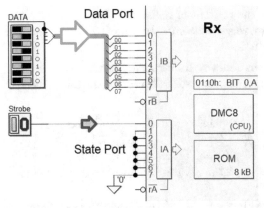

Now, let's focus on the receiver Rx; see the figure to the left.

We have used a "Data Port" to receive the information transmitted, and a "State Port" to detect the pulses sent on the Strobe line.

This port is generally called a "State Port" because it allows us to evaluate the state of the interface.

A potential solution for handling communication from the receiver's side is found below. After the usual definitions and initializations, the program zeros register A before entering the MAIN loop. This will zero the output port, which visualizes the number received from time to time, for testing purposes.

OUTP	EQU	00h	; OA output port
DATAP	EQU	01h	; IB input port (Data Port)
STATP	EQU	00h	; IA input port (State Port)
	ORG	0000h	; link to the reset
	JP	START	; jump to the program start
	ORG	0100h	
START:	LD	SP,0FFFFh	; initialize the Stack Pointer
	LD	A,00000000b	; Set A = '0'
MAIN:	OUT	(OUTP),A	; copy A to the output port

At the WAIT0 label, we find two consecutive wait loops, where we read the state port, looking for the rising edge of Strobe. In the first loop we check that the state goes to '0' or is already there. In the second loop, we wait until the line goes from '0' to '1'.

WAIT0:	IN	A,(STATP)	; read the State Port
	BIT	0,A	; check the Strobe line
	JP	NZ,WAIT0	; wait for a '0'
WAIT1:	IN	A,(STATP)	; read the State Port
	BIT	0,A	; check the Strobe line
	JP	Z, WAIT1	; wait for the transition from '0' to '1'

We retrieve the data as soon as the rising edge of Strobe is identified.

	IN	A,(DATAP)	; copy to A the received byte
	CALL	PROCESS	; simulate a latency time
	JP	MAIN	; repeat the sequence as of MAIN

We have inserted a call to the PROCESS subprogram to simulate a "latency time" in the data processing.

The subprogram doesn't change the content of register A. We go back to the MAIN label and copy the byte we received from A to port OUTP to visualize it, then we wait for the next data.

The code of the subprogram is as follows. Its structure is that of a normal delay loop that simulates the time the processor takes to use the data received (totally symbolic in our case).

```
PROCESS:    LD     C,10      ; simulate a latency time in processing data
PRO:        DEC    C         ; 4 +
            JP     NZ,PRO    ; 10 = 14 cycles; 14 x 10= 140 cycles +
            RET              ; 17 (call) +7 (ld C) +10 (ret) = 174 cycles
```

Now let's look at the whole system, shown in the figure below. It is made up of the two modules Tx and Rx that we have just now examined.

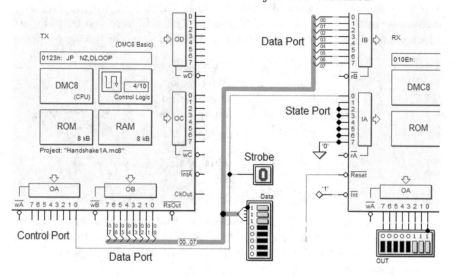

Let's try to think about the possibilities this interface offers, keeping in mind that the handling software in the two modules is what we have seen before.

Regarding Tx, all the transmission operations are executed by the program, including pulse generation on the Strobe line. The capacities of the Tx processor are mainly dedicated to handling the interface, since it needs to continually go and retrieve a number, then transmit it during the main loop.

If the processor performs other tasks, they should be executed in the main loop while the communication is being handled. We shall see further on how to save computational resources by using the processor less and separating interface management from the other tasks that it can execute.

As with Tx, all the communications management operations in Rx are executed by the program in the main loop, including detecting the Strobe pulse.

The most critical task from the perspective of wasting processing potential in the processor is definitely the double wait loop at the rising edge of the signal. The processor can certainly carry out other tasks in the main loop but with little residual computing capacity. This is because it must continue in each case to follow every point of the variations in the synchronization line since the Strobe signal has a limited duration.

Note that no mechanism allowing for a controlled slow-down of the transmission has been implemented. This means that the programmer must pay close attention to the execution times of the other tasks the processor carries out (aside from waiting for the Strobe pulse) to prevent data loss.

4.1.2 The bidirectional handshake

In cases where the receiving device isn't fast enough, the transmitter has to regulate its own transmission speed. Clearly, we don't achieve this by slowing down the transmitter's clock (that would be a real waste of resources). Rather, we add another line to the interface, that is oriented in the opposite direction from that of the data line. This is why we call this type of handshake "bidirectional".

The figure below shows a Tx device that transmits data in parallel to an Rx device, as in the previous case. In this block schematic, however, there is an additional wire called "Busy". It is generated by the receiver and sent back toward the transmitter.

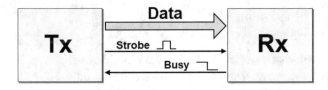

This new line allows Rx to inform Tx about its state. For example, we can make it so that when Busy is at '1', it indicates that Rx is working on a task and can't receive data. When Busy = '0', Rx is ready to receive the next number.

Example of a parallel interface with a bidirectional handshake

The following figure shows a potential connection from the new Busy line to the microcomputer, which transmits data. With a dedicated program, we can handle the sequence of operations involved.

Clearly, unlike the previous case, microcomputer Tx should check the Busy line before sending the number. One could say it is "asking the peripheral's permission".

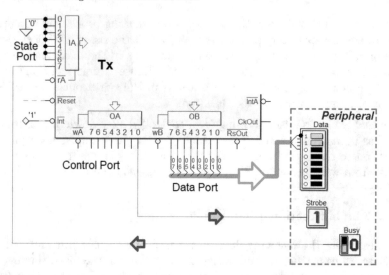

The schematic is similar to the one before but it has an added connection on an input port (the State Port) that evaluates the Busy line. Here, the line is connected to bit 7 (of the IA port in the schematic).

The handling program is similar to the one before but with an added test of the Busy line, obtained by reading the state port. Below, we'll show only the parts of the code that differ from the previous example of a transmitter. We have added the state port (IA) to the list of definitions.

```
CTRLP       EQU    00h          ; OA and OB output ports
DATAP       EQU    01h
STATP       EQU    00h          ; IA input port
```

The significant change is in the main cycle, highlighted below. Before transmitting the data, we go to read the state of the port on the WAIT label. Then we wait for the receiver to give us the permission to transmit. If the Busy line is already at '0' (or when it goes to '0'), we go ahead and transmit the number (the rest of the code is unchanged).

```
MAIN:       CALL   CREATE       ; simulate the time that is necessary
            INC    B            ; to 'create' a new number to transmit
WAIT:       IN     A,(STATP)    ; read the state of the Busy line
            BIT    7,A          ; wait for the peripheral to free itself
            JP     NZ,WAIT      ; go on if the peripheral is free
            LD     A,B          ; copy the number to register A
            OUT    (DATAP),A    ; and transmit it to the peripheral
            LD     A,00000001b  ; set the Strobe line to '1'
            OUT    (CTRLP),A    ; on the Control Port
            ...
```

The code presented here can accommodate some variations. For example, in the busy-waiting loop, we can carry out other tasks if required. This addition would make us slower to react to the variations of Busy.

Now, let's deal with the changes from the perspective of the receiver Rx.

See the figure to the right. We have the Data Port and the State Port, which were already present in the previous version and used to handle the data and Strobe lines.

Now we have added an output port that functions as a Control Port to generate the Busy line. The program is meant to handle the value of the Busy line.

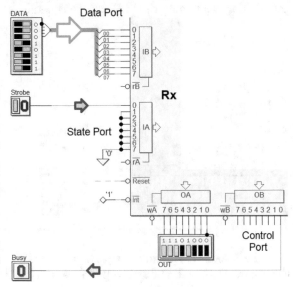

We have added the definition of the Control Port (OB) to the code:

```
OUTP    EQU    00h     ; OA output port (OUT)
BUSYP   EQU    01h     ; OB output port (BUSY)
DATAP   EQU    01h     ; IB input port (Data Port)
STATP   EQU    00h     ; IA input port (State Port)
```

Here, as before, we are only showing the differences from the previous receiver example. They are highlighted in the following list:

```
MAIN:       OUT    (OUTP),A     ; copy A to the output port
            LD     A,00000000b  ; set Busy = '0', to enable
            OUT    (BUSYP),A    ; the reception of new data
WAIT0:      IN     A,(STATP)    ; read the State Port
            BIT    0,A          ; check the Strobe line
            JP     NZ,WAIT0     ; wait for a '0'
WAIT1:      IN     A,(STATP)    ; read the State Port
            BIT    0,A          ; check the Strobe line
            JP     Z, WAIT1     ; wait for the transition from '0' to '1'
            LD     A,00000001b  ; activate Busy to avoid receiving new data
            OUT    (BUSYP),A    ; while processing
            IN     A,(DATAP)    ; copy to A the received data
            ...
```

As we can see in the main loop, we deactivate Busy before waiting for new data. We then reactivate it as soon as it is received. While the information is processed by the receiver, the transmitter will not be authorized to send new data. It can send only after data processing is completed, when the receiver indicates it is available for a new transmission. The following figure shows the complete system using the bidirectional handshake.

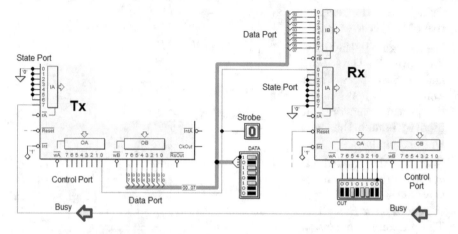

Now, let's look at the advantages of this version of the interface. Basically, the comments made about the previous version hold here as well, but we'll add two observations.

The first is that the synchronization between the transmitter and receiver is a clear advantage because it makes sure the receiver doesn't lose data if it is in a state where it can't receive new data.

The second is that the added wait cycle in the transmitter's code is a disadvantage. It creates another waste of processing capacity. The processor could also carry out other tasks aside from waiting for the receiver to authorize sending new data.

4.1.3 More complex handshake types

More complex devices may need to be handled by more control lines. For example, the figure below shows a new line called Ack ("Acknowledge"), which is used as another response by the receiver.

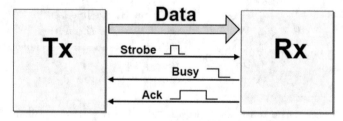

Here, the new line does not regulate the speed of transmission as the Busy signal does, but is used to confirm that the data has (or hasn't) been received and processed by the peripheral.

For example, if we didn't have this confirmation, the transmitter's software could decide to send the data again or tell the user, etc. We will not go further into the different forms of handshakes[2].

4.2 Hardware-supported handshake

Let's look at the parallel interface with bidirectional handshake that we introduced in Section 4.1.2. We have discussed the wasted processor potential, which is mainly caused by the wait loops it is involved in and the timing of signals. Since this is an introduction, we have so far exclusively referred to the available hardware in the microcomputer, that is its input and output ports.

To be clear, this interface, in its current state is not well designed. We need to relieve the processor of the low-level jobs that we have had to task it with to manage the lines.

Let's look for example, at an aspect of interface management that we haven't investigated. It regards the duration of the Strobe pulse, which has to be long enough so that the receiver can detect that the pulse has arrived. A shorter Strobe pulse requires the receiver to read the state port very often so that no pulse goes undetected. Also, under a certain minimum length, we can't even afford to go. In fact, let's consider the wait loop that we have used in the receiver to wait for the Strobe signal:

```
WAIT1:      IN     A,(STATP)    ; (11 clock cycles) read the State Port
            BIT    0,A          ; ( 8 cycles) check the Strobe line
            JP     Z, WAIT1     ; (10 cycles) wait for its activation
```

To be sure that we detect the pulse, the duration has to be greater than the loop's execution time, which is 29 cycles $(11 + 8 + 10)$.

We clearly need to add some hardware tricks aside from the simple input and output ports to prevent the receiver's processor from having to do nothing other than follow the interface signals to keep from losing data.

To start, it might be useful to use hardware to "capture" the arrival of Strobe, thus freeing the processor from keeping track of the situation. When the capture is done, the processor needs to be signaled immediately so that it will retrieve the data. Also, the generation of Busy can be automated.

This way, the receiver's processor can take care of everything else and dedicate only the necessary time to the interface.

---

[2] According to the system specifications, the types of handshakes can be very complex and go beyond the introductory scope of this book.

4.2.1 Example of a parallel interface with hardware handshake

The following figure shows an implementation of a parallel interface between two systems with a handshake managed by the hardware. Here we resolve some of the issues we had previously.

Both systems use a Data Port and a State Port. Control ports are unnecessary here since detecting the Strobe pulse and the function of the Busy signal are done by the added circuit between the two systems (the two NAND gates), with no direct participation by the processors.

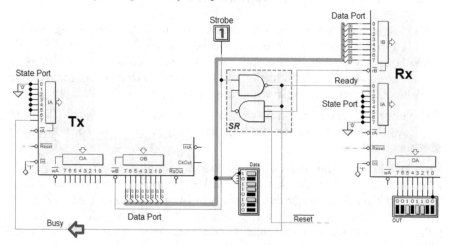

Regarding the receiver Rx, what is new is that the State Port is no longer used to keep tabs on the arrival of the Strobe signal (this is done by the hardware). The State Port has more time to check if data from the processor has arrived, by reading the Ready line (on bit 0).

As we can see in the schematic, the two NAND gates are connected in order to make an asynchronous flip-flop Set-Reset (SR) with "active low" inputs. Its output drives two lines: Ready (which means that the data is ready for the receiver), and Busy (which we know, and is meant for the transmitter). The other input comes from signal line $\overline{rB}$ on Rx's Data Port.

Also, $\overline{Reset}$ initializes Rx and also the flip-flop. So at the beginning, the Ready line is at '0', which indicates that Rx has not been sent any data yet. From the perspective of the transmitter, a '0' on the Busy line indicates that Rx is available. So the transmitter can send the first byte by writing its value on the Data Port (OB).

See the timing diagram shown in the figure below where the first two tracks at the top represent the instructions executed by the Tx and Rx processors. To make it easier to read, only the OUT instruction (executed by Tx) that sends the data byte to Rx, and the IN instruction (executed by Rx) that accepts it are included.

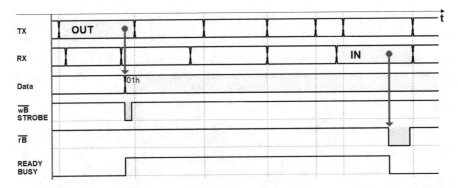

As shown by the red arrow on the left, the OUT instruction sends the data byte on the parallel OB port and also generates an active low short pulse on the associated output $\overline{wB}$ (see Section 2.3.2 on input and output ports).

We use this pulse to signal to the receiver that there is new data; for this we have connected $\overline{wB}$ to the Strobe line. Notice that, when it's done this way, Tx's software doesn't have to generate this pulse since that is automatic. We are also free from determining its duration since it isn't acquired by the receiver's program but by the SR flip-flop.

The Strobe pulse brings the output of the SR flip-flop to '1' (lines Ready and Busy), then tells the transmitter it can't send other data and that a data byte is ready to read from Rx. In the meantime, Rx first checks the Ready line on the state port and then goes to read the data lines, as required, by executing an IN instruction on port IB.

Executing an IN instruction also generates an active low pulse on the corresponding line $\overline{rB}$ (red arrow on the right). Activating this line zeroes the SR flip-flop and refreshes the functionality of the interface, which goes back to its initial state. Now the transmitter can send a new data byte.

Programming the parallel interface with handshake hardware

Now, let's analyze this interface's handling code. It derives from the code in the previous interface with bidirectional handshake (see Section 4.1.2). Let's first look at the transmitter's handling program.

Notice that there is one less port in the initial definitions; here we do not need the Control Port.

```
DATAP      EQU    01h          ; OB output port
STATP      EQU    00h          ; IA input port
           ORG    0000h        ; link to the reset
           JP     START        ; jump to the program start
           ORG    0100h
```

Before entering the main loop, we initialize the Stack Pointer and also register B, which we will use to create the data byte to transmit (on this subject, see the observations in Sections 4.1.1 e 4.1.2).

```
START:          LD      SP,0FFFFh    ; initialize the Stack Pointer and
                LD      B,0          ; the first number to be transmitted
```

The main loop includes: creating a new number, the wait loop for Busy to be zeroed and sending the new data byte. Here, we omit the subprogram code since it is identical to the one in Section 4.1.1.

```
MAIN:           CALL  CREATE     ; simulate the time that is necessary
                INC   B          ; to 'create' a new number to transmit
WAIT:           IN    A,(STATP)  ; read the state of the Busy line
                BIT   7,A        ; wait for the peripheral to free itself
                JP    NZ,WAIT    ; go on if the peripheral is free
                LD    A,B        ; copy the number to register A
                OUT   (DATAP),A  ; and transmit it to the peripheral
                JP    MAIN
```

The lines in the code about sending the number are highlighted. These lines are identical to the cited examples except that there is no part about generating Strobe since the hardware network has now made that automatic.

This means that the PTIME subprogram that timed the length of the pulse is also absent here. Now the transmitter code is streamlined even though we still have to cyclically check the level of the Busy line. We will soon resolve this problem as well by using interrupt techniques (see Section 4.4).

Here below is the receiver's code:

```
OUTP    EQU   00h          ; OA output port (OUT)
DATAP   EQU   01h          ; IB input port (Data Port)
STATP   EQU   00h          ; IA input port (State Port)
        ORG   0000h        ; link to the reset
        JP    START        ; jump to the program start
        ORG   0100h
START:  LD    SP,0FFFFh    ; initialize the Stack Pointer
        LD    A,00000000b  ; Set A = 0
```

After the usual definitions, we zero A and enter the main loop. This is similar to the code in Section 4.1.2, but there is no handling of the Busy line (which is now managed directly by the hardware).

The wait loop is also different. It now checks for the Ready line to be activated rather than searching for the arrival of the Strobe signal.

```
MAIN:    OUT   (OUTP),A     ; copy A to the output port OUTP
WAIT1:   IN    A,(STATP)    ; read the State Port
         BIT   0,A          ; check the READY line
         JP    Z, WAIT1     ; wait for its activation
         IN    A,(DATAP)    ; acquire in A the received data byte
         CALL  PROCESS      ; simulate a latency time
         JP    MAIN         ; repeat the sequence as of MAIN
```

As briefly explained for the transmitter, in Section 4.4 we will see how to use interrupt techniques to also eliminate the wait loops from the program code.

To summarize, in this version:

— much of the handshake management is resolved by the small hardware network that we added to the interface between the two systems.

— handshake management is largely invisible in the code. There is, in fact, no trace of it except for wait loops on Busy (Tx) and Ready (Rx).

— we can use a short Strobe pulse without forcing the receiver to waste time waiting for its arrival.

— the Ready signal value is maintained active by the circuit, and made available to the receiver's processor. This leaves the receiver free to continue executing other tasks while it waits to acquire a new data byte.

— the simple read/write of the data byte allows for the automatic activation/refresh of the handshake mechanism.

4.3 Polling

We have seen how Tx and Rx can talk to each other. However, there can be many more than two systems connected together.

The figure at the right shows an example where Rx (A) receives data from multiple Tx systems (B, C, D...).

Rx interacts with all the interfaces in the system; it continually checks for any data arriving. This cyclical, restless and time-consuming operation is called "polling".

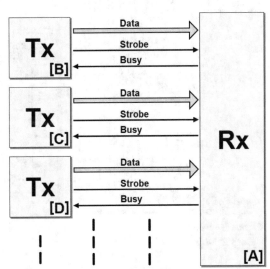

Rx polls the peripherals by reading the state ports connected to the handshake circuits. It generally needs to solve two problems:

— identify ("recognize") the device that needs an intervention.

— manage the "priority" for cases where more than one device makes a request at the same time.

Let's assume the device interfaces are designed according to the criteria from Section 4.2.1 for reasonably efficient communication.

Starting from this assumption, let's look at an example where three Tx units interface with one Rx as seen in the figure at the right.

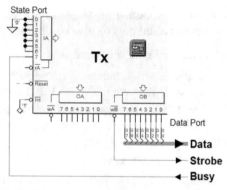

For simplicity's sake, let's imagine that the Tx units are all the same and they manage a Data Port (8 lines), a Strobe line directly retrieved from the write signal of that port and a Busy line received from a State Port.

Rx (see the figure) receives the data from three transmitters (B, C e D):

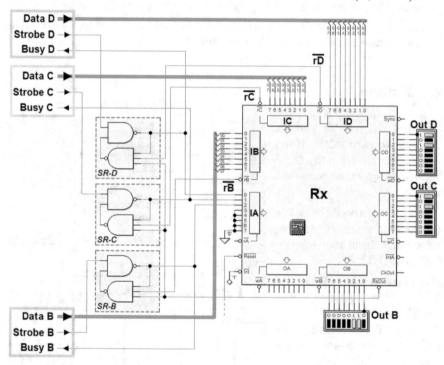

The data are acquired through the three input ports IB, IC and ID, respectively. As we have seen, each Tx manages its Strobe and Busy lines. The figure shows the three hardware handshake networks, which are all independent from each other and identical to those introduced in Section 4.2.1.

Port IA is used as a State Port and makes it possible to read the three flip-flops SR-B, SR-C and SR-D on lines 2, 1 and 0. If one or more are active, they indicate that the corresponding Tx has sent a data byte and it needs to be accepted.

The flip-flops $SR\text{-}B$, $SR\text{-}C$ and $SR\text{-}D$ are connected in order to receive outputs $\overline{rB}$, $\overline{rC}$ and $\overline{rD}$ that signal the reading of ports IB, IC and ID. Reading a data byte on these ports would therefore reset the corresponding flip-flop.

Now let's look at the individual devices' management programs. The software that checks Tx is identical to the one examined in the example in Section 4.2.1, so here, we will only look at the software for Rx. What follows is the definition of the Data Ports and the State Port:

DATA_B	EQU	01h	; IB input port (Data Port)
DATA_C	EQU	02h	; IC input port (Data Port)
DATA_D	EQU	03h	; ID input port (Data Port)
STATP	EQU	00h	; IA input port (State Port)

We also define the output ports that we use to visualize data after they are received for testing purposes.

OUT_B	EQU	01h	; OB output port (OUT_B)
OUT_C	EQU	02h	; OC output port (OUT_C)
OUT_D	EQU	03h	; OD output port (OUT_D)

We define the link to the reset, initialize the Stack Pointer, and then zero output ports OUT_B, OUT_C and OUT_D:

	ORG	0000h	; link to the reset
	JP	START	; jump to the program start
	ORG	0100h	
START:	LD	SP,0FFFFh	; initialize the Stack Pointer
	LD	A,00000000b	; set A = 0
	OUT	(OUT_B),A	; zero the three output ports
	OUT	(OUT_C),A	
	OUT	(OUT_D),A	

The main loop first calls the TASK_RX subprogram, which was introduced to simulate the receiver's execution of its primary task (which does not concern interface management). Right after, Rx focuses on its peripherals (goes to read the STATP port) and saves their states in register E.

MAIN:	CALL	TASK_RX	; simulate a generic and independent task,
			; executed in the program main loop
	IN	A,(STATP)	; read the State Port
	LD	E,A	; copy the state to E

Polling the devices implicitly resolves the "recognition" problem since it makes it possible to identify the device to serve through reading the state port. The "priority" issue is solved by testing high priority devices first.

We've chosen to focus on devices B, C and then D in order. As written in the comments, we first check the SR-B flip-flop and if there is no data from Tx B, we jump to the next test (TEST2).

If there is a data byte, we accept it from the DATA_B port, simulate a certain process time for it (by calling the PROCESS subprogram), and produce the received data on the test OUT_B output.

```
TEST1:    BIT    2,E              ; check the SR-B flip-flop
          JP     Z,TEST2          ; jump if there is no new data
          IN     A,(DATA_B)       ; read the byte from DATA_B
          CALL   PROCESS          ; simulate a latency time
          OUT    (OUT_B),A        ; copy the data byte to OUT_B
```

The other devices are managed in the same way. Finally, when all of the devices have been served, we go back to MAIN.

```
TEST2:    BIT    1,E              ; check the SR-C flip-flop
          JP     Z,TEST3          ; jump if there is no new data
          IN     A,(DATA_C)       ; read the byte from DATA_C
          CALL   PROCESS          ; simulate a latency time
          OUT    (OUT_C),A        ; copy the data byte to OUT_C

TEST3:    BIT    0,E              ; check the SR-D flip-flop
          JP     Z,ENDTEST        ; jump if there is no new data
          IN     A,(DATA_D)       ; read the byte from DATA_D
          CALL   PROCESS          ; simulate a latency time
          OUT    (OUT_D),A        ; copy the data byte to OUT_D

ENDTEST:  JP     MAIN             ; go back to MAIN
```

The subprograms called by the main loop are as follows. For testing purposes, they use delay loops to simulate the time it takes to execute a certain task (the set times are purely symbolic).

```
TASK_RX:  LD     C,55             ; simulate a generic task
TASK:     DEC    C                ; 4 +
          JP     NZ,TASK          ; 10 = 14 cycles; 14 x 55= 770 cycles +
          RET                     ; 17 (call) +7 (ld C) +10 (ret) = 804 cycles


PROCESS:  LD     C,10             ; simulate a latency time
PRO:      DEC    C                ; 4 +
          JP     NZ,PRO           ; 10 = 14 cycles; 14 x 10= 140 cycles +
          RET                     ; 17 (call) +7 (ld C) +10 (ret) = 174 cycles
```

The polling technique is a valid way to solve the problem of servicing the peripherals. It is used when the system needs the execution times to be highly predictable. The execution time of the main loop is documentable, as are the time and order in which the devices are polled. This makes it possible to know exactly when a certain action will be executed, which makes the behavior of the software predictable.

However, in systems where it is necessary to deal with the data sent by peripherals as quickly as possible, polling is not the best technique. Suppose the worst-case scenario when a new data byte is sent immediately after the poll. In this case, the peripheral has to wait for a repetition of the entire loop, which includes the next polls, the tasks to execute in the main loop and the previous polls. Only after all that will the new data be managed, and this is unacceptable for certain operations, for example reading an important or "critical" error code.

In these systems, questioning the peripherals can be more efficient if we use the interrupt technique, though this would mean sacrificing some of the predictability of system execution times.

4.4 Interrupt techniques

Interrupt techniques consist in peripherals being able to interrupt the normal functioning of the processor to ask for a specific "service". These techniques are different from polling because they don't require the processor to continuously control the peripherals.

For example, there are specific situations (such as receiving an alarm message in an industrial plant) where a delay with which the processor starts executing the requested service is unacceptable. This is the case of handling by polling. In these cases, we use interrupts.

Interrupts are possible because the processor is equipped with dedicated hardware. The method is based on one or more specialized inputs, which are usually called INT or IRQ ("Interrupt Requests"). They make it possible to use hardware to request the execution of specialized software, the "interrupt handlers". Interrupt inputs work directly on the processor's sequencer.

In the DMC8, the interrupt request occurs when an external device activates (at '0') at least one of the lines $\overline{\text{IRQ2}}$, $\overline{\text{IRQ1}}$ or $\overline{\text{IRQ0}}$ (as described in Section 2.1.5).

For simplicity's sake, when the situation permits, we will refer to the base version of the component available in the Deeds simulator library ("DMC8 Microcomputer", see Section 2.4.1), where lines $\overline{\text{IRQ2}}$, $\overline{\text{IRQ1}}$ and $\overline{\text{IRQ0}}$ are connected internally (see the following figure) and are accessible through input $\overline{\text{Int}}$ (see the arrow).

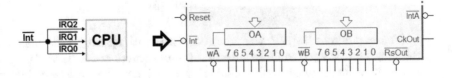

The processor's sequencer pays attention to the interrupt request input $\overline{\text{Int}}$ at the end of the execution of the current instruction (the one being executed when the request is made).

The following figure shows the sequence of operations that follow an interrupt request (represented by the red arrow). For the moment, we will not go into greater detail on this.

In this example, the moment the request is made, the RLA instruction is being executed. RLA execution is completed but, after this, the program's flow of execution is interrupted. The "interrupt handler" is executed instead, through a sort of forced jump. The handler includes the code (developed by the programmer) required to satisfy the service requested.

When it is finished, execution of the interrupted program resumes. There are clear similarities with the call and response subprograms.

It should be pointed out, however, that the interrupt handler was not launched by a CALL instruction inserted into the code. Rather it occurs through a hardware event, which by its nature can happen at any moment, asynchronously with the interrupted program.

A subprogram is called in synchronous mode because it is the program itself that operated it at a specific point in the algorithm's sequence.

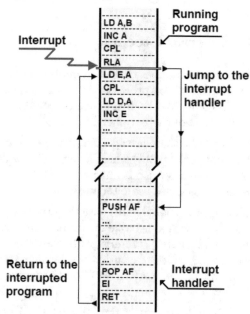

4.4.1 Enabling and disabling interrupts

The sequencer allows interrupt requests to be "masked" (ignored). To that end, the DMC8 has an internal IFF flip-flop (Interrupt Flip-Flop), which enables interrupts.

IFF = '0'	IFF = '1'
Disabled	Enabled

The system reset zeroes the IFF: the processor starts with interrupts disabled. It is the programmer's job to make sure that the interrupts are enabled if the system architecture requires it. Two enable/disable instructions are available and they change the value of IFF.

> EI   ; Enable Interrupts (IFF ← '1')
>
> DI   ; Disable Interrupts (IFF ← '0')

For reasons that will soon become clear, the EI instruction does not immediately enable interrupts. Rather their enabling is postponed until the next instruction has finished being executed.

The programmer inserts the EI instruction where necessary, usually after the system initialization instructions before entering the program's main loop.

We can also use these instructions to temporarily mask (and thus rehabilitate) interrupt requests. This can be useful when a group of instructions has to be executed without interrupts, that is in atomic mode[3].

For example, we might find ourselves in a situation where we need to respect specific time constraints and we can't interrupt the processor during the execution of certain sequences of instructions. So, let's insert a pair of DI and EI instructions as follows:

> DI   ; disable interrupts
>
> ... ; the atomic sequence starts here...
>
> ...
>
> ... ; and ends here
>
> EI   ; re-enable interrupts

In commercial processor models, we find different standards of interrupt requests. Some interrupt inputs ignore any disabling of the mechanism, which causes a "Not Maskable Interrupt". This is good for handling critical events like an alarm signal[4].

4.4.2 Interrupt mechanisms in detail

As we saw in Section 2.1.5, interrupt requests in the DMC8 are done through lines $\overline{IRQ2}$, $\overline{IRQ1}$ and $\overline{IRQ0}$. Let's assume the processor is executing a certain program when at least one of the inputs $\overline{IRQ2}$, $\overline{IRQ1}$ or $\overline{IRQ0}$ is activated (at '0') by a peripheral device requesting an interrupt.

Now let's examine DMC8's response mechanism to the interrupt. The following figure shows the steps of the "interrupt sequence", the launch of the interrupt handler and the return to the interrupted program.

1. While the processor is executing a program, it receives an interrupt request (at any time).
2. The current instruction is completed (RLA, in this example). The sequencer acknowledges the interrupt request only when RLA is ended.

---

[3] From the Greek "àtomos" which means "cannot be divided".

[4] The DMC8 only supports maskable interrupts.

3. If IFF = '0', the interrupts are disabled so the request is ignored. Then, as if nothing had happened, the processor goes on to the next instruction in the program without missing a clock cycle.

4. If interrupts are enabled, however, we start the sequence, but IFF is automatically zeroed to prevent further interrupts. This is necessary because an interrupt request is still active while the handler is being executed and another one would cause an interrupt of the interrupt.

5. The address of the instruction after the last one executed is saved on the top of the Stack (similarly to what happens with a CALL instruction). Let's keep in mind that at this moment, the processor only has the address of the next instruction to execute in the PC ('LD E,A' in this example).

6. The processor executes a jump to the location defined by the combination of lines $\overline{IRQ2}$, $\overline{IRQ1}$ and $\overline{IRQ0}$. The programmer needs to have allocated the specific interrupt handler at that address.

7. The processor executes the handler (following this, we assume that the peripheral removes the interrupt request).

8. An RET instruction returns the control to the interrupted program by retrieving the return address from the Stack (where it had been saved before the jump to the handler). RET is preceded by an EI, which enables the interrupts after the RET is executed. We have seen that the effect of EI is postponed to allow the handler to be entirely executed before any new interrupt comes.

9. The processor goes back to executing the interrupted program.

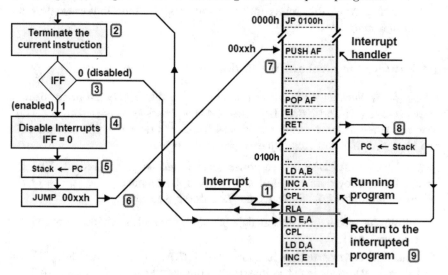

From step 3 to 6, this sequence lasts 11 clock cycles overall. Six of these cycles are used by the processor to save the return address in the Stack. In the first 5 cycles of the sequence, the processor informs the outside that it has accepted the interrupt request by activating the output (active low) $\overline{\text{INTA}}$ (Interrupt Acknowledgment).

The sequence is represented in the following timing diagram. Here we can see the activation of one of the interrupt request lines ($\overline{\text{IRQn}}$) at '0'.

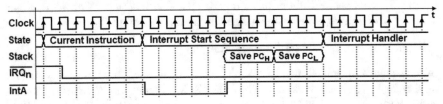

Notice that it isn't the transition of the line that requests the interrupt but simply the low level. This means that the device will keep the line active until the request is accepted. Assuming that the interrupts are enabled (IFF = 1), the processor starts the interrupt sequence when the execution of the current instruction is completed.

Output $\overline{\text{INTA}}$ is activated in the first 5 clock cycles of the interrupt sequence. During this time, the processor prepares the next operations internally. Line $\overline{\text{INTA}}$ is returned to the idle state and over the next 6 clock cycles, the processor saves the content of the Program Counter on the Stack.

At this time, the Program Counter has the address of the instruction that would have been executed if the processor had not been interrupted. Saving the content of the PC allows us to place the return address on the top of the Stack and so to restart the program as of the next instruction (that hasn't been executed yet).

After saving the return address, the sequencer overwrites the Program Counter with a jump address that depends on the combination of lines $\overline{\text{IRQ2}}$, $\overline{\text{IRQ1}}$ and $\overline{\text{IRQ0}}$, as explained in Section 2.1.5. To request an interrupt, we have up to seven different handlers available.

Interrupts are called "vectored", in the sense that the handlers' addresses (the "vectors") are arranged in the memory and are available in the DMC8 through an index made up of the combination of lines $\overline{\text{IRQ2}}$, $\overline{\text{IRQ1}}$ and $\overline{\text{IRQ0}}$. For ease of reference, we've reprinted here the table from Section 2.1.5.

$\overline{\text{IRQ2}}$	$\overline{\text{IRQ1}}$	$\overline{\text{IRQ0}}$	Interrupt Handler	Address
1	1	1	No Request	-
1	1	0	Interrupt 1	0008h
1	0	1	Interrupt 2	0010h
1	0	0	Interrupt 3	0018h
0	1	1	Interrupt 4	0020h
0	1	0	Interrupt 5	0028h
0	0	1	Interrupt 6	0030h
0	0	0	Interrupt 7	0038h

The first line corresponds to the case where there is no interrupt. In the other lines, we find the addresses of the seven possible handlers, which correspond to the different activations of $\overline{\text{IRQ2}}$, $\overline{\text{IRQ1}}$ and $\overline{\text{IRQ0}}$.

Notice that the area of memory reserved for the first six interrupt handlers is only 8 bytes each. This is just enough space to insert a jump to a larger memory area where we can memorize the handler's code.

Introductory example of an interrupt handler

Here, we show an initial example of an interrupt handler. For simplicity's sake, we'll refer to the base version of the "DMC8 Microcomputer" (Section 2.4.1), which has one input ($\overline{\text{Int}}$) that connects the three lines $\overline{\text{IRQ2}}$, $\overline{\text{IRQ1}}$ and $\overline{\text{IRQ0}}$ together. Therefore, the handler's code must be allocated as of location 0038h, based on the table.

Let's assume that a peripheral has sent data to the microcomputer and at the same time, activated input $\overline{\text{Int}}$. The processor executes the interrupt sequence that in order, disables the interrupts, saves the return address and jumps to the handler (at 0038h). The code of this example is as follows:

```
            ORG   0038h
IHANDLER:   PUSH  AF          ; save A and Flags contents on the Stack
            IN    A,(INP)      ; copy INP to OUTP (this is an example
            OUT   (OUTP),A     ; of operation requested by the peripheral)
            POP   AF          ; restore the contents of A and Flags
            EI                ; re-enable interrupts
            RET               ; return to the interrupted program
```

The processor executes the first instruction of the handler, PUSH AF. The instruction saves the content of register A and the flags on the Stack.

The next two instructions, IN and OUT, basically represent the operation requested by the interrupting device. For simplicity's sake, let's use the example of a request to copy an input port INP to an output port OUTP.

When this task has been completed, the handler restores the previous contents of register A and the Flags. Within the interrupt sequence the Program Counter is saved to be able to return to the calling program. Note that, aside from this, the sequence does not automatically save any other register. This means that it is the programmer's job to insert the right pairs of PUSH and POP to save and restore the registers that the handler modifies.

Here, the handler only uses the accumulator so the PUSH AF and POP AF pair has been inserted. This operation is necessary because the interrupt manager is executed asynchronously with respect to the program that was interrupted. This means that we need to make it so that, on returning to the interrupted program, this will find the same exact contents in the internal registers as there were before. The interrupt handler must leave the internal state of the processor perfectly unaltered.

The routine ends with the RET instruction, as in the case of subprograms but here there is an EI preceding it. This way, the interrupts, which were automatically disabled by the interrupt sequence are enabled again (as mentioned before, after RET is executed). RET brings the control back to the interrupted program, which goes back to work.

In this example, we have only considered the aspects related to the interrupt handler but ignored those related to activating and deactivating the request, which will be dealt with in the following pages.

4.4.3 Example of an interface with an interrupt request

Let's take another look at the interface circuit with handshake hardware that we studied in Section 4.2.1. Here, we have made changes so that it can be handled with the interrupt technique rather than by polling. The figure below highlights the additional parts in red boxes.

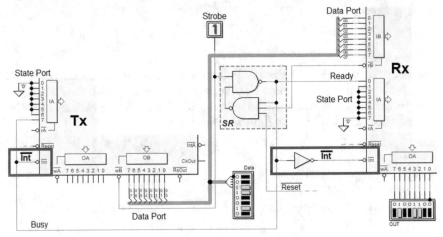

From the side of the receiver, the Ready line, which is already available to the processor for polling through the state port, is used here to request an interrupt (with the addition of a NOT since input $\overline{\text{Int}}$ is active low). When a data byte comes from Tx, it produces an interrupt request for Rx's processor, which can intervene in the interface only when necessary, without having to poll the state port.

From the other side, the transmitter receives an interrupt every time the Busy line goes to '0', that is when Rx shows it's ready to receive data. The data will no longer be sent in the main loop of Tx, but will be executed through interrupt requests.

The Tx and Rx subsystems' new way of working constitutes a great advantage. The receiver can continue to work on other tasks without having to periodically check for new data from the transmitter. It can handle requests only at the moment they are needed. The transmitter can also work on its own main tasks and only take care of transmitting data when Rx requests it.

Handling Rx through interrupts

Let's look at one example of Rx software management. We define the data port and the output port (the state port is not defined because it is unnecessary, as we will see in the code).

```
OUTP        EQU    00h          ; OA output port (OUT)
DATAP       EQU    01h          ; IB input port (Data Port)
```

We insert the usual connection to the reset at 0000h, which makes us jump to the start of the program (START). We'll add another jump that will let us go from the predefined location 0038h to the interrupt handler (IHANDLER). For our own convenience, we'll allocate that together with the rest of the code.

```
            ORG    0000h        ; link to the reset
            JP     START        ; jump to the program start
            ORG    0038h
            JP     IHANDLER     ; jump to the interrupt handler
            ORG    0100h
```

The program initializes the Stack Pointer and then zeroes the output port. Then before going into the main loop, it enables interrupts with the EI instruction. In the loop, we find a call to subprogram TASK_RX, which simulates a general task unrelated to the interface handling.

```
START:      LD     SP,0FFFFh    ; initialize the Stack Pointer
            LD     A,00000000b  ; initialize the output port to zero
            OUT    (OUTP),A
            EI                  ; enable interrupts
MAIN:       CALL   TASK_RX      ; simulate a generic task that
            JP     MAIN         ; is executed in the main loop
```

Note that the handshake network is identical to the one in Section 4.2.1, except that there is an added connection to line $\overline{\text{Int}}$. We've seen that a Strobe signal (which indicates that a new data byte has been received) activates the SR flip-flop output Ready. In that version, the receiver's program went to read the state port by polling, to check the state of SR.

In this circuit, however, activating SR directly produces an interrupt request to the processor. This interrupts the execution of the main program and "launches" the interrupt handler (its code is presented here below).

```
IHANDLER:   PUSH   AF           ; save A and Flags on the Stack
            IN     A,(DATAP)    ; acquire the received data byte
            OUT    (OUTP),A     ; copy it to the output port
            POP    AF           ; restore the registers saved before
            EI                  ; re-enable interrupts
            RET                 ; return to the interrupted program
```

The interrupt handler reads the data byte on the Data Port DATAP and produces it on output port OUTP. Because of the handshake hardware network, reading DATAP automatically resets the SR flip-flop, allowing it to log the arrival of new data again.

From the receiver's perspective, the important thing is that when the SR flip-flop is zeroed, it also deactivates the interrupt request. We can be sure that once the handler has been executed, the interrupt request has been deactivated and the processor can go back to carrying out the tasks that it had left undone without being interrupted again until new data arrive.

For the sake of completeness, the TASK_RX is shown below. As explained before, its purpose is only to simulate the execution of a generic task.

```
TASK_RX:   LD    C,22        ; simulate a generic task
TASK:      DEC   C           ; 4 +
           JP    NZ,TASK     ; 10 = 14 cycles; 14 x 22 = 308 cycles +
           RET               ; 17 (call) +7 (ld C) +10 (ret) = 342 cycles
```

Handling Tx through interrupts

The code for the transmitter is similar to that of the receiver. In the first couple lines, we define the output Data Port but not a State Port, which would serve no purpose here. The code also defines an 8-bit DATA variable, which simulates the data to transmit.

```
DATAP      EQU   01h         ; output Data Port (OB)
DATA       EQU   0FC00h      ; variable containing the data to transmit
```

The settings of the reset (0000h) and interrupt (0038h) locations is the same as that of Rx. They send us back to the START and IHANDLER labels.

```
           ORG   0000h       ; link to the reset
           JP    START       ; jump to the program start
           ORG   0038h
           JP    IHANDLER    ; jump to the interrupt handler
           ORG   0100h
```

We initialize the Stack Pointer and immediately after, we zero DATA (it is managed only by the interrupt handler), and then enable the interrupt mechanism (with the EI instruction). In the main loop, we simulate carrying out a general task by calling the TASK_TX subprogram. Notice that in the main loop we do not take care of data transmission. That is delegated entirely to the interrupt handler.

```
START:     LD    SP,0FFFFh   ; initialize the Stack Pointer
           LD    A,0         ; initialize the variable DATA with
           LD    (DATA),A    ; simulated data to transmit
           EI                ; enable interrupts
MAIN:      CALL  TASK_TX     ; simulate a generic task that
           JP    MAIN        ; is executed in the main loop
```

The interrupt is executed every time the Busy line is brought to '0', that is when Rx tells Tx to get ready to receive a new data byte.

```
IHANDLER:   PUSH  AF          ; save A and Flags on the Stack
            LD    A,(DATA)     ; read the variable DATA and increment it
            INC   A            ; to simulate a new data byte to transmit
            LD    (DATA),A     ; save the new value into the variable
            OUT   (DATAP),A    ; transmit it though the Data Port
            POP   AF           ; restore the register contents saved before
            EI                 ; re-enable interrupts
            RET                ; return to the interrupted program
```

We can see that the PUSH AF and POP AF instructions in the interrupt handler are necessary to preserve the internal state of the processor.

First, a new data byte is simulated by incrementing the DATA variable. Then, the new value is copied to the Data Port. The output writing launches the handshake mechanism, so Rx will acquire the new data. When the Busy line goes back to '0', it relaunches a new interrupt request in Tx.

The EI and RET instructions allow us to go back to the interrupted program and they enable the interrupts (which were disabled before the handler was launched) again.

Finally, for the sake of completeness, the code for the TASK_TX subprogram is shown below. Like the subprogram for Rx, this simulates a general transmitter task, independent from data transmission.

```
TASK_TX:   LD    C,10        ; simulate a generic task
TASK:      DEC   C           ; 4 +
           JP    NZ,TASK      ; 10 = 14 cycles; 14 x 10 = 140 cycles +
           RET                ; 17 (call) +7 (ld C) +10 (ret) = 174 cycles
```

4.5 Using vectored interrupts

Now, let's look at a general case where there are multiple devices that can request an interrupt from the processor. Here, we need to insert the appropriate hardware that will allow us to select an interrupt handler from those available, based on the specific device that requested the interrupt. It is also important to adopt criteria that will allow us to assign priority if we have simultaneous requests from multiple devices.

There are many solutions to this problem. For the "DMC8 Enhanced Microcomputer" component (included in the Deeds library, see Section 2.4.1), we decided to insert a priority encoder among the devices and the CPU, as shown in the following figure.

The circuit here is purely combinational. If none of the input lines to the encoder is active (i.e. if they are all high), the processor's interrupt requests will be high and there is no interrupt request for the CPU.

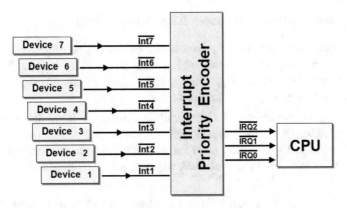

If one or more inputs is brought low, however, a configuration appears and corresponds to the highest priority input that is activated. The following truth table describes the encoder:

Int7	Int6	Int5	Int4	Int3	Int2	Int1	IRQ2	IRQ1	IRQ0	Interrupt
1	1	1	1	1	1	1	1	1	1	No request
1	1	1	1	1	1	0	1	1	0	Int. 1
1	1	1	1	1	0	-	1	0	1	Int. 2
1	1	1	1	0	-	-	1	0	0	Int. 3
1	1	1	0	-	-	-	0	1	1	Int. 4
1	1	0	-	-	-	-	0	1	0	Int. 5
1	0	-	-	-	-	-	0	0	1	Int. 6
0	-	-	-	-	-	-	0	0	0	Int. 7

For example, if only input $\overline{\text{Int5}}$ is active, the configuration '010' appears in the output. It corresponds to interrupt request 5. If another device also activates a line, for example $\overline{\text{Int2}}$, the encoder ignores it since it has a lower priority.

However, if we activate line $\overline{\text{Int7}}$ (the highest priority of all), the code goes to value '000' in the output, corresponding to interrupt request 7. All the other requests are ignored.

The "DMC8 Enhanced Microcomputer" (introduced in Section 2.4.1) has seven interrupt request lines (the detail is indicated by the arrow in the figure at the right).

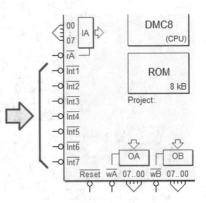

4.5.1 Considerations on recognition and priority

The network discussed above resolves two problems, recognizing the device that made the interrupt request and handling the priority among the devices making simultaneous requests. Then recognition happens formally since this network associates only one handler to each device.

The priority is embedded in the decoding system; a high priority request makes the network ignore lower priority requests. Note that having multiple pending requests at the same time means that only the highest priority will initially be served. Every device that requests an interrupt keeps the request active until it is fulfilled.

As soon as the highest priority device has been served, it must deactivate the request. Then the encoder provides the code corresponding to the device with the next highest priority (that is actively making a request).

4.5.2 Extending to a higher number of devices

Here, we analyze a system that has more than seven devices that can interrupt the processor. Since the processor has only 7 interrupt vectors, we need to find a way for multiple devices to share the same vector.

Here, we have chosen to change the connection of the lower priority devices and keep the better efficiency for those of higher priority.

As shown in the following figure, we have introduced a logic gate that conveys the requests of the lower priority devices (here, they are: 1A, 1B, 1C and 1D) into one single line. The AND gate activates line $\overline{\text{Int1}}$ if there is one or more interrupt request from devices 1A, 1B, 1C or 1D[5].

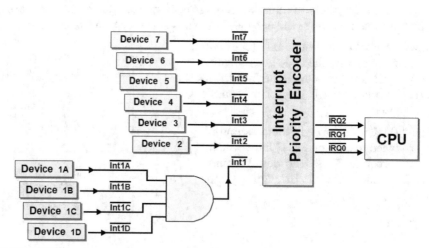

---

[5] Note that an AND gate, interpreted in "negated logic" (i.e., in terms of active low signals), generates '0' when at least one of the inputs is at '0'.

Clearly, there is no way to know through this encoding hardware which one of these four devices has requested the interrupt or which has priority over the others. We have to settle for doing this operation through software in interrupt handler 1.

First of all, this software solution should identify the device that requested the service by executing a test on the devices involved using the corresponding State Ports. So, we poll the devices after an interrupt request. Once the device is recognized, the interrupt routine executes the task requested.

The order of execution in the test allows us to assign a priority scale for the various devices (1A, 1B, 1C and 1D) so that the highest priority device's request is granted first if more than one request comes simultaneously.

4.5.3 Example of handling vectored interrupts

Here below is a possible outline of the code dealing with handling vectored interrupts. This example includes polling for the lower priority devices, which is in interrupt handler 1. At the beginning, we define a state port (STATP) and insert the usual jump to the start of the main program.

```
STATP      EQU    00h          ; input State Port (IA)
           ORG    0000h        ; link to the reset
           JP     START
```

Then we insert the definitions related to the vector table, i.e. the jumps to all the vectored interrupt handlers.

```
           ORG    0008h        ; interrupt 1
           JP     HINT1
           ORG    0010h        ; interrupt 2
           JP     HINT2
           ORG    0018h        ; interrupt 3
           JP     HINT3
           ORG    0020h        ; interrupt 4
           JP     HINT4
           ORG    0028h        ; interrupt 5
           JP     HINT5
           ORG    0030h        ; interrupt 6
           JP     HINT6
           ORG    0038h        ; interrupt 7
           JP     HINT7
```

The main program (not of interest to us in this explanation) is shown here in the form of a simple trace of the code. It includes the necessary definition of the Stack Pointer (it is important to remember that interrupts use the Stack). Also, interrupts are enabled (EI) before entering the main loop.

```
                  ORG    0100h           ; main program
START:            LD     SP,0FFFFh       ; initialize the Stack Pointer
                  ...    ...             ; ...omissis...
                  EI                     ; enable interrupts
MAIN:             ...    ...             ; ...omissis...
                  JP     MAIN
```

Now let's define the lines of code for the individual handlers in order of priority (just for convenience). Let's start with the handler of interrupt 7.

```
HINT7:            PUSH   AF              ; save the contents of the registers in use
                  ;...                   ; handler 7 (with the highest priority)
                  POP    AF              ; restore the registers saved before
                  EI                     ; re-enable interrupts
                  RET                    ; return to the interrupted program
```

This is clearly just an outline like those of the other interrupt handlers (identical in form) which we do not show up to interrupt handler 2.

```
HINT2:            PUSH   AF
                  ;...                   ; handler 2
                  POP    AF
                  EI
                  RET
```

To define the lowest priority interrupt handler, let's assume that devices 1D, 1C, 1B and 1A (shown in the previous figure) submit interrupt request lines $\overline{Int1D}$, $\overline{Int1C}$, $\overline{Int1B}$ and $\overline{Int1A}$ on bits 7, 6, 5 and 4, respectively, of the STATP state port.

After the register A and the Flags are saved, the code of the handler proceeds by polling the bits of the State Port in order of priority (we assume 1D has the highest priority and 1A the lowest).

Based on the state of lines $\overline{Int1D}$, $\overline{Int1C}$, $\overline{Int1B}$ and $\overline{Int1A}$, we jump to the corresponding part of the handler's code.

```
HINT1:            PUSH   AF              ; save A and Flags contents on the Stack
                  IN     A,(STATP)       ; read the State Port
                  BIT    7,A             ; check line !Int1D
                  JP     Z,Handle1D      ; jump if it requires to be served
                  BIT    6,A             ; check line !Int1C
                  JP     Z,Handle1C      ; jump if it requires to be served
                  BIT    5,A             ; check line !Int1B
                  JP     Z,Handle1B      ; jump if it requires to be served
```

The case of the handler of device 1A is evaluated by exclusion. In the following, we see the code outlines for the interrupt handler of each device. Each handler ends with a jump to the exit code (EXIT) where the previously saved register

A and the Flags are restored and the interrupts re-enabled. Finally, we go back to the interrupted program.

```
Handle1A:    PUSH  ...              ; save all the registers in use
             ;...                   ; handle device 1A
             POP   ...              ; restore the saved registers
             JP    EXIT

Handle1B:    PUSH  ...              ; save all the registers in use
             ;...                   ; handle device 1B
             POP   ...              ; restore the saved registers
             JP    EXIT

Handle1C:    PUSH  ...              ; save all the registers in use
             ;...                   ; handle device 1C
             POP   ...              ; restore the saved registers
             JP    EXIT

Handle1D:    PUSH  ...              ; save all the registers in use
             ;...                   ; handle device 1D
             POP   ...              ; restore the saved registers
;
EXIT:        POP   AF               ; restore A and Flags saved before
             EI                     ; re-enable interrupts
             RET                    ; return to the interrupted program
```

If more than one of these devices interrupts at the same time, the first one tested is served first. After leaving the handler, the hardware makes a new interrupt request from the devices that haven't been served yet, so the code is executed again until all the requests are satisfied in order of priority.

4.6 Interrupt timers

We have seen that if a system doesn't use interrupts, the processor executes one single program. This program can be made up of many different modules but they will always be executed under the strict control of the "main program". This method works for simple systems but we generally find that in real cases the processor needs to satisfy external needs that emerge at unpredictable times and are often urgent and irrevocable.

For example, in a system that receives "alarm signals" from a plant, the various input/output devices involved must request services from the processor through interrupts, so that they are satisfied as quickly as possible.

Another example is a system that acquires data from one or more sensors. Their output cannot be neglected by the processor otherwise important information could be lost.

There are other cases where the specifications require the execution of tasks at precise time intervals. Some examples are the request to see data on a panel at regular intervals or having them transmitted periodically to a control and supervision room.

If we tried to resolve the problem with a purely software approach, we would have to write a time-measuring program and in the meantime carry out other tasks such as cyclically checking a sensor, for example. We could write a delay subprogram and include the instructions for checking the sensor in the loop. However, this method is only applicable in very simple systems. If there were many sensors and many tasks to do with them or if the timelines were different among them, the source code would be complex and convoluted. Also, timing might not be reliable or simply approximate.

Clearly in these cases, we need to choose another path. This is why systems often include one or more hardware timing devices, which can request interrupts at pre-established times. From now on, we will use the term "timer", which is commonly used in practice and in the literature.

Timers are generally programmable (see the figure at the right), in that they have one or more control ports that the program uses to activate them, deactivate them, define the time interval or set a "one shot mode" or a "cyclic mode". Timers use pre-settable counters that are timed by the system clock and handled by a control logic that sets the counter according to their programming.

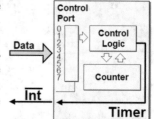

4.6.1 A specialized timer

For simplicity's sake, we will use pre-defined hardware timers that are not programmable by the processor. The component used here is called an "Interrupt Timer" and is available in the Deeds library among the counters (see below).

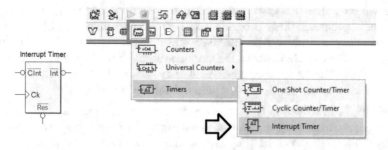

Let's assume we are working with a system that has only one interrupting device, the timer. We refer to a "DMC8 Microcomputer" (see Section 2.4.1) and its one interrupt request line $\overline{\text{Int}}$ (as shown in the following figure).

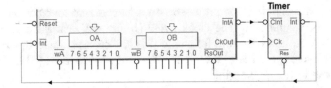

In this example, the microcomputer interacts with the timer by using lines $\overline{\text{Int}}$ and $\overline{\text{IntA}}$, which are connected to timer lines $\overline{\text{Int}}$ and $\overline{\text{CInt}}$ ("Clear Interrupt") respectively. The timer also has a reset input and a clock input ($\overline{\text{Res}}$ e Ck), which are connected to $\overline{\text{RsOut}}$ and CkOut of the microcomputer.

We won't go into the timer's internal network since it is enough to examine how it behaves. The timer functions cyclically with a period of ΔT (for example: $1mS$). At the end of each interval ΔT, the device activates request line $\overline{\text{Int}}$. This periodic request is called a "Timer Tick".

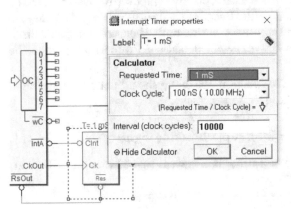

To set period ΔT in the design phase, we have to open the Interrupt Timer properties dialog box (see the figure at the left) by double clicking on the component.

Aside from being able to edit the label of the component, we can also define the desired time ΔT by selecting it from the list of available values.

It is important to declare the clock period the timer will use in the "Clock Cycle" field. The system calculates the number of clock cycles based on the ΔT that we want and the given clock period. This defines the "module" of the counter in the timer.

Notice that ΔT is rigorously guaranteed by the timer's internal hardware and is not influenced by the processor's interrupt response times. Cyclically, every ΔT, the timer activates line $\overline{\text{Int}}$.

Let's take a look at the following figure that shows the timing diagram of the signals exchanged between the timer and the microcomputer following the activation of the $\overline{\text{Int}}$ request, which happens in clock cycle (a).

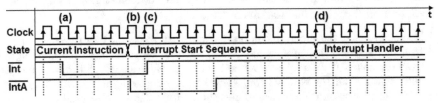

If interrupts are enabled, the processor stops the current instruction and starts the interrupt sequence in the clock cycle (b) by activating $\overline{\text{IntA}}$. The timer receives $\overline{\text{IntA}}$ on line $\overline{\text{CInt}}$ and following this, deactivates interrupt request $\overline{\text{Int}}$ in the next cycle (c). At the end of the interrupt sequence, in (d) the processor launches the interrupt handler.

Remember that interrupt request $\overline{\text{Int}}$ must be withdrawn by the device that made it before the corresponding handler has been fully executed. If it remained active, the processor would be interrupted again.

With the timer, the next interrupt should only come when the next ΔT interval expires. To achieve this, we have used the simplest, most immediate method: to reset the request as soon as the processor activates line $\overline{\text{IntA}}$. In real situations, we have many devices that can generate interrupts so requests should be deactivated differently and most importantly, selectively.

Now, let's look at a case where the interrupts are temporarily disabled[6], at the moment the timer activates line $\overline{\text{Int}}$. Because the processor doesn't memorize interrupt requests, they must be kept active until they are accepted. Since the timer's request remains active, it will be satisfied (albeit late) when the interrupts are re-enabled.

Note that for the mechanism to function, disabling should not last longer than a period ΔT, otherwise we lose a "Timer Tick".

4.6.2 Example of a timer interrupt: blinking lights

Our first example of the use of a timer is a system that flashes an LED light, as shown in the following schematic.

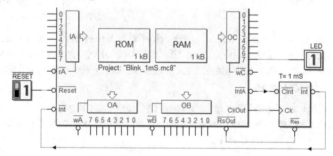

We have connected an LED light to bit 7 of the OC port of a "DMC8 Microcomputer" (see Section 2.4.1). The timer has been connected and set ($\Delta T = 1 mS$) as it was in Section 4.6.1. We want the LED light to stay on for $1 mS$, and off for the same amount of time[7].

---

[6] Remember that the program can disable and then re-enable interrupts.

[7] In a real system, this time would be too short to be able to see the LED light flash. However, if we set the clock animation to a frequency of $10 KHz$ (that is 1000 slower than the nominal frequency), and simulate in interactive mode, the LED light turns on for one second and turns off for the same amount of time.

In the first couple of lines of code, we define the output port (LED) and a variable (LEDMEM), which we use to store a software copy, in the RAM, of the last byte copied to the port. Then we define the jumps to the start of the program (START) and to the interrupt handler (HINT).

```
LED        EQU    02h          ; OC output port
LEDMEM     EQU    0FC00h       ; software copy of the output port
           ORG    0000h
           JP     START
           ORG    0038h
           JP     HINT
           ORG    0100h
```

After we initialize the Stack Pointer, the output port LED and the corresponding software copy LEDMEM. The LED light will be off at reset. Before entering the main loop, the EI instruction enables the interrupts.

```
START:     LD     SP,0FFFFh     ; initialize the Stack Pointer
           LD     A,00000000b   ; and the output port LED
           OUT    (LED),A       ; start with LED off
           LD     (LEDMEM),A    ; copy the LED state to the software copy
           EI
```

For our blinking light, it's possible for the main loop to be empty. However, it should be pointed out that a standard system could perform any task in the main loop and those tasks would be totally independent from what is carried out in the interrupt program. Therefore, we have inserted in the main loop a call to a standard subprogram called PROCESS.

```
MAIN:      CALL   PROCESS       ; execute a generic task (any one)
           JP     MAIN
```

To be complete, we show the PROCESS subprogram, which could execute any task but here, is left blank.

```
PROCESS:   NOP                  ; execute any task
           RET
```

Periodically, every $1mS$, the timer requests an interrupt, so the interrupt handler HINT is executed. The code also has our usual PUSH, POP, EI and RET, which have already been copiously described. The part on managing LED lights, however, has been highlighted.

In the following listing, we can see that the timer guarantees that the HINT handler is launched cyclically, every millisecond. So, upon each call we only need to invert the state (on/off) of the LED light. The state is stored in bit 7 of variable LEDMEM, so (as is written in the code) we read the variable, invert the value of bit 7, re-save it in the memory and transcribe it on the output port so it turns on (or off) the LED light.

```
HINT:        PUSH  AF            ; save A and Flags
             LD    A,(LEDMEM)    ; get the software copy in A
             XOR   10000000b     ; invert bit 7
             LD    (LEDMEM),A    ; save back the software copy
             OUT   (LED),A       ; copy the new state to the port
EXIT:        POP   AF            ; restore A and Flags
             EI                  ; re-enable interrupts
             RET                 ; return to the interrupted program
```

Note that the interrupt handler generally cannot trust the contents of the processor's registers. This is because they are normally used by the main program and the subprograms it calls. This is why if the interrupt handler has to save information, it has to rely on the variables in the memory (such as LEDMEM in this example). In doing so, it must always save and refresh the contents of the registers it uses.

Changes to the blinking light

Here we have a change in the software that manages the blinking light, which leaves the hardware and settings unchanged, especially the ΔT of the timer at $1mS$. However, we want to bring the on/off times to one second by making the microcomputer work with the clock at a frequency of $10MHz$, and the interrupt handler being called every $1mS$.

In order to invert the state of the LED light each second, when the handler is called it has to execute a count for 999 calls. Then on the thousandth call, it inverts the state of the LED light and then resumes counting.

In the following, we only show the changes to the previous code. Among the initial definitions, we now have an additional 16-bit variable (TIME), which is used to measure time, or more precisely, the number of interrupt calls.

```
LED          EQU   02h           ; output port OC
LEDMEM       EQU   0FC00h        ; software copy of the port
TIME         EQU   0FC01h        ; time count variable (16-bit)
```

Here we omit the link to the reset and the interrupt, since they are identical to the previous example. However we show the initializations in full; the new variable TIME has been added here and initialized with the number 1000. As we shall soon see, this is to count back the calls in the handler's code. The 16-bit constant is loaded in register HL and then in the memory.

```
START:       LD    SP,0FFFFh     ; initialize the Stack Pointer
             LD    A,00000000b   ; and the output port
             OUT   (LED),A       ; LED off on start
             LD    (LEDMEM),A    ; copy the LED state to the software copy
             LD    HL,1000       ; save the constant 1000 in TIME (16-bit)
             LD    (TIME),HL
             EI
```

For greater clarity, we've shown the main loop as well. It is actually identical to the previous case so the same observations apply here. We've omitted the code of the PROCESS subprogram, which is identical to the previous one.

```
MAIN:       CALL  PROCESS        ; execute a generic task (any one)
            JP    MAIN
```

Let's pay special attention to the new interrupt handler where the differences have been highlighted in another color. Among the registers saved in the beginning, we've added HL, which is used by the handler as a time counter.

```
HINT:       PUSH  AF             ; save A, Flags and HL
            PUSH  HL
```

Before updating the state of the LED light with the same instructions sequence as the previous example, we need to check whether it is time to execute it (i.e. if one second has passed).

The processor executes this check for every call, that is every 1 mS. It copies the TIME variable in HL, decrements it (on 16 bits), and then updates the memory with the new value. The count is executed backward from 1000. We need to check each time if it has reached 0.

Since the 16-bit decrement instructions do not change the flags (see Section 3.3.2.5), we insert two instructions (LD A,H and OR L) to evaluate whether HL goes to zero. If it hasn't reached 0 yet, the conditional jump will have us leave the handler by going to the "exit code" (EXIT).

```
            LD    HL,(TIME)      ; copy the TIME variable to HL
            DEC   HL             ; decrement it and
            LD    (TIME),HL      ; update it in memory
            LD    A,H            ; check if HL has been zeroed
            OR    L
            JP    NZ,EXIT        ; exit if it is not
```

Otherwise, we move on to re-initialize TIME to its initial value, and then invert the state of the LED light (note that the processor executes this part of the code once every 1000 handler calls).

```
            LD    HL,1000        ; re-initialize HL to 1000
            LD    (TIME),HL      ; and the TIME variable
            LD    A,(LEDMEM)     ; get the software copy in A
            XOR   10000000b      ; invert bit 7
            LD    (LEDMEM),A     ; save back the software copy
            OUT   (LED),A        ; copy the new state to the port
```

The exit code adds the content restoration of HL to the previous example.

```
EXIT:       POP   HL             ; restore A, the Flags and HL
            POP   AF
            EI                   ; re-enable interrupts
            RET                  ; return to the interrupted program
```

Closing observations

— The interrupt mechanism guarantees the tasks are carried out with no need to intervene in the logic of the main program.
— Interrupts that are associated to the timer guarantee a rigorous assessment of time, which is independent of the main program's tasks.
— Delay loops haven't been used to measure time. As a result,the processor's computing capacity is not wasted.

4.6.3 Timers and concurrent program execution

When one single program is in execution in our system, the sequence of operations is rigorously determined by the programmer. In a system with many devices managed through interrupts, there is generally one handler for each.

Each handler can be launched by its interrupt request at any time, that is totally asynchronously from the main program. As a consequence, the individual interrupts are asynchronous from each other. The handler call order can never be known ahead of time unless there are clear functional constraints. If an operation 'B' needs the results from an operation 'A', the order of execution of the operations must certainly be first 'A' and then 'B'.

When operations 'A' and 'B' are executed by two different asynchronous programs, we need to guarantee that the appropriate order of execution is respected. These operations can be synchronized through the use of shared variables called "semaphores", which let the various modules signal when they are finished executing their operations.

A timer lets us face this type of problem in an orderly way. Therefore, we will be able to regularly execute tasks that are totally independent of one another and independent of those carried out by the main program (except for some that are interrelated like the case described above).

In a way, it is as if we had separate processors, each executing its own program. In reality, there is only one processor and we will execute various programs a bit at a time concurrently by breaking down the processor's computational capabilities over time[8].

The programs that execute operations 'A' and 'B' described above are executed concurrently. We must consider the coherence of the data when exchanging data between concurrent and communicating processes. In other words, the synchronization mentioned before must allow for the individual processes to execute atomic sequences so that the interrupt mechanism itself doesn't cause partial updates of the data. We can achieve this through a judicious use of interrupt enable (EI) and interrupt disable (DI) instructions, as described in Section 4.4.1. In the following section, we will examine some programming examples including cases that require atomic operations.

---

[8] Multi-user and multi-tasking operating systems (such as Windows®, Linux®, etc.) are based on these concepts, but outside the scope of this book.

4.7 Examples of programming and interfacing

The examples in this section[9] will let us develop the programming and interfacing techniques that use interrupts.

4.7.1 Pulse generator (at system reset)

We want to generate a high level pulse with a duration of $5\,mS$ as of system reset by using the interrupt technique. The pulse is produced at the PULSE output, connected to bit 0 of the OC port on a DMC8 Microcomputer (see the following schematic). Line EN, which is connected to bit 0 of input port IA, checks if the pulse generation has been enabled.

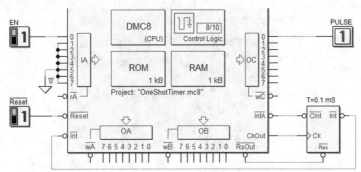

Assume that the system executes other (unspecified) tasks in the main program, which are summarized in a periodical call to the standard PROCESS subprogram. The "Interrupt Timer" component is programmed to request an interrupt every $0.1\,mS$, through line $\overline{\text{Int}}$. When the processor activates $\overline{\text{IntA}}$, this deactivates that request in the timer.

Solution

In the first part of the code, we define:

(a) the addresses of the ports (the same as default),
(b) the constant 'NCalls' at 50 (the number of calls to count to get $5\,mS = 50 \cdot 0.1\,mS$),
(c) the TIME and ENABLED variables, to count the time and to enable pulse generation, respectively,
(d) a jump to the interrupt handler at the reserved location 0038h.

ENP	EQU	00h	; IA input port (EN)
PULSEP	EQU	02h	; OC output port (PULSE)
NCalls	EQU	50	; number of timer ticks in 5 mS

---

[9] Remember that all the programs and networks in this text are available on the Deeds website, ready to analyze, simulate and change.

```
TIME        EQU    0FC00h          ; time count variable
ENABLED     EQU    0FC01h          ; count enable variable
            ORG    0000h
            JP     START
            ORG    0038h           ; link to interrupt handler
            JP     HINT
            ORG    0100h
```

Before entering the main loop, we initialize the Stack Pointer and the TIME variable at 50. Then we read input EN, copy it in the ENABLED variable and zero the bit we are not interested in. The fact that the position of EN on bit 0 corresponds with that of PULSE, also on bit 0, allows us to directly use the value that is now in ENABLED to immediately activate PULSE (or not). Finally, we enable the interrupts with the EI instruction.

```
START:      LD     SP,0FFFFh       ; initialize the Stack Pointer
            LD     A,NCalls        ; and the variable TIME
            LD     (TIME),A
            IN     A,(ENP)         ; read EN from the port
            AND    00000001b       ; masks the bits that don't interest us
            LD     (ENABLED),A     ; and copy EN to variable ENABLED
            OUT    (PULSEP),A      ; activate the output line
            EI                     ; enable interrupts
```

The main program cyclically calls the PROCESS subprogram, whose content is not relevant for the present aim (in this example, it's just a placeholder), but we assume it engages the processor in some task. Notice that pulse generation does not interfere with operations in PROCESS (except for periodic interrupts for a couple microseconds).

```
MAIN:       CALL   PROCESS         ; execute any task
            JP     MAIN
PROCESS:    NOP                    ; placeholder for any task
            RET
```

The HINT interrupt handler is executed every 0.1 mS thanks to the timer's periodic interrupt requests. Note that this time is an assurance for our system so we can take it as a reference point to measure the amount of time that has elapsed as of reset.

At the start of the handler, we insert a PUSH AF instruction to save the current content of register A and of the flags. Our program doesn't change the content of other registers so we don't need to save anything else.

```
HINT:       PUSH   AF              ; save A and the Flags
            LD     A,(ENABLED)
            BIT    0,A             ; if the pulse generation is not enabled...
            JP     Z,EXIT          ; exit immediately
```

Right after, we check bit 0 of the ENABLED variable. If it is at zero, we simply exit immediately from the handler. If it's not, the pulse generator is enabled so we read the TIME variable, decrement it and rewrite it in the memory. TIME is therefore decremented by one each time the handler is executed, every 0.1 mS. After 50 decrements, 5 mS will have passed so we can go on.

```
LD    A,(TIME)        ; count the interrupt calls
DEC   A
LD    (TIME),A
JP    NZ,EXIT         ; exit if a second has not elapsed
```

So when we get to zero, we stop the pulse by zeroing the PULSEP port. We also zero the ENABLED variable, so that we can simply exit the next time the handler is executed.

```
LD    A,00000000b     ; one second elapsed, zero
OUT   (PULSEP),A      ; the output and the variable ENABLED
LD    (ENABLED),A
```

Finally, before executing the EI and RET instructions, we restore the A content and the Flags.

```
EXIT:      POP   AF        ; restore A and the Flags
           EI              ; re-enable interrupts
           RET             ; return to the interrupted program
```

4.7.2 Finite State Machines

The specifications of this example are the same as that in Section 3.5.4 but are printed here again for ease of consultation. In this case, however, we need a solution based on the use of interrupts since we assume the processor is continually working on executing other unspecified tasks in the main loop.

We want to emulate the function of the synchronous sequential component that has the connections shown in the figure at the left.

Outputs W1 and W0 can assume binary values from '00' to '11'. The component is described in Finite State Machine terms (FSM) in the ASM (Algorithmic State Machine) chart on the right.

At each rising edge of the clock CK, the component increments or decrements number W1W0 based on the value of UP.

We are asked to write a program that creates the synchronous FSM described.

Use a timer that simulates the active edge of the clock by interrupting the processor at regular intervals of 0.4 mS.

The first solution

In the following schematic, we have connected UP to pin 0 of input port IA in a DMC8 Microcomputer, and outputs W1 and W0 to output port OA. The unused wires of IA are connected to '0'.

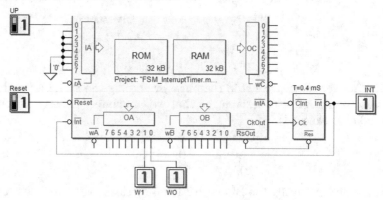

The microcomputer is paired with a timer that generates an interrupt request on $\overline{\text{Int}}$ every $0.4\,mS$. Since line $\overline{\text{IntA}}$ is connected to the timer's input $\overline{\text{CInt}}$, the processor's activation of $\overline{\text{IntA}}$ deactivates the timer's request. The timer emulates a clock with a period of $0.4\,mS$ ($2500 Hz$).

In this first solution, rather than reproducing the description of the states given by the ASM chart, we re-interpret the FSM in terms of a non-cyclical bidirectional counter. Based on the value of the input, we ascertain the direction of the count, so we increment or decrement the state code (which we make coincide with the number generated on outputs W1 and W0), and make sure we don't go over the limits of the count.

The algorithm described here is executed by the interrupt handler. However, in the main program, we have inserted a simple placeholder for a standard independent task as suggested in the previous examples.

Now, let's look at the following code. To begin, we have defined the two input and output ports and declared STATE, a variable used to memorize the code for the state of the FSM (and thus of the outputs). What follow are the definitions of the jumps to the start of the program and the interrupt handler.

```
UPINP      EQU    00h          ; IA input port (UP input line)
W1W0P      EQU    00h          ; OA output port (W1,W0 output lines)
STATE      EQU    0FC00h       ; state of the FSM
           ORG    0000h
           JP     START
           ORG    0038h
           JP     HINT
           ORG    0100h
```

We initialize the Stack Pointer, the STATE variable at state (a) and the port at the corresponding outputs (W1W0 = '00'). Then we enable the interrupts and enter the main loop, which as mentioned before, calls the PROCESS subprogram that we imagine is carrying out general tasks.

```
START:      LD      SP,0FFFFh       ; initialize the Stack Pointer
            LD      A,00h           ; set the FSM state to zero
            LD      (STATE),A
            OUT     (W1W0P), A      ; set outputs as W1 = '0', W0 = '0'
            EI                      ; enable interrupts
MAIN:       CALL    PROCESS         ; execute any task
            JP      MAIN
PROCESS:    NOP                     ; placeholder for any task
            RET
```

Now, let's look at the interrupt handler (HINT), which is called by the timer[10] every $0.4\,mS$. First, we save A and the Flags, then we see if input UP is requesting to send the count up or down.

```
HINT:       PUSH    AF              ; save A and Flags
            IN      A,(UPINP)       ; read line UP from the port
            BIT     0,A             ; and check its value
            JP      Z,GODOWN        ; '0' = previous state, '1' = next one
```

If it continues below, the count has to go up. It checks if the state is already at the maximum value. If it is, it exits the handler because neither the state nor the outputs should change. If the state is not at the maximum value, it increments and saves it in the memory, updates the outputs and then exits.

```
GOUP:       LD      A,(STATE)       ; read the current state
            CP      00000011b       ; if W1 = '1', W0 = '1' then do not
            JP      Z,EXIT          ; increment the state and exit
            INC     A               ; otherwise increment it
            LD      (STATE),A       ; save the new state code in memory
            OUT     (W1W0P),A       ; update the corresponding outputs
            JP      EXIT
```

To decrement the state, the logic is very similar. Finally, we find the code to get out of the handler at the EXIT label.

```
GODOWN:     LD      A,(STATE)       ; read the current state
            CP      00h             ; if W1 = '0', W0 = '0' then do not
            JP      Z,EXIT          ; decrement the state and exit
            DEC     A               ; otherwise decrement it
            LD      (STATE),A       ; save the new state code in memory
            OUT     (W1W0P),A       ; update the corresponding outputs
EXIT:       POP     AF              ; restore A and Flags
            EI                      ; re-enable interrupts
            RET                     ; return to the interrupted program
```

---

[10] Formally it is called on the active edge of the clock in a real machine.

The second solution

The hardware circuit is the same as in the first solution. From a software perspective, this solution is distinctive because of a very general and reusable algorithm. It makes this solution reasonably adaptable to other problems that can be solved in FSM terms. Despite its length and repetitiveness, it is actually very readable and easy to maintain.

The FSM algorithm is executed only by the interrupt handler, as in the previous case. We also use the STATE variable, as before, to memorize the FSM state. What is different is that the writing of the code does not take individual cases into account (that is, in our example the FSM is in fact a counter), but is intended to be as general as possible, applicable to any ASM chart.

Let's look at the code here below. We have the definitions of the ports and of STATE (identical to those on the first solution). Then we define the constants 'Code_a', 'Code_b', 'Code_c' and 'Code_d', which in turn define the codes assigned to states (a), (b), (c) and (d) of the ASM chart, respectively.

```
UPINP     EQU    00h        ; IA input port (UP input line)
W1W0P     EQU    00h        ; OA output port (W1,W0 output lines)

STATE     EQU    0FC00h     ; FSM state variable

Code_a    EQU    00000000b  ; state (a)
Code_b    EQU    00000001b  ; state (b)
Code_c    EQU    00000010b  ; state (c)
Code_d    EQU    00000011b  ; state (d)
```

The code was chosen so that outputs W1 and W0 correspond to bits 1 and 0 of the code itself. Therefore it wouldn't be necessary to define the constants of the outputs corresponding to the states. However, for the sake of completeness and generality, we're including them anyway. Then come constants 'Out_a', 'Out_b', 'Out_c' and 'Out_d', which correspond to the outputs of states (a), (b), (c) and (d).

```
Out_a     EQU    00000000b  ; state (a) outputs: W1 = '0', W0 = '0'
Out_b     EQU    00000001b  ; state (b) outputs: W1 = '0', W0 = '1'
Out_c     EQU    00000010b  ; state (c) outputs: W1 = '1', W0 = '0'
Out_d     EQU    00000011b  ; state (d) outputs: W1 = '1', W0 = '1'
```

Then we have the jumps to the program and the interrupt handler.

```
          ORG    0000h
          JP     START
          ORG    0038h
          JP     HINT
          ORG    0100h
```

The program begins by assigning the code of state (a) to the STATE variable and the corresponding outputs to the port. Then we enter the main loop.

```
START:      LD      SP,0FFFFh       ; initialize the Stack Pointer
            LD      A,Code_a
            LD      (STATE),A       ; set the initial state equal to (a)
            LD      A,Out_a
            OUT     (W1W0P),A       ; set the corresponding outputs
            EI                      ; enable interrupts
MAIN:       CALL    PROCESS         ; execute any task
            JP      MAIN
PROCESS:    NOP                     ; placeholder for any task
            RET
```

As mentioned before, tasks are carried in the main loop but this is not our focus. Rather, let's look at HINT, the interrupt handler. After saving the registers that are used, it acquires the FSM input and moves it to register B.

After that, it takes the current state (STATE) of the FSM and compares it linearly to the four possible codes. Then it jumps to the label corresponding to the current state of the FSM.

```
HINT:       PUSH    AF              ; save A, Flags, B and C
            PUSH    BC
            IN      A,(UPINP)       ; read the FSM input
            LD      B,A             ; move it to register B
            LD      A,(STATE)       ; read the current state of the FSM
            CP      Code_a          ; execute a "linear search"
            JP      Z,STATE_A
            CP      Code_b
            JP      Z,STATE_B
            CP      Code_c
            JP      Z,STATE_C
            JP      STATE_D         ; the last one is evaluated by exclusion
```

The following code is divided into four sections, which are very similar in terms of structure. The sequences are mnemonically labeled to remind us of the state they handle. What follows is the code for state (a):

```
STATE_A:    BIT     0,B             ; check input 'UP'
            JP      Z, EXIT         ; do not change state if UP = '0'
            LD      A,Code_b        ; but if UP = '1', go to state (b)
            LD      (STATE),A
            LD      A,Out_b         ; set state (b) outputs: W1 = '0', W0 = '1'
            OUT     (W1W0P),A
            JP      EXIT
```

As we can see, the sequence begins by evaluating the UP input (now in bit 0 of register B). Based on its value, the new code of state (b) is assigned (or not) in STATE (and the outputs W1 and W0).

Note that when the handler ends, STATE represents the new state assumed by the FSM. What follows is the code for state (b):

```
STATE_B:    BIT    0,B              ; check input 'UP'
            JP     Z,GO_a

            LD     A,Code_c         ; if UP = '1', go to state (c)
            LD     (STATE),A
            LD     A,Out_c          ; set state (c) outputs: W1 = '1', W0 = '0'
            OUT    (W1W0P),A
            JP     EXIT

GO_a:       LD     A,Code_a         ; if UP = '0', go to state (a)
            LD     (STATE),A
            LD     A,Out_a          ; set state (a) outputs: W1 = '0', W0 = '0'
            OUT    (W1W0P),A
            JP     EXIT
```

The code is, in fact, a loyal transcription of the ASM chart and the conditions that provoke changes of state. As we can see in the code, we can go to state (c) or state (a) depending on the value of UP. Note that changes of state consist in assigning a new code to STATE and then exiting the handler. The state's new code is considered only when we enter the handler again.

Below is the code that handles state (c), which is very similar to state (b):

```
STATE_C:    BIT    0,B              ; check input 'UP'
            JP     Z,GO_b

            LD     A,Code_d         ; if UP = '1', go to state (d)
            LD     (STATE),A
            LD     A,Out_d          ; set state (d) outputs: W1 = '1', W0 = '1'
            OUT    (W1W0P),A
            JP     EXIT

GO_b:       LD     A,Code_b         ; if UP = '0', go to state (b)
            LD     (STATE),A
            LD     A,Out_b          ; set state (b) outputs: W1 = '0', W0 = '1'
            OUT    (W1W0P),A
            JP     EXIT
```

The code of state (d), however, is similar to that of (a). In either one, if the input requests it, the state can remain unchanged.

```
STATE_D:    BIT    0,B              ; check input 'UP'
            JP     NZ,EXIT

            LD     A,Code_c         ; if UP = '0', go to state (c):
            LD     (STATE),A
            LD     A,Out_c          ; set state (c) outputs: W1 = '1', W0 = '0'
            OUT    (W1W0P),A
            JP     EXIT
```

The handler ends by re-enabling interrupts, restoring the contents of the registers and returning to the interrupted program.

EXIT: POP   BC
           POP   AF
           EI
           RET

4.7.3 Sinusoidal waveform generator

We want to generate a sinusoidal waveform by using an 8-bit virtual digital/analog converter (DAC) connected to an OF port. The network schematic, which uses the enhanced version of the "DMC8 Microcomputer", is shown in the following figure. The processor's clock frequency is 10 *MHz*, and the timer generates an interrupt every 100 μS, through line $\overline{\text{Int7}}$.

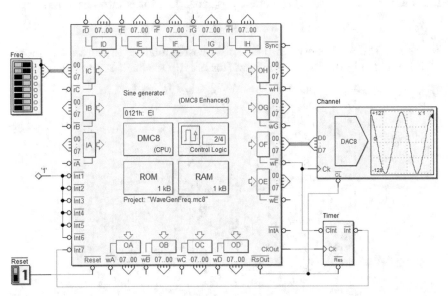

Note that in this case, input $\overline{\text{CInt}}$ is not connected to the microcomputer's output $\overline{\text{IntA}}$ but to the output port's write signal $\overline{\text{wF}}$. As a result, the interrupt request generated by the timer is stopped by a new value written on the port. The write signal $\overline{\text{wF}}$ is also used as a clock for the DAC.

The sinusoidal wave frequency has to be linearly proportionate to the parameter Freq, acquired by input port IC. The value of the parameter is limited to the interval of 0..31 (the value 0 stops the oscillation, generating a constant).

Finally, we calculate the relation between the parameter Freq and the frequency of oscillation. From this we can calculate the highest value that can be generated.

Solution

The interrupt handler retrieves the function values from a table of constants allocated in the ROM. We could, for example, describe a complete sinusoidal cycle with a table of 256 8-bit values.

Each value would correspond to an advancement of 1/256 of a 360 degree angle (a round angle), so it would make sense to measure the angle in 256ths of a round angle. The angle, would thus be between 0 and 255 and conveniently, this would be nothing other than the table index.

Thanks to the sinusoid's property of symmetry, we can save memory space by constructing a table with the 128 values of the positive half wave, while the negative half wave can be obtained by inverting the sign of the numbers in the table by a two's complement operation[11].

The values V to insert in the table should be calculated beforehand. The following expression takes the angle described in 256ths into account:

$$V(\theta) = \lceil 127 \cdot sin(\theta \tfrac{360}{256}) \rceil \qquad\qquad 0 \le \theta \le 127$$

where θ is the angle between 0 and 127, corresponding to the table's index. This is defined in the ROM as follows (for brevity's sake, it is partially shown):

SINTAB:	DB	000	; x = 0 (0 degrees)
	DB	003	; x = 1
	DB	006	; x = 2
		... omissis ...	
	DB	085	; x = 30
	DB	088	; x = 31
	DB	090	; x = 32 (45 degrees)
	DB	092	; x = 33
	DB	094	; x = 34
		... omissis ...	
	DB	127	; x = 63
	DB	127	; x = 64 (90 degrees)
	DB	127	; x = 65
		... omissis ...	
	DB	094	; x = 94
	DB	092	; x = 95
	DB	090	; x = 96 (135 degrees)
	DB	088	; x = 97
	DB	085	; x = 98
		... omissis ...	
	DB	006	; x = 126
	DB	003	; x = 127
	DB	000	; x = 128 (180 degrees, not used)

---

[11] We could also reduce the size of the table to just 64 values (1/4 wave) but this would overly complicate the code.

Now, let's describe the code of the solution. First we define the input and output ports then we define a variable ANGLE, which memorizes the last angle corresponding to the last value generated.

```
FREQ        EQU    02h          ; IC input port (Freq)
OUTWAV      EQU    05h          ; OF output port (Sinusoid)

ANGLE       EQU    0FC00h       ; current angle
```

What follows are the jumps to the program and the interrupt handler.

```
        ORG    0000h
        JP     START
        ORG    0038h        ; Int. 7
        JP     HINT7
        ORG    0100h
```

Before entering the main loop, we initialize the Stack Pointer, zero the output port and the current angle of the sinusoid and then enable interrupts.

```
START:      LD     SP,0FFFFh    ; initialize the Stack Pointer
            LD     A,00h        ; zero the waveform output
            OUT    (OUTWAV),A
            LD     (ANGLE),A    ; zero the current angle
            EI                  ; enable interrupts
```

The only action the main loop takes is to continually read the frequency parameter from the FREQ port (if > 31, it is reduced to 31), which is saved in register B. Note that the parameter is acquired from the main program but is then used by the interrupt handler. In the following, we will often encounter this type of communication between programs.

```
MAIN:       IN     A,(FREQ)     ; read the frequency parameter
            CP     32           ; limit its value to 31
            JP     C, NOLIMIT   ; if C = '1' then A < 32 and so skip
            LD     A,31         ; otherwise limit A to 31
NOLIMIT:    LD     B,A
            JP     MAIN
```

The timer makes sure that every $100\,\mu S$ the interrupt handler HINT7 is executed, by first saving register A and the Flags.

```
HINT7:      PUSH   AF           ; save A and Flags
```

Every time there is a call, we have to take a new value of the sinusoid from the table and copy it to the port. The index of the table is calculated by adding the Freq parameter to the previous angle (remember that Freq is continually updated in register B by the main program).

```
            LD     A,(ANGLE)    ; compute the new angle
            ADD    A,B          ; by adding the Freq parameter
            LD     (ANGLE),A    ; to the previous angle
```

The angle is passed to the subprogram WAVEFORM, which we delegate to read the value table and return the corresponding function value to us in register A. This value is copied to the output port where the Deeds virtual DAC allows us to visualize the waveform that is generated.

```
CALL  WAVEFORM   ; get the next wave value from the table
OUT   (OUTWAV),A  ; copy it to the output port (to the DAC)
```

As described in the specifications, writing the port has the effect of erasing the interrupt request in the timer. This happens because the $\overline{wF}$ signal of the port itself is connected to input $\overline{CInt}$ of the timer.

As always, the handler ends by restoring the saved registers, re-enabling the interrupts and returning to the interrupted program.

```
POP   AF            ; restore A and Flags
EI
RET
```

As mentioned before, the WAVEFORM subprogram provides the value of the sinusoid in function of the angle we send it in the accumulator.

This subprogram has to manage 256 values, the first half of which are positive and read directly from the table of 128 locations. The second half are negative and derived by calculating two's complement of the values in the table.

This is why we save the registers used then save the value of the angle in register C. A little further on, this makes its most significant bit available so that we can distinguish the first 128 values of the positive half wave from the 128 values of the negative side.

Then, the most significant bit is zeroed in register A so that we can translate the value of the requested angle in the table's index (which has 128 locations). This way, we also map the negative half wave on the positive one in the table.

```
WAVEFORM:  PUSH  HL            ; save register HL and BC
           PUSH  BC

           LD    C,A           ; save bit 7 of the angle in C, and mask
           AND   01111111B     ; it to avoid readings outside the table
```

The following instructions translate the table's index into the address of the location of interest then load the value we need in the accumulator.

```
LD    HL,SINTAB   ; get the base address of the table
ADD   A,L         ; add the index to it
LD    L,A         ; to obtain the address of the location
JP    NC,NoCarry  ; of interest in register HL
INC   H
NoCarry:  LD    A,(HL)    ; get the value
```

Now, we have the value read from the table in A, but we must assess whether the angle requested referred to the positive half wave or the negative one.

If bit 7, which was first saved in C, was 0, then the returned value must be positive, otherwise we need to invert the sign (with a NEG instruction). The function ends by restoring the previously saved registers[12].

```
           BIT   7,C            ; check if we are in second half wave
           JP    Z,Positive     ; if not, the value is positive
Negative:  NEG                  ; otherwise invert the sign of the value
Positive:  POP   BC             ; restore registers BC and HL
           POP   HL
           RET
```

Communication between the main program and the interrupt handler

As noted previously, this example shows a parameter (Freq) that is produced in the main program (reading a port) but used by the interrupt handler.

This is the first example of communication between programs. How reliable is passing parameters between two totally "asynchronous" processes since the interrupt can come at any moment with respect to the main loop?

We don't know beforehand when the interrupt will happen in the loop. For ease of reading, the loop code is shown again below.

```
MAIN:      IN    A,(FREQ)
           CP    32
           JP    C, NOLIMIT
           LD    A,31
NOLIMIT:   LD    B,A            ; ← the instruction is this
           JP    MAIN
```

We always have to go through the instruction that updates register B (marked by the arrow in the listing) regardless of the value read and the underlying calculations. We must concentrate on this instruction since the handler expects to find what it needs in register B.

Luckily, this updating operation is "atomic" (as we've seen, this means indivisible), so all that can happen is that the register will already be updated at the time of the interrupt or it will be immediately afterward.

There can be no partial update of its content; the register cannot assume incongruous values. Simply put, the interrupt handler reads either the old or the new value, never an incongruous value.

---

[12] Arguably, the pair of PUSH HL and POP HL instructions is not useful because registers H and L are not used by the main program. Nevertheless, we preferred to insert them to comply with the general rule that the interrupt program must preserve the contents of the registers in the processor.

The frequency of the generated sinusoid

Let's set the Freq parameter at 1. Each time the handler is called, the value right next to the previous one is retrieved from the table. This is to say that all 256 values that describe the entire sinusoid cycle are read one after the other.

Therefore, the sinusoid cycle is repeated after 256 calls to the handler, which all happen at a distance of $100\,\mu S$ from each other (thanks to the timer). So, independently of the processor's clock frequency, the generated waveform's period T and frequency F are:

$$T = 100\,\mu S \cdot 256 \qquad \text{and} \qquad F = \frac{1}{100\,\mu S \cdot 256}$$

Defining:

$$F_0 = \frac{1}{100\,\mu S} = 10\,KHz$$

If Freq = 1, the generated frequency is:

$$F = F_0 \frac{1}{256} \approx 39.1\,Hz$$

If Freq = N (> 1), each time the handler is called, the value at N positions after the previous one is retrieved from the table. In other words, the values are read by jumping N positions. This means that the table is read at a "speed" N times faster.

For example if N = 4, one out of every four values is read from the table, thus taking 1/4 the time to generate the whole cycle (the frequency is quadrupled). The relation deriving from this is the following[13]:

$$T = \frac{256}{N} \cdot 100\,\mu S \qquad \text{and} \qquad F = N \cdot \left(F_0 \frac{1}{256}\right)$$

Given that the specifications require that the highest value of Freq is 31, the highest frequency generated is:

$$F = 31 \cdot \left(10\,KHz\,\frac{1}{256}\right) \approx 1.21\,KHz$$

4.7.4 Dual sinusoidal waveform generator

We must design a two-channel (Right, Left) sinusoidal waveform generator using two 8-bit virtual DACs based on the solution to the previous exercise (see the following figure).

Connected to port OH, we have the second converter and also a second timer (Timer L), which is programmed to interrupt the processor every $130\,\mu S$. The first timer (renamed Timer R) is still programmed at $100\,\mu S$.

Another frequency controller has also been added so now FreqL (read on port IC) controls the left channel while FreqR (read on port IA), controls the right channel.

---

[13] We can verify that this relation stands even when N is not a power of 2.

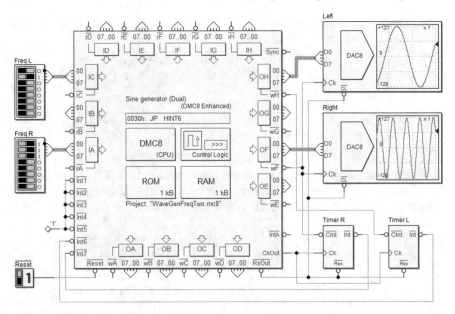

Channels Right and Left are updated independently from each other, each in relation to its own timer: Timer R and Timer L, respectively. Note that since the timers are set at different times, the two sinewaves have different frequencies if the FreqR and FreqL parameters remain the same.

Solution

The solution here is very similar to that of the previous example, so many of the same comments will not be repeated. The first part of the code has the definitions of the two input and two output ports, which are different between the left and right channels. Similarly, we use two different variables depending on the channel to store the last angle used to generate the value.

FREQR	EQU	00h	; IA input port (Right Frequency)
FREQL	EQU	02h	; IC input port (Left Frequency)
OUTWAVR	EQU	05h	; OF output port (Right Channel)
OUTWAVL	EQU	07h	; OH output port (Left Channel)
ANGLER	EQU	0FC00h	; current angle (Right)
ANGLEL	EQU	0FC01h	; current angle (Left)

Notice the addition of the second interrupt handler, HInt6, for the left channel, while HInt7 will now manage the right channel.

```
ORG    0000h
JP     START
ORG    0030h          ; Int. 6 (Left Channel)
JP     HINT6
ORG    0038h          ; Int. 7 (Right Channel)
JP     HINT7
ORG    0100h
```

There are no substantial changes regarding the initialization sequence except the higher number of ports and variables.

```
START:      LD     SP,0FFFFh        ; initialize the Stack Pointer
            LD     A,00h            ; zero the waveform outputs
            OUT    (OUTWAVR),A
            OUT    (OUTWAVL),A
            LD     (ANGLER),A       ; and the corresponding angles
            LD     (ANGLEL),A
            EI                      ; enable interrupts
```

Now, the main program reads the other input port as well. The codes for the two channels are practically identical apart from the difference in the destination of the frequency parameter (saved in register B for the right channel and in D for the left).

```
MAIN:       IN     A,(FREQR)        ; read the Right frequency parameter
            CP     32               ; and limit its value to 32
            JP     C, NOLIMR        ; if C = '1' then A < 32 and so skip
            LD     A,31             ; otherwise limit a to 31
NOLIMR:     LD     B,A              ; save it in B (used by the Right channel)

            IN     A,(FREQL)        ; do the same thing for the Left channel
            CP     32
            JP     C, NOLIML
            LD     A,31
NOLIML:     LD     D,A              ; save the parameter in D (Left channel)

            JP     MAIN
```

Every $100\,\mu S$ the right channel's interrupt handler is executed. The code is identical to that of the previous example except for the different names of the variables and output port.

Remember that the parameter contained in register B is added to the angle and then the WAVEFORM function is called. Depending on the new angle, the function returns the corresponding value of the waveform.

```
HINT7:      PUSH   AF               ; save A and Flags
            LD     A,(ANGLER)       ; update the current angle (Right),
            ADD    A,B              ; adding the control parameter to it
            LD     (ANGLER),A
            CALL   WAVEFORM         ; get the waveform sample from the table
            OUT    (OUTWAVR),A      ; and send it to the output port (Right)
            POP    AF               ; restore A and Flags
            EI
            RET
```

Every $130\,\mu S$ the left channel's interrupt handler is launched. Note that it is identical to that of the other channel except for the different names of the variable, output port and the increment of the angle of the value in D.

HINT6:	PUSH	AF	; save A and Flags
	LD	A,(ANGLEL)	; update the current angle (Left),
	ADD	A,D	; adding the control parameter to it
	LD	(ANGLEL),A	
	CALL	WAVEFORM	; get the waveform sample from the table
	OUT	(OUTWAVL),A	; and send it to the output port (Left)
	POP	AF	; restore A and Flags
	EI		
	RET		

The WAVEFORM function and the SINTAB table are identical to those of the previous example, so we will not repeat them here.

However, it makes sense to discuss the importance of the two interrupt handlers working independently, as if each were the only one present. They are launched by two different interrupt requests from two timers with different periods.

The only time these two "meet" is when the two interrupt requests periodically overlap. The priority dynamic solves this problem, however. Simply put, the left channel is slightly penalized by a small delay in the generation of a value. It is updated after the right channel request is satisfied.

4.7.5 Object counters

In the system represented below, input port IA is used to acquire the output of an (idealized) proximity sensor on bit 7. The sensor is inserted into an object handling apparatus (in an industrial plant), and generates '1' each time an object passes near the sensor.

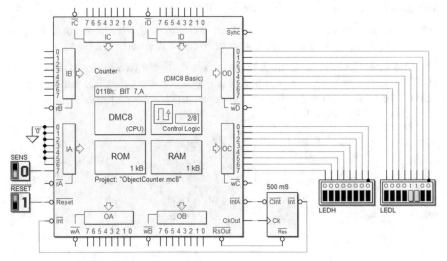

Output ports OC and OD drive a 16-LED array divided into two 8-bit groups (the high part is LEDH and the low part, LEDL). The lights are arranged horizontally on a panel visible to the user and they show a 16-bit binary number. The timer is set to activate the interrupt request line every $500\,mS$.

The program cyclically checks the sensor and counts the objects that have passed. Every 10 seconds, it updates the number of objects counted on the LED lights then restarts the count from zero. To call the user's attention, each time the number is updated on the LED lights, the lights are turned off for one second and then turned on again, showing the new number.

Let's assume there is a maximum of 50 objects per second that can pass in front of the sensor.

Solution

The last specification requires the system to assess well-defined time intervals during which it still needs to count objects. This need to double-task suggests it would be a good idea to use timer interrupts. In fact, using delay loops to assess the times given by the specifications is inconvenient. The resulting code would be very complex since the objects need to be counted in addition to the other tasks.

So let's divide the tasks between the main program and the interrupt handler. The main program will count the objects, while the interrupt handler will show the count to the user every 10 seconds.

Each time an object passes on front of the sensor, the main program increments the count. The object coming close activates the output of the sensor at '1'. To increment the count, we can wait for the object to move away and the sensor output go back to '0'. In other words, the main program should wait and count the falling edges on line 7 of the SENS port.

Let's look at the following code. After defining the addresses of the ports described in the text, we insert the definition of the two variables.

The first one, (COUNT), counts objects. We dedicate two bytes to that activity. The text says that the number of objects per second is 50, maximum. It follows that every 10 seconds, we can count up to 500 objects. This requires us to use a 16-bit variable.

```
SENS        EQU    00h        ; IA input port (sensor)
LEDH        EQU    02h        ; OC output port (LED15..LED8)
LEDL        EQU    03h        ; OD output port (LED7..LED0)

COUNT       EQU    0FC00h     ; object counter (16-bit)
TIME        EQU    0FC02h     ; time counter
```

The second one, (TIME), measures time by counting the timer's calls to the handler. To measure 10 seconds, we need to count 20 calls since the timer is set at a cycle of $500\,mS$.

After the usual jumps to the start of the program and the interrupt handler, the program does a series of initializations.

```
ORG    0000h
JP     START
ORG    0038h
JP     HINT
ORG    0100h
```

After initializing the Stack Pointer, we zero output ports LEDH and LEDL, so that we can start with all the lights off (the text of the exercise does not specify this but it is reasonable to do it this way).

```
START:    LD    SP,0FFFFh    ; initialize the Stack Pointer
          LD    A,00h        ; zero the output ports
          OUT   (LEDL),A
          OUT   (LEDH),A
```

As we have seen, the TIME variable allows us to count by 20. Let's count backward; we need to initialize it at 20. The COUNT variable, however, must be zeroed since we use it to count objects.

```
          LD    A,20
          LD    (TIME),A      ; initialize the time counter
          LD    HL,0000h
          LD    (COUNT),HL    ; initialize the object counter
```

At the start of the main loop (MAIN), we enable EI (we'll soon see why it is inserted here). Then we find two consecutive wait loops that assess objects moving in front of the sensor. The first loop waits for the sensor to produce a '1', which signals an object is present. The second waits for a '0', meaning the object has moved away.

```
MAIN:     EI                  ; enable interrupts
WAITH:    IN    A,(SENS)      ; loop to check if...
          BIT   7,A           ; the sensor output goes high
          JP    Z,WAITH
WAITL:    IN    A,(SENS)      ; loop to check if...
          BIT   7,A           ; the sensor output goes low
          JP    NZ,WAITL
```

Note that the double wait loop, the wait for the falling edge of SENS, cannot be substituted by a simple value check. The object, in fact, stays close to the sensor for a longer time than the processor takes to execute the main loop.

So if there were just a value check on the SENS output, this would probably detect the same object... thousands of times.

As soon as the object has moved away, we increment the count variable by one unit before repeating the main loop as of MAIN.

To guarantee that the updating sequence of the variable isn't interrupted, the interrupt disable instruction DI is placed before it. The need to prevent the sequence from being interrupted will be explained a bit further on in a description of the interrupt handler.

```
        DI

        LD      HL,(COUNT)    ; increment the object count
        INC     HL
        LD      (COUNT),HL

        JP      MAIN
```

The interrupts are then re-enabled right afterward by the EI instruction, which we found at the start of the main loop.

Note that if any interrupt request is made during the time when the interrupts are disabled, it is not lost. Rather, it is "pending" and will be satisfied soon after, when they are re-enabled.

The interrupt handler is totally dedicated to showing the count. It is executed every 500 mS and each time, it decrements the TIME variable.

```
HINT:       PUSH  AF              ; save the registers in use
            PUSH  HL

            LD    A,(TIME)        ; decrement the time count
            DEC   A
            LD    (TIME),A
```

When we get to the 20th interrupt, 10 seconds will have gone by and we jump to DISPLAY. However, when one second after the time is up, the count will have gone down to 2 so we'll jump to SWOFF, to turn the lights off (as required in the specifications). If we are in neither of these conditions, we go to EXIT, and return straight to the interrupted program.

```
            JP    Z,DISPLAY       ; jump if 10 seconds are gone by
            CP    2               ; are we at one second before time ends?
            JP    Z,SWOFF         ; jump if yes
EXIT:       POP   HL              ; restore the registers saved before
            POP   AF
            EI
            RET
```

We have seen that we jump to the DISPLAY label if 10 seconds have gone by. It is now the time to show the count by reading the COUNT variable and copying it to the output port (the high part on LEDH, the low on LEDL).

After it is shown, the count of the objects must be zeroed. Also, before jumping to the handler output, we re-initialize the time count as well.

```
DISPLAY:    LD      HL,(COUNT)    ; read the current object count
            LD      A,H           ; display it on the LED array
            OUT     (LEDH),A
            LD      A,L
            OUT     (LEDL),A
            LD      HL,0000h
            LD      (COUNT),HL    ; re-initialize the object counter
            LD      A,20
            LD      (TIME),A      ; and the timer counter
            JP      EXIT
```

Finally, we have seen that one second before the end of the 10-second period, a jump is executed to the SWOFF label where the two output ports are zeroed, the LED lights are turned off and we exit the handler.

```
SWOFF:      LD      A,00h         ; switch off the LED array
            OUT     (LEDL),A
            OUT     (LEDH),A
            JP      EXIT
```

One final observation: this is another example of communication between the main program and the interrupt handler, obtained through the COUNT variable. Every 10 seconds, the COUNT variable is incremented in the main loop while the handler is zeroed.

We have seen that the group of instructions related to the increment has been made non-interruptable. If it were not, the content of the COUNT variable could be corrupted because an interrupt could come at any time, even in the middle of the increment sequence.

Let's look at an example to demonstrate this. Let's assume we have received an interrupt request while we are executing the first instruction in the sequence.

```
            LD      HL,(COUNT)
```

First, the instruction is completed. Let's assume that COUNT contains the number 175 (just as an example); now this number is in register HL.

At this point, the interrupt handler is executed. Let's assume that the required 10 seconds have gone by so the COUNT variable is seen on the LED lights and then zeroed immediately after. When the handler's work is done, we go back to the interrupted program and execute the following instructions:

```
            INC     HL
            LD      (COUNT),HL    ; → ERROR!
```

The error is that the content of COUNT has been corrupted. The number 175, which was in HL, is now incremented and copied in the variable, which has actually just been zeroed.

4.7.6 Sensor evaluation in parallel

Let's work with the system in the figure below, which is based on a DMC8 Microcomputer. Port IA is connected to the outputs of three proximity sensors (SA, SB and SC), which are positioned on three outputs of a machine that makes metal washers. The sensors generate a high level when a washer comes near them.

Output ports OA and OB drive 16 LED lights that show the binary number N15..N0 (divided into the high part NUMH and the low part NUML).

A Timer activates interrupt request $\overline{\text{Int}}$ every $10\,mS$. When the request is accepted through output $\overline{\text{IntA}}$, this automatically deactivates line $\overline{\text{Int}}$.

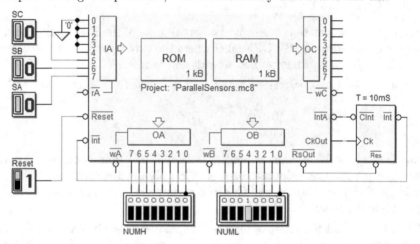

We need to write a program in assembly language that meets the specifications below.

The main program executes the required initializations after system reset, then enters an infinite loop where it continually updates number N15..N0 on outputs NUMH and NUML. It does so by copying the number from a 16-bit variable called NUMBER. However, it is the interrupt handler that updates NUMBER.

The interrupt program identifies the rising edges of sensors SA, SB and SC's signals in order to get an overall count of the washers produced every 20 seconds. The count comes in one single 16-bit variable called COUNT. Let's assume that no more than 35 washers come before each sensor per second.

Every 20 seconds, the count in COUNT is copied in the NUMBER variable, making it available to the main program. Then the variable COUNT is re-initialized at zero.

To let the operator know that the number is about to be updated, all 16 LED lights must be off for one second before the new value is shown.

Solution

First we define the input (SENS) and output (NUMH and NUML) port addresses, then we define the NUMBER and COUNT variables. As described in the text, we must assure that two bytes are allocated for each variable since they both need to be 16 bits.

The 8-bit TIME1 and TIME20 variables are used to count time. As we will see further on, the time will be counted so that we can assess one second and 20 seconds.

The PREV variable stores the previous state of the sensors so that they can be compared with the current ones every time the interrupt handler is called. Finally, we have the jumps to the start of the program and the interrupt handler.

SENS	EQU	00h	; IA input port (sensors A, B, C)
NUMH	EQU	01h	; OA output port (number, 15..8)
NUML	EQU	02h	; OB output port (number, 7..0)
NUMBER	EQU	0FC00h	; the number to copy to the LEDs (16-bit)
COUNT	EQU	0FC02h	; object counter (16-bit)
TIME1	EQU	0FC04h	; one second time counter
TIME20	EQU	0FC05h	; 20 seconds time counter
PREV	EQU	0FC06h	; previous sensor state
	ORG	0000h	
	JP	START	
	ORG	0038h	
	JP	HINT	
	ORG	0100h	

As always, at the start of the main program, we initialize the Stack Pointer, the variables in play and the output ports. We insert '1s' in the PREV variable to prevent the rising edges from being misidentified the first time the sensors are assessed (further on, we will see the description for the algorithm that identifies the edges).

START:	LD	SP,0FFFFh	; initialize the Stack Pointer
	LD	A,11111111b	; initialize the previous state
	LD	(PREV),A	

The timer interrupts the processor every $10\,mS$, so to make one second go by, we need to count to 100, that value we initialize TIME1 with. To assess the required 20-second interval, however, we simply need to count the seconds in TIME20 that have already passed in the TIME1 count.

	LD	A,100	; initialize the time counters:
	LD	(TIME1),A	; one second (100 x 10 mS)
	LD	A,20	
	LD	(TIME20),A	; 20 seconds (20 x 100 x 10 mS)

The two 16-bit variables, COUNT and NUMBER, are initialized at zero. Also, before entering the main loop, we enable interrupts.

```
LD      HL,0000h        ; initialize the object counter
LD      (COUNT),HL
LD      (NUMBER),HL   ; and the number to display

EI                      ; enable interrupts
```

As required by the specifications, the main loop expects the interrupt handler to update the NUMBER variable that is copied to the 16-bit HL register. The high part H is then written to the NUMH port and the low part to the NUML port. At the start, NUMBER is clearly still zero because the interrupt handler hasn't been called yet.

```
MAIN:     LD      HL,(NUMBER)   ; read the number to display
          LD      A,H           ; copy it to the output ports,
          OUT     (NUMH), A     ; separated into the high...
          LD      A,L
          OUT     (NUML), A     ; and low byte
          JP      MAIN
```

The interrupt handler HINT is called every $10\,mS$. When we enter, we save the registers involved in the processing (A, F, B, C and HL).

```
HINT:     PUSH  AF              ; save the registers in use
          PUSH  BC
          PUSH  HL
```

We need to find out whether there has been a rising edge at the output of the three sensors. We do this through a logic function that does a parallel test of all the sensors and shows us when the rising edges get to their lines (including the simultaneous ones). Look at the following figure which shows the trend of a standard signal on a line L.

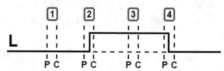

If we periodically execute a check on L, we will always have a current value 'C' and a previous value 'P'. The table below describes the four possible cases.

Case	P	C	Description	Transition	$F = \overline{P} \cdot C$
1	0	0	Constant value	('0' → '0')	0
2	0	1	Positive edge	('0' → '1')	1
3	1	1	Constant value	('1' → '1')	0
4	1	0	Negative edge	('1' → '0')	0

Our focus is on case 2. There is a rising edge on the line if 'P = 0' and 'C = 1', that is if the Boolean expression $F = \overline{P} \cdot C$ gives '1'.

To check the rising edges on all three sensors at the same time, we calculate this function for all the bits in register A simultaneously by reading the lines' current values on the SENS port and the previous values in the PREV variable.

$$F = \overline{PREV} \cdot SENS$$

The considerations above give rise to the following code. Initially, we read the state of the port of the sensors and copy it temporarily to B.

```
IN    A,(SENS)        ; read the sensors (in parallel)
LD    B,A             ; save their state in B
```

Then we copy the previous state (PREV) to the accumulator and we negate all its bits by executing a CPL instruction. The next AND instruction with B completes the above function, executed on all the bits in parallel. Then we save the result of the function in register C.

```
LD    A,(PREV)        ; get the previous state and invert all
CPL                   ; the bits... after the AND, a positive edge
AND   B               ; results in a '1' on the corresponding bit
LD    C,A             ; save this evaluation in register C
```

The value read on the SENS port, which we have moved to B, is then saved in the PREV variable (it is needed for the check on the next interrupt).

```
LD    A,B             ; save the sensors state as "previous"
LD    (PREV),A        ; for the check on the next interrupt
```

In bits 7, 6 and 5 of register C we have the result of the function regarding sensors SC, SB and SA, as shown in the following table (the remaining bits clearly don't interest us so we ignore them).

Value	SA (bit 7)	SB (bit 6)	SC (bit 5)
0	-	-	-
1	⌐	⌐	⌐

We copy the current 16-bit COUNT value to HL. We check bit 7 of register C (bit 7 corresponds to sensor SA). If it is '1', a rising edge has appeared so we need to increment the count (now in HL).

```
LD    HL,(COUNT)      ; get the current object count
BIT   7,C             ; check for a positive edge on sensor SA
JP    Z,TEST1
INC   HL              ; edge detected: increment the count
```

We repeat the same operation for sensors SB and SC:

```
TEST1:    BIT   6,C             ; check for a positive edge on sensor SB
          JP    Z,TEST2
          INC   HL              ; edge detected: increment the count
TEST2:    BIT   5,C             ; check for a positive edge on sensor SC
          JP    Z,ENDTEST
          INC   HL              ; edge detected: increment the count
ENDTEST:  LD    (COUNT),HL      ; update the count in memory
```

When the tests are finished, we update the object count in the COUNT variable. This operation is executed every $10\,mS$. Now we need to assess the time of one second. So, we decrement the time count in TIME1 and exit the handler if one second hasn't gone by yet. If it has, at the hundredth call, we go ahead and re-initialize the count.

```
LD    A,(TIME1)        ; decrement the TIME1 variable
DEC   A
LD    (TIME1),A
JP    NZ,EXIT          ; exit if not a second has passed
LD    A, 100           ; otherwise re-initialize TIME1 = 100
LD    (TIME1),A
```

If the processor has gotten here it means that one second has gone by since the last time we went through this sequence. Now we need to decrement the variable TIME20 to see if 20 seconds have gone by.

When its value gets to 1, we will have gotten to one second before the end of the count so we need to turn the LED lights off by zeroing the NUMBER variable (remember that the output ports are updated in the main loop).

```
         LD    A,(TIME20)       ; decrement the TIME20 variable
         DEC   A
         LD    (TIME20),A
         CP    1                ; check if time is 1 second before the 20th
         JP    NZ,CHECK0        ; second, jump if not
         LD    HL, 0            ; otherwise zero the variable NUMBER,
         LD    (NUMBER),HL      ; so all the LEDs will be switched off
         JP    EXIT
CHECK0:  CP    0                ; check if 20 seconds are passed
         JP    NZ,EXIT          ; jump if not
```

If the count of the seconds gets to zero, it's the time to re-initialize it but also to copy the count of the objects from COUNT to NUMBER and zero the COUNT as per specifications.

```
DISPLAY:  LD    A, 20          ; re-initialize the 20 seconds counter
          LD    (TIME20),A
          LD    HL,(COUNT)     ; copy the object count to NUMBER,
          LD    (NUMBER),HL    ; it will be displayed in the main loop
          LD    HL,0           ; re-initialize the object count
          LD    (COUNT),HL
```

Finally, we restore the contents of the previously saved registers, re-enable the interrupts, exit the handler and return to the interrupted program.

```
EXIT:     POP   HL             ; restore the registers saved before
          POP   BC
          POP   AF
          EI                   ; re-enable interrupts and
          RET                  ; return to the interrupted program
```

4.7.7 Push-button interface for a video game

The system in the following figure is based on a DMC8 Microcomputer component and implements an interface for a video game. Input port IB is connected to eight command buttons (U7..U0) that the user controls[14].

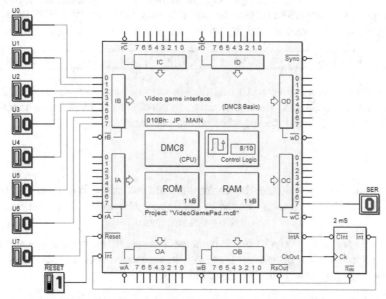

Output port OC connects our system to the game console through an asynchronous serial line SER connected on the port's bit 7. A Timer activates interrupt request $\overline{\text{Int}}$ every $2\,mS$. The processor accepts the request activating output $\overline{\text{IntA}}$. This event, through the $\overline{\text{CInt}}$ line, deactivates line $\overline{\text{Int}}$.

We need to write a game interface handling program with the following specifications.

For this project, the main program doesn't execute operations that are significant for our interface except for the required initialization of the variables and ports at system reset. Right after, the main program enters an infinite loop where is inactive.

The interrupt handler reads the push-buttons and does a "debouncing" operation on their state. Every time the state of the push-buttons changes, the information is sent to the game console in serial form on the SER line.

The push-buttons are sampled with a period of $12\,mS$ and the debouncing check consists of verifying that the configuration is identical to the readings from $12\,mS$ and $24\,mS$ before. This check prevents the mechanical bounces of the contacts from misleading the system.

---

[14] For educational purposes, the interface has been simplified and only push-buttons are present.

Once the current state of the push-buttons is confirmed by the debouncing check it is compared to the one confirmed by the previous check. If it has changed, the new state is prepared to be sent in serial format on line SER.

Bit serialization also has to be timed by the interrupt handler. The asynchronous, serial transmission protocol has a bit time with a duration of $2\,mS$ and a packet made up of a start bit at '1', the 8 data bits (the 'U7' button is transmitted first), and a stop bit at '0' (see the following figure).

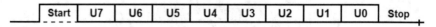

Notice that transmitting a (10 bit) packet takes $20\,mS$ overall, and that the push-button checks and data transmission checks don't need to be mutually exclusive, but rather done together where necessary.

Solution

In this solution, we define the input and output ports but then insert no definition of the variables allocated in the memory because we use the registers to keep the information handled by the program.

This is a very odd choice to make because it engages the registers, an important resource for the processor. However, in rare cases such as this, where the system does no other task, it is feasible.

USER	EQU	01h		; IB input port (user push-buttons)
SOUT	EQU	02h		; OC output port (serial line SER)
;	'H'	=		; time count
;	'B'	=		; push-buttons state (12 mS before)
;	'C'	=		; push-buttons state (24 mS before)
;	'D'	=		; push-buttons previous confirmed state
;	'E'	=		; transmission buffer

What follow are the usual jumps to the program and the interrupt handler.

```
ORG    0000h
JP     START
ORG    0038h
JP     HINT
ORG    0100h
```

Notice that the declaration of the registers comes in the form of a comment. While useful for the programmer, it is not essential to the program logic.

The main loop doesn't carry out tasks after the initializations. These consist of zeroing the registers used: B, C, D and E, and then assigning the value of 6 to register H, the time counter[15].

---

[15] $12\,mS$ correspond to 6 calls of the interrupt handler (generated by the timer).

After enabling the interrupts, the processor enters an empty, infinite loop where it does nothing.

```
START:      LD      SP,0FFFFh       ; initialize the Stack Pointer

            LD      H,6             ; time: 6 = 12 mS /(2 mS of Timer Tick)
            LD      B,00h           ; zero the other registers in use
            LD      C,B
            LD      D,B
            LD      E,B

            EI                      ; enable interrupts
MAIN:       JP      MAIN            ; empty main loop
```

The interrupt handler is called every $2\,mS$ and in the first couple rows, it handles the serial transmission of bits (one at a time, every $2\,mS$). Notice the absence of PUSH instructions. Since the main loop is inactive the registers are unused, so their content doesn't need to be saved or retrieved.

```
HINT:       LD      A,E             ; get data to transmit
            AND     10000000b       ; mask the bits we don't care
            OUT     (SOUT),A        ; send bit 7 to the serial line
            SLA     E               ; left shift data (for the next time)
```

The contents of the serial packet to transmit are defined in the next part of the code, which we will study a bit further along. While waiting to build the first packet, register E contains only zeroes.

Therefore, in the beginning, the instructions retrieve a '0' from bit 7 in E (every $2\,mS$) and send it on line SER. The left shift of register E by the SLA instruction doesn't change the situation since it inserts a zero from the right. At the beginning, in fact, this operation is not useful but the receiver that reads our SER line will not be disturbed. It will, in fact, continue to read a constant zero on the line (the "idle state").

The code prepares the packet, as we shall see, and then saves it in register E. The lines of code examined here serialize all its bits on line SER, starting with the bit in position 7. The SLA instruction allows us to left shift the bits contained in register E. One by one, a new bit will find itself in position 7 every $2\,mS$ (where the SER line is connected).

The left shifting also inserts zeroes to the right in the register so after transmitting the whole packet (through 8 shifts), we continue to send only zeroes, taking us back to the original situation.

Alternatively, we can introduce a variable that signals if we need to transmit a serial packet or not. However, every time the handler is called this choice requires us to check whether we have something to transmit and also to count the number of bits transmitted so we know when to stop sending them. To make the code simpler and more compact, however, we choose to proceed as described above.

After the transmission on the serial line, we have to make sure that $12\,mS$ have gone by before reading the push-buttons. So we decrement register H, which will be zeroed after 6 calls.

```
DEC    H              ; count the time (12 mS)
JP     NZ,EXIT        ; jump and exit if < 12 mS
LD     H,6            ; otherwise re-initialize the time count
```

If this time has not gone by yet, we jump to EXIT, re-enable the interrupts (with the EI and RET instructions) and go back to the interrupted program. Otherwise, at the sixth call, $12\,mS$ will have gone by. If so, we re-load 6 in register H and proceed ahead to handle reading the push-buttons.

Push-buttons and switches are electro-mechanic devices. The figure at the right shows one of the possible connection techniques. With these circuits, we are able to transform their manual operation into a two-level signal readable by a logic device.

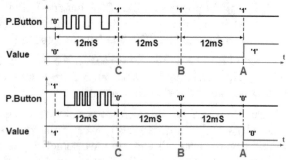

The upper part of the figure shows a push-button in the idle state, that is not pressed (represented with the electric contacts in the "open" position).

In this state, output PB of the NOT gate is at '0', in that the resistor (or "pull-up", in technical jargon) guarantees a high logical level at its input. When the push-button is pressed the contacts close so we see a '1' at the output of NOT. When the push-button is released, it goes back to the idle position (due to a spring). The contact is open again and PB is at '0'.

For the switch represented in the lower part of the figure, the circuit is identical. The difference is mechanical in nature, in the sense that a switch is a stable, two-position device so the value we set manually is maintained and to change it, we must press it again.

Pressing a push-button or flipping a switch produces a series of "mechanical bounces" in the electrical contact due to their mechanical properties. This causes a non-ideal trend in the signal produced.

For a short time interval, the signal shows a sudden, random fluctuation in values before reaching a stable state.

The figure on the right shows the possible level changes: '0→1' (above) and '1→0' (below).

The maximum duration of the bounces is ascertained and declared by the maker of the devices and is normally on the order of a few mS.

To overcome this, we should apply a "debouncing technique" in our program. Let's adopt the technique of "multiple readings". We acquire the line multiple times and we confirm its state only after three consecutive readings over a reasonable space of time have given the same result.

In the upper part of the figure above, we read the state of the push-button ('1') at time 'A'. The two previous readings, 'C' and 'B', which are $12\,mS$ apart, have given the same logical level. The new value (shown in red) is thus confirmed at time 'A'.

Now let's go back to writing the code. We read the input port and get the current state of all the push-buttons in register A. As defined at the beginning, we find the state of the push-buttons $12\,mS$ before in register B and $24\,mS$ before in register C.

```
IN    A,(USER)      ; read the push-buttons state
CP    C             ; compare it to the 24 mS ago reading
JP    NZ,SHIFT      ; jump if they differ
CP    B             ; compare it to the 12 mS ago reading
JP    NZ,SHIFT      ; jump if they differ
```

The two CP instructions allow us to discard a reading if it is different from the previous ones (A ≠ B ≠ C). This occurs because there has been a change in the state of the buttons. If this happens, we jump to the SHIFT label, where the "history" of the previous readings is updated when the content of B is moved to C and that of A is moved to B.

We move forward only if A = B = C, in which case the state of the push-buttons is confirmed. The new confirmed value must now be compared to the previously confirmed one (found in register D) to decide whether to send the new state of the push-buttons to the serial line. If the new confirmed value is the same as the previous one, we don't need to send anything and we jump to the SHIFT label.

```
CP    D             ; compare the state with the previous
JP    Z,SHIFT       ; confirmed value, jump if they are equal
```

If the values are different, we transmit the new push-button configuration to the game console. As we've seen, for this to be serialized one bit at a time on line SER, it needs to be set in register E (the transmission buffer). The new configuration is also saved in register D where it will serve as the "previous confirmed value" for the next comparison.

```
TRASM:    LD    E,A       ; copy the byte to transmit to E and
          LD    D,A       ; to D, as "previous confirmed value"
```

According to the specifications, the packet must begin with a start bit at '1' and end with a stop bit at '0'. However, in register E there are only the eight data bits of the packet. Before transmitting them we need to send a start bit, which we can do now.

It will be kept on the line for $2\,mS$, until it is substituted by the first data bit through the next call to the handler.

```
LD    A,10000000b      ; send the start bit on line SER...
OUT   (SOUT),A
JP    EXIT             ; and exit
```

The stop bit is guaranteed to arrive because register E gradually empties out and shifts its contents to the left, while it is filled with zeroes from the right.

Finally, the exit code executes the shift (A → B → C) of the "history" of the previous push-button readings and then it re-enables the interrupts and goes back to the interrupted program.

```
SHIFT:   LD    C,B      ; copy the 12 mS state into 24 mS state
         LD    B,A      ; and the current one into 12 mS state
EXIT:    EI             ; re-enable interrupts and
         RET            ; return to the interrupted program
```

4.7.8 Asynchronous serial communication

The system in the following figures is based on two DMC8 Microcomputer components. It demonstrates serial communication through two hardware components: a transmitter and a receiver.

In the serial data transmitter (see the following figure) port OD has a serialization device attached (ASTX, which will be described further on).

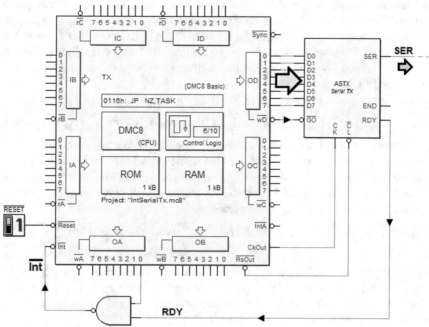

This added component transforms the OD port into a "serial output port". This means it allows for a connection to another system through a single wire (SER) where bits travel one at a time.

We have already met the concept of serial transmission in an example in Section 4.7.7. There, however, the serialization was handled internally via software. Here, we use hardware components dedicated to serialization and de-serialization that guarantee greater efficiency and save the processor's computational resources.

If we look at the schematic, we see that the ASTX component receives 8 data bits in parallel from the OD port it's connected to. While writing to the port, the synchronization signal $\overline{wD}$ allows us to start generating a serial sequence on line SER containing bits D0..D7, which are available at the port.

The serial transmission protocol used in this example is inspired by the main specifications of the classic standard RS-232[16] (see the following figure).

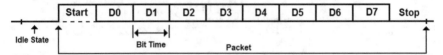

Here, we have defined a bit time of $1.6\,\mu S$, which corresponds to 16 clock cycles of the processor ($10\,MHz$). So the resulting bit rate is $625\,Kb$ (Kilo bits) per second. The packet is made up of 10 bits (a start bit at '1', the 8 data bits 'D0'..'D7', and finally a stop bit at '0').

Notice that the bits are transmitted without clock sync information, in the sense that this doesn't accompany the signal. As we will soon see, it's the job of the receiver to synchronize with the sequence received since it has a clock that is 16 times higher than the nominal velocity of the bit.

When the transmission is finished the ASTX component activates the RDY line (ready) to signal that it is ready to acquire a new number to serialize. To do this, the RDY signal commands interrupt request line $\overline{Int}$ to assure the processor's highest response readiness. If necessary, the processor will immediately send a new number to the OD port.

Notice that the interrupt request line is conditioned by a logic port connected to port OA, in order to disable it when we have no need to transfer data.

The following figure shows the receiving part of the system. Port IB has an added de-serialization component (ASRX, which is described below) that transforms it into a "serial input port".

The schematic shows the ASRX component, which can synchronize with the serially received sequences and extract the 8 data bits from them. The data bits are returned in parallel to the IB port it is connected to.

---

[16] EIA RS-232 (Electronic Industries Alliance Recommended Standard - 232), or simply RS-232, defines a type of synchronous serial connection. This is equivalent to the European standard CCITT V21/V24.

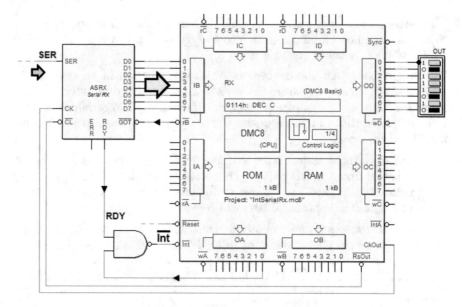

When the RDY (Ready) line is activated, the component signals that a new packet has been received. RDY produces an interrupt request through line $\overline{\text{Int}}$, so the processor goes to get the received data (by reading port IB). The request can be disabled through port OA.

We need to write a program that demonstrates the handling of communication both for the transmitter and for the receiver. Communication must be made through the interrupt mechanism. For test purposes, the transmitter continually sends an automatically generated data byte. The receiver receives the data and and sets them one by one on output port OD (OUT).

For the purposes of writing the code, assume that the main program of the two systems executes operations that are not significant for the communication interface, except for the necessary initializations.

Component specifications: the ASTX serial module

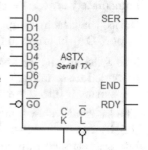

The ASTX transmitter block shown in the figure to the right, has 8 parallel data inputs (D0..D7). A transition from '1' to '0' at input $\overline{\text{GO}}$ loads the data byte into the component, which starts the serialization.

The serial packet is produced at the SER output, following the format described previously.

The bit time is 16 clock periods. The asynchronous clear input $\overline{CL}$ allows us to initialize the component. The RDY output is activated (high) when the component is inactive, that is when it isn't transmitting and is waiting for a new data byte to send. The END output, however, is activated at the end of the transmission, after the stop bit for the duration of one bit time.

The following figure shows the internal schematic of the ASTX component.

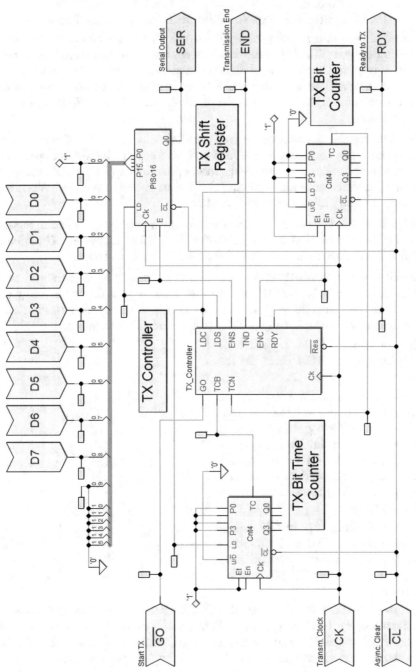

The schematic on the opposite page shows:

— a 'TX Bit Time Counter' on the left hand side,
— a 16-bit 'TX Shift Register' on the upper right hand side,
— a 'TX Bit Counter',
— a 'TX Controller' based on the FSM in the center of the figure.

The bit time counter generates a pulse every 16 clock cycles (1.6 μS, the bit time) on output TC. This line, which is connected to the controller's input TCB, is used to time the transmission of individual bits in the packet at the correct bit rate. The controller's LDC output allows us to make the counter start from value '1111' (set up through inputs P3..P0), at the start of a new transmission.

The shift register serializes the given parallel data byte D0..D7 on the SER output line. The controller's lines LDS and ENS govern the register operation. When the controller activates LDS, the register loads bits D0..D7, the start bit at '1' and the stop bit at '0' and the 6 unused bits in parallel. When it is loaded, the start bit appears on line SER.

The next bits in the packet are transmitted one by one every time the register shifts (ENS is activated by the controller every 16 clock cycles). If LDS and ENS are not activated, the state of the register remains the same.

The bit counter counts the number of bits to transmit and is initialized at 10 when the controller's line LDC is activated. The count is decremented by one each time the controller activates line ENC. When the counter reaches zero, it activates the controller input TCN just to end the transmission.

The controller's FSM algorithm is described in the ASM chart on the right.

Input $\overline{GO}$ controls the transmission. States (a) and (b) wait for its transition from '1' to '0', while they keep the two counters initialized by activating line LDC. In state (b), the RDY line is activated to signal that the system is waiting to transmit another data byte.

When the falling edge of $\overline{GO}$ comes, we go to state (c) where LDS orders the TX Shift Register to load data and ENC orders the TX Bit Counter to decrement.

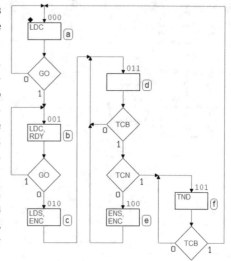

In state (d), the FSM waits for the TCB signal from the bit time counter. Activated every 16 clock cycles, TCB forces the controller to go to state (e) where ENS orders the register to right shift, while ENC orders the TX Bit Counter to decrement and counts the remaining bits to transmit.

When the number of remaining bits reaches zero, the activation of TCN forces the FSM to exit loop (d)-(e) and move to state (f) where it generates TND (that drives the output END), to signal the end of the transmission. Finally, after one more bit time, the FSM returns to state (a).

Component specifications: the ASRX serial receiver

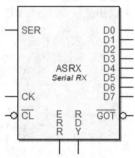

The ASRX receiver block (see the figure on the right) checks line SER, which waits for the start bit of a packet. When the start bit comes, the receiver synchronizes with the sequence and then begins acquiring all of its bits one by one until the stop bit comes. When it finishes receiving the packet, the data bits are presented in parallel on the 8 outputs D0..D7.

At this point, if the stop bit has been received correctly (at '0'), it activates the RDY line and then waits for the handshaking input $\overline{GOT}$ to activate. Input $\overline{GOT}$ will be activated from the outside to signal that the number has been acquired so the component can go back to waiting for the next packet on SER.

The ERR output is activated if the stop bit is at '1' (which can be due to "noise", that is to disturbances along the communication line). Also, ERR is kept active as long as the SER line is at '1' (which could depend on a malfunction of the line).

The internal schematic of the ASRX module is shown in the following figure. Some of the elements are similar to those of the transmitter. Note:

— the 'RX Bit Time Counter' on the left,
— the 8-bit 'RX Shift Register' at the upper right hand side,
— the 'RX Bit Counter',
— the 'RX Controller' based on the FSM, in the center.

The bit time counter is similar to that of the transmitter and is used for a similar purpose: to scan the bit time. It is set to count backward and it activates the controller's line TCB.

In the transmitter, the goal of the bit time counter is to regulate the bit generation rate, whereas in the receiver, it synchronizes bit acquisition. The details on how it works will be discussed further on.

The serial−parallel shift register in the receiver acquires bits on line SER and returns them in parallel on outputs D0..D7. The register operation is set by the controller through line ENS. If ENS = '0', the register stays in the previous state (its content is not modified). When ENS is activated by the controller (for one clock cycle, positioned at the center of the bit time), the register acquires the value currently on line SER and right shifts all the previously memorized values.

After eight acquisitions, all the bits (D0..D7) will be available in parallel on the output lines.

The controller initializes the counter of the received bits at 8 through line LDC. By activating line ENC every time a bit is received, it decrements the count. When it reaches zero, it activates line TCN. So, the controller stops receiving data since all the bits will have been received.

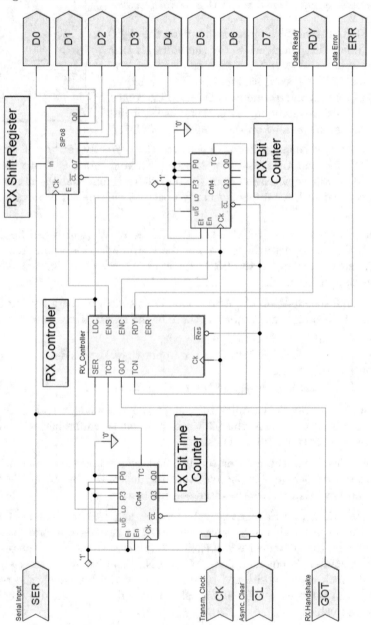

The controller's algorithm is described in the following ASM chart.

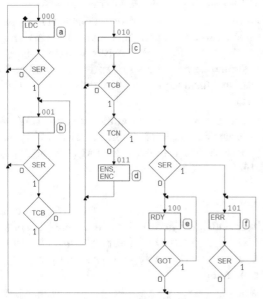

It is important to observe that the receiver does NOT have access to the transmitter's clock. Rather it has its own clock generator, even though its nominal frequency must be the same (in our case 10 *MHz*).

However, the two frequencies won't be precisely equal in the real world because of the different manufacturing tolerances of the different components.

To solve the synchronization problem, we have a bit time that is 16 times the clock period. This gives us the chance to correct the acquisition time of the packet's bits at increments of $1/16$ of the bit time.

Therefore we need to execute a synchronization sequence by verifying that the start bit has arrived in steps of one 16th of the bit time (i.e. one clock cycle). Also, to maximize the likelihood of all the following bit values being read correctly (at least for the duration of one packet), it would be smart to acquire them at the center of the corresponding bit times.

As we see in the ASM chart, to capture the start bit, the FSM reads line SER in state (a). When SER goes to '1', the FSM moves to (b), where it keeps monitoring SER, and also line TCB.

If we look at the network schematic, we see that the activation of line LDC in state (a), has initialized the bit time counter at the value of '0110'. This means that TCB will activate the first time at the center of the start bit time. Since the counter will continue to count cyclically, the next times that TCB activates will be every 16 clock cycles, i.e. at the center of the bit time for all the bits in the packet.

If SER stays at '1' before TCB activates, when TCB arrives, we validate the start bit and go to state (c). If, however, SER goes to '0' ahead of time, the FSM goes back to state (a) and waits for a new start bit (this "filter" helps reduce the chance of interpreting any disturbance on the line as a start bit).

In state (c) we wait for TCB, so we can synchronize with the center of the bit time. When TCB activates, it goes to state (d), where we activate ENS and ENC. ENS causes the bit currently on line SER to be acquired, and also all the other bits in the register to be right shifted.

ENC, however, decrements the bit counter. Notice that the pair of states (c) and (d) is repeated 8 times, i.e. for all the data bits, until TCN activates.

The algorithm's next task is to confirm the stop bit. When TCB activates in state (c), all the data bits have been received and rather than ordering the acquisition of the stop bit in the register, the FSM checks its value directly. If the stop bit is '0', the packet is assumed to be valid so RDY is activated and the FSM goes to state (e) where it stays until input $\overline{\text{GOT}}$ is activated (by the processor that reads the Data Port).

If the stop bit is at '1', the data bits received lose their significance and RDY is not generated. The FSM goes into a waiting state (f), which activates output ERR and waits[17] for SER to go back to '0'. Finally, the FSM goes back to state (a) to wait for a new start bit.

Solution (transmitter assembly code)

First let's deal with the transmitter code. We define the addresses of output ports OA (CTRLP) and OD (DATAP), and the DATA variable, used to simulate the data to be transmitted on line SER. Next, we define the jumps to the main program and the interrupt handler.

```
CTRLP      EQU    00h        ; OA output port (RDY inter. enable)
DATAP      EQU    03h        ; OD output port (serial TX)
DATA       EQU    0FC00h     ; simulated data to transmit

           ORG    0000h
           JP     START
           ORG    0038h
           JP     HINT
           ORG    0100h
```

The main program defines the initial content of the Stack Pointer and the DATA variable. The ASTX device is initialized at system reset so for now, on the program start, we don't need to write in the OD port since that would cause a packet to be sent on the SER line.

```
START:     LD     SP,0FFFFh   ; initialize the Stack Pointer
           LD     A,55h       ; initialize variable DATA with a first
           LD     (DATA),A    ; simulated data byte to transmit
           LD     A,00000001b ; enable interrupt coming from RDY line
           OUT    (CTRLP),A
```

Writing '1' to bit 0 of Control Port CTRLP (see the schematic of the transmitter) enables the ASTX component's interrupt request RDY. Since the ASTX component is already activating line RDY to signal that it is ready to transmit, the processor will receive an interrupt request immediately.

---

[17] Note that this is the simplest check possible that we can carry out and it doesn't pretend to fix all the possible errors.

This request, however, will only be satisfied after the interrupt mechanism is enabled by the EI instruction before going into the main loop.

```
            EI                      ; enable interrupts
MAIN:       CALL   TASK_TX          ; simulate a generic task
            JP     MAIN             ; executed in the main loop
```

In the main loop, we simulate the execution of a generic task by calling subprogram TASK_TX. For the sake of completeness, this subprogram code is shown below, but it is a simulation of a fictitious task done by the transmitter, independent of data transmission (as explained in previous examples).

```
TASK_TX:    LD     C,10             ; simulate a generic task
TASK:       DEC    C                ; 4 +
            JP     NZ,TASK          ; 10 = 14 cycles; 14 x 10 = 140 cycles +
            RET                     ; 17 (call) +7 (ld) +10 (ret) = 174 cycles
```

Aside from saving the registers used, the interrupt handler "creates" a trial data byte by incrementing the DATA variable.

```
HINT:       PUSH   AF               ; save A and Flags
            LD     A,(DATA)         ; get DATA variable content
            INC    A                ; create a new trial data byte
            LD     (DATA),A         ; update DATA in memory
```

The number is then copied to the Data Port and sent to the ASTX component. At the end of the handler, we restore the contents of the saved registers, re-enable the interrupts and go back to the interrupted program.

```
            OUT    (DATAP),A        ; transmit the data byte to the serial port
            POP    AF               ; restore the saved registers
            EI
            RET
```

When the number is written on the Data Port, the ASTX serializer:

— riceives the signal $\overline{GO}$ from line $\overline{wD}$ on the same port,
— deactivates line RDY and, in so doing, the interrupt request,
— starts transmitting data on SER,
— reactivates RDY when the whole packet is generated.

The interrupted program goes back in execution, only to be interrupted again as soon as the ASTX serializer stops generating the packet and reactivates the RDY line.

Solution (receiver assembly code)

At the start of the receiver code, we define the address of the DATAP port (connected to the ASRX de-serializer), and the CTRLP and OUTP output ports. Then we define the jumps to the program and the interrupt handler.

```
DATAP       EQU    01h            ; IB input port (serial receiver )
CTRLP       EQU    00h            ; OA output port (RDY inter. enable)
OUTP        EQU    03h            ; OD output port (received data output)
            ORG    0000h
            JP     START
            ORG    0038h
            JP     HINT
            ORG    0100h
```

The program initializes the Stack Pointer and output port OD (OUTP). The ASRX de-serializer component is initialized by system reset so it doesn't need software for this. However, we need to enable the interrupt request from the RDY line through the CTRLP port.

```
START:      LD     SP,0FFFFh      ; initialize the Stack Pointer
            LD     A,00000000b
            OUT    (OUTP),A       ; zero the received data output port
            LD     A,01h          ; enable interrupt coming from RDY line
            OUT    (CTRLP),A
```

The main loop, akin to that of the transmitter, simulates the execution of a generic task (subprogram TASK_TX).

```
            EI                    ; enable interrupts
MAIN:       CALL   TASK_RX        ; simulate a generic task
            JP     MAIN           ; executed in the main loop
TASK_RX:    LD     C,22           ; simulate a generic task
TASK:       DEC    C              ; 4 +
            JP     NZ,TASK        ; 10 = 14 cycles; 14 x 22 = 308 cycles +
            RET                   ; 17 (call) +7 (ld) +10 (ret) = 342 cycles
```

The interrupt handler saves and recovers A and the Flags, and also reads the received data byte from the DATAP Data Port then copies it to port OUTP.

```
HINT:       PUSH   AF             ; save A and Flags
            IN     A,(DATAP)      ; read the data byte from the receiver
            OUT    (OUTP),A       ; and copy it to the output port
            POP    AF             ; restore the saved registers
            EI                    ; re-enable interrupts
            RET                   ; return to the interrupted program
```

The code that read the received data is simple. This is because of the tasks executed in hardware by the ASRX de-serializer. When we read the Data Port DATAP:

— It receives the $\overline{\text{GOT}}$ signal from line $\overline{\text{rB}}$ from the same port.
— It deactivates the RDY line and so, the interrupt request.
— It goes back to wait for a new packet from SER.
— When it finishes receiving the packet, it reactivates RDY and the processor is interrupted again so that it goes to retrieve the new received data byte.

4.8 Exercises

The digital content pages of the book on the Deeds simulator website have outlines of the schematics, diagrams and/or programs to complete for each exercise. Those same web pages also have the files for the solutions, so that students can check their work.

4.8.1 Interrupt techniques

1. Assume we have the system in the figure below, which is based on the "DMC8 Microcomputer" component.

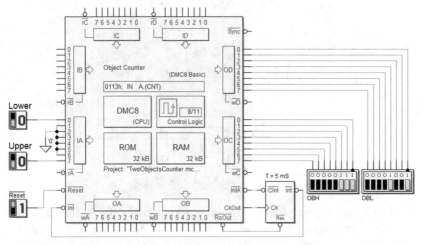

OC and OD, the two parallel ports, (at addresses 02h and 03h) drive two groups of 8 LED lights each (OBH and OBL, respectively).

The IA parallel input port (at address 00h) lets us read the outputs of the two optic sensors (LOWER and UPPER) on bits 7 and 0. The sensors are positioned at the same point next to a conveyor belt, one on top of the other. The different heights of the two sensors allow the system to distinguish between 'tall' and 'short' objects that move along the belt (when an object moves in front of a sensor, it generates a '1').

A timer activates the microcomputer's interrupt request line $\overline{\text{Int}}$ every $5\,mS$. When the interrupt is accepted, the processor's response $\overline{\text{IntA}}$ automatically deactivates line $\overline{\text{Int}}$.

We need to write a program in assembly language that keeps count of the number of tall and short objects that move along the conveyor belt.

The system counts the tall items that pass in front of the sensors separately from the short ones. Every second it updates the number of items on the LED lights in binary code. Then it restarts the count from zero (OBH = the number of tall objects, OBL = the number of short objects).

At most 50 items move along the conveyor belt per second. We must turn off the LED lights (for about one tenth of a second) before showing the new number. While the LED lights are off, the system must continue to count items.

2. Assume we have the system in the figure below, which is based on the "DMC8 Microcomputer" component.

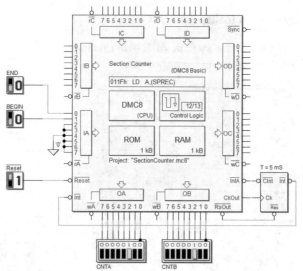

The parallel port of input IA (at address 00h) lets us read the outputs of the two proximity sensors (BEGIN and END) on bits 1 and 0. The sensors are positioned at the beginning and the end of a conveyor belt in an industrial plant.

OA and OB, the two output parallel ports, (at addresses 00h and 01h) drive two groups of 8 LED lights each (CNTA and CNTB, respectively).

A timer activates the microcomputer's interrupt request line $\overline{\text{Int}}$ every $5\,mS$. When the interrupt is accepted, the processor's response $\overline{\text{IntA}}$ automatically deactivates line $\overline{\text{Int}}$.

We need to write a program in assembly language that keeps count of the number of objects that move along the conveyor belt based on the following specifications.

The main program constantly checks the two sensors. Each time an object moves in front of a sensor, it generates a '1'.

Assume that when a sensor output goes from zero to one, this correctly indicates that an object is passing on front of it. Only one object can move before a sensor at a time (the objects are in single file on the conveyor belt). Obviously, two objects can be on the belt at the same time, one in front of the BEGIN sensor and another in front of the END sensor.

The main program counts the objects on the conveyor belt (that have passed from the BEGIN sensor but haven't reached the END sensor). Assume that there can be no more than 200 objects on the conveyor belt at the same time.

Every second, the system updates the number of items on the conveyor belt on the CNTA LED lights in binary code through the interrupt handler. At the same time, it shows the difference between the current number of items and the number shown one second before on the CNTB LED lights (as a signed integer coded in two's complement).

3. Assume we have the system in the figure below, which is based on the "DMC8 Microcomputer" component.

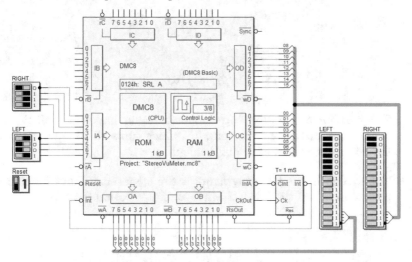

Input port IA acquires two 4-bit binary numbers, LEFT and RIGHT, connected to lines 7..4 and 3..0, respectively. The four output ports OD, OC, OB and OA drive two "columns" of 16 LED lights each.

A timer activates the microcomputer's interrupt request line $\overline{\text{Int}}$ every 1 mS. When the interrupt is accepted, the processor's response $\overline{\text{IntA}}$ automatically deactivates line $\overline{\text{Int}}$.

We need to write a program in assembly language that manages the system based on the following specifications.

The main program executes the necessary initializations at hardware reset and then enters an infinite loop where it acquires the numbers LEFT and RIGHT from the input port.

The data are divided into two variables VLEFT and VRIGHT, which will then be read by the interrupt handler. Every 5 mS, the interrupt handler converts the values of the variables VLEFT and VRIGHT to linearly represent the values received (see the figure below) on the two "LED light columns". Define a value table in the source code like the one below:

```
TABLE:   DB    00000000b, 00000000b  ; index 0 → all LED lights off
         DB    00000000b, 00000001b
         DB    00000000b, 00000011b
         DB    00000000b, 00000111b
         DB    00000000b, 00001111b
         DB    00000000b, 00011111b
         DB    00000000b, 00111111b
         DB    00000000b, 01111111b
         DB    00000000b, 11111111b
         DB    00000001b, 11111111b
         DB    00000011b, 11111111b
         DB    00000111b, 11111111b
         DB    00001111b, 11111111b
         DB    00011111b, 11111111b
         DB    00111111b, 11111111b
         DB    01111111b, 11111111b  ; index 15 → 15 LED lights on
```

The first line of the table corresponds to value 0 and turns all the lights off; the last corresponds to value 15, which turns all 15 lights on (notice the "diagonal" trend of the bit configuration).

4. The system in the following figure is the management unit for a remote control based on the "DMC8 Microcomputer" component.

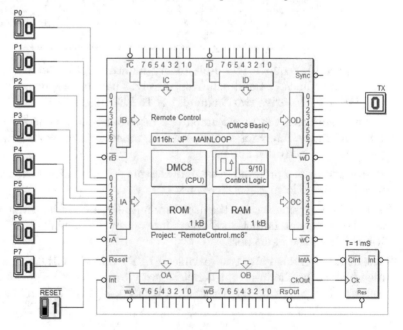

Input port IA lets us acquire the state of eight push-buttons (P7..P0) at the disposition of the user. Output port OD connects our system to an infrared transmitter (not shown in the figure) through a serial asynchronous line TX connected to bit 1.

Every 1 mS a Timer activates the microcomputer's interrupt request line $\overline{Int}$. When the processor accepts the interrupt, $\overline{IntA}$ automatically deactivates line $\overline{Int}$.

We need to write a program in assembly language that manages the remote control according to the following specifications.

For this project, the main program does not execute important operations except for the necessary initializations of the system at reset. Next, it enters an infinite loop where it is inactive, leaving the control of the system to the interrupt handler.

The interrupt handler reads the push-buttons and does a "debouncing check" on their state. The push-buttons are checked every 10 mS, and this consists of making sure the configuration read on the port is identical to that read 10 mS before.

This first check prevents mechanical bounces in the contacts from causing errors in reading the push-buttons. The state of the push-buttons is considered validated when the last two readings of the push-button port are identical regardless of whether the push-buttons are pressed.

Then, the new configuration is coded on 4 bits: T, C, B and A, based on the following table :

P7	P6	P5	P4	P3	P2	P1	P0	T	C	B	A
1	-	-	-	-	-	-	-	1	1	1	1
0	1	-	-	-	-	-	-	1	1	1	0
0	0	1	-	-	-	-	-	1	1	0	1
0	0	0	1	-	-	-	-	1	1	0	0
0	0	0	0	1	-	-	-	1	0	1	1
0	0	0	0	0	1	-	-	1	0	1	0
0	0	0	0	0	0	1	-	1	0	0	1
0	0	0	0	0	0	0	1	1	0	0	0
0	0	0	0	0	0	0	0	0	0	0	0

As the table shows, the coding works on "priority". If the user presses multiple push-buttons by mistake, only the one with the highest number is recognized. Bits C, B and A represent the code that is transmitted on serial line TX. When at '1', bit T indicates if that configuration corresponds to at least one pressed push-button. If none of the push-buttons is pressed the system sends no code.

The system transmits if at least one push-button is pressed and if the current code is different from that obtained previously. This is done by serializing bits C, B and A on line TX (bit T is only used to identify the "no push-button pressed" configuration, and is not transmitted).

The asynchronous serial transmission protocol has a bit time of $1\,mS$, and the packet is composed of a start bit at '1', bits A, B and C (in order), and a stop bit at '0' (see the following figure).

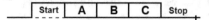

Suggestion: an example of software-managed serial transmission was given in Section 4.7.7.

5. Assume we have the system in the figure below, which is based on the "DMC8 Microcomputer" component. Input port IA lets us read the state of a push-button (PBUT) on bit 0. Output port OD shows an 8-bit number on 8 LED lights.

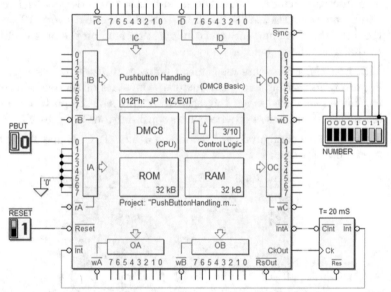

The system must show the value of a variable (NUMBER) to the outside through the LED lights connected to the OD port. The user can control the value of NUMBER through the push-button.

Every $20\,mS$, a timer activates the microcomputer's interrupt request line $\overline{\text{Int}}$. When the interrupt is accepted, the processor's response $\overline{\text{IntA}}$ automatically deactivates line $\overline{\text{Int}}$.

We need to write a program in assembly language that manages the value of the variable (copied on port OD), based on pressing the button.

The main program performs the necessary initializations and then enters an infinite loop where it is inactive.

The interrupt handler, however, checks the push-button periodically. To prevent mechanical bounces of the push-buttons from causing misreadings, the interrupt handler validates the state of the push-buttons through two consecutive readings $20\,mS$ apart.

When the push-button is pressed, it generates a high level. By pressing the push-button for half a second, we increment the variable NUMBER by one unit. If the button is kept pressed, the increment is repeated every half second. If the number is at the largest value that can be represented with 8 bits, the increment is not executed. When the push-button is not pressed, the variable NUMBER is decremented by one unit every half second, unless that number has reached zero.

6. The figure represents a system based on the "DMC8 Enhanced Microcomputer" component. The parallel port of input IA acquires data lines D3..D0. The four parallel output ports OA, OB, OC and OD drive 4 rows each with 8 LED lights, which make up the four sides of a square.

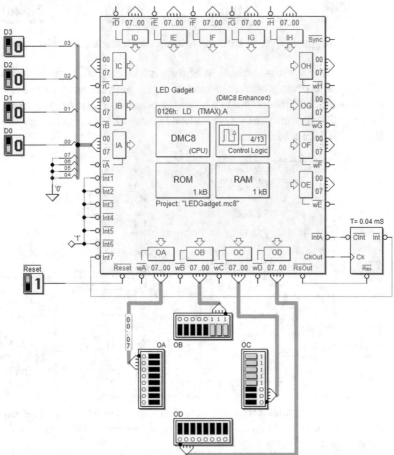

A timer activates the microcomputer's interrupt request line $\overline{\text{Int}}$ every $40\,\mu S$. When the processor accepts the interrupt, $\overline{\text{IntA}}$ automatically deactivates line $\overline{\text{Int}}$.

The main program executes the necessary initializations at reset (all LED lights on the OC port on, all the others off), and then enters an infinite loop where it cyclically acquires a 4-bit number from the input port's lines D3..D0. After incrementing the value of a unit, it saves it in the memory (in the TCYCLE variable).

When the interval lasting TCYCLE * $40\mu S$ is over, it makes the eight lit LED lights rotate counter clockwise by one position. The following figure shows the first three rotations (1)(2)(3) as of the initial position (0).

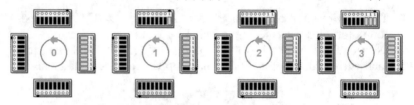

The LED lights go back to their initial position after 32 rotations.

7. Assume we have the system in the figure below, which is based on the "DMC8 Microcomputer" component. Input port IA lets us read the state of push-button PB on bit 0 (when it is pressed, it gives a low level). The two parallel output ports, OD and OC show a 12-bit number on the LED lights (NUMH and NUML).

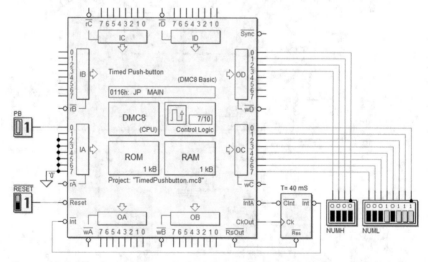

Every $40\,mS$ a timer activates the microcomputer's interrupt request line $\overline{\text{Int}}$. When the interrupt is accepted, the processor's response $\overline{\text{IntA}}$ automatically deactivates line $\overline{\text{Int}}$.

We need to write a program in assembly language that manages the system based on the following specifications.

For this project, the main program does not execute important operations except for the necessary initializations of the system at reset, including zeroing the number shown on the LED lights NUMH and NUML.

Next, the main program enters an empty, infinite loop and delegates system control to the interrupt handler.

Every 40 mS, the interrupt handler reads the state of the push-buttons (there is no request for a debouncing check, since we assume this has been solved by the hardware, not shown). The functionality of the push-button depends on how long the user presses it.

— If the push-button is pressed for less than 200 mS, the number on LED lights NUMH and NUML is incremented by one when it is released.
— If the button is pressed for more than 200 mS and less than 4 seconds, the number is incremented by eight when it is released.
— If the button is pressed for at least 4 seconds, the number goes to zero when it is released.

The number must be limited to the largest number that can be represented with 12 bits.

Note: for assessing times, an approximation of ±40 mS is acceptable.

8. The figure on the next page shows a sine wave digital generator based on the "DMC8 Enhanced Microcomputer" component.

Input port IC is connected to a serial ASRX receiver (described in the example in Section 4.7.8) that allows us to receive a command word from the SER serial line.

The serial packet is composed of 8 data bits plus the start and stop bits (see the following figure). The bit time is 1,6 μS, corresponding to 16 of the processor's clock cycles (10 MHz).

| Start | F0 | F1 | F2 | F3 | F4 | – | – | – | Stop |

Only bits F4..F0 are used to control the generator. They are given by the ASRX receiver on IC port lines D4..D0.

We need to use a virtual 8-bit DAC connected to port OF to generate a sinusoidal wave form (see the example in Section 4.7.3). The processor's clock frequency is 10 MHz, and the timer generates an interrupt every 100 μS, through line $\overline{Int7}$. The timer's input $\overline{CInt}$ is connected to write strobe $\overline{wF}$ of port OF (also used for the DAC's clock).

The sinusoid wave frequency has to be linearly proportional to the FREQ parameter acquired by the SER serial line. This parameter's value comes directly from the bits of the serial packet (after limiting the received value to an interval of 0..31). Note that a zero stops oscillation and generates a constant in the output.

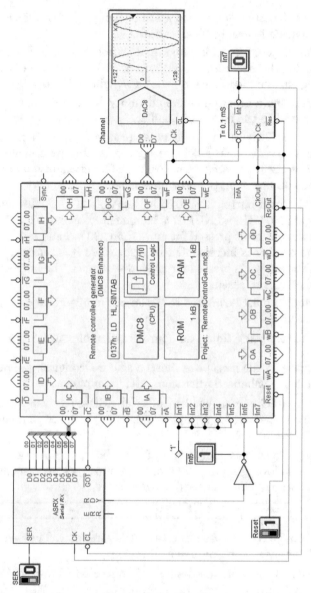

We need to write a program in assembly language that manages the system based on the following indications.

The main program executes the necessary initializations at hardware reset and then enters an infinite loop where it is inactive.

There are two interrupt handlers. One of them generates the sinusoidal wave form and responds to the interrupts generated by timer ($\overline{\text{Int7}}$). The other depends on interrupt $\overline{\text{Int6}}$, which is generated by the ASRX receiver when a new packet comes from the serial line.

4.9 Solutions

4.9.1 Interrupt techniques

1. Based on the system's specifications, we first see that we only need 8 bits to count at most 50 items per second. With the timer set at 5 mS, we should count 200 calls from the interrupt handler to allow one second to go by, so we also only need 8 bits to measure time.

   We define the names of the ports and the TIME variable, used to assess time, as well as the usual jumps to the program and the interrupt handler.

```
SENS        EQU     00h             ; IA input port (sensors)
OBH         EQU     02h
OBL         EQU     03h
TIME        EQU     0FC00h          ; time counter
            ORG     0000h
            JP      START
            ORG     0038h
            JP      HINT
            ORG     0100h
```

   Before entering the main loop, we initialize the TIME variable and registers B and C, which are used to count the low and high objects, respectively. Then we zero the output ports and enable the interrupts.

```
START:      LD      SP,0FFFFh       ; initialize the Stack Pointer
            LD      A,200           ; and the time counter
            LD      (TIME),A
            LD      B,0             ; zero B and C registers, used to count
            LD      C,0             ; the low and high objects, respectively
            LD      A,0
            OUT     (OBH),A         ; on start, all LED lights OFF
            OUT     (OBL),A
            EI                      ; enable interrupts
```

   In the main loop, the program keeps track of object movement by checking the low sensor. When it signals that an object is near, we check whether the high sensor is active too.

```
MAIN:       IN      A,(SENS)        ; wait for an object movement
            BIT     0,A             ; by checking the low sensor
            JP      Z,MAIN
            BIT     7,A             ; check if the object is high
            JP      NZ,ISHIGH       ; jump if it is
```

   According to the height of the object, we wait for the sensor to signal that it has left the field of view, then we increment the count.

```
ISLOW:    IN     A,(SENS)     ; check if the low object is passed by
          BIT    0,A
          JP     NZ,ISLOW
          INC    B            ; if yes, increment the low object count
          JP     MAIN         ; and return to wait for the next object
ISHIGH:   IN     A,(SENS)     ; check if the high object is passed by
          BIT    7,A
          JP     NZ,ISHIGH
          INC    C            ; if yes, increment the high object count
          JP     MAIN         ; and return to wait for the next object
```

The interrupt handling routine deals with visualization. It is launched every 5 mS, and each time, it decrements the count by 1 (which, as we know, was initialized at 200).

```
HINT:     PUSH   AF           ; save A and Flags
          LD     A,(TIME)     ; load the time counter in register A,
          DEC    A            ; decrement it and
          LD     (TIME),A     ; write back the time counter in memory
```

When the time count reaches 1/10 of a second (TIME = 20), the LED lights are turned off by a jump to LEDOFF, as per specifications.

```
          CP     20           ; if we are 1/10 second from the end...
          JP     Z,LEDOFF     ; jump and turn LED lights off;
          CP     0            ; if time count reached zero...
          JP     Z,DISPLAY    ; jump and display the object count
EXIT:     POP    AF           ; otherwise exit, restore A and Flags
          EI                  ; and re-enable interrupts
          RET
LEDOFF:   LD     A,00         ; turn off the LED lights
          OUT    (OBH),A
          OUT    (OBL),A
          JP     EXIT         ; and exit
```

However, when the count is zeroed, one second has gone by and we jump to DISPLAY, where the number of high and low objects is shown on their ports and then zeroed.

```
DISPLAY:  LD     A,C          ; get the number of high objects...
          OUT    (OBH),A      ; and display it
          LD     A,B          ; get the number of low objects...
          OUT    (OBL),A      ; and display it
          LD     C,0          ; zero the low and high object count
          LD     B,0
          LD     A,200        ; re-initialize the time counter to 200
          LD     (TIME),A
          JP     EXIT         ; and exit
```

Finally, the time count is re-initialized and starts again from 200.

2. We define the I/O ports and four variables. SPREC stores the previous state of the sensors and is used to identify the signal edges. TIME lets us assess the time required by the text. COUNT is used to count the number of objects and is incremented when an object comes onto the conveyor belt and is decremented when it leaves. PCOUNT records the object count from the previous interval.

```
SENS        EQU    00h              ; IA input port (sensors)
CNTA        EQU    00h              ; OA and OB output ports (counts)
CNTB        EQU    01h

SPREC       EQU    0FC00h           ; previous state of the sensors
TIME        EQU    0FC01h           ; time counter
COUNT       EQU    0FC02h           ; object counter
PCOUNT      EQU    0FC03h           ; previous object count

            ORG    0000h
            JP     START
            ORG    0038h
            JP     HINT
            ORG    0100h
```

We initialize the Stack Pointer and then zero the variables of COUNT and PCOUNT, as well as the two output ports. In SPREC, we insert a value so as not to identify edges during the first reading of the port. The timer interrupts the processor every 5 mS, so to make one second go by, we count by 200 (which we insert in TIME).

We enable interrupts with an EI instruction before entering the main loop. Here, checks on the rising edges are executed in parallel because they could come simultaneously on the two sensors. We determine if there is a rising edge through a logic operation between the current state of the sensors and the previous one.

```
START:      LD     SP,0FFFFh        ; initialize the Stack Pointer
            LD     A, 00h
            LD     (COUNT),A
            LD     (PCOUNT),A
            OUT    (CNTA), A
            OUT    (CNTB), A
            LD     A, 00000011b     ; value useful to discard the first reading
            LD     (SPREC), A
            LD     A, 200           ; initialize the time counter to 200
            LD     (TIME),A         ; (1 second = 200 x 5 mS)
            EI
```

As explained in Section 4.7.6, if we define the value read previously on one of the sensors as 'P', and the current one as 'C', on that line there is now a rising edge if C = '1' and P = '0', that is if the Boolean function $F = (C \cdot \overline{P})$ gives us '1'.

In the processor, this function can be calculated on all the bits of the accumulator simultaneously, so we can detect rising edges on two sensors at the same time.

```
MAIN:     IN    A, (SENS)      ; get the current state of sensors
          AND   00000011b      ; masks the bits that don't interest us
          LD    B, A           ; save the current sensors state in B
          LD    A, (SPREC)     ; get the previous state
          CPL                  ; negate it and execute a bitwise AND
          AND   B              ; between it and the current state
          LD    C, A           ; copy the result to register C
          LD    A, B           ; save the current as "previous state" to
          LD    (SPREC), A     ; SPREC variable, for the next check
```

We acquire the current state of the sensors, zero the bits that don't interest us and then save the state in register B. In the accumulator, we retrieve the previous state from SPREC. We use the CPL instruction to invert the previous state of the sensors as per the Boolean function described above. The result of the AND is copied to register C, while the current state of the sensors is saved in SPREC for the next check.

We have the results of the function for both sensors in register C. Let's check bit 1. If it is not '0', a rising edge has come to the sensor, so we need to increment the COUNT of the objects present. The test of bit 0 is similar, but this bit comes from the sensor placed at the end of the conveyor belt. If an edge has occurred on it, COUNT must be decremented[18]. After these operations, we go back to the MAIN label.

```
S_BEGIN:   BIT   1, C          ; check the sensor at the beginning
           JP    Z,S_END       ; of the conveyor belt
           LD    A,(COUNT)     ; if active increment the count
           INC   A             ; of the objects that have entered
           LD    (COUNT),A
S_END:     BIT   0, C          ; check the sensor at the end of the belt
           JP    Z,MAIN        ; of the conveyor belt
           LD    A,(COUNT)     ; if active decrement the count
           DEC   A             ; of the objects that have come out
           LD    (COUNT),A
           JP    MAIN
```

---

[18] We can also use two separate variables to count entering and exiting objects. This would not, however, be a good choice because some complications would arise, such as the need to subtract the two values at the end of one second and to keep count of any overflow of the counts. To prevent this, the two values would be re-initialized when the time was up, while keeping count of the number of objects on the conveyor belt at that time, and loading the difference between the two counts in the variable of the objects that entered. In any case, an assumption on the conveyor belt's maximum velocity is needed.

Now, let's look at the interrupt handler, which is called every 5 *mS*. We immediately save all the registers involved (A, Flags, B and C) on the Stack. We assess if one second has passed by decrementing TIME. If one second has not gone by, we exit the interrupt handler; if it has, we proceed by re-initializing the time count.

```
HINT:       PUSH  AF            ; save the registers involved
            PUSH  BC

            LD    A,(TIME)      ; assess the time that has passed
            DEC   A
            LD    (TIME),A
            JP    NZ, EXIT      ; exit if one second has not gone by

            LD    A, 200        ; otherwise, it has passed, so
            LD    (TIME),A      ; re-initialize the time counter
```

One second has gone by and we need to show the results on the LED lights. We temporarily move the previous count PCOUNT in B. Then we take the current count and show it on the CNTA lights, as per specifications. This number is also copied in the PCOUNT variable to be used next time.

```
            LD    A,(PCOUNT)  ; copy the previous count to register B
            LD    B,A

            LD    A,(COUNT)   ; display the current count
            OUT   (CNTA), A   ; on port CNTA and also save it
            LD    (PCOUNT),A  ; as previous count for the next time
```

So we subtract the number of previous objects from the number of current ones and copy that on the CNTB port, as per specifications.

```
            SUB   B           ; subtract the previous count from the
            OUT   (CNTB), A   ; current one and display it
```

The handler finishes by restoring the registers used and returning to the interrupted program.

```
EXIT:       POP   BC          ; restore the registers
            POP   AF
            EI                ; re-enable interrupts
            RET               ; return to the interrupted program
```

3. We define the addresses of the input and output ports used in the project.

```
DATAIN    EQU   00h           ; IA input port
LED_RH    EQU   03h           ; OD output port
LED_RL    EQU   02h           ; OC output port
LED_LH    EQU   01h           ; OB output port
LED_LL    EQU   00h           ; OA output port
```

Then we define the variables VLEFT and VRIGHT put forth in the text, and we introduce a variable TIME as well, to assess the required $5\,mS$. The variables are all 8-bit integers. This is followed by the usual links to the reset and the interrupt request.

```
TIME       EQU    0FC00h        ; time counter
VLEFT      EQU    0FC01h        ; left channel variable
VRIGHT     EQU    0FC02h        ; right channel variable
           ORG    0000h
           JP     START
           ORG    0038h
           JP     HINT
           ORG    0100h
```

After the Stack Pointer, we initialize the TIME variable at 5 to count the $5\,mS$, since the timer interrupts us every $1\,mS$. We zero variables VLEFT and VRIGHT, as well as all the output ports (to turn the lights off). Then we enable the interrupts and enter the main loop.

```
START:     LD     SP,0FFFFh     ; initialize the Stack Pointer
           LD     A,5
           LD     (TIME),A      ; set the time counter at 5
           LD     A, 00h
           LD     (VLEFT),A     ; zero the left and right variables
           LD     (VRIGHT),A
           OUT    (LED_LL),A    ; and zero all the output ports
           OUT    (LED_LH),A
           OUT    (LED_RL),A
           OUT    (LED_RH),A
           EI
```

The main loop reads the DATAIN port (IA), and saves it in register B. It masks bits 7 through 4 with an AND instruction, while it copies the value of data lines 3..0 in RIGHT. It takes what was saved in register B and right shifts it four positions to save the values in LEFT. Then it repeats these operations infinitely.

```
MAIN:      IN     A,(DATAIN)    ; read input port DATAIN
           LD     B,A
           AND    00001111b     ; let there the low part only
           LD     (VRIGHT),A    ; save it in the right channel variable
           LD     A,B           ; now process the high part
           SRL    A             ; shift right it 4 times
           SRL    A
           SRL    A
           SRL    A
           LD     (VLEFT),A     ; and save it in the left channel variable
           JP     MAIN
```

Every 1 mS the interrupt handler is executed. We save the registers used, i.e., A, the flags and HL, on the stack. Right after, we assess the time that has elapsed. If the TIME variable has gone to zero, 5 mS have gone by and we proceed. If it has not, then we simply exit.

```
HINT:      PUSH  AF          ; save A, Flags and HL on the Stack
           PUSH  HL

           LD    A,(TIME)     ; assess the time that has passed
           DEC   A
           LD    (TIME),A
           JP    NZ,EXIT      ; exit if 5 mS have not elapsed

           LD    A,5          ; otherwise, re-initialize variable TIME
           LD    (TIME),A     ; and goes on
```

5 mS have gone by and we have re-initialized the time count. Now we need to update the LED columns consistently with variables VLEFT and VRIGHT. We delegate to the LINEAR subprogram the task of translating the value of the variable (passed through register A) to the bit pattern to write to the LED ports (returned to H and L).

```
           LD    A,(VLEFT)    ; convert the left channel variable
           CALL  LINEAR       ; in the corresponding LED patterns
           LD    A,L          ; which are returned in L and H registers
           OUT   (LED_LL),A   ; and display them on the output ports
           LD    A,H
           OUT   (LED_LH),A
```

We call the function twice: the first time, we pass the VLEFT variable in A, and the second, the VRIGHT variable. Once we've got the corresponding patterns of the LED lights in H and L, we update the corresponding ports.

```
           LD    A,(VRIGHT)   ; do the same for the right channel
           CALL  LINEAR
           LD    A,L
           OUT   (LED_RL),A
           LD    A,H
           OUT   (LED_RH),A
```

The exit code restores the contents of the registers saved before, re-enables the interrupts and goes back to the interrupted program.

```
EXIT:      POP   HL           ; restore HL, Flags and A from the Stack
           POP   AF
           EI
           RET
```

The LINEAR subprogram uses the table from the text. The address of the first row in the table (referenced by the label TABLE) is set in HL.

Every row in the table stores the pattern of 16 lights to turn on or off according to a certain index passed from the calling program through A.

The SLA instruction multiplies the given index by 2, thus offsetting the row we are interested in since the table is made up of 2 bytes per row. So we add this offset to the address in HL.

```
LINEAR:   LD    HL,TABLE      ; read the table item specified by the
          SLA   A             ; index passed through register A
          ADD   A,L           ; calculate HL = HL + (2 · A)
          LD    L,A
          JP    NC,GETHIGH    ; take into account the carry, if any
          INC   H
```

In HL, we now have the address of the first byte of the row corresponding to the given index. We use HL to read that byte from the table and copy it in A. Then we increment HL to point to the second byte that interests us. After the second access to the table, the address in HL will no longer be useful so we can reuse L and H for another purpose. Therefore, we copy the second byte to L.

After transferring A in H, the function returns to the calling program as the two bytes with the LED patterns are in registers H and L.

```
GETHIGH:  LD    A,(HL)        ; read the two corresponding patterns
          INC   HL
GETLOW:   LD    L,(HL)
          LD    H,A           ; and return them in registers H and L
          RET
```

For easy readability, the table described in the text is shown here.

```
TABLE:    DB    00000000b, 00000000b  ; index 0 → all LED lights off
          DB    00000000b, 00000001b
          DB    00000000b, 00000011b
          DB    00000000b, 00000111b
          DB    00000000b, 00001111b
          DB    00000000b, 00011111b
          DB    00000000b, 00111111b
          DB    00000000b, 01111111b
          DB    00000000b, 11111111b
          DB    00000001b, 11111111b
          DB    00000011b, 11111111b
          DB    00000111b, 11111111b
          DB    00001111b, 11111111b
          DB    00011111b, 11111111b
          DB    00111111b, 11111111b
          DB    01111111b, 11111111b  ; index 15 → 15 LED lights on
```

4. We define the addresses of the input and output ports and the variables TIME, TXBUFF, PREV and CONV. This is followed by the link to the reset and the interrupt request.

```
USER        EQU     00h             ; IA input port (push-buttons)
TXOUT       EQU     03h             ; OD output port (serial output TX)

TIME        EQU     0FC00h          ; time counter
TXBUFF      EQU     0FC01h          ; transmission buffer
PREV        EQU     0FC02h          ; the previous state of the input port
PTCBA       EQU     0FC03h          ; the previous validated code

            ORG     0000h
            JP      START
            ORG     0038h
            JP      HINT
            ORG     0100h
```

We initialize all the variables and the output port at zero; we set TIME at 10. We enable the interrupts and enter the main loop which is empty.

```
START:      LD      SP,0FFFFh       ; initialize the Stack Pointer
            LD      A,10
            LD      (TIME),A        ; 10 * 1 mS (timer tick) = 10 mS
            LD      A,0             ; zero the other variables
            LD      (TXBUFF),A
            LD      (PREV),A
            LD      (PTCBA),A
            OUT     (TXOUT),A       ; and the output port
            EI
MAIN:       JP      MAIN
```

On entering the interrupt handler, we save the registers used and, on exit, we restore them before returning to the interrupted program. We do this even though the main loop is empty, in anticipation of any future modification.

```
HINT:       PUSH  AF                ; save registers A, Flags, D and E
            PUSH  DE
```

The interrupt handler is launched every $1\,mS$ and it executes serial bit transmission (one every $1\,mS$) in the first couple lines. We have defined a transmission buffer (the TXBUFF variable), which is even used when the serial line is in the idle state. In the absence of a packet, the buffer contains zeroes. The value of the line remains idle, and the handler continues to send zero after zero, right shifting the register.

```
            LD      A,(TXBUFF)      ; get the transmission buffer
            SRL     A               ; shift right it of one bit position
            LD      (TXBUFF),A      ; and write back it in memory
            AND     00000010b       ; mask all the bits, except bit 1
            OUT     (TXOUT),A       ; send its value to the output port
```

When we have a packet to send, we copy it in TXBUFF so that the handler right shifts it one position at each call and sends its bits out one by one. When the packet is done being transmitted, TXBUFF will be totally at zero again.

Considering the position of the TX serial line on bit 1 of the output port and the specifications in the text, the packet initially set in the buffer has this format:

— bit 7 = '0' (idle state of the line)
— bit 6 = '0' (stop bit)
— bits 5, 4 and 3 = C, B and A (data bits, C is transmitted last)
— bit 2 = '1' (start bit)
— bit 1, 0 = '0'

For the transmission, we have chosen to shift the buffer before sending a bit, therefore the packet contains the bits arranged so that the start bit will be in the position of line TX at the first sending.

We send a bit on the serial line and then assess the time that has elapsed, decrementing the TIME variable. We move on only if 10 mS have passed, otherwise we exit the handler.

```
LD    A,(TIME)      ; assess the time (10 mS)
DEC   A
LD    (TIME),A
JP    NZ,EXIT        ; exit if 10 mS have not elapsed,
LD    A,10           ; otherwise re-initialize the time count
LD    (TIME),A
```

10 mS have elapsed, so we need to read the state of the push-buttons after loading the configuration read 10 mS before (PREV) in register D.

We get rid of the bounces by exiting the handler. We move forward if the two configurations are equal, thus validating the new configuration. In any case, we save the current configuration in the PREV variable to use it in the next 10 mS.

```
LD    A,(PREV)      ; copy the previous reading
LD    D,A           ; to register D
IN    A,(USER)      ; get the current state of push-buttons
CP    D             ; and compare it with the previous one
LD    (PREV),A      ; save the current as previous for next
JP    NZ,EXIT        ; exit if the current state is different
```

Then we validate the new value and move on to encode the key pressed in the four bits T, C, B and A, as described in the text (for easy readability, the table of the "priority encoder" is shown in the following).

To immediately exclude the possibility expressed in the last row of the table, we do a preliminary check to see that no key has been pressed.

P7	P6	P5	P4	P3	P2	P1	P0	T	C	B	A
1	-	-	-	-	-	-	-	1	1	1	1
0	1	-	-	-	-	-	-	1	1	1	0
0	0	1	-	-	-	-	-	1	1	0	1
0	0	0	1	-	-	-	-	1	1	0	0
0	0	0	0	1	-	-	-	1	0	1	1
0	0	0	0	0	1	-	-	1	0	1	0
0	0	0	0	0	0	1	-	1	0	0	1
0	0	0	0	0	0	0	1	1	0	0	0
0	0	0	0	0	0	0	0	0	0	0	0

In the case of the last row, we jump to CODED, taking advantage of the fact that the accumulator is zeroed (with the meaning of TCBA = '0000').

```
OR    A              ; check if at least a button is pressed
JP    Z,CODED        ; avoid encoding if no button is pressed
```

For priority encoding, we chose the "by calculations" method (we could also have used the "decision tree" method). To do this, we begin with the attempt code TCBA = '1111' in the accumulator and check the most significant bit of the push-buttons (whose state is found in D). If the push-button is not pressed, we shift the next push-button to position 7, decrement the TCBA attempt code and repeat the loop.

```
         LD    A,00001111b    ; code TCBA (and loop counter)
DECODE:  RL    D              ; rotate through Carry flag the buttons
         JP    C,CODED        ; if the flag is set, we have the code in A
         DEC   A              ; otherwise, repeat the loop
         JP    DECODE         ; until the code TCBA is found
```

The algorithm necessarily converges because we have excluded the situation where no push-button is pressed. The new, validated TCBA code that we have in the accumulator, now has to be compared with the previous one (PTCBA), as per specifications, to determine whether we send a new packet. Next, we copy the new TCBA code to register E and compare it with the previously validated one. If they are equal, it does not need to be transmitted.

```
CODED:   LD    E,A            ; copy in E the new obtained code
         LD    A,(PTCBA)      ; get the previous validated code in A
         CP    E              ; compare it with the new code
         JP    Z,EXIT         ; exit if they are equal
```

The specifications also forbid data transmission if no key is pressed. Therefore, we save the validated code in the PTCBA variable and then make sure it is not zero. If that is the case, we transmit nothing and exit.

```
         LD    A,E            ; get the new validated code from E
         LD    (PTCBA),A      ; save it into the PTCBA variable
         OR    A              ; check if at least a button is pressed
         JP    Z,EXIT         ; exit if no button is pressed
```

Bits C, B and A are to be transmitted and are found now in the accumulator, but we must eliminate bit T, which is not required in the transmission. Bits C, B and A also need to be moved to the correct position through three shifts to format the packet according to the specifications described above. We add the start bit and load the packet in the TXBUFF transmission buffer.

```
TXCODE:   AND    00000111b     ; leave there only CBA bits
          SLA    A             ; shift the code in the correct position
          SLA    A
          SLA    A
          OR     00000100b     ; set the start bit at '1'
          LD     (TXBUFF),A    ; and save the packet in the TX buffer
```

Finally, we restore the registers and go back to the interrupted program.

```
EXIT:     POP    DE            ; restore registers D, E, A and Flags
          POP    AF
          EI                   ; re-enable interrupts
          RET                  ; return to the interrupted program
```

5. We define the input and output ports, the variables in play and the jumps to the beginning of the program and the interrupt handler.

PRECB stores the previous state of the push-button for a debouncing check, while CONFB records its confirmed state downstream of the check. NUMBER keeps the value that is incremented/decremented whereas TIME counts the 500 mS.

The constant MaxTIME is used to initialize the time count.

```
BUTP      EQU    00h           ; IA input port (push-button)
NUMP      EQU    03h           ; OD output port (number display)

PRECB     EQU    0FC00h        ; used for push-button debouncing
CONFB     EQU    0FC01h        ; previous-confirmed push-button value
NUMBER    EQU    0FC02h        ; the number that we handle
TIME      EQU    0FC03h        ; time counter

MaxTIME   EQU    25            ; time constant corresponding to 50 mS

          ORG    0000h
          JP     START
          ORG    0038h
          JP     HINT
          ORG    0100h
```

We initialize output port NUMP and variables PRECP, CONFB and NUMBER at zero. TIME is loaded at 25 to assess the half second (20 mS · 25 = 500 mS). Then we enable the interrupts and enter the main loop that does nothing (it is empty).

```
START:    LD    SP,0FFFFH      ; initialize the Stack Pointer
          LD    A,0            ; initialize the following variables
          LD    (PRECB),A      ; and the output port at zero
          LD    (CONFB),A
          LD    (NUMBER),A
          OUT   (NUMP),A
          LD    A,MaxTIME      ; initialize the time counter (500 mS)
          LD    (TIME),A
          EI                   ; enable interrupts
MAIN:     JP    MAIN           ; empty main loop
```

The interrupt handler saves and restores registers A, Flags, B and C, even though the main loop is inactive (to allow for future changes).

```
HINT:     PUSH  AF             ; save the used registers
          PUSH  BC
```

The following sequence of instructions reads the new value of the push-button and compares it to that stored in PRECB.

If the values are different, a transient or a bounce is in progress so we immediately discard the value and exit[19]. If the values are equal, we go on saving the confirmed value in C.

The current value is copied in PRECB for the comparison that will be executed during the next call to the handler.

```
          LD    A,(PRECB)      ; get the previous push-button value
          LD    B,A            ; copy it to register B
          IN    A,(BUTP)       ; read the push-button current value
          AND   00000001b      ; mask the unused bits and save the new
          LD    (PRECB),A      ; value as 'previous' (for the next check)
          CP    B              ; compare values for debouncing
          JP    NZ,EXIT        ; exit if they are different
          LD    C,A            ; otherwise copy the new confirmed
                               ; value to register C
```

So let's see if the confirmed value has changed. If it has, we need to reset the time count so that the next increment (or decrement) happens as of 500 mS after the change. We also save the new state of the push-button in CONFB for the next check.

```
          LD    A,(CONFB)      ; get the previous confirmed value
          CP    C              ; compare it to the current value
          JP    Z,COUNT        ; jump if they are equal
```

---

[19] When we exit the handler this way, the time count is interrupted. There will be an error of 20 mS for every bounce or transient identified. However, since we are dealing with manually pressing buttons, the user will not notice such a small error compared to 500 mS.

```
        LD      A,MaxTIME    ; otherwise, they are different, so
        LD      (TIME),A     ; re-initialize the time count and
        LD      A,C          ; save the new confirmed value as the
        LD      (CONFB),A    ; 'previous' confirmed (for the next time)
```

When the 500 mS are up, we need to change the number in function of the confirmed value of the push-button. We count the time in the TIME variable, in terms of the number of calls to the handler, and we exit if the 500 mS haven't elapsed yet. Otherwise, it is time to increment or decrement the variable NUMBER (after re-initializing the time counter).

```
COUNT:  LD      A,(TIME)     ; assess the time that passes
        DEC     A
        LD      (TIME),A
        JP      NZ, EXIT     ; exit if 500 mS have not elapsed,
        LD      A,MaxTIME    ; otherwise re-initialize the time
        LD      (TIME),A     ; counter and go on
```

When the 500 mS have elapsed, we assess the confirmed value of the push-button that we had saved in register C. If we have to increment, we jump to INCR, if not, we move on to DECR.

```
        BIT     0,C          ; check the push-button current value
        JP      NZ,INCR      ; jump if it is high
```

The two sequences are very similar. They start with a compare to the limit value, coherent with the direction of the count, so the content of NUMBER and thus that of output port NUMP are updated.

```
DECR:   LD      A,(NUMBER)   ; decrement the number, but we check
        CP      0            ; beforehand if it is possible
        JP      Z,EXIT
        DEC     A
        LD      (NUMBER),A   ; write back the new variable value
        OUT     (NUMP),A     ; and write it to the output port
        JP      EXIT
INCR:   LD      A,(NUMBER)   ; increment the number, but we check
        CP      0FFh         ; beforehand if it is possible
        JP      Z,EXIT
        INC     A
        LD      (NUMBER),A   ; write back the new variable value
        OUT     (NUMP),A     ; and write it to the output port
```

Finally, we restore the contents of the saved registers, re-enable the interrupts and go back to the interrupted program.

```
EXIT:   POP     BC           ; restore the contents
        POP     AF           ; of the saved registers
        EI                   ; re-enable interrupts
        RET                  ; return to the interrupted program
```

6. Among the definitions, we find the input port (DATAP) and output port (LED_RT, LED_UP, LED_LF and LED_DN); the TIME variable, used to count time; the TCYCLE variable, alluded to in the text of the exercise; and four variables to copy the current state of the LED lights (M_RT, M_UP, M_LF and M_DN). This is followed by the connections to reset and the interrupt handler.

```
DATAP    EQU    00h        ; IA input port (Inputs D3..D0)
LED_RT   EQU    02h        ; OC output port (LEDs, right side)
LED_UP   EQU    01h        ; OB output port (LEDs, upper side)
LED_LF   EQU    00h        ; OA output port (LEDs, left side)
LED_DN   EQU    03h        ; OD output port (LEDs, lower side)
TIME     EQU    0FC00h     ; time counter
TCYCLE   EQU    0FC01h     ; duration of the time step
M_RT     EQU    0FC02h     ; software copies of the output ports
M_UP     EQU    0FC03h
M_LF     EQU    0FC04h
M_DN     EQU    0FC05h
         ORG    0000h
         JP     START
         ORG    0038h
         JP     HINT
         ORG    0100h
```

At reset we jump to START and initialize the Stack Pointer. We have chosen to set the time counter to 1 so we can immediately execute the first shift, as we will see further on. Other choices, which are equal from a functional perspective, would have been good as well. We continue with the initialization of the ports and the variables which record their values. They are set based on what is described in the text (all LED lights on the port LED_RT are set on, all others off). Then we enable the interrupts and enter the main loop.

```
START:   LD     SP,0FFFFh       ; initialize the Stack Pointer
         LD     A,1             ; the time counter,
         LD     (TIME),A
         LD     A,11111111b     ; and the output ports
         LD     (M_RT),A
         OUT    (LED_RT),A
         LD     A,00000000b
         LD     (M_UP),A
         OUT    (LED_UP),A
         LD     (M_LF),A
         OUT    (LED_LF),A
         LD     (M_DN),A
         OUT    (LED_DN),A
         EI
```

In the main loop we read lines D3..D0 from the input port and we mask the bits that are not in use with an AND. To comply with the specifications, we increment the number read and save this value in TCYCLE. In the text, TCYCLE is multiplied by $40\,\mu S$, but this time enters into play simply because the timer calls the interrupt handler at regular intervals of $40\,\mu S$.

```
MAIN:     IN     A,(DATAP)      ; read D3..D0 from the input port
          AND    00001111b
          INC    A
          LD     (TCYCLE),A
          JP     MAIN
```

First of all, the interrupt handler saves all the used registers on the Stack even through the main program doesn't strictly need them[20].

```
HINT:     PUSH   AF             ; save the used registers
          PUSH   BC
          PUSH   DE
```

Right after, we decrement the time count. If the required time has not gone by, we jump directly to the exit code. Otherwise, it is time to change the on/off state of the LED lights.

Before that, however, we need to re-initialize the time counter with the value set by the main program in TCYCLE (the time count will start again at the next call to the handler).

```
          LD     A,(TIME)       ; assess the time
          DEC    A              ; to regulate the rotation speed
          LD     (TIME),A
          JP     NZ,EXIT        ; exit it time has not elapsed

          LD     A,(TCYCLE)     ; re-initialize the time counter
          LD     (TIME),A
```

Then, we calculate the new state of the LED lights by reading it from the software copy stored in the four variables in memory M_RT, M_UP, M_LF and M_DN.

```
          LD     A,(M_RT)       ; now is the time to rotate the LEDs lit
          LD     E,A            ; copy their current state from memory
          LD     A,(M_UP)       ; to registers E, D, C and B
          LD     D,A
          LD     A,(M_LF)
          LD     C,A
          LD     A,(M_DN)
          LD     B,A
```

---

[20] It is always a good idea to insert the appropriate PUSH and POP, so that if there are changes to the main program, we do not need to remember to change the interrupt handler as well.

The current state, copied into registers B, C, D, E of the CPU, is shown in the figure below as one single 32-bit register.

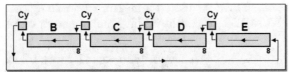

We need to do a left rotation of these 32 bits. Since we cannot do this in one operation, we break it up into four 8-bit shifts that go through the Carry flag (Cy in the figure).

The first RL A instruction in the sequence takes advantage of the fact that A still contains the value of the eight farthest left bits of the 32 bit set (those now in B). It shifts them in order to copy bit 7 in the Carry flag. The next RL E inserts that bit at the right of bit 0 in E.

```
RL    A        ; copy bit 7 of B to the Carry flag
RL    E        ; and rotate left 32 bits, in four steps:
RL    D        ; B ← C ← D ← E ← B7
RL    C
RL    B        ; close the 32 bits rotation
```

So the next RL intructions left shift all the bits one register at a time but the bit that exits from the left of a register is inserted at the right through the Carry flag in the register on the left.

When the rotation is complete, we need to update the state of the LED lights in the memory in their software copies and the output ports.

```
OUTPUTS: LD    A,E           ; display the new state on the ports
         LD    (M_RT),A      ; and update their software copies
         OUT   (LED_RT),A
         LD    A,D
         LD    (M_UP),A
         OUT   (LED_UP),A
         LD    A,C
         LD    (M_LF),A
         OUT   (LED_LF),A
         LD    A,B
         LD    (M_DN),A
         OUT   (LED_DN),A
```

Finally, we take back the previous content of the registers, re-enable the interrupts and go back to the interrupted program.

```
EXIT:    POP   DE            ; restore the saved registers
         POP   BC
         POP   AF
         EI                  ; re-enable interrupts
         RET                 ; return to the interrupted program
```

7. We define the addresses of the input port (PBUT) and output ports (NUMH e NUML). The PREV variable stores the previous state of the push-button. TIME is used to measure how long the push-button is pressed. NUMBER stores the 12-bit number shown on ports NUMH and NUML.

```
PBUT      EQU    00h              ; IA input port (push-button)
NUMH      EQU    03h              ; OD output port (NUMH, bit 11..8)
NUML      EQU    02h              ; OC output port (NUML, bit 7..0)

PREV      EQU    0FC00h           ; previous state of the push-button
TIME      EQU    0FC01h           ; how long the push-button is pressed
NUMBER    EQU    0FC02h           ; current displayed number (2 bytes)
```

There are also two more definitions: those of constants TIME4000 and TIME200, the time thresholds required by the text (measured in the 40 mS increments marked by the timer).

```
TIME4000  EQU    100              ; 4 S = 100 · 40 mS
TIME200   EQU    5                ; 200 mS = 5 · 40 mS
```

This is followed by the links to the reset and the interrupt handler.

```
          ORG    0000h
          JP     START
          ORG    0038h
          JP     HINT
          ORG    0100h
```

Then we initialize the Stack Pointer, do the first reading of the push-button port and save that state in the PREV variable.

```
START:    LD     SP,0FFFFh        ; initialize the Stack Pointer and
          IN     A,(PBUT)         ; the push-button previous state
          AND    00000001b
          LD     (PREV),A
```

The time counter TIME is zeroed (it will be incremented when the button is pressed next). We also zero the NUMBER variable and the corresponding output ports, by means of the OUTPUT subprogram. Then we enable the interrupts and enter the (empty) main loop.

```
          LD     A,0              ; initialize the time counter
          LD     (TIME),A
          LD     HL,0             ; zero the variable NUMBER
          CALL   OUTPUT           ; and display it on the ports
          EI                      ; enable interrupts
MAIN:     JP     MAIN             ; empty main loop
```

The OUTPUT subprogram receives the number to save in the NUMBER variable in HL, copies it on output ports NUMH and NUML, and makes that visible on the LED lights.

```
OUTPUT:  LD    (NUMBER),HL   ; update the variable NUMBER
         LD    A,H           ; and display its contents on the ports
         OUT   (NUMH), A
         LD    A,L
         OUT   (NUML), A
         RET
```

The interrupt handler saves all the used registers in anticipation of any extensions of the code, even though the main loop doesn't use them.

```
HINT:     PUSH  AF            ; save the used registers
          PUSH  BC
          PUSH  DE
          PUSH  HL
```

The following sequence of instructions acquires the current state of the push-button from the port and compares it with the state memorized in PREV, which was moved to C beforehand (AND only masks the bits that are not of interest).

The current state is copied in B and saved in PREV for the comparison that will be executed the next time the handler is called.

```
          LD    A,(PREV)      ; get the previous push-button state
          LD    C, A          ; and copy it to register C
          IN    A,(PBUT)      ; read the current state from the port
          AND   00000001b     ; mask the bits that are not of interest
          LD    B,A           ; copy the current state in B
          LD    (PREV),A      ; and save it for the next call
```

Notice that we are not simply looking for the edges of the signal; the assessment is more complex. The specifications require us to calculate the amount of time the push-button is pressed, and to act differently once it is released.

So we check if the button is being pressed. If it is, we check how long it has been pressed. If it isn't, we jump to UPTEST, where we check whether the button has been released now.

```
          BIT   0,B           ; is the push-button pressed?
          JP    NZ,UPTEST     ; jump if it is not
```

To understand the code, however, let's assume for now that the push-button is pressed and we move ahead without executing the jump. If the push-button remains in this state, every time the handler is called, we pass from here and go on to increment the time count.

If fewer than 4 seconds have gone by, we continue to count, otherwise we stop. Once 4 seconds have passed, in fact, we are no longer interested in the exact time value. By stopping the time count we also prevent a possible overflow. After assessing the time, we exit the handler since nothing more is required if the button is pressed.

```
          LD    A,(TIME)      ; because the button is pressed,
          CP    TIME4000      ; check if four seconds are elapsed
          JP    Z,NOINCR      ; if so do not increment the time count
          INC   A             ; otherwise, increment it and
          LD    (TIME),A      ; write back it to the memory
NOINCR:   JP    EXIT          ; and exit from the handler
```

If the push-button is at rest, however, we check its previous state to show if the button has been released at this moment. If it has been a rest since before, however, we simply exit. But it has just been released now, so let's check how long the button was pressed and act according to the specifications.

We have copied the previous state of the push-button to register C, so if the previous state is also high, the button was not released and we jump straight to the exit code.

```
UPTEST:   BIT   0,C           ; check if the button has been released
          JP    NZ,EXIT       ; exit if it is not
```

Otherwise, the button has been released and we need to act according to how long it was pressed. So, let's read the TIME variable, which was being updated the whole time the button was kept pressed.

If it is at the maximum count, it was pressed for at least 4 seconds so we jump to CLEAR. If it is not, we check if the time is greater than 200 mS (between 200 mS and 4 seconds), and if that is the case, we jump to the MS200 label. Notice that after the second CP instruction, the conditional jump assesses the Carry flag for the equal-to or greater-than condition.

```
          LD    A,(TIME)      ; the button has been released,
          CP    TIME4000      ; so assess the elapsed time...
          JP    Z,CLEAR       ; jump if pressed for at least 4 seconds

          CP    TIME200       ; check if time is less than 200 mS
          JP    NC,MS200      ; jump if it is equal or greater
```

If we continue, then the time has been assessed as lesser than 200 mS by exclusion, so we need to increment NUMBER by one unit, inserting a 1 in register D and jumping to the ADDNUM label. Similarly, 8 is loaded in D at the MS200 label. In fact, a little further on in the code, the sequence beginning with the ADDNUM label increments the NUMBER variable by the quantity passed in D.

```
ONE:      LD    D,01h         ; by exclusion, time is less than 200 mS,
          JP    ADDNUM        ; increment NUMBER by 1

MS200:    LD    D,8           ; time is between 200 mS and 4 seconds,
          JP    ADDNUM        ; increment NUMBER by 8

CLEAR:    LD    HL,0          ; otherwise, it is greater than 4 seconds,
          JP    UPDATE        ; zero the NUMBER variable
```

Notice that at the CLEAR label we want to zero NUMBER, so we zero HL and jump directly to UPDATE without passing through ADDNUM.

As we have seen, we enter the ADDNUM label after loading the constant to add (1 or 8) in register D. The content of NUMBER is copied in HL and the D content is added to it. Any carry toward the high part of the result will be taken into account.

```
ADDNUM: LD    HL,(NUMBER)  ; copy the NUMBER variable to HL
        LD    A,L
        ADD   A,D          ; add the D content to the low part L
        LD    L,A
        JP    NC,NOCY      ; take into account the carry
        INC   H            ; if any, increment the high part H
```

Right after, and before we proceed to the UPDATE label, the code limits the number to the largest that can be represented with 12 bits (0FFFh), as per specifications, assessing if its high part has gotten to 10h.

```
NOCY:   LD    A,H          ; check if the number is greater than
        CP    10h          ; the maximum 12-bit number
        JP    C,UPDATE     ; if the high part is ≥ 4096
        LD    HL,0FFFh     ; limit the number to the maximum
```

We get to the UPDATE label after updating the number in HL. When we call the OUTPUT subprogram, the value is transcribed in the NUMBER variable and on the LED lights NUMH and NUML. To finish managing releasing the push-button, we zero the time count again (TIME), in anticipation of the next button being pushed.

```
UPDATE: CALL  OUTPUT       ; update NUMBER and the output ports
        LD    A,0          ; re-initialize the time counter
        LD    (TIME),A     ; for the next time
```

Finally, the handler retrieves the content of the saved registers, re-enables the interrupts and goes back to the interrupted program.

```
EXIT:   POP   HL           ; restore the saved registers
        POP   DE
        POP   BC
        POP   AF
        EI                 ; re-enable interrupts
        RET                ; return to the interrupted program
```

8. In large part, this solution falls in among the examples in Sections 4.7.8 and 4.7.3 cited in the text of the exercise. Apart from some necessary changes, it copies almost all of their code and integrates the two examples' solutions. For this reason, many explanations have been omitted or summarized here, so we suggest going back to those examples ahead of time for more information.

We define the address of input port PFREQ, which is connected to the serial receiver, and the address of the output port OUTWAV, that drives the virtual DAC.

The ANGLE variable stores the current angle in the sine calculation. FREQ is used to record the parameter of the frequency. What follow are the link to the reset and the two different interrupt requests.

```
PFREQ      EQU   02h          ; IC input port (frequency)
OUTWAV     EQU   05h          ; OF output port (sinusoidal wave)
ANGLE      EQU   0FC00h       ; current angle
FREQ       EQU   0FC01h       ; frequency parameter
           ORG   0000h
           JP    START
           ORG   0030h        ; Int. 6
           JP    HINT6
           ORG   0038h        ; Int. 7
           JP    HINT7
           ORG   0100h
```

Downstream of reset, we initialize the Stack Pointer. Then we zero the OUTWAV output port, the current sine angle and the frequency parameter (before getting to any serial line command, the output will be at zero). Finally, we enable the interrupts and enter the empty main loop[21].

```
START:     LD    SP,0FFFFh    ; initialize the Stack Pointer
           LD    A,00h        ; zero the waveform output,
           OUT   (OUTWAV),A
           LD    (ANGLE),A    ; the angle and
           LD    (FREQ),A     ; the frequency parameter
           EI                 ; enable interrupts
MAIN:      JP    MAIN         ; empty main loop
```

The interrupt handler launched by the receiver's request, as usual, saves and recovers the used registers. Its main task is to read the number from the serial receiver (through the PFREQ input port) and to copy it to the FREQ variable. The number is stored after a check is executed on its value, as required in the text (it should not be over 31). Reading the port deactivates the interrupt request inside the ASRX component.

```
HINT6:     PUSH  AF           ; save A and Flags on the Stack
           IN    A,(PFREQ)    ; get the data byte from the serial port,
           CP    32           ; limit its value to 31
           JP    C, NOLIMIT   ; jump if C='1' (A < 32)
           LD    A,31         ; otherwise A ≥ 32, load 31 in A
```

---

[21] Although the the main loop is inactive, we still insert a save and restore of the registers in the interrupt handlers so we won't have to correct the code later on if we change the main loop.

```
NOLIMIT:      LD    (FREQ),A        ; copy it to the FREQ variable
              POP   AF              ; restore the saved registers
              EI                    ; re-enable interrupts
              RET                   ; return to the interrupted program
```

Interrupt handler HINT7 is launched by the timer every $100\,\mu S$, and it deals with the generation of the wave form. This part of the code is identical to what we find in Section 4.7.3, so we will not go into the detail found there.

```
HINT7:        PUSH  AF              ; save the used registers
              PUSH  BC

              LD    A,(FREQ)        ; copy the frequency parameter
              LD    B,A             ; to register B

              LD    A,(ANGLE)       ; compute the new angle
              ADD   A,B             ; by adding the frequency parameter
              LD    (ANGLE),A       ; to the previous angle

              CALL  WAVEFORM        ; get the next value from the table
              OUT   (OUTWAV),A      ; copy it to the DAC output port

              POP   BC              ; restore the saved registers
              POP   AF
              EI
              RET
```

The WAVEFORM function generates the waveform through a table of values, as described in Section 4.7.3.

```
WAVEFORM:     PUSH  HL              ; save register HL and BC
              PUSH  BC

              LD    C,A             ; save bit 7 of the angle in C, and mask
              AND   01111111B       ; it to avoid readings outside the table

              LD    HL,SINTAB       ; get the base address of the table
              ADD   A,L             ; add the index to it
              LD    L,A             ; to obtain the address of the location
              JP    NC,NoCarry      ; of interest in register HL
              INC   H
NoCarry:      LD    A,(HL)          ; get the value

              BIT   7,C             ; check if we are in second half wave
              JP    Z,Positive      ; if not, the value is positive
Negative:     NEG                   ; otherwise invert the sign of the value
Positive:     POP   BC              ; restore registers BC and HL
              POP   HL
              RET
```

The table that describes the values of the positive half sine wave is also identical so it will not be shown here.

5

Microprocessor systems on FPGA

Abstract In this chapter, we will learn what a Field Programmable Gate Array (FPGA) device is and how we can use it to physically implement what we have just simulated until now. In fact, Deeds supports both the simulation and deployment on FPGA of the microprocessor-based systems. After a brief introduction to the FPGA, we will present a few examples of FPGA-based experimentation boards that we will use to implement our projects. Then we will present several practical project examples (i.e. LED light dimmer, special sound effects, a musical box, a stepper motor controller, and an LCD display-based stopwatch). We will start from the specification, we will continue with the phases of conception, hardware design, and assembly language programming, and we will conclude by showing how quickly the Deeds allows their physical implementation on different FPGA boards.

5.1 Introduction to FPGAs

All the examples in the previous chapters were given with the aim to facilitate understanding of the architecture and programming techniques of microprocessor systems. These are essential capabilities for any designer/programmer but we have not yet focused on their practical implementation. Now, readers who have acquired the basics can work on programming and simulating.

This chapter offers readers a process of experimentation. These examples are inspired by ones from the previous chapters in terms of approach and complexity. The difference here is that readers can work on them and test how they work in practice.

The physical implementation of a system depends greatly on the state of the art of the technologies in use and is subject to rapid changes and consequent obsolescence. A good understanding of the basic concepts from the previous chapters will allow designers/programmers to easily approach the changes coming along the way.

Right now, designers can take advantage of the programmable components called FPGAs[1], and a wide range of boards for prototypes called "FPGA boards", based on these components. All the practical examples in this chapter are made on FPGA boards.

An FPGA component contains a large number of basic logical elements (logical ports, flip-flops and more complex circuits), that can be connected through a network of connections to create a system. We choose the connections by using specific development software provided by the producers of the FPGA.

Their programming is loaded in the chip and kept in the internal memory. The following figure shows two examples of FPGA devices, produced by *Intel/Altera FPGA*[2], on the left, and by *Xilinx*[3], on the right.

FPGA components are the younger descendants of the the large family of PLDs[4], a term for all the chips that can be specialized for a specific application. Their connections can be internally established when the system is produced or "in the field", i.e., when it is already in use. Since the 1980s, PLD components have profoundly changed the world of complex system design.

FPGAs are very valuable from an educational perspective as well. They are suited to developing prototypes designed quickly and economically for educational purposes. Finally, an FPGA component should not be confused with a microcomputer, where the hardware is pre-defined and programming consists of setting a sequence of instructions to be executed.

5.1.1 Creation of prototypes with FPGA

Not long ago, "prototyping" (creating a prototype) of a circuit required the connection of a large number of discrete components through soldered wires. This process was very time-consuming and quite prone to connection errors or bad wiring. It was often hard to tell if the system malfunction was due to design error or faulty connections.

---

[1] Field Programmable Gate Arrays

[2] https://www.intel.com/content/www/us/en/products/details/fpga.html

[3] https://www.xilinx.com/products/silicon-devices/fpga.html

[4] Programmable Logic Devices

In laboratories, it was very common to have boards with solderless connections called "breadboards" with a fixed grill made of internally connected holes. Students used them to insert components and the connections were stabilized by wires. For example, the following figure shows an 8-bit parallel-serial interface implemented on a breadboard. This was done with commercial integrated circuit connections that carry out the various logical functions required (logical ports, flip-flops, registers and counters).

The problems with this prototyping system are similar to those of the traditional soldered systems. Two advantages are the fact that it is easier to change the connections and there is no risk associated with using a soldering iron. A disadvantage is the problem of bad contacts due to wear and tear on the boards or oxidation of the wires.

Currently, solderless breadboards are very useful for quick prototyping of systems based on FPGAs. For example, simple interfaces or support circuits connected to the FPGA board in use can be mounted on a breadboard.

Many types of prototyping boards based on FPGA components are found on the market. They are developed for educational use and run from the simplest and cheapest to more complex versions. They include various types of interfaces in input and output, and allow for the creation of system prototypes, often without needing additional components aside from the board itself.

If it has the appropriate capacity, an FPGA component can be programmed so that it implements a microcalculator using the large number of logic elements at its disposal. When a processor is loaded on an FPGA, it is called a "soft-processor", in that its hardware is assembled from the software during the design development. A soft-processor behaves exactly like a "hard-processor", that is, like those physically manufactured on a chip.

5.1.2 Some examples of FPGA boards

Many models of FPGA boards are available commercially and they have constantly evolving features and capabilities. They are made for a broad spectrum of applications: from cheap, simple boards for educational use to fast, complex boards for professional use. Designers and experimenters can find the right board on the market for their applications, in terms of capabilities, software availability and budget.

As an example, we'll give a brief description of some FPGA boards suitable for implementing the microprocessor systems developed in the book. It should be noted that each of these has the capacity to host much more complex systems and could allow for a natural transition to professional design. Some of these boards will be described in more detail in Section 5.4, regarding elements on it that could be put to advantage in conjunction with the Deeds environment. For now, we will only give some general information about them.

The following figure shows the DE2 board, produced by *Terasic*[5] for *Intel/Altera FPGA*[6]. This board is supported by Deeds.

This was conceived for mid/high complexity digital circuit experimentation and includes numerous devices, from the simplest (switches, push-buttons, LED lights, seven-segment displays) to more complex (LCD matrix displays,

---

[5] https://www.terasic.com.tw/en/
[6] https://www.intel.com/content/www/us/en/products/details/fpga.html

Ethernet network interfaces, A- and B-type USB 2.0 interfaces, connectors for SD memory cards, analog audio inputs and outputs, analog video input, VGA video output, etc.).

The core of the board is the chip EP2C35 "Cyclone® II" family from *Intel/Altera FPGA*, which has more than 33,000 logic units (the basic logic block will be explained a but further ahead in Section 5.2). The DE2 board allows for the creation of simple, introductory projects (that use a small portion of its potential) to complex systems that can include specialized microcomputers and interfaces.

The following figure shows the DE0-CV board produced by *Terasic* for *Intel/Altera FPGA*. As the image shows, it has push-buttons, switches, LED lights, seven-segment displays, connectors and other devices, although fewer than the previous board had. This board is also supported by Deeds.

The core of the board is the big, black square in the center. This is the FPGA chip 5CEBA4, part of the "Cyclone® V" family, produced by *Intel/Altera FPGA*. This chip contains a matrix of about 49,000 logic units and 3,080 Kbits of integrated RAM.

Other, larger members of the same chip family include the *ARM Cortex*[TM] dual-core microprocessor by ARM (Advanced RISC Machine)[7], which is used in many mobile phones.

---

[7] https://www.arm.com/products/silicon-ip-cpu

The following figure shows another example: the ARTY S7-50 board produced by *Digilent*[8], which uses the XC7S50 chip from the "Spartan®-7" family, by *Xilinx*[9]. The FPGA contains over 52,000 basic combinational blocks and over 65,000 flip-flops. The FPGA chip is the one positioned at a 45 degree angle on the board[10].

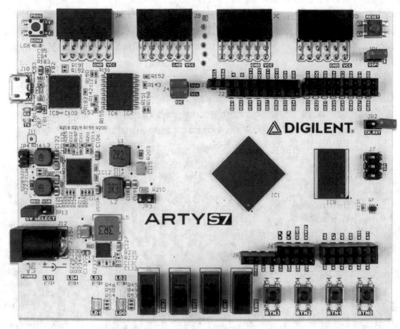

As in the previous example, this board has push-buttons, switches, LED lights and other interfaces. More specifically, we have connectors designed to host input and output boards ("shields"), which were originally designed for *Arduino* microcontroller boards[11].

The last example we will show on these pages is a very inexpensive FPGA board available online and supported by Deeds (see the following figure). It is based on a "Cyclone® II EP2C5T144C8" chip from *Intel/Altera FPGA*.

The user has four connectors positioned around the chip, only three LED lights and one push-button. The two 10-pin connectors on the left, however, are used to program the FPGA chip. For brevity's sake, we will refer to this board as "EP2C5".

To implement our designs, which often need multiple input and output devices, we need to use the four connectors to externally connect all the necessary push-buttons, switches, LED lights and displays.

---

[8] https://store.digilentinc.com
[9] https://www.xilinx.com/products/silicon-devices/fpga.html
[10] This board is not yet supported by the Deeds environment.
[11] https://www.arduino.cc/

The chip is big enough to implement a microcomputer based on the Deeds DMC8 processor if it is configured with a small RAM and ROM (once the DMC8 is loaded on the FPGA component, it uses about half of the 4,600 logic units available and only 400 of the 4,600 flip-flops).

5.2 The architecture of FPGA components

FPGA makers offer a wide array of devices classified into "component families", that differ in their complexity and their fields of application.

Some typical examples are audio/video signal processing, radar, automobile systems and generally, all the applications that require high performance but don't have the volume to justify the cost of a "full custom" chip.

Despite the wide variety, all the devices have the same base architectural structure in common.

An FPGA chip is essentially a big matrix of logic blocks set in rows and columns as shown in the figure at the right.

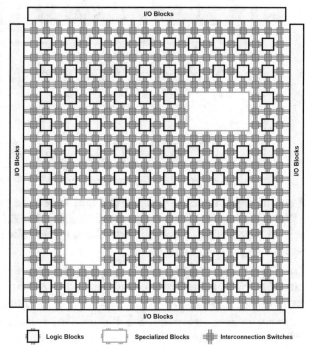

Every block contains one or more flip-flops and programmable combinational networks. A matrix of connections goes across the whole chip, using up most of its area.

At every intersection of the matrix, programmable electronic switches allow for the local connection between rows and columns, and the interconnections with the blocks. Local sub-matrices of connections can be made available to improve the speed of communication between nearby blocks.

There can also be special blocks inside the matrix intended for specific functions like read/write RAM memory, arithmetic circuits (often multipliers), etc. The matrix is also surrounded along its four edges, by input and output logic blocks responsible for the interface of the chip and external devices.

The previous figure shows only the elements of the chip that are available for the design. It does not show the memory elements that program the connections and configure the logic blocks. There is a high number of flip-flops connected in cascade, that make up a very long shift register as shown in the following figure.

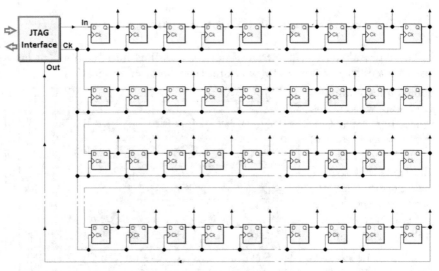

A specific serial interface (called JTAG, which will be explained later) is tasked with writing the flip-flops in the programming phase of the FPGA chip (in the normal functioning, however, these flip-flops are not accessible, they cannot be used as part of our projects).

The FPGA component must be re-programmed every time the system is turned on[12]. To solve this problem, we add to the circuit board a Flash ROM memory chip (see Appendix A.1) that stores the programming information.

At every system power up, the Flash ROM will transfer the programming information into the FPGA, through dedicated pins.

---

[12] As we know, a flip-flop does not store information when the power is off.

5.2.1 Logic blocks

The following figure shows the basic schematic of one generic "logic block" in an FPGA. The block has a type-E flip-flop (edge-triggered), driven by a combinational logic network.

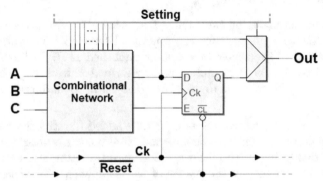

The combinational network's functionality is controlled by the configuration shift register ("Setting", in the upper part of the figure), and also by the 2-to-1 multiplexer on the right, which gives us the option to use the flip-flop to register the output of the combinational network.

It might be interesting to more closely examine the combinational network, that is implemented in the form of an LUT (Look-Up Table). Its values are selected by a multiplexer driven by the network inputs (see the figure below).

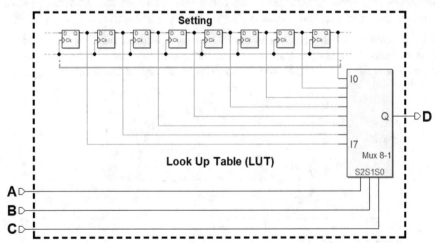

During the FPGA chip programming operations, the values of the desired table are stored in the flip-flops of the above-mentioned shift register. Once they are stored, these values remain the same during the normal operations of the FPGA, and the multiplexer copies them on output D, according to the combination of A, B and C, giving us the requested Boolean function.

Notice that according to the FPGA family, the logic block might be more complex than what is presented here. For example, it might also contain an adder, XOR networks, other flip-flops and multiple combinational networks.

5.2.2 JTAG programming

JTAG is the acronym for the (Joint Test Action Group) consortium that defined a standard protocol for the functional testing of integrated circuits toward the end of the 1980s. It was later published as IEEE 1149.1 ("IEEE Standard Test Access Port and Boundary-Scan Architecture"). In the following, this will be referred to as JTAG (the term "Boundary-Scan" is also used)[13].

The version of this protocol released in 1994 added the capacity to program memory, microcontrollers and other devices. It also made it possible to execute functional debugging of the firmware and activate automatic tests ("Built-In Self-Test"), defined by the maker of the component. To achieve this, a standard language ("Boundary Scan Description Language") was developed to access the components using the JTAG interface.

Currently, the JTAG interface is the only method used to access the internal hardware of electronic systems such as mobile phones, tablets, wireless "access points" etc. both for testing during production and for fault diagnostics.

Let's examine its basic operations. Briefly, the standard offers the possibility to stop the normal system operations and going to a modality where the JTAG interface is activated. The interface takes control of all the external pins of the components and the test and programming circuits in the system itself.

The physical JTAG interface is composed of a limited number of standard connections. The simplest set allows us to communicate with the circuit by using the TCK, TMS, TDI and TDO lines as shown in the following figure.

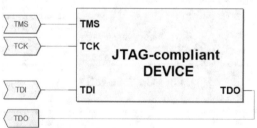

Pin	Name	Function
TCK	*Test Clock*	Data Clock Pin
TMS	*Test Mode Select*	Mode Control and Operation Selection
TDI	*Test Data In*	Serial Data Input Pin (toward the device)
TDO	*Test Data Out*	Serial Data Output Pin (from the device)

---

[13] https://www.jtag.com

All the signals of the JTAG interface are serial and synchronized by the clock TCK (usually in the $10..100\ MHz$ interval). Activating TMS signals the system to turn on the JTAG-compliant mode. So, through the same line, we can execute the required operation by using a specific state algorithm, the "JTAG State Machine" (not described here).

The standard also defines an optional control line, Test Reset (TRST), but its functionality can be obtained by controlling TMS following a specific sequence and it is often not used as in the example above.

The standard furthermore defines the chain connection of the TDI and TDO pins of more than one device so that we can access all the JTAG-compliant devices located on the same board (see the following figure).

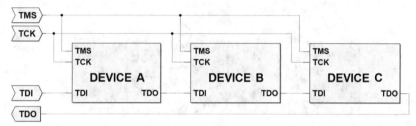

This method makes it possible to run a "chain integrity test", for example. Each JTAG-compliant device has its own ID code. All the ID codes can be read and checked against the design project ID to see if the JTAG chain works as it should.

5.2.3 Devices for programming FPGAs

Many programmable devices like FPGAs use JTAG for programming, and not only for testing purposes.

It should be noted that FPGA chips are programmable after they have been soldered on the board. This brings many advantages including simplifying the programming phase, avoiding the use of external programmers and adding the possibility of updating and changing the networks programmed inside the chip. This all makes FPGA systems ideal for the implementation of prototypes and experimental circuits, including those for educational use.

There are different standards for the physical JTAG interface.

The simplest uses a 10-pin connector, as shown in the figure on the right, which refers to an EP2C5 board.

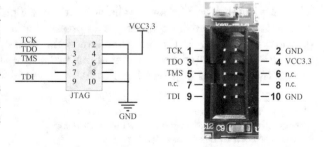

In the following figure, we see a JTAG programmer and its 10-pin cable (on the right) as well as a standard USB cable to connect it to a PC. The software is provided by the maker of the FPGA.

Often, especially in more complete boards, the programmer is integrated and available through a dedicated USB interface. This is the case for the first three boards described before.

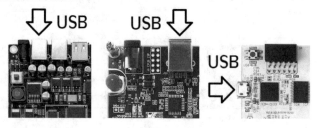

5.3 FPGA development tools

The makers of FPGAs sell proprietary development tools for professional designers to implement complex systems. The makers provide reduced versions of the same tools usually for free, to educators.

Clearly, the goal is to promote their products to future designers by influencing their choices later on in the world of work. This is why makers provide a great deal of documentation on their software.

In the following, we present a brief panorama of what is available for free online. All the tools mentioned offer similar functional capabilities, such as schematic editors, source code editors, compilers, pin planners and optimization tools. They make it possible to design digital systems by using logical schematics or languages to describe the hardware (HDL) such as the Verilog[14], the VHDL[15], or the System-C[16].

The large number of functions available that make a professional's work more productive, may give a beginner the impression of a complex arena that is difficult to manage. In the following, we will see that Deeds allows for the use of FPGAs for rapid prototyping of our projects without entering into the typical technicalities of professional tools.

---

[14] https://standards.ieee.org/standard/1800-2017.html
[15] https://standards.ieee.org/standard/1076_6-2004.html
[16] https://accellera.org/community/systemc

Regarding boards based on *Xilinx* FPGA devices, at the moment this book is being written, there are free tools available such as: *Vivado® Design Suite HL WebPACK™* and *ISE® WebPACK™*. The following screen-shot shows *Vivado®*, which is made for the most recent families of devices.

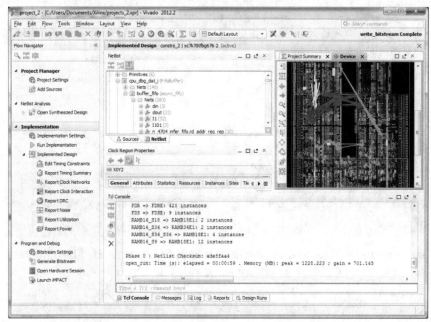

Below, the *ISE®*, which supports the less recent FPGA families.

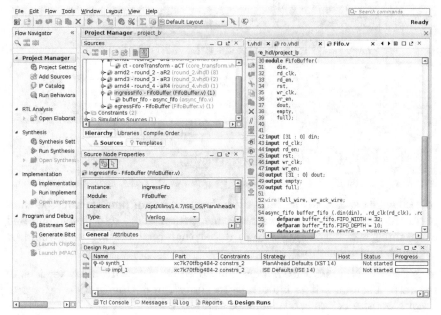

Intel/Altera FPGA offers two free tools: *Quartus® Prime Lite Edition™* and *Quartus® II Web Edition™*. The following screenshot shows the main window of the first. It supports the most recent families of FPGAs.

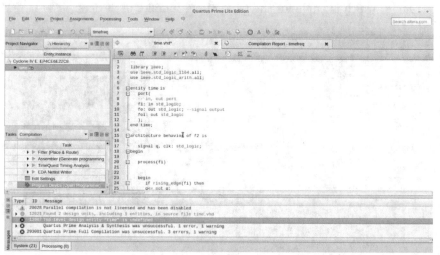

The previous versions of this software are called *Quartus® II*. The main command window for project management is found below.

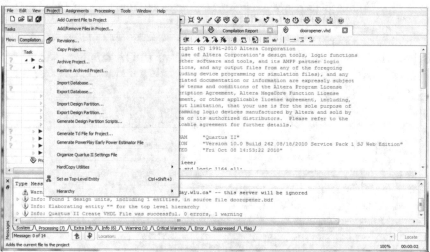

In recent years, due to the increase in the number of families and the complexity of chips FPGA makers have put greater effort into developing and maintaining newer generation tools rather than guaranteeing they would be compatible with older ones.

In general, less recent FPGA boards need to be used with older versions of software. Therefore, it is important to study the documentation on a producer's website before choosing chips and tools.

A useful function (*Software Selector*) associates an FPGA family to the corresponding software to use.

The screenshot below, from the *Intel/Altera FPGA* website shows that the *Cyclone®* II family is supported by version "*13.0 - ServicePack 1*" and previous versions from *Quartus® II Web Edition™*.

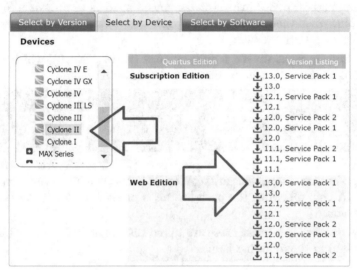

5.4 The FPGA boards used in the examples

5.4.1 The DE2 board

As introduced before, in Section 5.1.2, the Terasic/Altera board DE2[17] makes it possible to do simple, basic projects (using few of the boards resources), as well as systems including a microcomputer and its interfaces.

The DE2 board has various types of components[18]. Those interested in learning more about what it can do should consult the exhaustive documentation on the manufacturer's website[19], which also contains the user manual. In our examples, we use those components of the board that can be directly interfaced with the projects generated by Deeds.

For example, there are 18 slide switches (Sw17 .. Sw0) connected to 18 pins of the FPGA chip (see the following figure).

---

[17] http://www.terasic.com.tw/cgi-bin/page/archive.pl?Language=English&CategoryNo=39&No=30

[18] https://www.terasic.com.tw/cgi-bin/page/archive.pl?Language=English&CategoryNo=39&No=30&PartNo=2#section

[19] https://www.terasic.com.tw/cgi-bin/page/archive.pl?Language=English&CategoryNo=39&No=30&PartNo=4#section

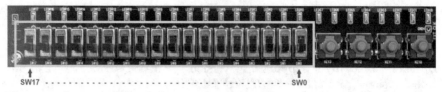

SW17 - SW0

The manufacturer defined the switch connection so as to provide a '0' to the FPGA pins when their cursors are brought low, toward the edge of the board and a '1' when they are high.

On the bottom right we see 4 push-buttons (Key3 .. Key0), which are also connected to the FPGA input pins (see the following figure).

KEY3 - - - - - - - - - - - KEY0

The push-button connections are set so as to generate a '0' on the corresponding FPGA chip pin when the push-button is pressed, and a '1' when it is in idle position.

Directly above the switches, there are 18 red LED lights (LEDR17 .. LEDR0), as indicated in the following figure.

LEDR17 - LEDR0

There are also 9 green LED lights (see the following figure), 8 of which are placed on the push-buttons (LEDG7 .. LEDG0), and one (LEDG8) positioned near the seven-segment display.

LEDG8   LEDG7 - - - - - - - - - - - - - LEDG0

The eight seven-segment displays (HEX7 .. HEX0, see the following figure), can manage the individual segments by driving them directly (the native modality of the board).

HEX7 - HEX0

Deeds allows for the use of the "decoded" displays in the library.

These components make it possible to use the
board display by providing the 4-bit number in
binary to show as a hexadecimal number.

The board also has two 40-pin "expansion connectors" (GPIO_0 e GPIO_1, see the figure on
the right). The connector pins are directly connected to the FPGA chip and can be programmed individually as inputs or as outputs.
See the above-mentioned user manual for more
information on connectors' pinouts (a few power
supply lines are also available on them).

Deeds allows the user to connect the connector
pins freely to external networks, as we will see
later in this chapter.

There is a special issue about clock generators on the board, where we find
two crystal oscillators, one at 27 MHz, and the other at 50 MHz.

Deeds supports both generators. From the second one, it derives (by division)
our projects' clockwork frequencies. Therefore, we can define clock frequencies
of 50 MHz, 27 MHz, 10 MHz, 5 MHz, 2 MHz, 1 MHz, 500 KHz, 200 KHz, 100 KHz,
etc. down to the lowest, 1 Hz).

5.4.2 The DE0-CV board

In this chapter, we will also use the Terasic/Altera DE0-CV board[20]. As
mentioned in Section 5.1.2, this board makes it possible to develop projects
of varying levels of complexity, similar to what we saw for the DE2 board. It
has an FPGA chip of the newest generation "Cyclone® V" (which is more
powerful) but has fewer switches, push-buttons, LED lights and displays.

The DE0-CV board also has a wide variety of components[21]. The manufacturer's website has exhaustive documentation[22]. The user manual is a good
starting point to learn more about its components.

Now, let's look at components of the board that will allow us to interact
with our networks. The figure below shows 10 slide switches (Sw9 .. Sw0),
connected to 10 FPGA inputs.

---

[20] https://www.terasic.com.tw/cgi-bin/page/archive.pl?Language=English&CategoryNo=167&No=921

[21] https://www.terasic.com.tw/cgi-bin/page/archive.pl?Language=English&CategoryNo=167&No=921&PartNo=2

[22] https://www.terasic.com.tw/cgi-bin/page/archive.pl?Language=English&CategoryNo=167&No=921&PartNo=4

They are connected in order to provide the chip pins with a '0' when the cursor is low, toward the edge of the board and a '1' when we move it high.

At the lower right hand side of the following figure, there are 5 push-buttons (Key3 .. Key0 e RESET), which are also connected to the input pins of the FPGA. The push-buttons generate a '0' on the corresponding pin of the FPGA chip when the push-button is pressed, and a '1' when it is not.

There are 10 red LED lights (LEDR9 .. LEDR0) in line above the switches, as shown in the figure below. There are no green LED lights on this board.

As shown in the following figure, there are six seven-segment displays. We can manage their individual segments. Deeds allows for the use of the "decoded" displays here as well (These are the same as those belonging to the DE2 board).

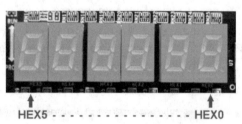

The board also has two 40-pin "expansion connectors" (GPIO_0 e GPIO_1, see the following figure), which are identical to those in the DE2 board.

The connectors, here as well, are directly connected to the FPGA pins and can be defined as inputs or outputs. We can also connect the connector pins to external networks.

See the board user manual for more information on connectors' pinouts (a few power supply lines are also available on them).

We see only one $50MHz$ quartz clock generator on this board. With the support of Deeds, we can use this to obtain the clock work frequencies of our projects ($50MHz$, $10MHz$, $5MHz$, $2MHz$, $1MHz$, $500KHz$, $200KHz$, $100KHz$, etc. down to $1Hz$).

5.4.3 The EP2C5 board

The third board used in the examples of this book is called the "EP2C5". It is an economical, unbranded FPGA board that is easy to find online. It is based on the "Cyclone® II EP2C5T144C8" chip from *Intel/Altera FPGA*. As mentioned in Section 5.1.2, this board allows us to create networks that are smaller in size. Yet, we are still able to load a DMC8 microcomputer on it, if configured with a small sized ROM and RAM system.

Documentation for this board, including the electrical schematics can be found online[23]. It has no switches and offers the user only one push-button. There are only three red LED lights and no display. Essentially, any interface component that we would need would have to be connected to the board.

The one available push-button (KEY0) is shown in the following figure.

On the same side of the board, we find the three LED lights (LED2..LED0).

---

[23] http://land-boards.com/blwiki/index.php?title=Cyclone_II_EP2C5_Mini_Dev_Board

The board has four 28-pin connectors (P4, P3, P2 and P1, see the figure on the right) connected directly to the FPGA chip pins.

They can be set as inputs or outputs and given the scarcity of devices on the board, they'll necessarily be used to connect to external interfaces, as shown in Section 5.5.

The EP2C5 board has a $50\,MHz$ clock generator, like the boards we have studied before. For our projects, Deeds offers the use of the native clock frequency and many of its submultiples (those listed for the DE0-CV board).

5.5 Microprocessor system prototypes on FPGA

In this section, we will show how to implement a microprocessor system on FPGA that is designed through the Deeds software suite. In the examples, we will use the 3 FPGA boards described in Sections 5.1.2 and 5.4). In any case, the concepts presented here can be reused on every FPGA board supported by Deeds to create not only microprocessor systems but also any digital network.

5.5.1 The steps to take

To put a project created with Deeds on an FPGA board, we need to follow the procedure below, which is applicable for all the boards that Deeds supports. After simulating our system to check that it works correctly, we take the following steps:

— Associate the input and output devices on the Deeds schematic with those available on the chosen FPGA board,
— Launch the automatic conversion of the Deeds project in VHDL code,
— Program the FPGA with the *Quartus® II Web Edition™* software.

In the following, we will go through this procedure step by step to provide a practical example to refer to. First, we will present a microprocessor system to implement on FPGA. Then, we will provide an example for each board presented in Sections 5.1.2 and 5.4.

If none of these boards is available, we suggest following this procedure regardless since it is usable in all the boards supported by Deeds. By studying the DE2 and DE0-CV boards, we will learn to use the I/O devices already on the board, while studying the EP2C5 will show us how to connect input and output devices that are not on the board through the board's connectors.

5.5.2 A system to implement on FPGA: An example

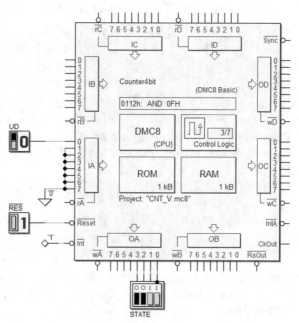

To facilitate learning, we will begin with a simple example: emulating a 4-bit bidirectional binary counter.

The image on the right is a schematic that includes a "DMC8 microcomputer" component (in its basic version, introduced in Section 2.4.1), configured with 1 kB of ROM and 1 kB of RAM.

An input checks the direction of the count (UD), and a push-button is connected to the reset input $\overline{RES}$.

A bar of 4 LED lights (STATE) is connected to output port OA. The interrupt line is set at '1' since it is not used.

When UD is at '0', the count goes up; when it is at '1', it goes down. Notice the input port IA wiring, where we connected[24] the unused lines at '0'.

Now, let's look at the program loaded in the microcomputer. First, we see the label definition for input port CNTRL, where input UD is read from. Next, we see the label definition for output port STATE, where the internal state of the counter is displayed. Then we find the usual link to the reset.

CNTRL	EQU	00h	; IA input port (UD control line, bit 0)
STATE	EQU	00h	; OA output port (the counter output)
	ORG	0000h	; link to the reset
	JP	0100h	
	ORG	0100h	

---

[24] If we don't connect the unused lines to '0' (or '1'), Deeds signals the reading of unknown values during the simulation.

Before entering the program's main loop, we zero output port STATE and register B, which contains the internal state of the counter. We do not use the Stack in the code, so we do not initialize the Stack Pointer.

```
INIT:        LD      A,00h          ; initialize the counter state
             OUT     (STATE),A      ; and the outputs to zero
             LD      B,A
```

Every time the main loop is executed, the UD input is first checked (by reading the input port CNTRL), so register B is then incremented or decremented.

```
MAIN:        IN      A,(CNTRL)      ; read the input port CNTRL
             BIT     0,A            ; check the value of input line UD
             JP      NZ,DN          ; jump and decrement the state if it is at '1'
UP:          INC     B              ; otherwise, increment the state
             JP      UPDATE         ; jump to update and display it
DN:          DEC     B              ; decrement the state
```

Then bits 7, 6, 5 and 4 are set to zero through bit masking to reduce the count to 4 bits. Finally, the new value of the count is shown on output port STATE.

```
UPDATE:      LD      A,B            ; mask the bits in position 7,6,5 and 4
             AND     0Fh            ; to reduce the count to 4 bits
             LD      B,A            ; update the state in B
             OUT     (STATE),A      ; and copy it to the output lines
             JP      MAIN
```

Note that bit masking is not strictly necessary since the bits connected to the bar of LED lights are only those in positions 3, 2, 1 and 0, but it is done anyway to make the code more legible.

5.5.3 Implementing the network on an FPGA board

Now let's look at the procedure required for implementing the system above on an FPGA board. The following explanation goes more into detail on DE2, DE0-CV and EP2C5 boards[25].

Associating input and output devices

Let's study how to associate the input and output devices in the Deeds schematic with those available on the FPGA board.

The following figure shows the Deeds-DcS main page with our project open in the editor. We press the "Test on FPGA" button on the main tool bar (highlighted by the red box on the upper right hand side).

---

[25] There are extensive tutorials available about this procedure on the Deeds website.

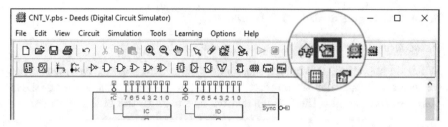

A new window, shown in the following figure, will open. This allows us to associate the push-buttons, switches, clock generators, LED lights, etc. in our schematic with their respective components on the FPGA board. By using expansion connectors, we can connect other devices. In some boards, such as the EP2C5, this operation is necessary since the components available to the designer are not enough to create our system.

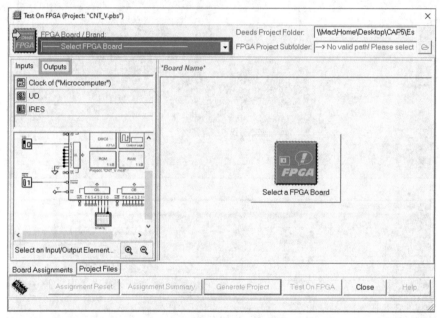

In the box at the left, we can see the input devices in the schematic. By clicking on "Outputs" (in the green box) we can see the output devices. When we click on the upper left hand side menu (in the red box), we can select the FPGA board where we want to implement the project defined in the schematic.

When we select an FPGA board, its inputs (or outputs, depending on the page selected) will be made available to the designer. When we click on one of the schematic's inputs (or outputs), the inputs (or outputs) of the board that we can associate with it will be shown.

The following pages will show the associations chosen for each board.

5.5.4 Settings for the DE2 board

Associating input devices

When we select the DE2 board, the window below will appear. We have clicked on input UD in order to associate it with one of the switches on the DE2. Here, we have chosen the SW[00] switch, as shown in the blue rectangle.

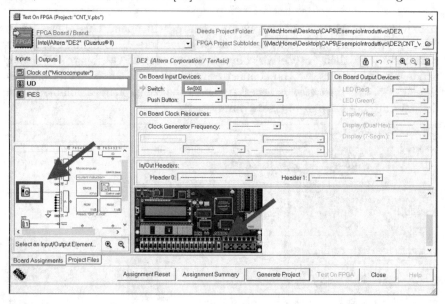

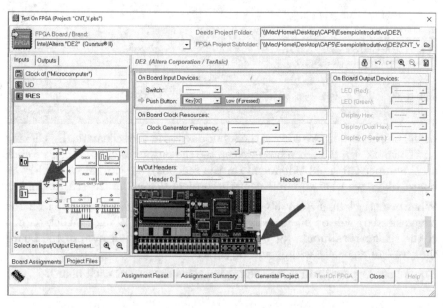

This window shows both the position of the component selected in the Deeds schematic and that of the one associated to it on the FPGA board, on the right. Both are highlighted in red (see the arrows).

The same procedure was followed for input $\overline{RES}$, which was associated to push-button KEY[00] and set to generate a low level when pressed (see the figure at the bottom of the opposite page).

We have chosen the DMC8 Microcomputer component in order to set its clock frequency (see the following figure).

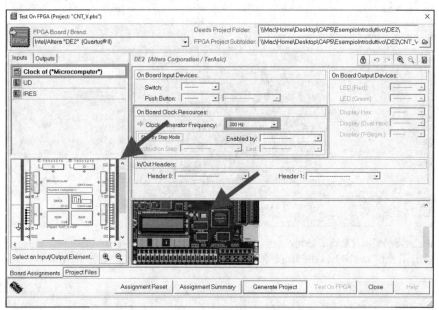

Although the clock is fixed at 10 MHz in the simulation, when creating it on FPGA, we can select a lower or higher frequency clock source. See the figure at the right. However, if we choose a frequency that is different from 10 MHz, we need to consider the fact that the calculations for any delay loops will have to be redone for the new frequency.

As shown in the blue rectangle in the center of the figure above, a 200 Hz clock was chosen in order to make human interaction with the device possible. There are other options for function checks on the clock for debugging, but they have been omitted here for simplicity's sake.

All the clock frequencies we can choose in the box ensure the proper operations of the microcomputer. As highlighted by the red arrows in the figure above, the clock generator on the board corresponds to the selected DMC8 component. This association is shown by the red outlines.

Associating output devices

Once the input devices are defined, we go on to associating the outputs by clicking on the "Outputs" palette outlined in green in the figure.

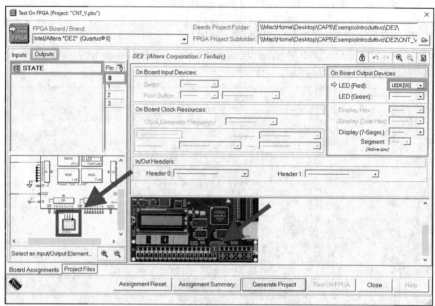

We select the STATE component (outlined in blue), then we associate the red LED lights available on the FPGA board to its four pins (one by one).

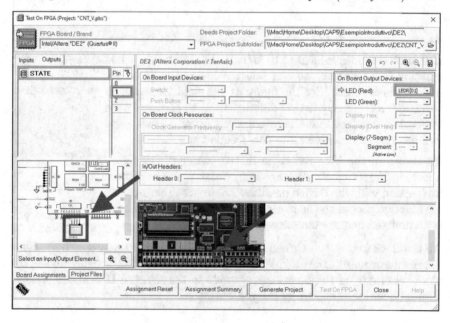

Now, we focus on the figures on the opposite page. In the window shown at the top, the STATE component and index pin 0 (the least significant bit) were selected (highlighted by the blue outline). Then we select the red LED light that we want to associate to it (LEDR[00]) using the menu at the upper right hand side (also outlined in blue).

In the example at the bottom, we select index pin 1 (the bit of weight 2) and associate LEDR[01] to it, as shown by the blue box. Using the same procedure, we then associate the LED lights belonging to STATE of weight 4 and 8 to the LED lights LEDR[02] and LEDR[03].

Notice that for output components as well, the window shows the device selected in the Deeds schematic and the FPGA device associated to it in a red box (as highlighted by the red arrows in the figure).

The following picture shows a useful overview of the function of the devices.

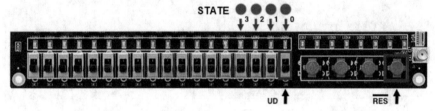

5.5.5 Settings for the DE0-CV board

Associating input devices

When we select the DE0-CV board (see the yellow box), the window will appear as follows. With a click, we also select input UD.

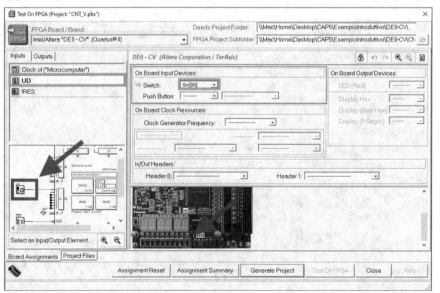

We have associated input UD to the SW[00] switch on the DE0-CV board, as shown by the red boxes and arrows in the figure.

We'll do the same thing (see the following figure) with input $\overline{\text{RES}}$: associate it to KEY[00] (set to generate a low level if pressed).

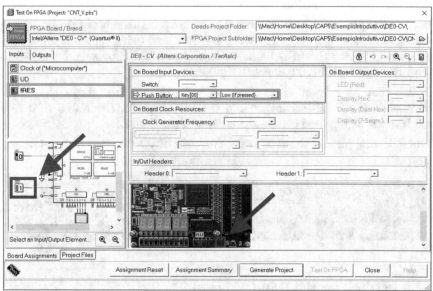

We have associated a $200\,Hz$ clock source to the microcomputer to make the behavior of the device observable with the human eye (see more on the clock options in Section 5.5.4).

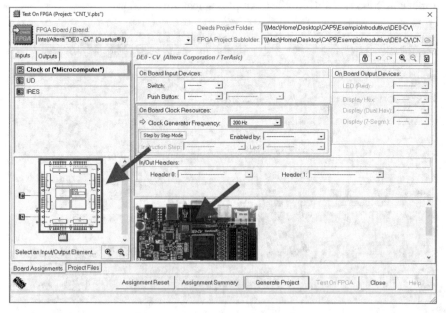

Associating output devices

Once the input devices are associated, we will deal with the output devices by clicking on the "Outputs" palette (see the green box in the figure).

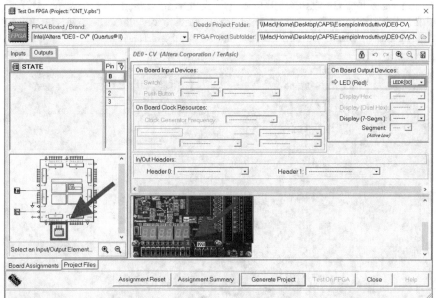

As before, the window lets us select the output device in the Deeds schematic and associate it to an output device on the FPGA board (as shown in the figure, in blue).

We follow the same procedure that we did for input devices and associate the four STATE outputs to the LED lights available on the FPGA board. We select STATE and the index of pin 0 (see the boxes on the upper left hand side of the window), then we use the list box control at the upper right hand side to associate that pin to the red LED light LEDR[00] on the board.

Once this is done, the Deeds component and the corresponding physical device will be marked with red boxes (indicated here by the arrows). We repeat the procedure for the other indexes and associate the bits of weight 2, 4 and 8, to the LED lights LEDR[01], LEDR[02] and LEDR[03].

The following figure shows a summary of the associations, useful to test the system on the board.

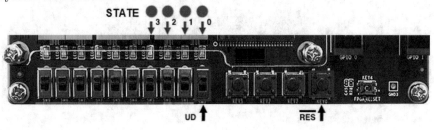

5.5.6 Settings for the EP2C5 board

Associating input devices

When we select the EP2C5 board, the following screen will appear. As considered in Section 5.4.3, the only input device on the board is the KEY0 push-button, which we will use for the reset input $\overline{\text{RES}}$.

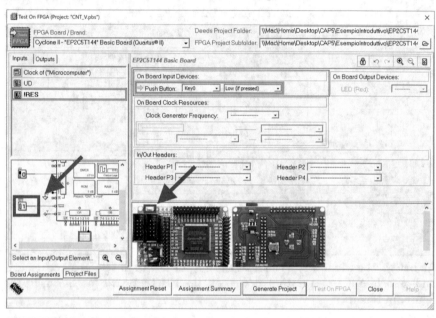

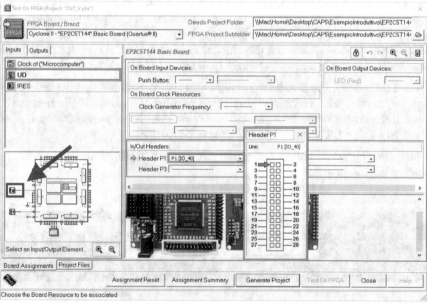

In the figure at the upper part of the opposite page, we selected line $\overline{\text{RES}}$ in the list of inputs on the left, then we associated it to that of the KEY0 push-button (see blue boxes). The arrows point out this association in the schematic and on the board.

To connect the remaining input and output devices, we need to rely on the expansion connectors available on the board.

The figure at the bottom of the opposite page shows the association between input line UD and connector pin P1 (chosen arbitrarily among those available). When we select the connector pin, the drawing of the connector appears with its pins numbered (as seen in the figure). This will be useful when we need to physically connect a wire to the pin because the red arrow points out its position and number.

The clock generator is similar to that of the DE0-CV board (50 Mhz) and is managed in the same way by Deeds. As shown in the following figure, we assign a frequency of 200 Hz to the microcomputer to allow for interaction with the user (for more on this, see the clock options in Section 5.5.4).

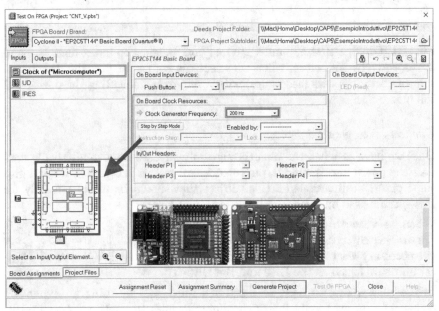

Associating output devices

The following figure shows that when we click on "Outputs" (in the green box) we can go on to assign output devices.

On the screen in the blue boxes, we see that pin P1[IO_71] on connector P1, has been associated to index pin 0 on the STATE lines.

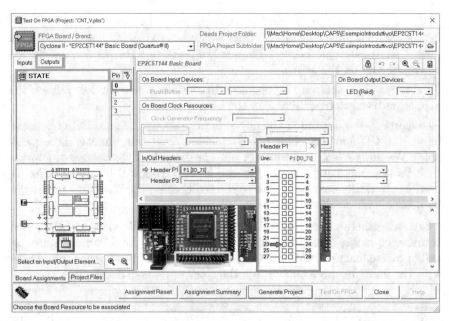

Pins P1[IO_63], P1[IO_53] and P1[IO_44] are assigned to index lines 1, 2 and 3, respectively. As before, when we select the connector pin, the drawing of the connector appears (see figure) with a red arrow indicating the position of the pin.

Connection to physical devices

Once we have assigned the pins through software, we need to physically connect the board and the input/output devices. In this subsection, we will give some practical directions on this subject that should be useful not only for this specific case but also for all the examples given.

When we want to make an input component like an "Input Switch" correspond to a physical slide switch, the necessary electrical connections have already been set on boards such as the DE2 and the DE0-CV.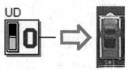

In the case where we need to reach the same goal by using an expansion connector pin, we should follow some simple rules to electrically connect it. We have already seen some points on this in the example in Section 4.7.7).

Switches and push-buttons are electromechanical devices and we need to transform their mechanical action into a two-level physical quantity that can be read by a logical device.

The following figure shows the electrical symbols of four of these single-pole devices (there are also other types). They are, from left to right: (1) an "on-off switch", (2) a "double throw switch", (3) a "normally open push-button" and (4) a "normally closed push-button".

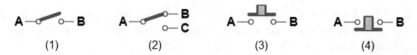

(1)   (2)   (3)   (4)

(1) The "on-off switch" makes it possible to open or close an electric circuit between electrodes A and B. For example, to turn the light in a room on or off, we use this switch. The device has two stable positions, so the state that we set manually is kept and to change it, we have to push it again.

(2) The "double throw switch" allows us to re-route the current into two distinct connections (A ↔ B, or A ↔ C). This type of switch has two stable positions[26]. Here, for reasons of availability, we use this kind of switch but without connecting one of the two electrodes (B or C), so they will actually be used as on-off switches.

(3) The "normally open push-button" behaves like an on-off switch electrically, but it has only one mechanically stable position, that is it closes the contact if we press it and the return spring brings it back to the open position. Other devices that behave this way are keyboard keys, and the buttons on elevators, remote controls or televisions.

(4) The "normally closed push-button" behaves like the open one mechanically, but the contact remains closed when at rest and it opens only when we press it (we will not use this model in our examples).

The left side of the following figure shows the connections of an on-off switch to a connector pin, which is connected to an FPGA chip input (the NOT gate is purely illustrative and represents a logical input).

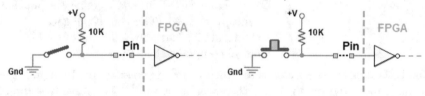

In the figure on the right, the electrical network is identical but we are connecting a push-button. The difference is only mechanical, in the way they are activated, as described before (and so, in the way they are used).

If the contact is open, the logical input is kept high by the "pull up" resistor[27]. When the contact is closed, however, the logical input is forced low by the electrical connection to the ground (Gnd).

As discussed in the example in (Section 4.7.7), the mechanical properties of electromechanical devices make them susceptible to "mechanical contact

---

[26] Note that when the contact is moved, it is unconnected for an instant from either contact before closing on the chosen side.

[27] The 10 $K\Omega$ (Kilo-Ohm) value is used as an example and needs to be adapted to the electrical parameters of the input.

bounces". These are generally resolved by doing multiple reads through software. However, in our elemental example, we will ignore this problem for simplicity's sake.

On the left of the following figure, we see a double throw slide switch, whose contact happens by sliding a cursor. The central contact A gets connected by sliding the cursor to contact B or C.

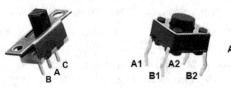

The movement is stable, that is, the cursor remains where we move it. In our examples, we will employ it as an on-off switch, using the contact pairs AB or AC arbitrarily.

In the middle and on the right, we have a "tactile" push-button (shown right side up and on its side). The internal contact is kept open by a spring, but it closes when pressed. When it is released, it opens again.

As the figure shows, the push-buttons often have 4 pins and we can choose those most convenient for connecting to the circuit. A1 and A2 are connected together internally, as are B1 and B2 (so for example we can use only the A1/B1 pair and ignore the other).

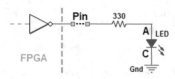

In the figure on the left, we see the connection between an LED light and a connector pin that comes from the FPGA (the NOT gate is there as a formality and represents the output of any logical component).

The LED light must be connected with the right polarity. In the figure, the anode is indicated by an A and the cathode by a C[28]. The resistance limits the working current[29] of the LED light.

LED lights on the market come in a wide variety of colors, shapes, sizes and power characteristics. The figure on the left shows a green LED light. Following convention, the longer terminal is always the anode (A) and the shorter is the cathode (C).

When the FPGA output is low, the tension generated by the logic gate is not enough to turn the light on. When the output is high, it can provide enough current to turn on a low-power light, suitable for our purposes.

---

[28] The terms "anode" and "cathode", come from the field of Electrochemistry; the anode is at a higher tension than the cathode.

[29] The value of 330 Ω is just an example that can be applied here. It would be reduced according to the electrical parameters of the LED lights and the output.

Let's go back to our example. For ease of consultation, the figure below provides a table summarizing the associations that have been made (as reported in Deeds).

In/Out Name	VHDL Name	Board Resource	Aux. Info
Inputs:			
Ck of ("Microcomputer")	—	Clock: 200 Hz	
UD	iUD	Header 0: P1 [IO_40]	
!RES	inRES	Push-Button: Key0	Low (if pressed)
Outputs:			
STATE.0	oSTATE(0)	Header 0: P1 [IO_71]	
STATE.1	oSTATE(1)	Header 0: P1 [IO_63]	
STATE.2	oSTATE(2)	Header 0: P1 [IO_53]	
STATE.3	oSTATE(3)	Header 0: P1 [IO_44]	

The result of these connections will be similar to what we see in the figure on the right.

We have attached the central pin of the slide switch to input UD, that is pin P1[IO_40] of the connector at the bottom (the pin is simply reported as '40' on the silk-screen printing of the board).

We have also connected one of the two wires of a 10 $K\Omega$ resistor to the same pin of the slide switch, while we brought the other one to the power supply (Vcc, 3.3V, pin on the connector at the left).

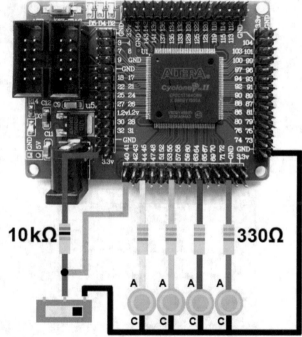

Either of the opposite pins of the switch will be connected to the ground (GND, pin on the connector at the right).

Finally, we connect the 4 LED lights to the outputs that we assigned to connector pins P1[IO_44], P1[IO_53], P1[IO_63] and P1[IO_71] at the bottom (shown as '44', '53', '63' and '71' on the silk-screen printing of the board). We connect these pins to the anodes of the LED lights, taking care to insert the 330 Ω resistor. The cathodes (the shorter terminals) will then be connected to the ground (GND, the same as the pin above).

5.5.7 Converting the Deeds project into VHDL

When we have finished associating the input and output devices of the board to the Deeds project, we generate the VHDL code by clicking on the "Generate Project" button (highlighted by the blue box in the figure below).

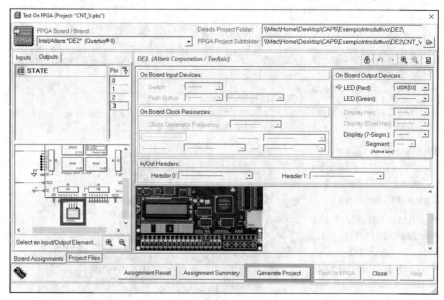

After the short time it takes for the VHDL code to be generated automatically, we get to the following window:

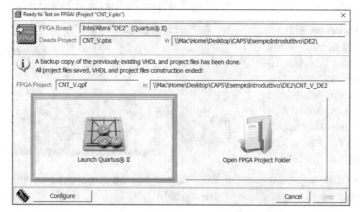

Clicking on the "Launch Quartus® II" button will run the software tool of the same name[30] (introduced in Section 5.3), with which we will go on to program the board.

---

[30] The Deeds website tutorials on using FPGAs, have useful information on installing the software.

5.5.8 Programming the FPGA board

Once Quartus® II is open, the following window will appear. If we click on the "Files" command in the blue box in the figure, we can examine the VHDL files that come from the Deeds schematic and the association of the input/output components. This may be interesting for those who want to learn more, but it is not necessary for those who want to simply program the circuit on the FPGA board.

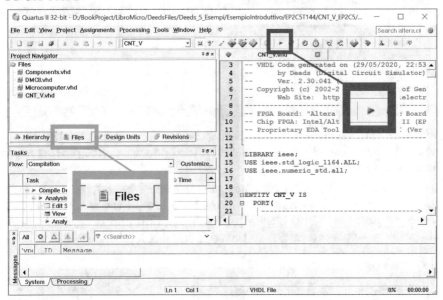

Before going on to actually programming the board, we need to process the program's VHDL file by pressing the "Compile" command (see the red box). After a few minutes, an overview on the project compilation will appear in the window. This indicates that the project is now ready to be loaded on the FPGA board. Usually, many warnings are generated but for the educational scope of these projects, we do not need to take them into account unless they are explicit error messages.

To load the compiled project on the FPGA board, we need to click on the "Programmer" command to open the programmer tool window (highlighted by the red box in the following figure).

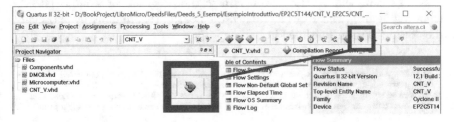

The programmer window will open (see the following figure).

When we click on the "Hardware Setup" button (see the red box), a dialog box will open. This allows us to select the USB port where the programming hardware is connected.

In the "Hardware Setup" dialog box (see the figure below), when we open the "Currently Selected Hardware" drop-down list (in the red box) we can select (for example) "USB Blaster [USB-0]", the USB port recognized by the operating system as connected to the programming hardware in use[31].

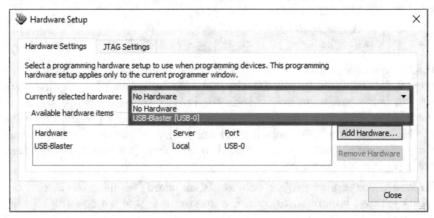

When we close this window, we return to the programmer window (see the following figure), where we can launch the board programming by clicking on the "Start" button (outlined in red).

---

[31] If this option weren't there, we would have to check that the FPGA is connected to the PC and powered. If that option were still missing, we would have to check that the QUARTUS drivers were updated and installed correctly. For more on this, consult the tutorials on the Deeds website: https://www.digitalelectronicsdeeds.com/learningmaterials/labtopics.html#fpgatutorial

During the programming phase, we can see how far along we are by looking at the progress bar (see the blue outline in the figure). Once the process is over, we can test the functionality of our project on the FPGA board.

5.6 Project examples

As mentioned in the beginning of this last chapter, we offer some projects developed with the Deeds simulator that are easy to replicate and experiment with on an FPGA board.

Each project comes with an implementation on all three boards that we have shown in this chapter. This will allow the reader who has one of these boards on hand to immediately reproduce these projects.

All the projects shown here are available in their entirety on the Deeds website on the pages regarding this book. Readers can redesign and extrapolate on this material as they like, using the skills they have developed and also that bit of creativity that any future microprocessor system designer/programmer should have. Readers are encouraged to check that the system is working correctly, go through the steps shown here, and then to try to modify it in a creative way.

With the first example, we will learn to control the luminosity of an LED light and with the second we will attempt to create a light gadget. Then we will focus on producing sounds, first by creating a special effects generator and then a music box that can play a famous tune.

From these projects with fun applications, we will move on to more technical/industrial examples. We will first design and build a stepper motor controller and then we will provide two examples of the use of a small alphanumeric and graphical display that had been used in a very popular mobile phone.

As mentioned at the beginning of this chapter, the physical implementation of a system depends greatly on the technologies in use. Cutting edge technologies evolve and become obsolete rapidly, often after only a few years.

This is why it is important to focus on the programming techniques in the previous chapters. The device interfaces will certainly change over time, but the approaches we use to study them and the techniques we use to program them will remain. These approaches and techniques can be reused to design the systems of the future and they will help those who have mastered them to face technological changes with success.

5.6.1 Light dimmer

We need to design and build a light dimmer that can progressively control an LED light from completely off to completely on and vice versa, through the PWM (Pulse Width Modulation) technique.

The dimmer needs an UP and a DOWN push-button. When UP is kept pressed, the light gets progressively brighter and when DOWN is pressed, progressively dimmer. If both buttons are pressed simultaneously, DOWN has priority. When the buttons are released, the level of brightness stays fixed.

5.6.1.1 The system (Version 1)

In this version (see the following figure), the system carries out the task requested by the specifications using a software-only approach.

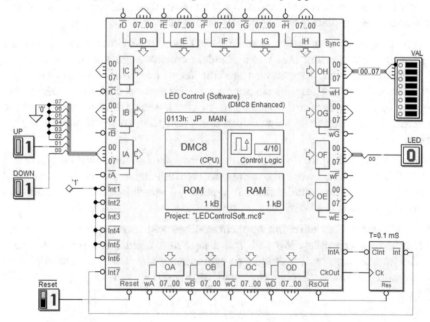

The system is based on a "DMC8 Enhanced Microcomputer" (see Section 2.4.1) with an added interrupt timer.

The LED has been connected to bit 0 of port OF, while two lines of port IA are dedicated to reading the UP and DOWN buttons (when pressed, the buttons generate a low level). The schematic also shows an 8-LED display (connected to port OH) that we will only use for simulations to assess the state of the LED control.

The timer interrupts the processor every $100\,\mu S$. The requested task is executed exclusively by the interrupt handler, aside from the necessary initializations, which are executed by the main program.

We were introduced to the PWM technique in an example from Section 1.5.3.3. It makes it possible to generate programmable average voltage on one line thanks to a succession of fixed period pulses but variable duration. The average voltage that is generated depends on the ratio between the duration of the high part and the whole period.

Here, the PWM period is $25.6\,mS$ $(= 256 \cdot 100\,\mu S)$, that is 256 timer interrupts. The pulse duration is established by the 8-bit variable VALUE. If it has a value of zero (VALUE = 0) the LED light is off; if it has the maximum value (VALUE = 255), the light remains on for 255 calls and is turned off by the 256th.

The continual on/off flashing cannot be perceived by the human eye due to retinal persistence. This makes it so that our brains do not see variations but rather stable images whose brightness is regulated by the ratio VALUE/256.

The reading of the buttons should be confirmed by two consecutive reads $25.6\,mS$ apart to eliminate contact bounces (see Sections 4.7.7 and 5.5.6).

The program

First we define ports USER, PLED and PVAL (read the comments in the code), then we define three variables in the memory.

VALUE memorizes the PWM pulse duration (a number from 0 to 255), which controls the LED light. PUSER records the previous reading of the push-button port so the debouncing check can be done. TIME, on the other hand, is used to assess the time that passes in the period of the PWM output.

```
USER        EQU    00h           ; IA input port: user push-buttons
PLED        EQU    05h           ; OF output port: LED control
PVAL        EQU    07h           ; OH output port: value monitor

VALUE       EQU    0FC00h        ; LED control value
PUSER       EQU    0FC01h        ; the previous state of USER port
TIME        EQU    0FC02h        ; time counter
```

We define the link to the reset and the interrupt handler. Then, we initialize the Stack Pointer and the variables. We enable the interrupts and enter the infinite loop MAIN where no operation is executed. Here, no operations are executed because everything will be done by the interrupt handler.

```
            ORG    0000h
            JP     START
            ORG    0038h
            JP     HINT
            ORG    0100h
START:      LD     SP,0FFFFh      ; initialize the Stack Pointer
            LD     A,00h
            LD     (VALUE),A      ; zero the LED control value
            OUT    (PLED),A       ; and the corresponding output port
            LD     (PUSER),A      ; no button previously pressed
            LD     (TIME),A       ; initialize also the time counter
            EI                    ; enable interrupts
MAIN:       JP     MAIN           ; main loop (empty)
```

Every $100\,\mu S$, the interrupt handler is launched. The code begins with the usual PUSH instructions and ends with the required POPs so that we can preserve the content of the registers used by the handler. This is done even though the main program is empty for now, leaving space for any future development of the code.

```
HINT:       PUSH   AF             ; save the used registers
            PUSH   BC
```

As we will see further on, the TIME variable is decremented in the exit code of the handler (that is every $100\,\mu S$). This means that TIME behaves like a cyclical 256 module down counter.

Since we need to turn the LED on for the duration defined by VALUE, we compare VALUE with TIME at every call. If TIME $\geq$ VALUE we jump to NOPULSE and put the output at '0'; if not, we activate the output at '1'.

In other words, the output goes to '1' when TIME, while decrementing, reaches VALUE. The output goes back to '0' when TIME starts back again from 255.

```
            LD     A,(VALUE)      ; copy the control value
            LD     B,A            ; to register B
            LD     A,(TIME)       ; compare the current time
            CP     B              ; with the control value
            JP     NC,NOPULSE
PULSE:      LD     A,00000001b    ; PWM = '1' , if TIME < VALUE
            JP     WPORT
NOPULSE:    LD     A,00000000b    ; PWM = '0', if TIME ≥ VALUE
WPORT:      OUT    (PLED),A       ; update the PWM output port
```

Cyclically, every 256 calls (25.6 mS), TIME is zeroed. So the next check verifies if it is now time to assess the push-buttons.

```
LD    A,(TIME)        ; check if it is time
CP    0               ; to assess the push-buttons' state
JP    NZ,EXIT         ; EXIT if it is not
```

If 25.6 mS have gone by, we check the state of the push-buttons and compare that to the state saved the last time in PUSER. The new state is saved in this variable to be used for the next comparison and in register C.

```
LD    A,(PUSER)       ; copy the previous port state
LD    B,A             ; to register B
IN    A,(USER)        ; read the current push-buttons' state
CP    B               ; compare it with the previous state
LD    (PUSER),A       ; save the new state in the variable
LD    C,A             ; and in register C
JP    NZ,EXIT         ; debouncing: exit if they are different
```

If the values are different we assume that the cause is a contact bounce or a transient, so we exit. If the values are equal, we validate the state that's just been read. Then, we go on to act on the basis of activation or push-buttons. DOWN has priority over UP, so we test it first and if it is pressed, we do not test the state of UP.

```
TESTDN:   BIT   0,C         ; 'DOWN' is pressed? (it has priority)
          JP    NZ,TESTUP   ; jump to the other test if it is not
```

If DOWN is pressed (it is at '0'), we decrement the VALUE variable by 1 unless it is already at zero.

```
LD    A,(VALUE)       ; get the control value
CP    0               ; check if it is already at zero,
JP    Z,EXIT          ; if it is so do not decrement, exit
DEC   A               ; otherwise decrement the value
LD    (VALUE),A       ; and save back it in memory
JP    TESTV           ; jump because 'DOWN' has priority
```

If DOWN is not pressed, we check to see if UP is pressed. We go on if it is.

```
TESTUP:   BIT   1,C         ; check if 'UP' is pressed
          JP    NZ,EXIT     ; jump to exit if it is not
```

UP is being pressed (it is at '0'), so we increment the VALUE variable by 1 unless it is already at the maximum value.

```
LD    A,(VALUE)       ; get the control value
CP    0FFh            ; check if it is at the maximum value,
JP    Z,EXIT          ; if it is so do not increment, exit
INC   A               ; otherwise increment the value
LD    (VALUE),A       ; and save back it in memory
TESTV:    OUT   (PVAL),A    ; display the new value for test
```

Finally, as mentioned before, we decrement the TIME variable and then exit the interrupt handler. The count is cyclical, so if TIME is at zero, it returns to 255 by decrementing.

```
EXIT:       LD    A,(TIME)        ; count cyclically the time
            DEC   A
            LD    (TIME),A

            POP   BC              ; restore the used registers
            POP   AF
            EI                    ; re-enable interrupts and
            RET                   ; return to the interrupted program
```

The version we have now analyzed can be changed if we introduce specialized hardware that automatically executes the generation of the PWM output, and therefore can remove that task from those of the microprocessor.

This hardware feature is integrated into microcontrollers. In version 2, which is discussed below, we will go further into the functionality of a specially designed PWM converter, which we add to the microcomputer.

5.6.1.2 The system (Version 2)

In this new version, we leave it to the microcomputer to handle the function of the push-buttons and the content of VALUE (see version 1 since we've taken some of the code from it).

The following figure shows that the microcomputer no longer controls the LED light directly. Rather, it copies the number in VALUE to the OF port, leaving the external PWM component to generate the correct pulse sequence. The component uses the 10 *MHz* processor clock.

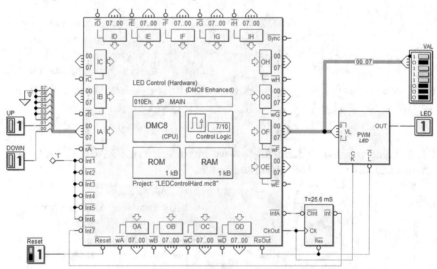

In this version, the timer interrupts the processor only to read the state of the push-buttons, which it does on a regular basis. Its interval is now defined as $25.6\,mS$, as per specifications.

The PWM component (see the figure at the right) achieves the same thing through hardware as version 1 does through software, but can do it much faster. More importantly, the added component allows us to save the processor's computational capacity. It accepts the number VL in the input and produces the corresponding PWM signal in the output, with a period 256 times the clock period.

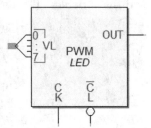

With a $10\,MHz$ clock, the PWM signal period is $25.6\,\mu S$ (thousands of times faster than the software version).

The inside of the component is described in the following schematic. The 8-bit counter "Cnt8" cyclically produces the decreasing sequence from 255 to 0, which is brought to inputs A7..A0 of the magnitude comparator "Cp8".

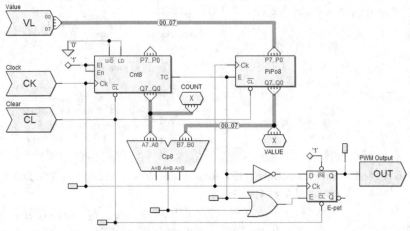

Each time the counter reaches zero, its TC (Terminal Count) output enables loading the number VL in parallel register "PiPo8". The output of this is brought to the comparator's inputs B7..B0. The comparator signals the moment when the number generated by the counter becomes equal to the number set at input VL. When this happens, the OUT output (the PWM signal) is brought to '1'; it will be brought to '0' when the counter starts the count again from 255.

The program

The assembly code derives from that of version 1.

USER	EQU	00h	; IA input port: user push-buttons
PLED	EQU	05h	; OF output port: LED control
VALUE	EQU	0FC00h	; LED control value
PUSER	EQU	0FC01h	; the previous state of USER port

The TIME variable has been eliminated from the definitions since generating the PWM signal is no longer the job of the software. After the jumps to the start of the program and the interrupt handler, we initialize the Stack Pointer and the variables. The main program is similar to that of version 1.

```
            ORG    0000h
            JP     START
            ORG    0038h
            JP     HINT
            ORG    0100h
START:      LD     SP,0FFFFh      ; initialize the Stack Pointer
            LD     A,00h
            LD     (VALUE),A      ; zero the LED control value
            OUT    (PLED),A       ; and the corresponding output port
            LD     (PUSER),A      ; no button previously pressed
            EI                    ; enable interrupts
MAIN:       JP     MAIN           ; main loop (empty)
```

As mentioned before, the interrupt handler only manages debouncing and incrementing/decrementing the VALUE variable.

```
HINT:       PUSH   AF             ; save the used registers
            PUSH   BC
            LD     A,(PUSER)      ; copy the previous port state
            LD     B,A            ; to register B
            IN     A,(USER)       ; read the current push-buttons' state
            CP     B              ; compare it with the previous state
            LD     (PUSER),A      ; save the new state in the variable
            LD     C,A            ; and in register C
            JP     NZ,EXIT        ; debouncing: exit if they are different
```

With regard to this functioning, the code is identical to version 1. The DOWN push-button is checked first (priority).

```
TESTDN:     BIT    0,C            ; 'DOWN' is pressed? (it has priority)
            JP     NZ,TESTUP      ; jump to the other test if it is not
```

If the button is pressed, the content of VALUE is decremented. If it is not already at the minimum value, then it jumps to the SEND label.

```
            LD     A,(VALUE)      ; get the control value
            CP     0              ; check if it is already at zero,
            JP     Z,EXIT         ; if it is so do not decrement, exit
            DEC    A              ; otherwise decrement the value
            LD     (VALUE),A      ; and save back it in memory
            JP     SEND           ; jump because 'DOWN' has priority
```

If DOWN is not pressed, the check moves to the state of the UP push-button. If UP is not pressed either, we exit the handler without changing VALUE.

```
TESTUP:     BIT    1,C            ; check if 'UP' is pressed
            JP     NZ,EXIT        ; jump to exit if it is not
```

If, however, UP is pressed, the content of VALUE is incremented unless it is already at the maximum value. Then at the SEND label, we copy the new content of VALUE to the PLED port and pass this number to the PWM module.

```
          LD    A,(VALUE)     ; get the control value
          CP    0FFh          ; check if it is at the maximum value,
          JP    Z,EXIT        ; if it is so do not increment, exit
          INC   A             ; otherwise increment the value
          LD    (VALUE),A     ; and save back it in memory
SEND:     OUT   (PLED),A      ; send the value to the PWM generator
EXIT:     POP   BC            ; restore the used registers
          POP   AF
          EI                  ; re-enable interrupts and
          RET                 ; return to the interrupted program
```

5.6.1.3 Implementation on FPGA

From the functional perspective, versions 1 and 2 carry out the same operations. To implement them on an FPGA board, however, we have used version 2, which is available on the Deeds website in the online content for this section. This allows interested readers to implement version 1 on their own. The following paragraphs will offer synoptic images that summarize the associations chosen for each board[32].

The DE2 board

The following figure provides a visual indication of the devices used on the DE2 board, which is useful when we test the system.

Scheda DE0CV

The assignment of devices for the DE0-CV board is very similar to that for the DE2.

---

[32] For more information on the associations, see the online content for this section.

The EP2C5 board

For the EP2C5 board, since some of the connections have to be made by hand, it is a good idea to use the summary generated by Deeds (see the following figure). Notice that the outputs that were just set for the simulation have not been mapped.

In/Out Name	VHDL Name	Board Resource	Aux. Info
Inputs:			
Ck of ("LED	—	Clock: 10 MHz	
!Reset	inReset	Push-Button: Key0	Low (if pressed)
UP	iUP	Header 0: P1 [IO_40]	
DOWN	iDOWN	Header 0: P1 [IO_47]	
Outputs:			
!Int7	onInt7		
VAL.0	oVAL(0)		
VAL.1	oVAL(1)		
VAL.2	oVAL(2)		
VAL.3	oVAL(3)		
VAL.4	oVAL(4)		
VAL.5	oVAL(5)		
VAL.6	oVAL(6)		
VAL.7	oVAL(7)		
LED	oLED	LED (Red): LED0	

The following figure shows these connections and highlights the network for the two push-buttons to attach to the board.

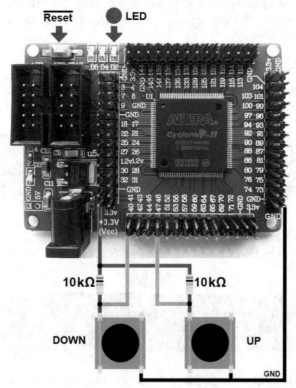

5.6.2 LED gadget

We need to design and build a small gadget that "rotates" the on/off state of three LED lights in a cyclical sequence. The three lights have to gradually turn on and off like a sine wave. The waves are offset from each other by 120 degrees.

The gadget has two push-buttons (UP and DOWN), that control the speed of rotation between a high and low point (that can be perceived by the human eye). When UP is kept pressed, the speed of rotation increases progressively; when DOWN is kept pressed, it decreases progressively. When neither button is being pressed, the speed remains the same. If both buttons are being pressed at the same time, DOWN has priority over UP.

5.6.2.1 The system

Some elements of this project are reminiscent of the previous example (see Section 5.6.1). The push-button handling is similar, so some of the code can be reused in this project. Also, to control the LEDs' brightness, it would be a good idea to use the PWM component described in the previous example.

As we can see in the following figure, a "DMC8 Enhanced Microcomputer" uses output ports OH, OG and OF to drive three PWM generators connected to LED lights 0, 1 and 2, respectively. The generators use the same 10 MHz processor clock.

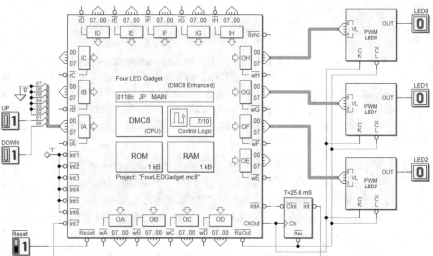

Following the previous example, we use the two lines of the IA port to read the UP and DOWN buttons (when pressed, they generate a low level).

We also add a timer that will interrupt the processor every 25.6 mS, and will be used both for debouncing the reading of the buttons and to handle the on/off rotation of the LED lights.

The same time has been chosen as that of the previous example, but this is not critical and could be changed (in which case the speed of rotation of the lights would change as well).

The example from Section 4.7.3 also inspires the technique of generating outputs with a sinusoidal shape in this case. Here, we reuse the waveform table in Section 4.7.3 and the method to read it. Every time the handler is called, the read index is incremented by the size the user sets through the UP and DOWN buttons, thus controlling the resulting oscillation frequency.

The three sine waves that control the PWM modules are phase-shifted from each other by 120 degrees. So, the read index of the table is offset by a value corresponding to that angle for each light (as required by the specifications).

All the required tasks are carried out exclusively by the interrupt handler. The main program only does the necessary initializations of the ports and the variables used.

The program

First we define ports USER, PLED0, PLED1 and PLED2, then we declare the PHASE120 constant. This will allow us to offset the three sine waves by 120 degrees from each other (1/3 of a round angle but in 256ths).

```
USER        EQU    00h          ; IA input port: user push-buttons
PLED0       EQU    07h          ; OH output port: LED0
PLED1       EQU    06h          ; OG output port: LED1
PLED2       EQU    05h          ; OF output port: LED2
PHASE120    EQU    85           ; 85/256 = about 120 degrees
```

Among the variables, we have PUSER, which memorizes the state of the buttons (for debouncing checks). FREQ is the multiplication parameter of the oscillation frequency, which the user increments/decrements through the push-buttons. Finally, the ANGLE variable records the current angle of the generation of the sine wave related to LED0 (the others use the same angle, with the addition of the PHASE120 offset).

```
PUSER       EQU    0FC00h       ; USER port previous state
FREQ        EQU    0FC01h       ; frequency parameter (0..31)
ANGLE       EQU    0FC02h       ; current angle
```

Then we define the jumps to the program and the interrupt handler.

```
            ORG    0000h
            JP     START
            ORG    0038h        ; Int. 7
            JP     HINT7
            ORG    0100h
```

When the program is launched, we initialize the Stack Pointer, the variables and the output ports. The FREQ parameter is set at an intermediate value.

```
START:      LD      SP,0FFFFh       ; initialize the Stack Pointer
            LD      A,00h           ; zero the output ports
            OUT     (PLED0),A       ; LED0, LED1 and LED2
            OUT     (PLED1),A
            OUT     (PLED2),A
            LD      (ANGLE),A       ; zero the current angle
            LD      (PUSER),A       ; no button previously pressed
            LD      A,5             ; default frequency
            LD      (FREQ),A
```

Then we enable the interrupts and enter the main loop (which is empty since everything is done by the interrupt handler).

```
            EI                      ; enable interrupts
MAIN:       JP      MAIN            ; main loop (empty)
```

The interrupt handler is organized into different subprograms. The contents of the registers in use are saved at the beginning and then restored at the end. This allows us to leave the handler unchanged if functions are added to the main program (which is empty now).

```
HINT7:      PUSH    AF              ; save the used registers
            PUSH    BC
```

We copy the FREQ parameter to register B, and increment the ANGLE variable by this value. Every time the timer is called, that is, we move forward in reading the waveform table by using ANGLE as an index. The new value is also saved in register C.

```
            LD      A,(FREQ)        ; copy the frequency parameter
            LD      B,A             ; to register B
            LD      A,(ANGLE)       ; update the current angle
            ADD     A,B             ; by adding the parameter to it
            LD      (ANGLE),A
            LD      C,A             ; copy the new angle to register C, too
```

The angle is passed through A to the WAVEFORM function, that reads the value table and returns the corresponding sample of the function, which is then sent to the PWM component that drives LED0.

```
            CALL    WAVEFORM        ; read the sample from the table
            OUT     (PLED0),A       ; send it to the port of LED0
```

This same operation is repeated twice more for the other two lights, but with 120 degrees added to the current index (what we saved in register C).

```
            CALL    PHASE           ; shift the phase of 120 degrees
            CALL    WAVEFORM        ; read the sample from the table
            OUT     (PLED1),A       ; send it to the port of LED1
```

After we send the value to the PWM component that drives LED1, we deal with the one connected to LED2.

```
            CALL   PHASE         ; shift the phase of 120 degrees
            CALL   WAVEFORM      ; read the sample from the table
            OUT    (PLED2),A     ; send it to the port of LED2
```

Then we call UPDOWN, which assesses the state of the push-buttons and increments (or decrements) the FREQ parameter, and we exit the handler.

```
            CALL   UPDOWN        ; assess the state of push-buttons
            POP    BC            ; restore the saved registers
            POP    AF
            EI
            RET
```

The PHASE subprogram increments the current angle by 120 degrees and puts it back into register A (as we have seen, it is called before the WAVEFORM function in relation to LED lights 1 and 2).

```
PHASE:      LD     A,C           ; get the current angle from register C
            ADD    A,PHASE120    ; move it 120 degrees ahead
            LD     C,A           ; save back it in C and also return it in A
            RET
```

As mentioned before, the UPDOWN subprogram manages the buttons. It checks their state and compares that with what is saved in PUSER. The new state is then saved in this variable and in register C to be used in the next comparison. If the two states are different, we assume a transitory or contact bounce, so we exit the subprogram with the RET NZ instruction.

```
UPDOWN:     LD     A,(PUSER)     ; copy the push-buttons previous state
            LD     B,A           ; to register B
            IN     A,(USER)      ; read the current push-buttons state
            CP     B             ; compare it with the previous
            LD     (PUSER),A     ; save back the new state in memory
            LD     C,A           ; and in register C
            RET    NZ            ; debouncing: if they are different, exit
```

If they are the same, the state is valid. We then go ahead and assess the DOWN button first (it has priority). If DOWN is pressed (= '0'), we decrement the FREQ parameter unless it is not already at the lowest value[33], and we return to the calling program.

```
TESTDN:     BIT    0,C           ; is the DOWN button pressed?
            JP     NZ,TESTUP     ; (it has priority) jump if it is not
            LD     A,(FREQ)      ; get the previous parameter value
            CP     1             ; if it is already at the lowest value
            RET    Z             ; do not decrement it and exit,
            DEC    A             ; otherwise decrement
            LD     (FREQ),A      ; and save back it in memory
            RET
```

---

[33] The lowest value must be greater than 0, otherwise the advance of the angle stops.

If DOWN is not being pressed, we jump to TESTUP to check the UP button. If it is not being pressed either, we exit the function with the RET NZ instruction; if it is being pressed, we increment the FREQ variable (if it is not already at the highest value[34]), and we return to the calling program.

```
TESTUP:    BIT    1,C              ; is the UP button pressed?
           RET    NZ               ; exit if it is not

           LD     A,(FREQ)         ; get the previous parameter value
           CP     31               ; if it is already at the highest value
           RET    Z                ; do not increment it and exit,
           INC    A                ; otherwise increment
           LD     (FREQ),A         ; and save back it in memory
           RET
```

As mentioned before, the WAVEFORM function provides the value of the sine in function of the angle that we pass in the accumulator. The function uses the SINTAB table, which contains the values of the positive half cycle of the sine wave. The negative values are retrieved from the positive ones by two's complement. This is almost identical to what is used in the programming example in Section 4.7.3). Therefore, we'll bypass any explanation of the details on how it functions. The only difference is that the returned values are offset into the positive range (0..254) and constant +127 is added.

```
WAVEFORM:  PUSH   HL               ; save register HL and BC
           PUSH   BC
           LD     C,A              ; save bit 7 of the angle in C, and mask
           AND    01111111B        ; it to avoid readings outside the table

           LD     HL,SINTAB        ; get the base address of the table
           ADD    A,L              ; add the index to it
           LD     L,A              ; to obtain the address of the location
           JP     NC,NoCarry       ; of interest in register HL
           INC    H
NoCarry:   LD     A,(HL)           ; get the value

           BIT    7,C              ; check if we are in second half wave
           JP     Z,Positive       ; if not, the value is positive
Negative:  NEG                     ; otherwise invert the sign of the value
Positive:  ADD    A,127            ; move the samples in the range 0..254
           POP    BC               ; restore registers BC and HL
           POP    HL
           RET
```

The SINTAB table is defined in the ROM and has been calculated previously. It is identical to the one used in the example cited above. For convenience, part of it is re-printed here.

```
SINTAB:    DB     000              ; x = 0    (0 degrees)
           DB     003              ; x = 1
           DB     006              ; x = 2                        (cont.)
```

---

[34] The highest value was determined experimentally and can be changed.

```
                ... omissis ...
    DB      088              ; x = 31
    DB      090              ; x = 32   (45 degrees)
    DB      092              ; x = 33
                ... omissis ...
    DB      127              ; x = 63
    DB      127              ; x = 64   (90 degrees)
    DB      127              ; x = 65
                ... omissis ...
    DB      092              ; x = 95
    DB      090              ; x = 96   (135 degrees)
    DB      088              ; x = 97
                ... omissis ...
    DB      006              ; x = 126
    DB      003              ; x = 127
    DB      000              ; x = 128  (180 degrees, not used)
```

5.6.2.2 Implementation on FPGA

The following sections will show images that summarize the connections chosen for each board[35], which are useful for testing the system.

The DE2 board

The following figure shows the choice of devices for the DE2 board.

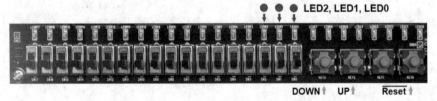

The DE0CV board

The assignment of devices for the DE0-CV board is very similar to that set for the DE2.

---

[35] For more information on the associations, see the online content for this section.

The EP2C5 board

The connections on the EP2C5 board require two external push-buttons to be connected. Among the resources on the board, we use the three red LED lights and the push-button (for reset). In any case, it should be useful to have access to the Deeds summary (see the following figure).

In/Out Name	VHDL Name	Board Resource	Aux. Info
Inputs:			
Ck of /"Four	—	Clock: 10 MHz	
IReset	inReset	Push-Button: Key0	Low (if pressed)
DOWN	iDOWN	Header 0: P1 [IO_40]	
UP	iUP	Header 0: P1 [IO_47]	
Outputs:			
IInt7	onInt7		
LED0	oLED0	LED (Red): LED0	
LED1	oLED1	LED (Red): LED1	
LED2	oLED2	LED (Red): LED2	

Below is a photograph of the board with the physical connections to make for the two push-buttons superimposed on it. Pay attention to the connections of the 10 KΩ pull-up resistors (highlighted in red and orange).

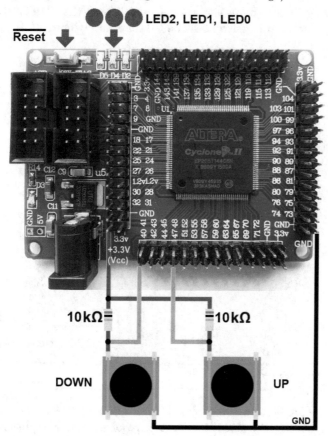

5.6.3 Special sound effects

We want to design and build a sound effects generator that imitates the typical sounds of 1980s video games. We will use a small piezoelectric speaker (called a "buzzer", see the photo at the right) as an acoustic transducer. This speaker has the benefit of being able to directly connect to the logic circuit without the need of an amplifier.

Piezoelectric materials are able to deform when an electrical field is applied. If the electrical field varies over time, we will get a transformation into acoustic vibrations.

5.6.3.1 The system

The generator requires a push-button (ON) and a switch (FAST). The sound is generated when we press the ON button. FAST allows us to select the type of effect we want (fast or slow pace).

The following figure shows the system, which uses the "DMC8 Enhanced Microcomputer" component with an added timer set to interrupt the processor every $100\,\mu S$.

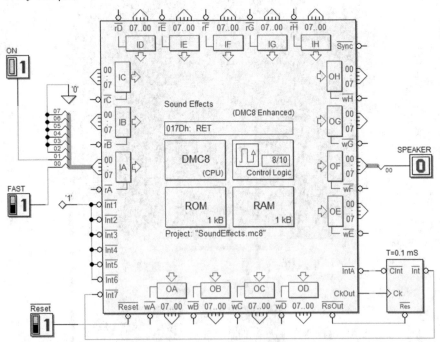

Through the IA port, we read the state of ON and FAST (we will not do debouncing checks here since they are not required by the application).

The piezoelectric speaker is directly connected to the SPEAKER line (bit 0 of output port OF). Since the audio signal sent by the transducer is generated by a logic output, it will always have exactly two levels. To generate the variety of sounds that we expect, we need to work on the oscillation frequency of the logic levels and above all, on the variation of the frequency itself, i.e., on its "modulation".

The main program acquires the inputs and translates them into two variables: PLAY and TMAX (used by the interrupt). PLAY enables sound generation and is obtained by the state of the ON button. TMAX, however, depends on the FAST switch. It contains the value used to re-initialize the time count that the frequency variation (fast or slow) and the type of effect generated depends on, as we shall see.

The interrupt handler primarily inverts the level of the SPEAKER output at the right time. To do this, it decrements a counter every time it is called and executes the inversion each time it zeroes. The count is therefore re-initialized with the value contained in the PERIOD variable.

If PERIOD contained a constant, the frequency of the output would be fixed. However, PERIOD is regularly incremented so it gradually lowers the frequency of the signal generated. This increment is cyclical (once it gets to the highest value, it restarts at the lowest).

This way, the frequency is modulated by a signal with a "descending sawtooth" trend. How quickly the frequency is decremented is in turn controlled by the TMAX variable, which as we remember, depends on the FAST switch.

Therefore, we get a fast or slow decline of the frequency generated that produces two types of sounds that are very different from each other from a psychoacoustic perspective[36]. The following timing diagram shows the rate of the signal in the output, which we obtain by setting FAST to '1'.

The program

We define the addresses of ports PSEL and PSOUND, and then the variables.

```
PSEL       EQU   00h        ; IA input port: user commands
PSOUND     EQU   05h        ; OF output port: audio output
```

The COUNT variable assesses the time that elapses between two output transitions. PERIOD, TMAX and PLAY have already been discussed. The SOUND variable is the software copy of the PSOUND output port. It records the last value written on the port in order to invert the value of the output line when requested.

---

[36] Obviously, we suggest creating the system and listening to the result!

TIME assesses the time elapsed between two variations of PERIOD and so allows us to modulate the frequency, making the two different modulations (fast or slow) possible.

```
COUNT      EQU   0FC00h        ; time counter (between two transitions)
PERIOD     EQU   0FC01h        ; requested time between two transitions
SOUND      EQU   0FC02h        ; software copy of the output
TIME       EQU   0FC03h        ; time counter (between two variations)
TMAX       EQU   0FC04h        ; requested time between two variations
PLAY       EQU   0FC06h        ; sound generation flag (ON/OFF)
```

The following constants determine the final result of the sound effect and can be changed as desired, even experimentally.

```
PHIGH      EQU   50            ; maximum time between two transitions
PLOW       EQU   7             ; minimum time between two transitions
TMLONG     EQU   255           ; time between variations (long and short)
TMSHORT    EQU   25
```

We define the jumps to the main program and the interrupt handler.

```
           ORG   0000h
           JP    START
           ORG   0038h
           JP    HINT
           ORG   0100h
```

After the initialization of the Stack Pointer, the main program calls the CLEAR subprogram, which defines the default values of the variables and the output port. After that, we enable the interrupts.

```
START:     LD    SP,0FFFFh     ; initialize the Stack Pointer
           CALL  CLEAR         ; and all the variables and the output port
           EI                  ; enable interrupts
```

The main program translates the state of the push-button and the switch into the proper values of variables PLAY and TMAX, as mentioned before. If the content of PLAY is not zero, the sound is generated. The FAST switch determines if the TMSHORT or the TMLONG constant is loaded in TMAX (corresponding to the quick or slow modulation, respectively).

```
MAIN:      IN    A,(PSEL)      ; read the user commands
           LD    B,A           ; save their state

           CPL                 ; invert all the bits
           AND   00000010b     ; if the push-button is pressed, bit 1 = '1'
           LD    (PLAY),A      ; save the 'ON' command flag

           BIT   0,B           ; assess the line 'FAST' and assign
           LD    A,TMLONG      ; the update time of the time between
           JP    Z,LONG        ; two transitions (long or short time)
           LD    A,TMSHORT
LONG:      LD    (TMAX),A

           JP    MAIN
```

The subprogram below, CLEAR, initializes all the default values of the variables and the output port (for more on this, read the comments in the assembly code). As we have seen, it is called at the start of the main program, but the interrupt handler uses it as well, as we will see further on.

```
CLEAR:    LD    A,PHIGH        ; this subprogram initialize:
          LD    (PERIOD),A     ; 1) the time between two transitions
          LD    (COUNT),A      ; 2) and its counter
          LD    A,00h
          LD    (PLAY),A       ; 3) zero the play enable
          LD    (SOUND),A      ; 4) zero the output software copy
          OUT   (PSOUND),A     ; and the output port
          LD    A,TMLONG       ; define a default
          LD    (TIME),A       ; to the time count
          RET
```

The interrupt handler is launched by the timer every $100\,\mu S$. The only registers used in it are the accumulator and the flags, so we save them with a PUSH AF instruction. On exit, we restore its original value with the corresponding POP AF.

```
HINT:     PUSH  AF             ; save the used register
```

We immediately check if the generation is enabled. If it isn't, we exit without generating anything, but we re-initialize the variables in play by calling CLEAR (even though it is called all the time, this poses no problem).

```
          LD    A,(PLAY)       ; is the generation enabled?
          CP    0
          JP    NZ,SING        ; jump if it is, otherwise
          CALL  CLEAR          ; re-initialize all the variables and exit
          JP    EXIT
```

If generation is enabled, we jump to the SING label where we count the time that passed since the last inversion to check if it is time to invert the output again. If the count is not at zero yet, we jump to the UPDATE label.

```
SING:     LD    A,(COUNT)      ; assess the time passed
          DEC   A              ; since the last output inversion
          LD    (COUNT),A
          JP    NZ,UPDATE      ; jump if time has not elapsed
```

If the time count is at zero, we make it start again, but we take the new duration from the PERIOD variable (which may have been changed).

```
          LD    A,(PERIOD)     ; re-initialize the time counter
          LD    (COUNT),A
```

We do the level transition by inverting the bit in position 0 of the SOUND variable, which is then copied to output port PSOUND.

```
LD      A,(SOUND)      ; invert the bit 0 of the output
XOR     00000001b
LD      (SOUND),A      ; save back the new state in memory
OUT     (PSOUND),A     ; and copy it to the output port
```

As explained at the beginning, if PERIOD were a constant, the frequency in the output would not change. At the next UPDATE label, however, the new value of PERIOD is calculated (if the time has come to do it).

We decrement the TIME variable and, if it is not at 0, we exit the handler without updating PERIOD.

```
UPDATE:   LD      A,(TIME)      ; during the generation, assess
          DEC     A             ; if it is time to change the duration
          LD      (TIME),A      ; of the next half-period
          JP      NZ,EXIT       ; exit if it is not
```

If TIME is zeroed, we re-initialize it with time TMAX.

```
LD      A,(TMAX)      ; otherwise re-initialize the counter
LD      (TIME),A      ; of the time to change the period
```

So, if TMAX contains the constant TMLONG, PERIOD will be changed less often, giving us the "slow" sound. If TMAX contains the constant TMSHORT, PERIOD will be changed more often, giving us the "fast" sound.

The following instructions produce a progressive increment in the PERIOD variable until it reaches its maximum: PHIGH. When it gets to PHIGH, PERIOD is re-initialized to the minimum value: PLOW.

```
          LD      A,(PERIOD)    ; assess the duration of the half-period
          CP      PHIGH         ; check if it has reached the maximum
          JP      NZ,INCP       ; jump if it is not
          LD      A,PLOW        ; otherwise restart the count from the
          LD      (PERIOD),A    ; minimum half-period
          JP      EXIT
INCP:     INC     A             ; increment the half-period by one
          LD      (PERIOD),A    ; save back the new value
```

Finally, we exit the handler and return to the interrupted program.

```
EXIT:     POP    AF             ; restore the saved registers
          EI                    ; re-enable interrupts
          RET                   ; return to the interrupted program
```

5.6.3.2 Implementation on FPGA

The following sections have figures that summarize the connections chosen for each board[37] used to test the system.

---

[37] For more information on the associations, see the online content for this section.

The DE2 board

The following figure shows the devices chosen for the DE2 board.

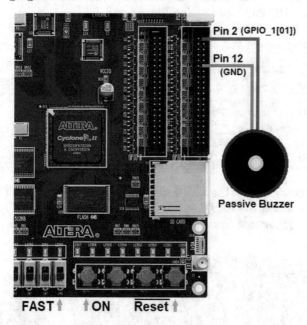

The DE0CV board

The assignment of devices for the DE0-CV board is very similar to that set for the DE2.

The EP2C5 board

The connections on the EP2C5 board require the connection of an external push-button (ON) and switch (FAST).

We will use the push-button already on the board for reset. The table below shows the summary of the connections as generated by Deeds.

In/Out Name	VHDL Name	Board Resource	Aux. Info
Inputs:			
Ck of ("Soun	—	Clock: 10 MHz	
!Reset	inReset	Push-Button: Key0	Low (if pressed)
FAST	iFAST	Header 0: P1 [IO_47]	
ON	iON	Header 0: P1 [IO_40]	
Outputs:			
!Int7	onInt7		
SOUND	oSOUND	Header 0: P1 [IO_71]	

Below is a photograph of the board with the physical connections to make for the ON button and the FAST switch. The connections of the $10\,\mathrm{K}\Omega$ pull up resistors are highlighted in red and orange.

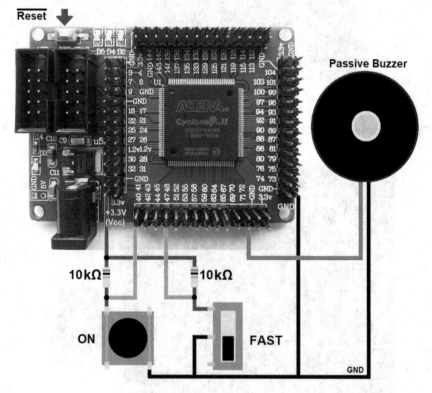

5.6.4 Music box

We need to build a music box that can cyclically play a short tune. The traditional music box is mechanical, box-shaped and decorated, and plays when the cover is lifted.

In our prototype, we will use the system reset button to simulate this behavior. We will make it work in the opposite sense from the usual. The reset will then be active with the push-button at rest. For us, "lift the cover" means pressing the reset button, and for this purpose, 'reset' should be called 'PLAY'.

For the transducer, we can use the same piezoelectric device described in Section 5.6.3. Then we can connect it directly to the circuit with no need for an amplifier.

A horn loudspeaker like the one photographed at the left is a potential alternative. It functions along the same principle.

Like most piezoelectric acoustic transducers, it can be directly connected to the logic circuit.

It is affordable, easy to repair, and produces a decidedly higher volume and better sound quality.

Among all the tunes that could be played, we chose a part of Bourrée in E minor (BWV 996) from the great composer J.S.Bach[38]. Below is the musical notation for the first 8 measures, which will repeat cyclically.

To make it possible to set some variations to the sound, we have added three switches in our system. Two of the switches (OCT1 and OCT0) allow us to "transpose" the execution an "octave".

From a Physics perspective, this means that the note's frequency f_N is multiplied by a certain factor in function of the setting of OCT1 and OCT0 (see the table below).

OCT1, OCT0	Frequency	Transposition
0 0	f_N	None
0 1	$f_N \cdot 2$	One octave above
1 0	$f_N \cdot 4$	Two octaves above
1 1	$f_N \cdot 8$	Three octaves above

---

[38] Johann Sebastian Bach (1685-1750) was a German composer and musician of the Baroque period. Originally composed for the lute, this song was covered in 1969 by Jethro Tull on the album "Stand Up".

The third switch (GLIDE) allows us to give a special touch to the tune by adding a "glide" effect between the notes. The glide between two consecutive notes consists in progressively raising or lowering the frequency from the first note to the second[39].

5.6.4.1 The system

The figure below shows the system, which uses a "DMC8 Enhanced Microcomputer" with an added timer (that interrupts every $50\,\mu S$) for the OCT0, OCT1 and GLIDE switches. The sound is generated on the output WAVE when the PLAY button is pressed, as described before.

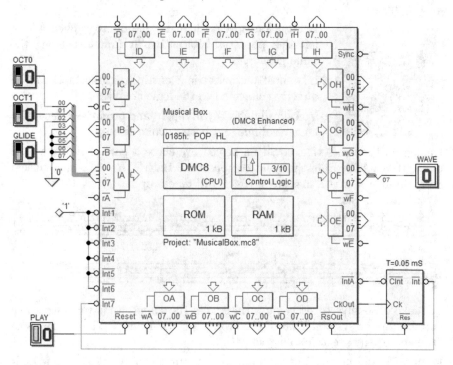

Generating notes

Before examining the assembly code, it is useful to describe the principle of operation behind the generation of notes. We'll approach the subject in steps, ignoring for the moment the specifications for the glide and transposing the octave.

Generating a note in our music box means producing a square wave signal, i.e., a two-level periodic wave on the WAVE output line.

---

[39] This can be done with the voice or various types of instruments (synthesizers, for example).

The notes are differentiated on the basis of their "pitch". In physical terms this means we simply need to control their frequency of oscillation[40].

Here, it makes sense to assess the half-period of the signal rather than the frequency because at the end of the half-period, we will have to invert the signal logical level on the output. We use a counter (COUNT), which is incremented by one every time the timer is called. When COUNT is equal to the half-period set in the CPERIOD variable, we invert the WAVE output and go back to counting from the top (see the following figure).

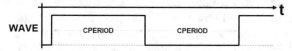

We should then define a table of constants, one for every note to generate (FTABLE). For a certain note, we consult the table and load the value of the desired half-period in the CPERIOD variable. The duration of the half-period T_H corresponding to the note's frequency f_N is:

$$T_H = \frac{1}{2 \cdot f_N}$$

Keeping in mind that here, time is marked by the interrupts (that come every $\Delta T = 50\,\mu S$), the number N_H to insert in the table for every note, is calculated by the following expression:

$$N_H = \frac{1}{2 \cdot \Delta T \cdot f_N}$$

Then we round off the result to the nearest integer number. To get a better approximation of the frequency generation, and thus an optimal pitch, we clearly need to reduce ΔT, but for our music box, the timer's $50\,\mu S$ are enough for an acceptable sound.

Reading the music

The sequence of notes to generate is read on the "musical score", which is just another table (MSCORE). MSCORE shows the codes of the notes that have to be executed one after the other.

---

[40] The frequency of notes is not universal but depends on the culture, the musical system and the historical period it is developed in. Just to cite a couple examples, within Europe, the Ancient Greeks tuned their instruments differently from those in the Baroque period, who in turn tuned them differently from the musicians of today. In modern "equal temperament" (introduced after the Baroque period), note frequencies are calculated starting from a reference point (the "central A note", generally defined as $440\,Hz$) and deriving the others by multiplying by a factor of $\sqrt[12]{2}$ (about 1.059). For example, the frequency of Bb, the note after A is $(440 \cdot \sqrt[12]{2}) \simeq 466.164$. Moving on to the note-by-note calculation, we do 12 multiplications and then get the frequency of A at the octave above, which is $880\,Hz$, simply double of the A at the beginning, since $\left(\sqrt[12]{2}\right)^{12} = 2$.

The codes to insert in the table were chosen according to the numbering defined by the MIDI standard[41], which identifies notes with numbers from 1 to 127. The lowest note we generate is fourth octave C, which corresponds to MIDI 60. Then, to index the FTABLE half-period table as of 0 (whose first line corresponds to our note), we need to subtract 60 from the code that we get by reading MSCORE.

The MSCORE table is read in a timed fashion; the new note is read when the previous one 's time elapses. The notes are divided into "eighths" i.e., one eighth of the time of a whole "measure". Let's look below at the figure of a part of the score. We will insert two identical codes (= 71) one after the other for note B, which must last for a quarter measure, then the codes for the notes A (= 69), G (= 67), and then two identical codes for F$_\#$ (= 66), for a quarter measure, and so on.

In the program, a one eighth note duration is assessed by counting the interrupts and is defined by the NOTETIME constant (the larger it is, the slower the execution and vice versa). When a note must last more than an eighth on the score, multiple identical codes are inserted consecutively in the table.

Octave transposition

Now let's add the specification for octave transposition. If the time interval between one interrupt and another were a lot shorter, we would be able to afford to extend the table of half-periods FTABLE.

Here though, if we included three higher octaves (36 notes), we would be forced to insert the values of the progressively smaller half-periods, which become ever more approximated in proportion. This would cause a drastic decline in pitch quality because the resulting frequencies would be too discordant from the nominal frequencies.

An acceptable solution from a musical perspective consists in a multiplication parameter (O_{ct} = 1, 2, 4 or 8) for the expression:

$$N_H = O_{ct} \cdot \frac{1}{2 \cdot \Delta T \cdot f_N} \ ,$$

so as to obtain higher frequencies with the same N_H in the denominator:

$$f_N = O_{ct} \cdot \frac{1}{2 \cdot \Delta T \cdot N_H} \ .$$

We can obtain this result by changing the way in which we calculate the half-period.

---

[41] Musical Instrument Digital Interface, https://www.midi.org/

Each time the interrupt handler is called, we continue to increment the counter (COUNT), but by the amount dictated by the parameter O_{ct} (OCTAVE, in the program). When COUNT is greater than or equal to the half-period set in the CPERIOD variable, we invert the WAVE output and instead of zeroing COUNT, we make it equal to the difference (COUNT - CPERIOD).

This means we bring COUNT back to zero after (CPERIOD $\cdot O_{ct}$) interrupt calls. This gives us a half-period only more or less what we want but that manages the pitch acceptably for our application.

The glide

Finally, we now add the glide between two consecutive notes. We need to take the CPERIOD variable (the current half-period used to calculate transition times) and place another PERIOD variable next to it, the "desired" half-period, that is, that of the next note.

Every time we take a new note from the MSCORE table, the corresponding half-period is calculated in PERIOD. If glide is enabled, rather than assigning the content of PERIOD directly to CPERIOD, we gradually increment (or decrement) CPERIOD in a time-controlled way until it reaches the value set in PERIOD. The resulting effect is that the frequency gradually shifts from one note to the next. As we will see, we can change the timing of this shift by changing the value of a constant.

The program

In the beginning, we define the input and output ports (PCTRL and PWAVE).

PCTRL	EQU	00h	; IA input port: control inputs
PWAVE	EQU	05h	; OF output port: square wave output

There are many variables so it is better to discuss them where they are used.

COUNT	EQU	0FC00h	; half-period time counter
PERIOD	EQU	0FC01h	; nominal duration of the half-period
CPERIOD	EQU	0FC02h	; current duration of the half-period
WAVE	EQU	0FC03h	; output's state
SINDEX	EQU	0FC04h	; index in the musical score
CNOTE	EQU	0FC05h	; current note in execution
OCTAVE	EQU	0FC06h	; octave transposition
TIME	EQU	0FC07h	; time count (16 bit)
GLITIME	EQU	0FC09h	; glide duration counter
GLIDEON	EQU	0FC0Ah	; glide mode On/Off flag

The NOTETIME constant determines the duration of an eighth note. The GLIDESET constant defines the duration of the glide between the notes (the shorter it is the less it is perceived).

NOTETIME	EQU	3700	; duration of an eighth note
GLIDESET	EQU	90	; duration of the glide

These constants can be changed as desired. What follow are the definitions of the jumps to the start of the program and the interrupt handler.

```
ORG     0000h
JP      START
ORG     0038h
JP      HINT
```

We allocate the FTABLE table in the ROM area that comes before the main program. As explained before, FTABLE contains the durations of the half-periods corresponding to each note in terms of units of time (50 μS, the interval defined by the timer).

The comments of each line have the names of the notes, their nominal frequencies and their MIDI codes.

```
         ORG     00C0h          ; Note frequency table
FTABLE:  DB      153            ; C4 = 261.626 Hz (MIDI: 60)
         DB      144            ; C#4 = 277.183 Hz (MIDI: 61)
         DB      136            ; D4 = 293.665 Hz (MIDI: 62)
         DB      129            ; Eb4 = 311.127 Hz (MIDI: 63)
         DB      121            ; E4 = 329.628 Hz (MIDI: 64)
         DB      115            ; F4 = 349.228 Hz (MIDI: 65)
         DB      108            ; F#4 = 369.994 Hz (MIDI: 66)
         DB      102            ; G4 = 391.995 Hz (MIDI: 67)
         DB      96             ; Ab4 = 415.305 Hz (MIDI: 68)
         DB      91             ; A4 = 440.000 Hz (MIDI: 69)
         DB      86             ; Bb4 = 466.164 Hz (MIDI: 70)
         DB      81             ; B4 = 493.883 Hz (MIDI: 71)
         DB      76             ; C5 = 523.251 Hz (MIDI: 72)
         DB      72             ; C#5 = 554.365 Hz (MIDI: 73)
         DB      68             ; D5 = 587.330 Hz (MIDI: 74)
         DB      64             ; Eb5 = 622.254 Hz (MIDI: 75)
         DB      61             ; E5 = 659.255 Hz (MIDI: 76)
         DB      57             ; F5 = 698.457 Hz (MIDI: 77)
         DB      54             ; F#5 = 739.989 Hz (MIDI: 78)
         DB      51             ; G5 = 783.991 Hz (MIDI: 79)
         DB      48             ; Ab5 = 830.609 Hz (MIDI: 80)
         DB      91             ; A5 = 880.000 Hz (MIDI: 81)
         DB      43             ; Bb5 = 932.328 Hz (MIDI: 82)
         DB      40             ; B5 = 987.767 Hz (MIDI: 83)
         DB      38             ; C6 =1046.502 Hz (MIDI: 84)
```

First of all, the main program initializes the Stack Pointer.

```
         ORG     0100h
START:   LD      SP,0FFFFh      ; initialize the Stack Pointer
```

Then it reads the first note found in the "score" (the MSCORE table, which we see at the end of the code), sets the execution and immediately after, calls the GPERIOD subprogram to take the corresponding half-period from the FTABLE table.

```
LD      A,(MSCORE)      ; get the first MIDI note from the score
LD      (CNOTE),A       ; and save it as current note
CALL    GPERIOD         ; get the corresponding half-period
LD      (CPERIOD),A     ; copy it to the current duration variable
LD      A,00h           ; zero the counter of the half-period
LD      (COUNT),A
```

We also zero output port PWAVE, its software copy WAVE, the read index for the notes from the MSCORE table and the octave transposition parameter (OCTAVE).

```
LD      (WAVE),A        ; zero the wave output port
OUT     (PWAVE),A
LD      (SINDEX),A      ; zero the read index for the notes and
LD      (OCTAVE),A      ; the octave transposition parameter
```

Before entering the main loop, we initialize the parameters for the glide and the metronome. After, we enable interrupts.

```
LD      A,GLIDESET      ; initialize the glide duration counter
LD      (GLITIME),A
LD      A,0             ; set glide mode OFF
LD      (GLIDEON),A
LD      HL,1            ; initialize the metronome glide counter
LD      (TIME),HL
EI                      ; enable interrupts
```

The main loop continually acquires the switches (this type of application doesn't necessarily require a debouncing check so it has been omitted for simplicity's sake).

Based on the state of the switches, it first defines the GLIDEON variable, which determines if the notes have been generated with or without a glide effect.

```
MAIN:      IN    A,(PCTRL)      ; read the input switches' state
           LD    C,A            ; and copy it to register C
           AND   00000100b      ; assess the glide control: glide mode
           LD    (GLIDEON),A    ; is ON if GLIDEON is not zero
```

Then, based on the configuration of the OCT1 and OCT0 switches, we load 1, 2, 4 or 8 in the OCTAVE variable.

```
LD      A,C             ; get the OCT1 and OCT0 bits from C,
AND     00000011b       ; encode them in a number ranging
INC     A               ; from 1 to 4 and copy it to register B
LD      B,A             ;                            (cont.)
```

```
                LD      A,00000001b     ; calculate (as power of 2) the octave
POWER:          DEC     B               ; transposition parameter
                JP      Z,SAVE          ; when finished, jump and save it
                SLA     A               ; multiply register A by two as many
                JP      POWER           ; times as specified by register B
SAVE:           LD      (OCTAVE),A      ; save the octave transposition parameter
                JP      MAIN            ; and repeat the main loop
```

The interrupt handler is called every $50\,\mu S$. As mentioned before, the shortest time possible was chosen to get the best approximation of the frequencies generated. We can verify, even with time simulation, that all the possible handler sequences are executed correctly within the time frame of $50\,\mu S$.

In order to understand the algorithms in the interrupt handler, it is important to note that it does different tasks all with the goal of generating sound.

These tasks produce results that are stored in variables to be then read and used by other modules, not necessarily written in subsequent order. In some cases, the results are not immediately used by a module executed right in the same interrupt call, but rather they are destined for a module that will be executed during the next interrupt.

At the start of the handler, we save the contents of the registers in use.

```
HINT:           PUSH  AF                ; save the registers in use
                PUSH  HL
                PUSH  BC
```

In the first part of the code, we check if a glide between notes is desired or not. Then we choose either the NOGLIDE, or the GLIDE subprogram. The NOGLIDE subprogram immediately assigns the half-period required by the current note for the generator. GLIDE, on the other hand, calculates the progressive approach from the current half-period to the half-period of the required note, little by little (as we will see further on).

```
                LD      A,(GLIDEON)     ; check if glide is enabled
                OR      A
                CALL    Z,NOGLIDE       ; OFF: assign directly the half-period
                CALL    NZ,GLIDE        ; ON : approach the desired half-period
```

The function IsBEAT tells us if it is time to take the next note from the score (it assesses if the time of one eighth note has elapsed, but the details will be addressed in the following pages).

```
                CALL    IsBEAT          ; is it time to read a new note?
                JP      NZ,NONEW        ; jump if it is not, otherwise
                CALL    NEXTNOTE        ; get the next note code and convert
                CALL    GPERIOD         ; it in the corresponding half-period
```

If it is time, we read the next note on the score and retrieve the corresponding half-period (this is an example of parameters that are not used immediately but will be used at the next call).

Whether a new note is taken or not, we check if the current code is a "rest note" (00h). If it is, we exit the handler because we do not generate transitions on the output during a rest note.

```
NONEW:    LD    A,(CNOTE)    ; check the current note code,
          OR    A            ; if we are during a rest note, exit
          JP    Z,EXIT
```

If it is not a rest note, we continue; we call the subprogram that generates the transitions of the output square wave (described a bit further on).

```
          CALL  GENERATE     ; call the output transition generator
```

Then the handler restores the content of the registers, re-enables the interrupts and goes back to the interrupted program.

```
EXIT:     POP   BC           ; restore the saved registers
          POP   HL
          POP   AF
          EI                 ; re-enable interrupts
          RET                ; return to the interrupted program
```

Now let's look at the subprograms that the interrupt handler calls.

The IsBEAT function counts time and if an eighth note has gone by, it gives authorization to take the next note to the calling program.

Specifically, it decrements the TIME variable (every $50\,\mu S$), and exits if the count is not zeroed. If it is zeroed, it reloads the NOTETIME constant[42] in the TIME counter and exits with the zero flag active.

```
IsBEAT:   LD    HL,(TIME)    ; assess the increment of time
          DEC   HL           ; (by steps of 50 microseconds)
          LD    (TIME),HL
          LD    A,H
          OR    L            ; exit if it's not time to read the next note
          RET   NZ           ; on the score, otherwise
          LD    HL,NOTETIME  ; the time of an eight note has passed,
          LD    (TIME),HL    ; re-initialize the time counter
          RET                ; and return to the calling program
```

The NOGLIDE subprogram is called when we need to execute notes without glides. It simply copies the PERIOD half-period, (which had been retrieved before from the FTABLE table), directly to the CPERIOD variable, which will be used by the generator.

```
NOGLIDE:  LD    A,(PERIOD)   ; use directly the requested
          LD    (CPERIOD),A  ; half-period value, without gliding
          RET
```

---

[42] To execute the piece more quickly we need to reduce the NOTETIME constant.

The GLIDE subprogram produces the glide by gradually moving the content of CPERIOD closer and closer to the desired half-period contained in PERIOD. As we have seen before, the note frequency will take a certain amount of time to move from one to the next so that the transition is clearly audible.

When we enter GLIDE, we immediately decrement the GLITIME ('glide time') variable, but if it is not zeroed yet, we simply leave the function.

```
GLIDE:      LD    A,(GLITIME)    ; assess the glide time
            DEC   A              ; decrementing the variable GLITIME
            LD    (GLITIME),A    ; exit if it is not time to modify the
            RET   NZ             ; half-period of the generated note
```

If it has been zeroed, we refresh the count by loading the GLIDESET constant in the GLITIME[43] variable.

```
            LD    A,GLIDESET     ; otherwise, re-initialize GLITIME
            LD    (GLITIME),A    ; to be able to restart the count
```

We check to see if the contents of the two variables are already equal, and if they are, we exit (the goal has been reached).

```
            LD    A,(CPERIOD)    ; copy the current half-period in B
            LD    B,A
            LD    A,(PERIOD)     ; get the requested half-period in A
            CP    B              ; compare the two values
            RET   Z              ; exit if they have become equal
```

Otherwise, the current half-period CPERIOD is still different from the requested PERIOD. So, we decide to increment or decrement CPERIOD by one based on the Carry flag in order to move the value closer to PERIOD.

```
            LD    A,B            ; move CPERIOD in A
            JP    C,GLIDEDN      ; jump if PERIOD < CPERIOD
GLIDEUP:    INC   A              ; otherwise, increment CPERIOD
            LD    (CPERIOD),A
            RET
GLIDEDN:    DEC   A              ; decrement CPERIOD
            LD    (CPERIOD),A
            RET
```

Now, let's examine the code of GENERATE, which inverts the output when it is time to do so. It increments COUNT by the amount in the OCTAVE variable (remember that it can contain 1, 2, 4 or 8). If the number in COUNT is not larger than CPERIOD, it is not time to invert the output yet.

```
GENERATE:   LD    A,(CPERIOD)    ; get the current note half-period
            LD    B,A            ; and copy it to B
            LD    A,(COUNT)      ; then, copy the time counter to C
            LD    C,A                                          (segue)
```

---

[43] We can change the entity of the glide by redefining the GLIDESET constant.

```
            LD      A,(OCTAVE)    ; get the octave transposition (1, 2, 4, 8)
            ADD     A,C           ; add it to the time counter:
            LD      C,A           ; COUNT ← COUNT + OCTAVE
            CP      B             ; the count is larger than CPERIOD?
            JP      NC,INVERT     ; jump to INVERT if it is, otherwise
            LD      (COUNT),A     ; save back the COUNT variable
            RET                   ; and leave the function
```

If COUNT is larger than CPERIOD, we jump to INVERT, complement the state of the output and save a copy of that in the WAVE variable. Before that, however, we refresh the COUNT variable by subtracting CPERIOD from it (as described before).

```
INVERT:     SUB     B             ; re-initialize the time counter to
            LD      (COUNT),A     ; the difference (COUNT - CPERIOD)

            LD      A,(WAVE)      ; do a transition on the output WAVE,
            XOR     10000000b     ; inverting the MSB of the software copy
            OUT     (PWAVE),A     ; and coping the new value to the port
            LD      (WAVE),A      ; save back the new port state
            RET
```

Now consider the reading of the musical score, that is, the MSCORE table (shown further on). In the MSCORE table we find:

— The codes for the *note* (MIDI, limited here to the interval of 60..84).
— A code for *rest note* (00h), which is not used in the song chosen here.
— A code for *refrain* (80h), which makes everything start from the top[44].

The table is read by using the SINDEX index, which was zeroed at the start. The index is added to the base address of the table. We obtain an address in the HL register that we use to take the code of the note from the table. The code is saved in the CNOTE variable.

```
NEXTNOTE:   LD      HL,MSCORE     ; copy the address of the table in HL
            LD      A,(SINDEX)    ; get the index of the note code to read
            ADD     A,L           ; add it to the table base address,
            LD      L,A           ; to obtain the address of the note code
            JP      NC,GETCODE
            INC     H             ; (handle the carry, if any)
GETCODE:    LD      A,(HL)        ; get the note code from the table
            LD      (CNOTE),A     ; copy it to the CNOTE variable
```

Before incrementing the SINDEX index (in order to handle taking the next note from the table), we check to see if we have read the code for refrain.

```
            CP      80h           ; check if it is a code for refrain
            JP      NZ,NEXTIND    ; jump if it is not
```

---

[44] This is only in our simplified project. In real musical notation, "refrain" signs would only repeat a part of the song not everything from the top.

If we have read a code for refrain, we have gotten to the end of the music described by the score so we zero the index and jump back to NEXTNOTE, then take the first note in the table and go back to the calling subprogram.

```
        LD      A,0             ; this is a refrain, zero the index
        LD      (SINDEX),A
        JP      NEXTNOTE        ; and jump back to get the first code
```

If we have not read it, we increment the read index of the table and exit.

```
NEXTIND:  LD    A,(SINDEX)      ; increment the index for the next time
          INC   A
          LD    (SINDEX),A
          RET                   ; and exit
```

The last subprogram to study is GPERIOD, which translates the code of the note into the value of the corresponding half-period. In GPERIOD, the initial check is to confirm that the code taken is for the rest note, in which case we exit (without producing a value for the half-period).

```
GPERIOD:  LD    A,(CNOTE)       ; get the current code and
          OR    A               ; check if it is a code for rest note
          RET   Z               ; exit if it is (no matter the return value)
```

Then we check just to be sure that the code of the note is within the interval that we handle and if it is not, we exit.

```
        SUB     60              ; calculate the index from the MIDI code
        RET     C               ; exit if the index is < 0 (not valid)
        CP      25              ; (A ≥ 25)? Cy = 1 if the index is < 25
        RET     NC              ; exit if ≥ 25 (not valid)
```

We continue if the index is valid (00..24). Then we add it to the base address of the table, to take the desired half-period.

```
        LD      HL,FTABLE       ; get the table base address in HL
        ADD     A,L             ; add the index to that address
        LD      L,A
        LD      A,(HL)          ; get the period
        LD      (PERIOD),A
        RET
```

The following is the entire musical score for the song, which has been transcribed note by note in the MSCORE table.

```
MSCORE:  DB  76     ; E5      (measure 0)
         DB  78     ; F#5

         DB  79     ; G5      (measure 1)
         DB  79     ; G5
         DB  78     ; F#5
         DB  76     ; E5
         DB  75     ; Eb5
         DB  75     ; Eb5
         DB  76     ; E5
         DB  78     ; F#5
```

```
DB  71        ; B4      (measure 2)
DB  71        ; B4
DB  73        ; C#5
DB  75        ; Eb5
DB  76        ; E5
DB  76        ; E5
DB  74        ; D5
DB  72        ; C5
```

```
DB  71        ; B4      (measure 3)
DB  71        ; B4
DB  69        ; A4
DB  67        ; G4
DB  66        ; F#4
DB  66        ; F#4
DB  67        ; G4
DB  69        ; A4
```

```
DB  71        ; B4      (measure 4)
DB  69        ; A4
DB  67        ; G4
DB  66        ; F#4
DB  64        ; E4
DB  64        ; E4
DB  76        ; E5
DB  78        ; F#5
```

```
DB  79        ; G5      (measure 5)
DB  79        ; G5
DB  78        ; F#5
DB  76        ; E5
DB  75        ; Eb5
DB  75        ; Eb5
DB  76        ; E5
DB  78        ; F#5
```

```
DB  71        ; B4      (measure 6)
DB  71        ; B4
DB  73        ; C#5
DB  75        ; Eb5
DB  76        ; E5
DB  76        ; E5
DB  74        ; D5
DB  72        ; C5
```

```
DB  71        ; B4      (measure 7)
DB  71        ; B4
DB  69        ; A4
DB  67        ; G4
DB  66        ; F#4
DB  66        ; F#4
DB  66        ; F#4
DB  69        ; A4
```

```
DB  67      ; G4      (measure 8)
DB  67      ; G4
DB  67      ; G4
DB  67      ; G4
DB  67      ; G4
DB  67      ; G4

DB  80h     ; code for refrain
```

Note: the last two notes (F#5 and G5) are not in the table because they are already at the beginning of the song and will be played when we go back to the top.

5.6.4.2 Implementation on FPGA

The following sections have figures that summarize the connections chosen for each board[45], used to test the system.

The DE2 board

The following figure shows the devices chosen for the DE2 board.

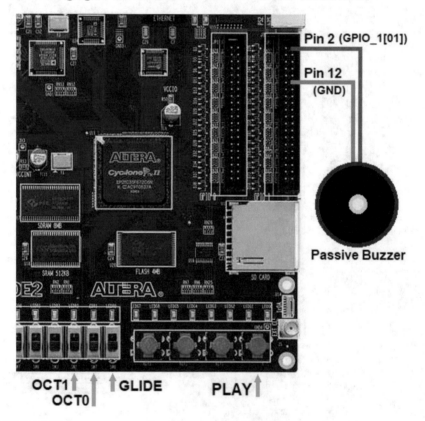

---

[45] For more information on the associations, see the online content for this section.

The DE0CV board

The assignment of devices for the DE0-CV board is very similar to that set for the DE2.

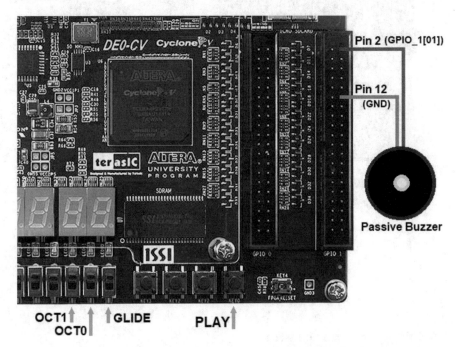

OCT1 OCT0 GLIDE PLAY

The EP2C5 board

The connections on the EP2C5 board require the external connection of three switches (OCT1, OCT0 and GLIDE). For the PLAY command, however, we will use the one push-button on the board. The table below shows the summary of the connections as generated by Deeds.

In/Out Name	VHDL Name	Board Resource	Aux. Info
Inputs:			
Ck of ("Musi	—	Clock: 10 MHz	
!Reset	inReset		
OCT0	iOCT0	Header 0: P1 [IO_57]	
GLIDE	iGLIDE	Header 0: P1 [IO_57]	
PLAY	iPLAY	Push-Button: Key0	High (if pressed)
Outputs:			
!Int7	onInt7		
WAVE	oWAVE	Header 0: P1 [IO_71]	

Below is a photograph of the board with the physical connections to make for switches OCT1, OCT0 and GLIDE. The connections of the three $10\,K\Omega$ pull up resistors are highlighted in red and orange.

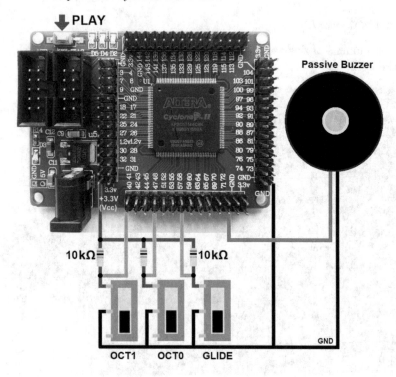

5.6.5 Stepper motor control

In this example, we will build a microprocessor-based system that continuously rotates a "stepper motor". After a brief introduction to stepper motor operations, we will analyze the system and implement it on FPGA.

The stepper motor

Here, we discuss the 28BYJ-48 component[46], shown in the figure on the right. It is easy to find and economical.

Stepper motors are electro-mechanical components that can execute small rotations of a predefined angle on command.

To move the motor we need electronic circuits whose operating principle goes beyond the scope of this book. We will use a small electronic "power" driver board as an interface between the motor and the output logic lines of the FPGA board we will use.

---

[46] https://datasheetspdf.com/pdf-file/1006817/Kiatronics/28BYJ-48/1
https://lastminuteengineers.com/28byj48-stepper-motor-arduino-tutorial

On the right, we have a photo of the electronic interface circuit that we will use.

Its job is to allow for driving the lines of the motor, given that they require higher current and tension than those available on the typical outputs of an FPGA component.

Further on, the instructions to connect this board between the chosen FPGA and the lines of the motor will be explained.

The operating principle

On these pages, we will show an abstract model of the motor and focus on the logical and operative aspects. The physical and electro-mechanical aspects will not be dealt with here. We use an arrow to represent the position of the angle of the motor shaft at a given time (see the following figure).

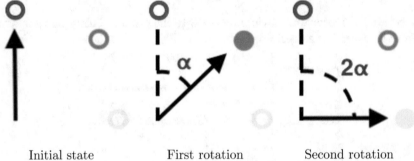

| Initial state | First rotation | Second rotation |

In the initial state the arrow is pointing up. There is an electromagnet to the right of the arrow[47] (pink circle) that attracts the arrow when activated. Once the electromagnet is activated, a certain amount of time elapses and the arrow completes the rotation of a certain angle α, whose value depends on the physical parameters of the motor (in the figure, it is represented at 45° just for clarity's sake). Until the electromagnet is active, the arrow will continue to point at it.

Now we activate another electromagnet (yellow circle in the figure) and make sure to deactivate the first one. As before, we will make the motor rotate again by attracting the arrow, giving us a total rotation of $2 \cdot \alpha$.

If we continue to activate the electromagnets along the circumference we can execute a full stepwise rotation of the motor. We proceed as before: each time, we activate the next electromagnet and deactivate the previous one. Clearly, if we invert the activation order of the electromagnets, we make the motor rotate in the opposite direction.

---

[47] An electromagnet produces an electric field "on command", with the passage of an electric current.

The stepper motor: an abstract model

Now let's look at the following figure, an abstract model of the 28BYJ-48 stepper motor. We see 32 electromagnets grouped by color. Four inputs (active high) are available to the software programmer. Each input is connected to a different group of electromagnets (the four groups are represented by the colors blue, pink, yellow and orange).

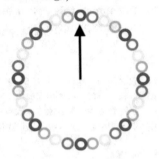

To rotate the motor we individually activate the groups of electromagnets one after the other in the desired direction.

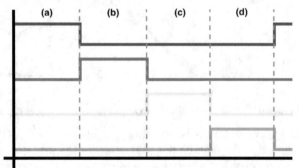

When we activate the blue, pink, yellow and orange electromagnets in sequence, the arrow rotates clockwise. When we change the sequence to orange, yellow, pink and blue, the arrow rotates counter-clockwise.

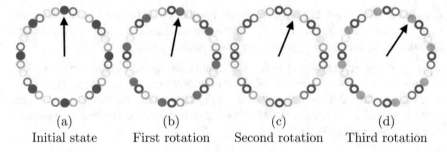

| (a) | (b) | (c) | (d) |
| Initial state | First rotation | Second rotation | Third rotation |

In the operating mode described, we see that each step corresponds to a rotation of 11.25 degrees (360/32) of the motor tree.

The component has a mechanical reduction gear set (as sketched in the figure on the right).

The gear on the motor shaft is shown in green. The gear set reduces the rotation by a ratio of about 1:64.

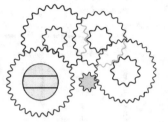

Each time a new electromagnet is activated in the sequence described, the external axis (shown in yellow) executes a rotation of 0.18 degrees downstream of the gear ratio.

Three different stepper motor control modes

We have already looked at the simplest control mode for the stepper motor. There are (at least) two other ways to drive this component.

One makes it possible to rotate the arrow by an angle half the standard step (that is 0.09, taking into account the gear set) degrees, by activating not only the electromagnet at the right of the arrow but also the one on the left. This way, the arrow will be attracted between the two electromagnets, passing from (a) to (b).

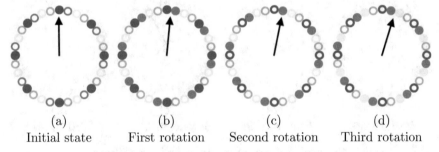

(a)	(b)	(c)	(d)
Initial state	First rotation	Second rotation	Third rotation

Then by deactivating the one on the left and leaving the one on the right active (c) we can execute a second 0.09 degree rotation. This way we get a complete step rotation, that is 0.18 degrees and we have doubled the motor's precision without changing the mechanics.

By continuing as shown in (d), (e), (f), (g), (h) and so on, we can make a whole round angle by 0.09 degree steps.

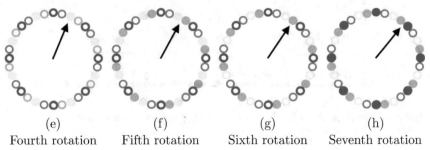

(e)	(f)	(g)	(h)
Fourth rotation	Fifth rotation	Sixth rotation	Seventh rotation

So, the sequence to activate for a clockwise rotation is the following:

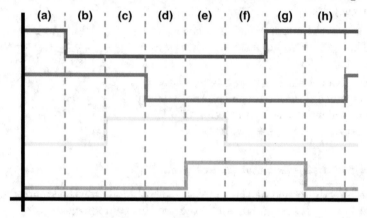

The other driving mode allows us to raise the angular momentum generated by the motor by activating not just the electromagnet closest to the arrow but also the next one, as shown in the following figure. It increases the strength of the attraction on the arrow.

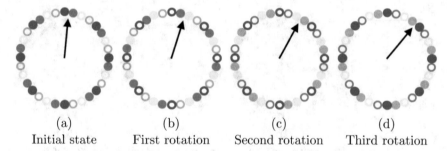

(a)	(b)	(c)	(d)
Initial state	First rotation	Second rotation	Third rotation

What follows is the timing diagram for a clockwise rotation.

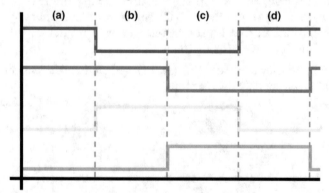

This mode will be used in the following programming example.

5.6.5.1 The system

The control system for the stepper motor is based on the "DMC8 Microcomputer" component (see the following figure). Pin 0 of port IA is connected to switch DIR, which can control the motor's direction of rotation. If it is set at '1' it is counter-clockwise but clockwise if it is at '0'.

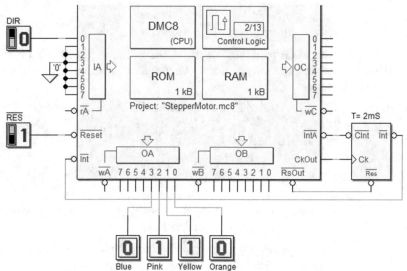

Output port OA drives the motor. Specifically, the blue, pink, yellow and orange control wires are driven by port bits 3, 2, 1 and 0, respectively. A timer activates interrupt requests $\overline{\text{Int}}$ every $2\,mS$. When the request is accepted a pulse on output $\overline{\text{IntA}}$ automatically deactivates line $\overline{\text{Int}}$.

Once it has executed the necessary initializations after system reset, the main program enters an infinite loop where it reads input port IA and memorizes the value in the DIR variable, which is read by the interrupt handler.

The interrupt handler updates the value of the blue, pink, yellow and orange commands in line with what is memorized in the DIR variable, which is updated by the main program. For every call, it must show a new value on output port OA for the sequence of commands needed to rotate the motor in the direction required by the DIR input.

The program

We declare the input port (CNTRL) and output port (MOTOR).

```
CNTRL      EQU    00h              ; IA input port: direction (bit 0)
MOTOR      EQU    00h              ; OA output port: motor control
```

We define the DIR and ANGLE variables (each one byte sized). The DIR variable stores the direction of rotation required. The use of ANGLE will be explained further on.

```
DIR          EQU    0FC00h        ; direction of rotation
ANGLE        EQU    0FC01h        ; index in the SEQUENCE table
```

We insert the jumps to the start of the program and the interrupt handler.

```
             ORG    0000h
             JP     START
             ORG    0038h
             JP     HINT
             ORG    0100h
```

At the start of the main program, as usual, we initialize the Stack Pointer, the variables used and the output ports. Then before entering the main loop, we enable interrupts.

```
START:       LD     SP,0FFFFh      ; initialize the Stack Pointer
             LD     A,00h          ; zero the variables DIR and ANGLE
             LD     (DIR),A
             LD     (ANGLE),A
             OUT    (MOTOR),A      ; and the output port MOTOR
             EI
```

As mentioned previously, the main loop does nothing more than update the required direction of rotation in the variable DIR.

```
MAIN:        IN     A,(CNTRL)      ; read the input port
             AND    00000001b      ; mask the unused bits
             LD     (DIR),A        ; update the rotation direction
             JP     MAIN
```

The interrupt handler HINT, which is called every $2\,mS$, updates the control lines of the motor.

The program saves the registers in use on the Stack and then reads the required direction of rotation from the DIR variable and decides whether to increment or decrement an index (ANGLE). As we will see in more detail further on, by incrementing the ANGLE index, we get a clockwise rotation and by decrementing it, we get a counter-clockwise rotation. The count is made cyclical on two bits (so that the value of ANGLE can assume only values from 0 to 3), and the resulting value is saved back in ANGLE.

```
HINT:        PUSH   AF
             LD     A,(DIR)        ; get the rotation direction
             OR     00000000b      ; modify the flags
             LD     A,(ANGLE)
             JP     Z,RIGHT        ; jump, or not, according to the direction
LEFT:        DEC    A              ; counter-clockwise rotation
             JP     UPDATE
RIGHT:       INC    A              ; clockwise rotation
UPDATE:      AND    00000011b      ; make the count cyclical (on two bits)
             LD     (ANGLE),A
```

The CONTROL subprogram, analyzed further on, returns to the accumulator the configuration of commands that are needed to rotate the motor, given the ANGLE index in the accumulator. The configuration we get is then transferred to the MOTOR output port. Then we retrieve the saved registers and the handler re-enables the interrupts and returns the control to the interrupted program.

```
CALL    CONTROL         ; get the motor commands from the table
OUT     (MOTOR),A       ; write the commands to the driver
POP     AF
EI                      ; re-enable interrupts
RET                     ; return to the interrupted program
```

What follows is the SEQUENCE table, which contains the sequence of commands needed to rotate the stepper motor clockwise, by driving it in the mode that generates the greatest angular momentum. If the table is read in reverse, it can also rotate the motor counter-clockwise.

```
SEQUENCE:   DB    00001100b    ; active commands: blue, pink
            DB    00000110b    ; active commands: pink, yellow
            DB    00000011b    ; active commands: yellow, orange
            DB    00001001b    ; active commands: orange, blue
```

The CONTROL subprogram reads the table. It takes an index passed through register A and returns the desired value back into the same register.

```
CONTROL:    PUSH  HL
            LD    HL,SEQUENCE ; add the index in register A
            ADD   A,L         ; to the table base address
            LD    L,A
            JP    NC,NOCARRY
            INC   H
NOCARRY:    LD    A,(HL)      ; get the desired value in A
            POP   HL
            RET
```

5.6.5.2 Implementation on FPGA

The following subsections have figures that summarize the connections chosen for each board, which are useful for testing the system.

The DE2 board

The following figures show the devices chosen for the DE2 board, both in table form (as generated by Deeds), and in terms of physical connections.

In/Out Name	VHDL Name	Board Resource	Aux. Info
Inputs:			
Ck of /"Micro	—	Clock: 10 MHz	
!RES	inRES	Push-Button: Key[00]	Low (if pressed)
DIR	iDIR	Switch: Sw[00]	
Outputs:			
Orange	oOrange	Header 1: GPIO_1[11]	
Yellow	oYellow	Header 1: GPIO_1[13]	
Blue	oBlue	Header 1: GPIO_1[17]	
Pink	oPink	Header 1: GPIO_1[15]	

Take care to also connect the power supply wires (+5V and GND) of the motor control board, as shown in the following figure.

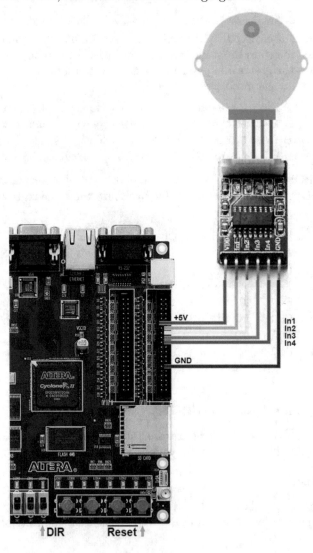

The DE0CV board

The assignment of devices for the DE0-CV board is very similar to that set for the DE2.

In/Out Name	VHDL Name	Board Resource	Aux. Info
Inputs:			
Ck of ("Micrd	—	Clock: 10 MHz	
!RES	inRES	Push-Button: Key[00]	Low (if pressed)
DIR	iDIR	Switch: Sw[00]	
Outputs:			
Orange	oOrange	Header 1: GPIO_1[11]	
Yellow	oYellow	Header 1: GPIO_1[13]	
Blue	oBlue	Header 1: GPIO_1[17]	
Pink	oPink	Header 1: GPIO_1[15]	

Likewise, pay close attention when connecting the motor control power supply (+5V e GND) as shown in the figure.

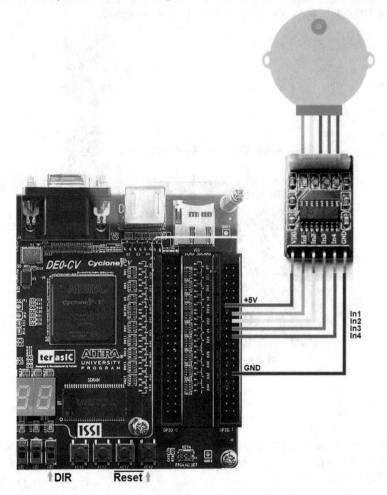

The EP2C5 board

The connections on the EP2C5 board require the connection of an external slide switch to manage the direction (DIR). We will use the only push-button on the board for manual system reset. The following table shows a summary of the connections as generated by Deeds.

In/Out Name	VHDL Name	Board Resource	Aux. Info
Inputs:			
Ck of ("Micro	—	Clock: 10 MHz	
!RES	inRES	Push-Button: Key0	Low (if pressed)
DIR	iDIR	Header 0: P1 [IO_47]	
Outputs:			
Orange	oOrange	Header 1: P2 [IO_104]	
Yellow	oYellow	Header 1: P2 [IO_101]	
Blue	oBlue	Header 1: P2 [IO_96]	
Pink	oPink	Header 1: P2 [IO_99]	

What follows is the photo of a board with the physical connections to make for the DIR slide switch and the $10K\Omega$ pull up resistors.

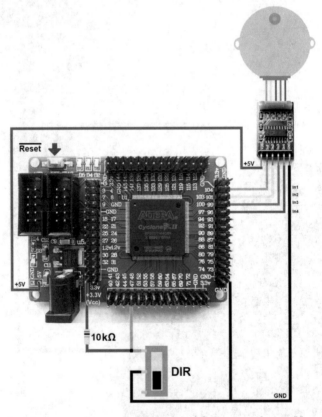

The motor's power supply can be connected by using the +5V on the board next to the power supply connector (as shown in the figure) even if there is no soldered pin.

5.6.6 Using a liquid crystal display (LCD)

In this sample project, we will first introduce the interface of the LCD (Liquid Crystal Display) that was mounted on a "vintage" Nokia 5110 mobile phone[48] and many others. Then we will analyze and build a microprocessing system that can show the classic "Hello World" message on this display.

A brief introduction to graphic displays

Any single-color display like the one here, can be understood as an ordered collection of small point sources of light that can be controlled electronically. When they are lit they have only one color depending on their physical characteristics. They are commonly called "pixels" (from "PIcture - ELement"), in that they represent the part of the image that is the smallest and cannot be divided.

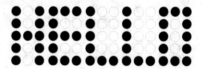

By turning some pixels on and others off, we can create any type of visual information (text, drawings, graphs, sheet music, photographs, etc.).

The term pixel does not only refer to single-color images; they can also be multi-colored.

If we put three small lights in red, green and blue (RGB) in the space for one single-color pixel, we get a tri-color display. By regulating the intensity of the three lights, we can represent most of the visible color spectrum.

This is the color model currently used for screens. There are others, however, such as the RGBY which adds yellow to the red, green, blue combination, and the CMYK based on cyan, magenta, yellow and black, which is commonly used in printers.

The lights discussed here can be produced with different technologies like LED, plasma and others. The differences come in the physical characteristics of the displays, which have an effect on the range of colors displayed, the luminosity, energy consumption, durability, resistance to mechanical stress... the basic features of an electronic system.

Aside from production technology, the screens differ in the density of pixels per inch (PPI), the higher it is, the less we see the outline of the pixels. Above 300 PPI we cannot see the outline of the pixels at a distance of 10-12 cm. Another difference that is important from the commercial perspective is the width and length of the pixel array.

---

[48] https://it.wikipedia.org/wiki/Nokia_5110

The Nokia 5110 display

The display used in the Nokia 5110 is single-color and made up of an array of 48 × 84 liquid crystals for a total of 4032 pixels (the component is shown at the right).

The pixel array is managed by an electronic circuit inside the display, designed specifically by the manufacturer to handle all the physical details autonomously. This allows the system designer to deal only with some settings and what is shown on the display.

The electronic circuit in question is an integrated component called the PCD8544[49] made by Philips in the 1990s. It offers programmers a serial interface for handling internal parameters and for sending the content to show on the display. Before analyzing the communication interface in detail, we will discuss the structure of the chip's internal memory.

The structure of the PCD8544 chip's internal memory

Inside the PCD8544 component, we have an 8-bit, 48 x 84 location RAM memory that contains the program for the image on the screen. The figure at the top of the opposite page shows the relation between the position of the RAM's flip-flops and that of the liquid crystals.

Up front, we see the array of liquid crystals. In the background, we see the memory divided into 6 banks (numbered 0 to 5). Each bank corresponds to an 84 pixel-wide, 8 pixel-high strip (in total, 6 strips × 8 pixels = 48 vertical pixels).

As an example, on the left of the figure, the flip-flops that make up the first location in every bank of the RAM are highlighted. An arrow shows the relation between the position of the first flip-flop of the memory location and the corresponding liquid crystal (a pixel).

Each flip-flop addresses a single liquid crystal. A logical '1' activates the liquid crystal to black. A '0' turns the crystal off. This allows the background color to pass through.

When the system is powered, the RAM contains random values and then needs to be initialized. To write the values in the RAM memory, the chip uses the data received on the serial line. Every byte that is sent writes 8 flip-flops, i.e., one RAM memory location.

The second figure on the same page shows the locations that make up the RAM memory.

---

[49] Datasheet PCD8544: https://www.sparkfun.com/datasheets/LCD/Monochrome/Nokia5110.pdf

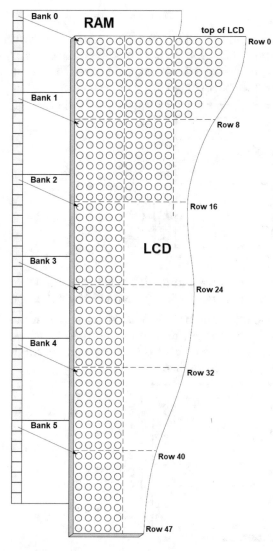

As shown in the figure at the bottom of this page, the RAM is organized into two dimensions (the X axis and Y axis).

The figure below also highlights the order in which the bits are organized in an individual location.

We start with the least significant bit, which is stored in the highest flip-flop and keep going down until we get to the most significant bit at the lowest flip-flop.

Consider again the figure at the left showing the relation of the flip-flop position to the corresponding liquid crystal. We see how the least significant bit of the first RAM location of Bank 0 drives the first liquid crystal of row 0. Continuing down we get to the most significant bit of the first RAM location of Bank 5, which drives the first crystal of row 47.

Two internal chip registers store the x and y coordinates of the memory cell currently in use. At system reset, they are both initialized to zero.

These registers can be set manually through the serial interface, although normally they are handled automatically by the chip according to the addressing method that's been chosen.

Addressing methods

There are two methods for addressing the RAM: "row addressing" and "column addressing". In row addressing (this will be used in our example and is shown in the following figure), the memory locations are addressed row by row from left to right, as we are used to writing with pen and paper.

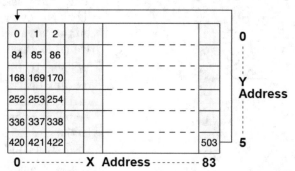

After each write operation, the registers are automatically incremented. Once the last row is written, the process restarts cyclically from the beginning.

For column addressing (see the following figure), first, all six locations in the first column (having index zero) are written from top to bottom. Then we increment the column and repeat the process. Once the last column is written the process restarts cyclically from the beginning.

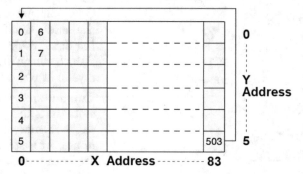

Now that we have analyzed the structure of the chip's internal memory, we can see how to interact with it.

Communication interface

The PCD8544 integrated circuit offers the designer a programming interface for the RAM memory based on five connections: SDIN, SCLK, D/$\overline{\text{C}}$, $\overline{\text{SCE}}$ and $\overline{\text{RES}}$. The interface provided requires serial synchronous communication with 8-bit words where the most significant bit is sent first (see the following figure).

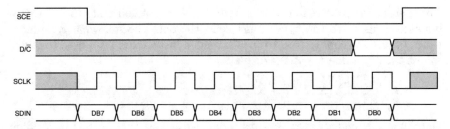

The five lines control the serial data (SDIN), the communication clock (SCLK) and the format of the data (D/$\overline{\text{C}}$), the chip enable ($\overline{\text{SCE}}$) and the device reset ($\overline{\text{RES}}$).

Both the control words that act on the internal settings and the data to show on the display can be sent to the chip through the interface. The use of the D/$\overline{\text{C}}$ line allows us to make this distinction.

To start communication with the chip, we need to activate line $\overline{\text{SCE}}$. This enables the action of the SCLK line, the serial communication clock.

The chip acquires the logic value presented on line SDIN at every rising edge of line SCLK. A data packet sent on this serial line is made up of 8 bits, starting from the most significant. The maximum frequency of the communication clock is 4 MHz.

The D/$\overline{\text{C}}$ line defines the content of the packet. If D/$\overline{\text{C}}$ is set at '0' while the last bit of the packet (the least significant) is sent, the word will be interpreted as a command. Otherwise the word will be interpreted as information to write on the display.

Since this explanation is introductory, we have chosen not to show the complete set of chip commands. This section only contains the basic commands[50]. Therefore, we will continue with a presentation of a brief display initialization sequence and then with useful instructions for writing the display contents into its RAM memory.

Initializing the display

This section will present a short series of instructions to make the display operative. As explained, we have left some instructions aside, such as those for handling some physical details. Yet, these are not essential for the display to function in closed environments with temperature control.

After we reset the system and activate the $\overline{\text{RES}}$ line for a short time, we can begin to interact with the display.

The first thing to do is set the contrast[51] of the display. If we omit this step, the information displayed might not be visible to the human eye.

---

[50] For a complete analysis, please refer to the PCD8544 datasheet cited previously.

[51] The contrast is regulated through the tension the liquid crystals work on.

This parameter can assume values from 0 (minimum contrast) to 127 (maximum contrast). To set it, we must send the chip the following command, which orders the chip to pass to the "extended instruction set", used to define the physical parameters of the display.

D/$\overline{C}$	DB7	DB6	DB5	DB4	DB3	DB2	DB1	DB0
0	0	0	1	0	0	0	0	1

Once in the extended instruction set mode, we can set the contrast by sending the byte below:

D/$\overline{C}$	DB7	DB6	DB5	DB4	DB3	DB2	DB1	DB0
0	1	C6	C5	C4	C3	C2	C1	C0

The protocol requires the most significant bit at '1'. The remaining bits represent the selected contrast value. A good contrast value to set for closed rooms is 16 (C6..c0 = 0010000_2). Once the contrast is set, we can go back to the instructions' "normal mode" with the following byte.

D/$\overline{C}$	DB7	DB6	DB5	DB4	DB3	DB2	DB1	DB0
0	0	0	1	0	0	0	V	0

In this step, we can set the desired addressing mode on bit V (a '0' sets the mode for rows and a '1', the mode for columns). If we want to use the display to show text, it is best to use row addressing as suggested in the previous example. We can change the addressing mode at any time by using the same command.

Lastly, we set the normal mode for the display and then send the data.

D/$\overline{C}$	DB7	DB6	DB5	DB4	DB3	DB2	DB1	DB0
0	0	0	0	0	1	1	0	0

Now we can finally move forward and send the data to be shown.

Transmitting the data to the display

The system reset has zeroed registers X and Y, which point to the memory.

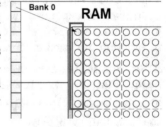

Therefore the currently selected RAM cell is the first one in Bank zero, which drives the vertical strip of pixels at the upper right (see the figure at the right). At every write operation, registers X and Y are incremented according to the chosen addressing method with no need for external intervention. So as we send more data bytes, contiguous locations will be written.

At the beginning, the RAM contains unknown values, so it is best to initialize it at a known value by sending 504 (84 × 6) data bytes. The automatic handling of write addresses makes it so that after 504 writes, the contents of register X and Y go back to the initial value (0 in this case).

To send the data, we need to select the "data mode" on the interface by setting the $\mathrm{D}/\overline{\mathrm{C}}$ line to '1'. At this point, we can start to send the bytes to the display, one by one. The following timing diagram shows the signals involved in sending two consecutive data bytes to the chip.

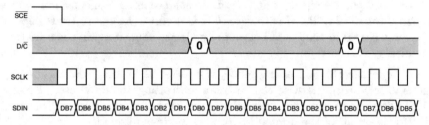

Changing the write flow of the RAM

If we wanted to skip to writing an arbitrary memory location, without following the established order of the selected writing modality, we could re-write the X and Y addressing registers. The next instruction writes register X, which addresses the X axis.

$\mathrm{D}/\overline{\mathrm{C}}$	DB7	DB6	DB5	DB4	DB3	DB2	DB1	DB0
0	1	X6	X5	X4	X3	X2	X1	X0

The following byte writes register Y, which addresses the Y axis.

$\mathrm{D}/\overline{\mathrm{C}}$	DB7	DB6	DB5	DB4	DB3	DB2	DB1	DB0
0	0	1	0	0	0	Y2	Y1	Y0

Once these registers are set, we can continue sending data, taking care to notify the interface by setting the $\mathrm{D}/\overline{\mathrm{C}}$ line to '1'.

Writing will start again as of the newly selected memory location. Registers X and Y will continue to be automatically managed by the chip according to the previously set addressing mode.

Other display operating modes

Aside from the normal mode, set at initialization, there are 3 others that we can select through the following command.

$\mathrm{D}/\overline{\mathrm{C}}$	DB7	DB6	DB5	DB4	DB3	DB2	DB1	DB0
0	0	0	0	0	1	D	0	E

The first mode turns off all the liquid crystals (D = '0', E = '0'), the second turns them all on (D = '0', E = '1') and the third inverts the colors of the display (D = '1', E = '1'). Normal mode is set by D = '1' and E = '0'.

Managing the display state

We can manage the display state by sending the following command byte:

D/$\overline{\text{C}}$	DB7	DB6	DB5	DB4	DB3	DB2	DB1	DB0
0	0	0	1	0	0	PD	V	H

Bit PD ("Power Down") active high, turns off the display and leaves the internal state of the registers unaltered. To achieve minimum consumption we must fill the display RAM with zeroes. It is useful to keep a software copy of what was stored in the RAM before to be able to show it again the next time the display is turned on.

Bit V regulates the addressing mode. If it is set at '0' the "row addressing" mode is selected, otherwise the "column addressing" mode is used. Finally, bit H allows us to enter the "extended mode" of the instructions, when set at '1'. If it is set at '0' the normal instruction mode is selected.

5.6.6.1 The system

The system is based on the "DMC8 Microcomputer". At output port OA the display serial interface lines $\overline{\text{SCE}}$, DC and SDIN are connected to bits 2, 1 and 0 respectively. The port's $\overline{\text{wA}}$ signal strobe has been connected to the SCLK line, as the serial communication clock. The reset of the $\overline{\text{DRES}}$ display has been connected to the microcomputer's reset output $\overline{\text{RsOut}}$.

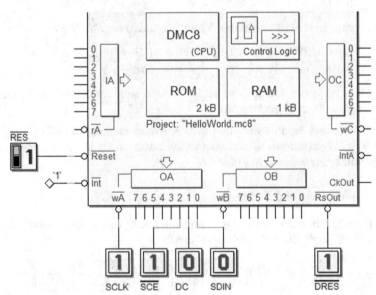

After the main program executes the necessary initializations, it shows the "Hello World!" message on the display and enters an infinite cycle where it is inactive.

The program

The address of the CNTRL output port is declared and this is followed by the link to the reset and the initialization of the Stack Pointer.

```
CNTRL        EQU    00h            ; IA port: SDIN (bit 0), DC (1), !SCE (2)
             ORG    0000h
             JP     0100h
             ORG    0100h
INIT:        LD     SP,0FFFFh      ; initialize the Stack Pointer
```

As seen in the previous section, before being able to see something on the display, we need to set the contrast. To set this parameter, we first need to use the "extended instruction" mode by sending the following command through the SENDC function, which will be analyzed later.

```
             LD     B,00100001b    ; PD = '0', V = '0', H = '1'
             CALL   SENDC          ; set the extended instruction mode
```

Then we send the next command, which sets the contrast to 8.

```
             LD     B,10001000b
             CALL   SENDC          ; set the contrast to 8
```

Finally we go back to normal instruction mode and set normal working mode.

```
             LD     B,00100000b    ; PD = '0', V = '0', H = '0'
             CALL   SENDC          ; go back to normal instruction mode
             LD     B,00001100b    ; D = '1', E = '0'
             CALL   SENDC          ; set normal working mode
```

The display's reset hardware has initialized all the internal registers to zero, so the pointers to the X and Y axes have been zeroed. We know that after reset, the state of the RAM is unknown. So the CLRSCR function is called to zero all the display's memory locations. This function leaves the state of the pointers intact.

```
MAIN:        CALL   CLRSCR         ; initialize all the RAM locations to zero
```

At this point, the address of the "Hello World!" string is loaded in index register IX. Then the X axis and Y axis where we want to start writing are loaded in registers B and C, respectively. After having loaded these parameters into the registers, we call the PRTSTR function, which shows the string on the display at the specified point (X,Y). Finally, the program enters an infinite loop where it is inactive, as per specifications.

```
             LD     IX,STR         ; copy the string's address to register IX
             LD     B,10           ; set X axis = 10 (through register B)
             LD     C,2            ; set Y axis = 2 (through register C)
             CALL   PRTSTR         ; finally, display the string
STOP:        JP     STOP           ; and enter in an inactive infinite loop
```

The CLRSCR function saves the used registers in the Stack then initializes the display RAM to zero. To do this, it enters a loop where it sends 504 data bytes through the SENDD function (analyzed below). After the saved registers are recovered, it goes back to the calling program.

```
CLRSCR:    PUSH  AF
           PUSH  BC
           PUSH  HL
           LD    HL,504      ; 84 x 6 = 504 RAM locations
           LD    B,0         ; set B = 0 to zero each location
CLRAM:     CALL  SENDD       ; of the display RAM
           DEC   HL          ; count the locations
           LD    A,L
           OR    H           ; set the flags according to HL value
           JP    NZ,CLRAM    ; exit if finished
           POP   HL
           POP   BC
           POP   AF
           RET
```

The next PRTSTR function shows on the display the string stored in memory at the address in register IX starting from the X axis indicated in B and the Y axis in C. After the used registers are saved as usual, the function sets the display's X and Y axes by calling the XSET and YSET subprograms analyzed below.

```
PRTSTR:    PUSH  AF
           PUSH  HL
           PUSH  BC
           PUSH  DE
           LD    A,B
           CALL  XSET        ; set X axis
           LD    A,C
           CALL  YSET        ; set Y axis
```

Then we read the first ASCII character of the string and make sure it is not equal to the string terminator (0). If it is, we exit with no further ado.

```
READ:      LD    A,(IX)      ; read the character pointed by IX
           CP    00h         ; check if it is the string terminator
           JP    Z,EXIT      ; exit if it is
```

If it is not equal, we go ahead and make sure the character we read is printable[52].

```
           CP    20h         ; if the character is not included
           JP    C,PINC      ; among the printable ones
           CP    7Fh
           JP    NC,PINC     ; jump to the label PINC
```

---

[52] That is can be shown and isn't a "control character", such as the previously mentioned terminator of the string.

If it is not printable, we jump to the PINC label, where we move on to reading the next character without showing anything.

The ASCII code that has been read is then adjusted to make index 0 coincide with the first printable character, to appropriately address the table containing the fonts[53], shown below. For now, it suffices to know that it is made up of 5 bytes for each printable code and it takes up a total of 480 bytes.

To get the address of the first byte representing the character graphically, we need to multiply the adjusted index by 5 to calculate the needed offset. Since there are more than 256 bytes making up the table, we have used the 16-bit HL register to process the index as described.

```
        SUB    20h          ; adjust the index excluding
                            ; the not-printable characters
        LD     L,A          ; copy the 8-bit index to HL
        LD     H,00h        ; and multiply it by 5
        ADD    HL,HL        ; x 2
        ADD    HL,HL        ; x 2
        ADD    A,L          ; +1
        JP     NC,NCY
        INC    H            ; if Carry, increment the high byte
NCY:    LD     L,A          ; now we have the offset in HL
```

Then, the resulting offset is added to the base address of the ASCII table, giving us the address of the first byte to send.

```
        LD     DE,ASCII     ; add the ASCII table base address
        ADD    HL,DE        ; to the offset
```

In a loop, all the data bytes are read and sent. We point to the data bytes using the HL register, which we increment on every loop repetition.

```
        LD     D, 5         ; set the number of bytes for each character
READ2:  LD     A,(HL)       ; read the byte to send (pointed by HL)
        LD     B,A
        CALL   SENDD        ; send data
        INC    HL           ; point to the next byte
        DEC    D            ; decrement the number of bytes
        JP     NZ,READ2     ; still to send
```

Finally, we go ahead and increment the IX register, to read the next character in the string on the next repetition as of the label READ.

```
PINC:   INC    IX           ; point to the next string character
        JP     READ         ; repeat loop as of READ
```

We only exit the subprogram and restore the used registers when we get to the terminator of the string.

---

[53] "Font", typically refers to typeface. Here, it refers to the sequence of bytes to send to the display that make up the characters in graphical terms.

```
EXIT:          POP    DE
               POP    BC
               POP    HL
               POP    AF
               RET
```

The XSET function sets the display's X axis pointer by using the value passed from the calling program to the accumulator. After saving the register in use, the function limits the value passed by the calling program to 83.

```
XSET:          PUSH   AF
               PUSH   BC
               CP     84          ; limit to 83 the value set for the X axis
               JP     C,XNCLP     ; if it is greater than 83 (when Carry = 0)
               LD     A,83
```

Then it sets bit 7 of the accumulator to '1', as required by protocol, and it sends everything to the display in the form of a command. In the end, we restore the previously saved registers and exit the function.

```
XNCLP:         SET    7,A         ; set bit 7 to '1',
               LD     B,A         ; as required by the protocol
               CALL   SENDC       ; send the command to the display
               POP    BC
               POP    AF
               RET
```

A similar operation is executed by the YSET function, which sets the address of the Y axis with the value passed to A, limiting it to 5.

```
YSET:          PUSH   AF
               PUSH   BC
               CP     6           ; limit to 5 the value set for the Y axis
               JP     C,YNCLP     ; if it is greater than 5 (when Carry = 0)
               LD     A,5
YNCLP:         SET    6,A         ; set bit 6 to '1',
               LD     B,A         ; as required by the protocol
               CALL   SENDC       ; send the command to the display
               POP    BC
               POP    AF
               RET
```

The SEND subprogram sends the byte in register B to the display in the form of a command if register C is set to 0, or in the form of data if it is not. We save the used registers on the Stack as usual, then counter E of the sent bits is initialized to 8.

```
SEND:          PUSH   AF
               PUSH   DE
               LD     E,8         ; initialize the bit counter to 8
```

After that, the $\overline{\text{SCE}}$ line is set to '0', while the DC line is set according to the content of register C.

```
          BIT   0,C
          JP    NZ,DATA      ; if C = '0'
COMMAND:  LD    A,00000000b  ; set DC = '0' (bit 1)
          JP    SLOOP        ; and go on to send a "command"
DATA:     LD    A,00000010b  ; otherwise send a "data byte" (DC = '1')
                             ; note that bit 2 (!SCE) is activated at '0'
```

We then go on to retrieve the most significant bit from register B and insert it in the bit in position 0, which commands data line SDIN. The configuration we get in the accumulator is copied on output port OA. Writing on the port causes the $\overline{\text{wA}}$ line to activate, which produces a clock pulse. On the clock rising edge, the data line is acquired by the peripheral.

```
SLOOP:    RLC   B            ; retrieve the next bit to send
          JP    NC,DRES      ; if it is high
DSET:     SET   0,A          ; set SDIN to '1' (on bit 0)
          JP    GO
DRES:     RES   0,A          ; otherwise set it to '0'
GO:       OUT   (CNTRL),A    ; send the bit on the serial line
```

This operation is repeated for all 8 bits in B. Then, we disable the display interface setting $\overline{\text{SCE}}$ to '1', restore the registers and exit the subprogram.

```
          DEC   E            ; decrement the number of bits to send
          JP    NZ,SLOOP     ; leave the loop if all bits have been sent

          LD    A,00000100b  ; set !SCE = '1' (bit 2)
          OUT   (CNTRL),A    ; to disable the display interface

          POP   DE
          POP   AF
          RET
```

The SENDD and SENDC versions of the SEND subprogram do nothing more than set the parameter of register C before calling it. SENDC puts C = 0 since it sends a command on the serial line. Because it sends a data byte on the serial line, however, SENDD puts C = 1.

```
SENDC:    PUSH  BC
          LD    C,0          ; send a command
          CALL  SEND
          POP   BC
          RET
SENDD:    PUSH  BC
          LD    C,1          ; send a data byte
          CALL  SEND
          POP   BC
          RET
```

What follows is the definition in the ROM memory of the message string to show and the table of the pixels corresponding to the ASCII characters (the table is shown only in part due to space restrictions).

STR:	DB	"Hello World!", 00h	
ASCII:	DB	00h, 00h, 00h, 00h, 00h	; 20 = (space)
	DB	00h, 00h, 5fh, 00h, 00h	; 21 = !
	DB	00h, 07h, 00h, 07h, 00h	; 22 = "
	DB	14h, 7fh, 14h, 7fh, 14h	; 23 = #
	DB	24h, 2ah, 7fh, 2ah, 12h	; 24 = $
	DB	23h, 13h, 08h, 64h, 62h	; 25 = %
	DB	36h, 49h, 55h, 22h, 50h	; 26 = &
	DB	00h, 05h, 03h, 00h, 00h	; 27 = '
	DB	00h, 1ch, 22h, 41h, 00h	; 28 = (
	DB	00h, 41h, 22h, 1ch, 00h	; 29 =)
	DB	14h, 08h, 3eh, 08h, 14h	; 2A = *
	DB	08h, 08h, 3eh, 08h, 08h	; 2S = +
	DB	00h, 50h, 30h, 00h, 00h	; 2C = ,
	DB	08h, 08h, 08h, 08h, 08h	; 2D = -
	DB	00h, 60h, 60h, 00h, 00h	; 2E = .
	DB	20h, 10h, 08h, 04h, 02h	; 2F = /
	DB	3eh, 51h, 49h, 45h, 3eh	; 30 = 0
	DB	00h, 42h, 7fh, 40h, 00h	; 31 = 1
	DB	42h, 61h, 51h, 49h, 46h	; 32 = 2
	DB	21h, 41h, 45h, 4bh, 31h	; 33 = 3
	DB	18h, 14h, 12h, 7fh, 10h	; 34 = 4
	DB	27h, 45h, 45h, 45h, 39h	; 35 = 5
	DB	3ch, 4ah, 49h, 49h, 30h	; 36 = 6
	DB	01h, 71h, 09h, 05h, 03h	; 37 = 7
	DB	36h, 49h, 49h, 49h, 36h	; 38 = 8
	DB	06h, 49h, 49h, 29h, 1eh	; 39 = 9
	DB	00h, 36h, 36h, 00h, 00h	; 3A = :
	DB	00h, 56h, 36h, 00h, 00h	; 3B = ;
	DB	08h, 14h, 22h, 41h, 00h	; 3c = <
	...	....	; ...
	...	....	; ...

5.6.6.2 Implementation on FPGA

The following sections have figures that summarize the connections chosen for each board, which are used to test the system.

Note: the LCD components on the market are produced by different companies. They may differ in the position of the connections on the connector. If any component is unlike the type indicated in the following figures, it will be necessary to review the connections according to the maker's indications, which are often shown on the silk-screen printing of the board itself.

The DE2 board

The following figures show the devices chosen for the DE2 board, both in table form (as generated by Deeds), and in terms of physical connections.

In/Out Name	VHDL Name	Board Resource	Aux. Info
Inputs:			
IRES	inRES	Push-Button: Key[00]	Low (if pressed)
Ck of ("Micrc	—	Clock: 10 MHz	
Outputs:			
ISCE	onSCE	Header 1: GPIO_1[03]	
DC	oDC	Header 1: GPIO_1[05]	
SDIN	oSDIN	Header 1: GPIO_1[07]	
SCLK	oSCLK	Header 1: GPIO_1[09]	
IDRES	onDRES	Header 1: GPIO_1[01]	

Take care to also connect the power supply wires (+3,3V and GND) of the display, as shown in the following figure.

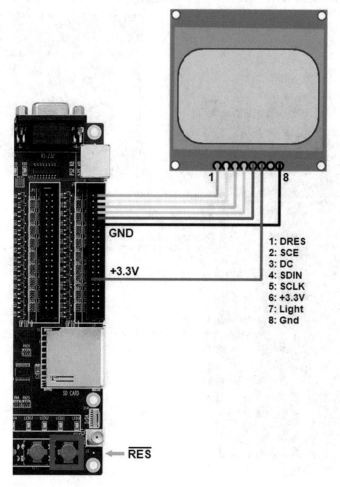

The DE0CV board

The assignment of devices for the DE0-CV board is very similar to that set
for the DE2.

In/Out Name	VHDL Name	Board Resource	Aux. Info
Inputs:			
IRES	inRES	Push-Button: Kev[001	Low (if pressed)
Ck of ("Micrc	—	Clock: 10 MHz	
Outputs:			
ISCE	onSCE	Header 1: GPIO_1[03]	
DC	oDC	Header 1: GPIO_1[05]	
SDIN	oSDIN	Header 1: GPIO_1[07]	
SCLK	oSCLK	Header 1: GPIO_1[09]	
IDRES	onDRES	Header 1: GPIO_1[01]	

As before, pay close attention when connecting the power supply for the display (+3.3V and GND) as shown in the figure.

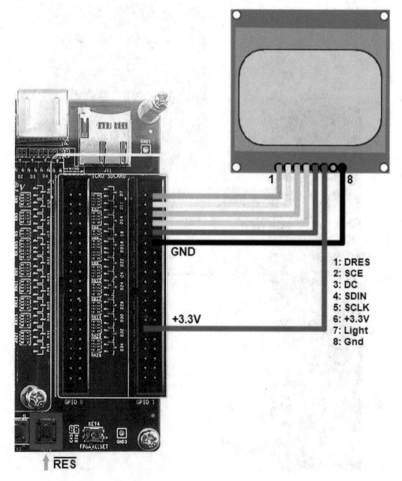

The EP2C5 board

We will use the only push-button on the board for manual system reset. The following table shows a summary of the connections as generated by Deeds.

In/Out Name	VHDL Name	Board Resource	Aux. Info
Inputs:			
IRES	inRES	Push-Button: Kev0	Low (if pressed)
Ck of ("Micro	—	Clock: 10 MHz	
Outputs:			
ISCE	onSCE	Header 2: P3 [IO_139]	
DC	oDC	Header 2: P3 [IO_136]	
SDIN	oSDIN	Header 2: P3 [IO_134]	
SCLK	oSCLK	Header 2: P3 [IO_132]	
IDRES	onDRES	Header 2: P3 [IO_142]	

What follows is the photo of a board with the physical connections to make. Again, pay close attention when connecting the power supply for the display (+3.3V and GND) as shown in the figure.

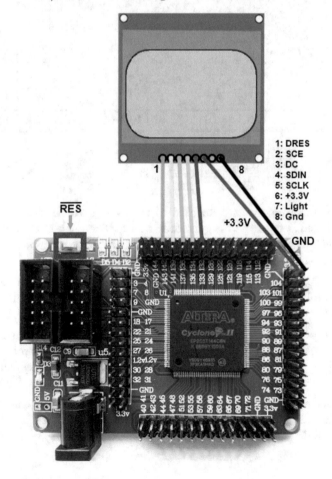

5.6.7 LCD stopwatch

In this project, we will analyze and build a digital stopwatch based on the DMC8 microcomputer that shows the time count on an LCD graphic display.

5.6.7.1 The system

This system, which is based on a "DMC8 Microcomputer" component, is shown in the following figure.

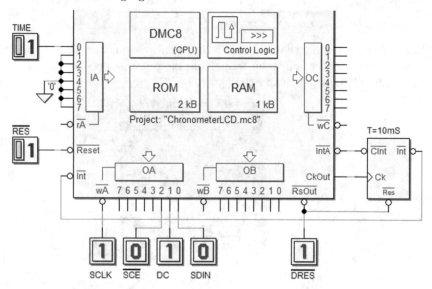

The TIME push-button is connected to the system through line 0 of input port IA. When it is pressed it generates a zero. The push-button also starts, pauses and restarts the time count. Pressing the $\overline{\text{RES}}$ push-button restarts the whole system and zeroes the count.

The counted time is shown on the LCD display of the Nokia 5110 phone used in the example in Section 5.6.6, much of whose code is borrowed to drive the component.

On output port OA, the display's interface lines $\overline{\text{SCE}}$, DC and SDIN are connected to bits 2, 1 and 0, respectively. The port's $\overline{\text{wA}}$ strobe signal is connected to the SCLK line (the serial communication clock). $\overline{\text{DRES}}$, the display reset, is connected to the microcomputer's reset output $\overline{\text{RsOut}}$.

A timer activates interrupt requests $\overline{\text{Int}}$ every 10 mS. When the request is accepted a pulse on output $\overline{\text{IntA}}$ automatically deactivates line $\overline{\text{Int}}$. The timer allows the system to precisely measure time.

The program

The code starts with the declaration of input port BUTTON and output port CNTRL. CNTRL drives the display's serial interface. Read the comments in the code.

```
BUTTON      EQU    00h         ; IA port: push-button (bit 0)
CNTRL       EQU    00h         ; OA port: SDIN (bit 0), DC (1), !SCE (2)
```

This is followed by the declaration of the variables. EN activates the count (when set at '1', the count is enabled, otherwise it is not).

BSTATE and PBSTATE store the current and previous states of the TIME button downstream of the debouncing checks.

```
EN          EQU    0FC00h      ; count enable
BSTATE      EQU    0FC01h      ; TIME push-button current
PSTATE      EQU    0FC02h      ; and previous states
```

Then the variables for the BCD (Binary Coded Decimal) time count are declared. BCD facilitates the translation of numbers into ASCII characters.

More specifically, CSEC counts the hundredths of second needed to make one tenth of a second elapse; DSEC counts tenths of a second; SEC1 and SEC2 count seconds and finally, MIN1 and MIN2 count minutes. Each of the variables above takes up one byte of memory.

```
CSEC        EQU    0FC03h      ; counter for the hundredths of second
MIN2        EQU    0FC04h      ; BCD most significant minutes digit
MIN1        EQU    0FC05h      ; BCD least significant minutes digit
SEC2        EQU    0FC06h      ; BCD most significant seconds digit
SEC1        EQU    0FC07h      ; BCD least significant seconds digit
DSEC        EQU    0FC08h      ; BCD tenths of a second digit
```

Finally, the STR variable is declared. It contains the string to show on the display. The format chosen for displaying the time is MM:SS:D, where MM represents the two digits for the minutes, SS the two digits for the seconds and D the digit for the tenths of a second.

The string is composed of 5 digits, two separators (':') and one character that terminates the string (00h). Therefore, it occupies a total of 8 bytes.

```
STR         EQU    0FC09h      ; string to show on the display
```

Registers B, C and D are dedicated to storing the intermediate readings of the debouncing check. B stores the last reading, C stores the second-to-last and D stores the third-to-last.

After the usual links to the reset and the interrupt handler, the Stack Pointer is initialized. The LCD component is initialized by the INITD subprogram that zeroes the graphic RAM and defines the working parameters.

```
                ORG    0000h
                JP     0100h
                ORG    0038h
                JP     HINT
                ORG    0100H
INIT:           LD     SP,0FFFFh    ; initialize the Stack Pointer
                CALL   INITD        ; and the LCD display
```

Then the variables and the registers used are zeroed.

```
                LD     A,00h         ; zero all the other variables
                LD     (EN),A
                LD     (BSTATE),A
                LD     (PSTATE),A
                LD     (CSEC),A
                LD     (MIN2),A
                LD     (MIN1),A
                LD     (SEC2),A
                LD     (SEC1),A
                LD     (DSEC),A
                LD     B,A           ; zero the registers in use
                LD     C,A
                LD     D,A
```

When the DISPLAY subprogram is called, a zeroed time count is shown on the display screen. The time count has been set to zero by the previous initialization. This subprogram's code will be analyzed next. After enabling the interrupts, we enter an empty main loop and wait for an interrupt request.

```
                CALL   DISPLAY       ; a zeroed time is shown on the display
                EI
MAIN:           JP     MAIN
```

The INITD subprogram initializes the LCD display, sets its physical parameters and zeroes the graphic RAM. The sequence of instructions is taken from the first rows of the code in the example in Section 5.6.6.

```
INITD:          PUSH   BC
                LD     B,00100001b   ; PD = '0', V = '0', H = '1'
                CALL   SENDC         ; set the extended instruction mode
                LD     B,10001000b
                CALL   SENDC         ; set the contrast to 8
                LD     B,00100000b   ; PD = '0', V = '0', H = '0'
                CALL   SENDC         ; go back to normal instruction mode
                LD     B,00001100b   ; D = '1', E = '0'
                CALL   SENDC         ; set normal working mode
                CALL   CLRSCR        ; initialize all the RAM locations to zero
                POP    BC
                RET
```

Every 10 *mS* the timer generates an interrupt, which causes the execution of its handler at the HINT label. Before acquiring the state of the TIME button, HINT shifts the previous readings in registers B and C into registers C and D, respectively. This makes room for the new reading that is stored in B (read the comments in the code).

```
HINT:      LD    D,C            ; second-to-last reading into third-to-last
           LD    C,B            ; last reading into second-to-last
           IN    A,(BUTTON)     ; get the current push-button state
           AND   00000001b      ; mask the bits not of our interest
           LD    B,A            ; copy the current state to B
```

The new reading is then compared to the previous ones. If they all show the same value, the current one is confirmed and saved in the BSTATE variable.

Before saving the new reading in the BSTATE variable, we need to transfer the previous content of BSTATE into PSTATE, where we store the reading that was confirmed before.

Then, a logic operation between the two values will allow us to detect the falling edges of the push-button line. To do this, the previous read confirmed is copied also to register L.

```
           CP    C              ; compare the state with the previous
           JP    NZ,TUPDATE
           CP    D
           JP    NZ,TUPDATE     ; if the reading is confirmed
           LD    E,A            ; copy it to register E
           LD    A,(BSTATE)     ; get the last confirmed reading
           LD    L,A            ; and copy it to register L
           LD    (PSTATE),A     ; and to the PSTATE variable
           LD    A,E
           LD    (BSTATE),A     ; save the confirmed current reading
```

We execute an XOR operation to invert the new reading and put the result in AND with the previous one. Then, if bit 0 is at '1', a falling edge is detected and the content of EN, which enables the time count, is inverted.

```
           XOR   00000001b      ; logic operation to detect a falling edge
           AND   L              ; on bit 0
           JP    Z,TUPDATE      ; if it is detected, go on and
           LD    A,(EN)         ; invert the enable variable (EN)
           CPL
           LD    (EN),A
```

Regardless of the result, we go on to evaluate the EN flag. If it is zero, we exit the handler without changing the state of the display. If it is not, we move on to the time count, starting from the hundredths of a second.

```
TUPDATE:   LD    A,(EN)         ; check if the time count is enabled
           OR    A
           JP    Z,EXIT         ; if it is, go on, otherwise exit
```

Then the variable that counts hundredths of a second (CSEC) is incremented. If 10 hundredths have elapsed, we proceed by incrementing the tenths of a second and zeroing the hundredths. Otherwise, we exit without changing the state of the display.

```
LD    A,(CSEC)    ; increment the hundredths of a second
INC   A
LD    (CSEC),A
CP    10          ; if 10 hundredths have not elapsed,
JP    NZ,EXIT     ; jump and exit, otherwise
LD    A,0         ; zero the hundredths of a second
LD    (CSEC),A    ; and go on, incrementing the tenths
```

If the tenths of a second are at 10, it means we need to increment the second counter. We zero the tenths of a second counter and go on to increment the least significant digit of the seconds. Otherwise, we jump to update the state of the display and exit.

```
LD    A,(DSEC)
INC   A           ; increment the tenths of a second
LD    (DSEC),A
CP    10          ; if a second have not elapsed,
JP    NZ,EXITD    ; jump to update the display and exit
LD    A,0
LD    (DSEC),A    ; otherwise zero the tenths of a second
LD    A,(SEC1)    ; and increment the least significant digit
INC   A           ; of the seconds
LD    (SEC1),A
```

If the least significant digit of the seconds has gotten to 10, it means that it is time to increment the most significant digit and zero the least significant. If not, we jump to update the display and exit.

```
CP    10          ; if 10 seconds have not elapsed,
JP    NZ,EXITD    ; jump to update the display and exit
LD    A,0         ; otherwise zero the least significant
LD    (SEC1),A    ; digit of the seconds
LD    A,(SEC2)    ; and increment the most significant digit
INC   A
LD    (SEC2),A
```

If the most significant digit of the seconds has gotten to 6, we need to update the minutes, so we proceed by zeroing that number[54] and incrementing the minutes. If not, we update the state of the display and exit.

```
CP    6           ; if 60 seconds have not elapsed,
JP    NZ,EXITD    ; jump to update the display and exit
```

---

[54] The seconds' least significant digit was already zeroed and passed from 59 to 60.

```
LD    A,0          ; otherwise zero the seconds (the least
LD    (SEC2),A     ; significant digit was already zeroed)
LD    A,(MIN1)     ; and increment the least significant digit
INC   A            ; of the minutes
LD    (MIN1),A
```

Through the same process we update the minutes: first, we increment the least significant digit and if it has not gotten to 10, we jump to update the display and exit. Otherwise, we zero it and increment the most significant digit. When this gets to 6, it is zeroed but the hours are not incremented since we will not see them[55]. Thus the count restarts from zero.

```
CP    10           ; if 10 minutes have not elapsed,
JP    NZ,EXITD     ; jump to update the display and exit

LD    A,0          ; otherwise zero the least significant
LD    (MIN1),A     ; digit of the minutes
LD    A,(MIN2)
INC   A            ; and increment the most significant digit
LD    (MIN2),A
CP    6            ; if 60 minutes have not elapsed,
JP    NZ,EXITD     ; jump to update the display and exit

LD    A,0          ; otherwise zero the minutes (the least
LD    (MIN2),A     ; significant digit was already zeroed)
```

Finally, we update the display by calling the DISPLAY subprogram, which will be analyzed further on. Then we re-enable the interrupts and go back to the calling program.

```
EXITD:    CALL  DISPLAY    ; update the display
EXIT:     EI
          RET
```

The DISPLAY subprogram updates the time count on the LCD display. The individual decimal digits stored in the BCD code are first translated into the corresponding ASCII code. To do this, we add their value to the hexadecimal constant 30h (the first code in the ASCII table section representing the numbers). The resulting characters are then linked in a string with a colon (':') as a separator between the minutes and seconds and another between the seconds and tenths of seconds, giving us the classic MM:SS:D format.

The subprogram begins by saving the registers in use. Then it enters a counting loop where every variable representing time is read in the order they are stored in, starting from the most significant digit of the minutes and ending with the tenths.

```
DISPLAY:    PUSH  AF       ; save the register in use on the Stack
            PUSH  DE
            PUSH  IX
            PUSH  IY
```

---

[55] With two extra variables, we could extend the code to count the hours.

Before entering the loop, register E is initialized at 5 (to count the number of digits used). Then the address of the STR string is assigned to index register IX (the chain of characters is stored in STR). IY is set at the address of the first digit of the measure to convert.

```
LD    E,5         ; set the counter of the 5 digits
LD    IX,STR      ; set the string address in register IX
LD    IY,MIN2     ; set the address of the first digit in IY
```

In the loop, the current digit of the time measure, whose address is contained in IY is read so that its value can be translated into the corresponding ASCII code. The resulting character is then saved in the STR string, concatenated to the previous characters.

```
DLUP:    LD    A,(IY)     ; get the current digit and add 30h
         ADD   A,30h      ; to translate it into the corresponding
         LD    (IX),A     ; ASCII code and add it to the string
```

Let's use a little trick. If the loop counter is not zero and if it is even, it means that we are between the minutes and the seconds or between the seconds and the tenths so we need to insert a colon (':') to separate the groups of digits.

```
LD    A,E         ; get the digit counter
CP    0
JP    Z,NOSEP     ; if this is not the last digit,

BIT   0,A         ; and if the counter index is even
JP    NZ,NOSEP

INC   IX          ; add the separator character ':' to
LD    A,3Ah       ; the string to separate minutes from
LD    (IX),A      ; seconds and seconds from tenths
```

In any case, we update indexes IX and IY, and the loop counter. If all the digits have been processed, we add the terminator (00h) to the sting end.

```
NOSEP:    INC    IY        ; otherwise, update the pointers
          INC    IX        ; to the string and to the digit
          DEC    E         ; and decrement the digit counter
          JP     NZ,DLUP   ; if all the digits have passed,

          LD     A,00h     ; add the string terminator
          LD     (IX),A
```

Lastly, the address of the STR string is reloaded in IX and then passed to the PRTSTR subprogram, which shows it on the display at the coordinates indicated in registers B and C. For a description of the PRTSTR subprogram, refer to the example in Section 5.6.6.

```
LD    IX,STR      ; copy the string's address to register IX
LD    B,22        ; set X axis = 22 (through register B)
LD    C,2         ; set Y axis = 2 (through register C)
CALL  PRTSTR      ; display the string
```

Finally, we restore the registers used and exit.

```
POP    IY
POP    IX
POP    DE
POP    AF
RET
```

5.6.7.2 Implementation on FPGA

Here, we show images that summarize the connections for each board, which are useful for testing the system (refer to the observations on Page 540 regarding the different versions of the LCD component).

The DE2 board

In/Out Name	VHDL Name	Board Resource	Aux. Info
Inputs:			
Ck of ("Chro)	—	Clock: 10 MHz	
TIME	iTIME	Push-Button: Key[03]	Low (if pressed)
IRES	inRES	Push-Button: Key[00]	Low (if pressed)
Outputs:			
ISCE	onSCE	Header 1: GPIO_1[03]	
DC	oDC	Header 1: GPIO_1[05]	
SDIN	oSDIN	Header 1: GPIO_1[07]	
SCLK	oSCLK	Header 1: GPIO_1[09]	
IDRES	onDRES	Header 1: GPIO_1[01]	

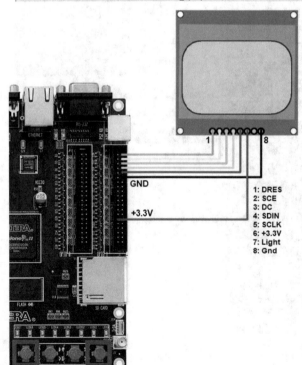

The table above lists the devices chosen for the DE2 board (as generated by Deeds).

The figure on the left shows the physical connections to make between the board and the LCD component.

Take care to connect the display's power supply wires (+3,3V e GND) as well, as suggested in the figure.

The DE0CV board

The assignment of devices for the DE0-CV board is very similar to that set for the DE2.

In/Out Name	VHDL Name	Board Resource	Aux. Info
Inputs:			
Ck of /"Chro(	—	Clock: 10 MHz	
TIME	iTIME	Push-Button: Key[03]	Low (if pressed)
IRES	inRES	Push-Button: Key[00]	Low (if pressed)
Outputs:			
!SCE	onSCE	Header 1: GPIO_1[03]	
DC	oDC	Header 1: GPIO_1[05]	
SDIN	oSDIN	Header 1: GPIO_1[07]	
SCLK	oSCLK	Header 1: GPIO_1[09]	
IDRES	onDRES	Header 1: GPIO_1[01]	

As before, take care to connect the display's power supply wires (+3.3V and GND) as well, as shown in the following figure.

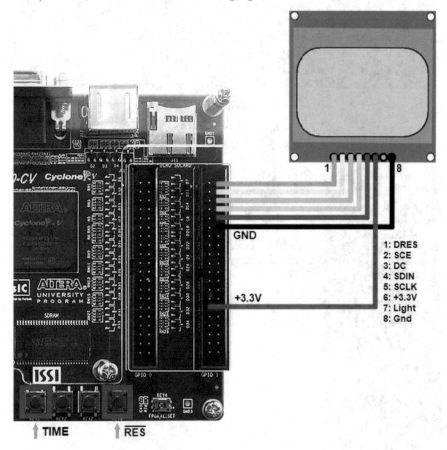

The EP2C5 board

In/Out Name	VHDL Name	Board Resource	Aux. Info
Inputs:			
Ck of f"Chro(	—	Clock: 10 MHz	
TIME	iTIME	Header 0: P1 [IO_40]	
IRES	inRES	Push-Button: Key0	Low (if pressed)
Outputs:			
!SCE	onSCE	Header 2: P3 [IO_139]	
DC	oDC	Header 2: P3 [IO_136]	
SDIN	oSDIN	Header 2: P3 [IO_134]	
SCLK	oSCLK	Header 2: P3 [IO_132]	
!DRES	onDRES	Header 2: P3 [IO_142]	

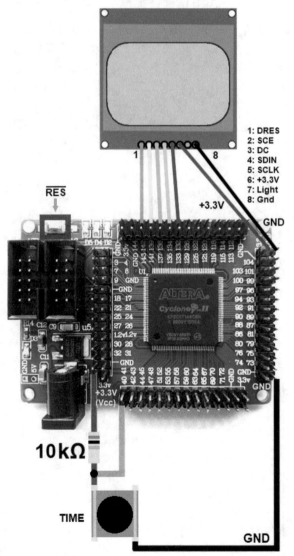

1: DRES
2: SCE
3: DC
4: SDIN
5: SCLK
6: +3.3V
7: Light
8: Gnd

The table above lists the devices chosen for the EP2C5 board (as generated by Deeds).

The figure on the left shows the physical connections to make between the EP2C5 board and the LCD display component.

We use the only push-button on the board to reset the system.

In the figure, the externally added TIME button and the 10K Ω pull-up resistor are highlighted.

Take care to also connect the power supply wires (+3,3V and GND), as shown in the figure.

A

Memories and busses

A.1 ROM memory

The term Read Only Memory (ROM) refers to a type of persistent memory that stores information even with no power supply. This type of memory is designed for long-term storage of information that does not change during system operation.

In early versions of ROM memory, the content could not be changed after it was made. Afterward, thanks to the evolution of technology, designers were able to change the contents of ROM memory directly in the lab in the system development phase. They were also able to do updates on equipment that was already operating.

In terms of principle, we do not want the content of the ROM to be changed in a system's normal working conditions. Despite this, many systems today allow for the contents of their ROMs to be changed even while they are working, but it is important to note that the reprogramming phase arrests the normal activity of the system. When the system starts again, the ROM goes back to being unchangeable and will keep the information even when the system is turned off.

A.1.1 A bit of history

As we can imagine, the ROM has undergone large-scale evolution, in part due to increasing technological capacities and in part due to the changing system requirements of ever more complex specifications.

The first forms of read-only memory were created with discrete components[1] and they were introduced to give computers the capacity to activate autonomously when the systems were turned on. They contained the first instructions computers would execute as of reset. The instructions served to

---

[1] Diodes, transistors and resistors, all encased separately and each less than one centimeter in size.

© The Editor(s) (if applicable) and The Author(s), under exclusive license
to Springer Nature Switzerland AG 2022
G. Donzellini et al., *Introduction to Microprocessor-Based Systems Design*,
https://doi.org/10.1007/978-3-030-87344-8

make the computer operative and able to load, in the main memory, the program to execute from the then-existing mass storage systems (punch cards, punch tape and magnetic tape at the start of the 1960s).

Later, with the development of solid state integrated circuit technology, large companies were able to build and sell integrated ROM memory on a single chip. Over time, different manufacturing technologies have been developed thanks to the success of microprocessors in reducing costs and improving capacity in terms of the number of storable bits.

MASK ROM

In the beginning, the first ROMs were made directly by producers of integrated circuits and programmed in the production phase according to the specifications of the system designer (MASK ROM). This was a very costly solution that was acceptable only to be sold in large-volumes. In fact, this technique required adapting the production line every time it was necessary to change the memory programming.

After this first solution, the industry began to offer ever more versatile and economical ones that could be programmed directly in the lab, allowing designers to develop and test programs and systems reasonably quickly. The programmable devices that have followed are: PROMs (Programmable ROMs), EPROMs (Erasable Programmable ROMs), EEPROMs (Electrically Erasable Programmable ROMs), EAROMs (Electrically Alterable Programmable ROMs) and finally Flash ROMs.

PROMs

Since the contents of PROMs could be defined after they are built, they were mass produced, leaving the definition of the content to the buyers. This type of memory was then programmed by the designer using laboratory hardware equipment called a "PROM programmer". Once the content was defined, it could not be changed again. Every time there was a programming error, a new component needed to be used[2].

Components like these that have to be inserted and re-placed very often in the design phase were normally mounted on a "zero insertion force" socket (see the image on the right), to make it easy to replace them by opening and closing the contacts with the lever.

From the industrial perspective, while an apparatus was being built, the PROMS were programmed on the production line before being mounted.

---

[2] A standard device that can be programmed only once is called OTP (One Time Programmable)

EPROMs

To make the components reusable, industry developed EPROMs, which are erasable and re-programmable in the lab. One feature of these components is the glass window that shows the chip and its connections (see the figure on the right).

Memory was erased by exposing the chip to ultraviolet rays (UVB) through the glass window, using a dedicated device. This was a rather long process (10-20 minutes) and could only be done a limited number of times.

Also in this case, chip programming was done with a dedicated EPROM programmer.

EEPROMs and EAROMs

The next technological development was EEPROM memory, which was not only programmable, but also electrically erasable through a dedicated programming circuit that could even be installed in the system itself. This made it possible to reprogram the component without physically removing it from the system.

Bear in mind that these components were still a type of ROM in that erasing and reprogramming components could not be done by the ordinary microprocessor connections but only at higher voltages (about 25 volts), that dedicated programming circuits can provide. Also, this was not done location by location but relatively slowly by entire cell blocks at a time.

Subsequently, EAROMs were introduced. They are very similar in their use to EEPROMs but they feature the option of changing individual locations rather than reprogramming entire blocks of data.

Flash ROMs

A "Flash" is a type of programmable and electrically erasable memory that is also used as mass storage (read and write) due to its high performance. The photo on the right shows an example (it does not seem different from many other integrated components that we can find on a board).

A "Flash ROM" is a "Flash" type memory used as a ROM (in the literal sense, i.e. preventing writing by the microprocessor).

Flash technology makes it possible to write and erase data in a single operation, unlike the previous technologies, giving us much greater speed.

Since the data is kept even when the computer is off, we find this used in cameras, mobile phones, network music players, USB keys, personal computers, tablets and wherever a lot of data storage is required.

A.1.2 Operating principle

Originally, the term ROM referred only to MASK ROM, but it has been extended and is used for all types of memory that do not erase when the power source is cut (they are also called "non volatile") and when the system is working in its normal mode, can only be read. Hereafter, we will use the term ROM with this meaning.

In this book, we will not go into the merits of the technological aspects of manufacturing ROM memory or the components available. Rather, let's consider the logical and functional perspective. In the design phase, we will virtually program the ROMs according to our needs. We will leave it to the Deeds environment to take care of the physical details of this operation without getting into the particulars of the actual programming technique. ROM programming will take shape in the synthesis of ROM functions directly in the combinational blocks available in the FPGA device mounted on a specific board prototype.

Let's look at the figure on the right that shows a ROM component available in the Deeds library. Only the connections needed for reading the data inside it have been made available (for example, to be connected to a microprocessor). As mentioned before, the programming has been done virtually thanks to the development environment. Therefore the pins that would be necessary in a real component, depending on the type, for its programming, are absent here.

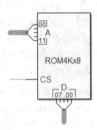

This is a 4-KByte ROM memory. Its 12 address wires (A11..A00) address 2^{12} one-byte locations inside it. The location selected by the address can be read through lines D7..D0.

When CS (Chip Select) is at '1' makes it possible to activate the functionality of the component. In this class of ROM components taken from the Deeds library, if CS = '0', the outputs are zeroed.

The figure on the right shows an idealized version of the inside of a generic ROM.

It appears in the form of a table where each location is identified by a unique address. Each location contains a constant (set in the programming phase).

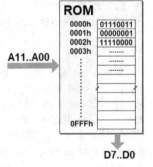

When a ROM is used in a microprocessor-based system, it contains the programs that need to be executed in the form of binary codes.

In any case, it should be stressed that a ROM behaves, from a logical perspective, like a purely combinational network. Once it has been programmed and is in use, it always provides the same outputs (the stored data) if the inputs (the address lines) are the same.

An example of ROM application: a sine wave generator

A ROM can be used to store different types of information. For example, it can contain a sequence of numbers that describes the shape of a "digital signal", such as the samples of a complete period of a sine wave.

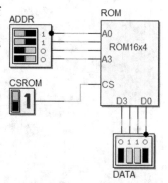

Let's take the smallest ROM memory available from the Deeds library and connect it as in the figure at the right, for test purposes. This component has only 16 4-bit locations (D3..D0), addressed by lines A3..A0. It also has the CS input to enable its functionality.

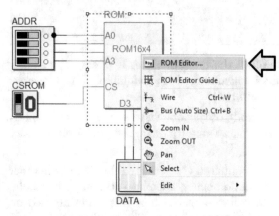

If we activate the component's context menu (see the yellow arrow in the figure on the left), we can open the "ROM Editor/Programmer" window that allows us to program the memory contents.

The figure below shows what appears in the window that will open. Among the various items, we see the ROM editor grid.

In the "ROM Editor/Programmer" window, the grid allows us to manually insert and/or change the content of the ROM, location by location.

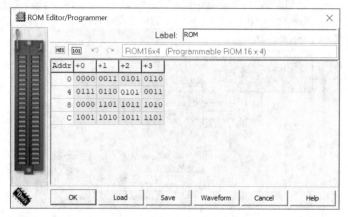

In this example, however, the ROM contents have been generated automatically. If we click on the "Waveform" button, shown at the bottom of the window, another window appears, as shown in the following figure.

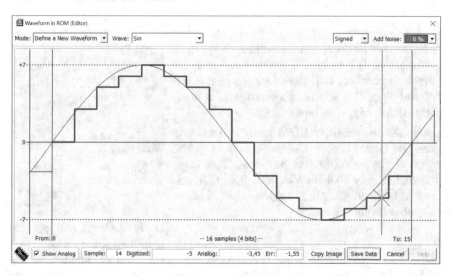

Without going into the merits of all the functionalities it offers, let's simply say that we can choose a waveform among those available. Here, we have chosen a sine wave (see the curve in red).

The software calculates the amplitude of the sine wave at regular time intervals along the wave period, giving us a value (or "sample") for each memory location (16 in this case).

Then these sample values are approximated ("discretized"), with the number of bits available in each memory location being taken into account. The results are shown graphically on the same figure on the blue line superimposed on the red curve of the original continuous sine wave. The resulting values are the ones that we have seen in the ROM Editor/Programmer grid.

In the network in the following figure, a binary counter addresses all the ROM locations one after the other in sequence. The ROM returns the corresponding values memorized inside.

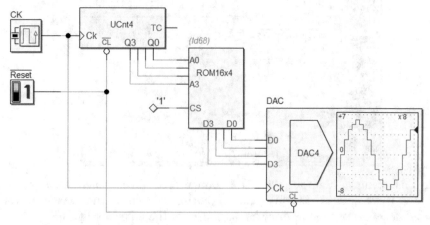

The digital-analog converter (DAC) translates the generated values into analog values shown graphically over time in the DAC internal pane[3]. By changing the clock frequency, we can change the period of the sine wave.

A.1.3 Internal architecture

The internal architecture of a ROM memory is formally based on what is called a "Programmable Logic Array" (PLA). The PLA is used in integrated circuits to conveniently implement combinational logic networks[4] of a certain level of complexity.

The PLA is based on Shannon's expansion theorem[5], which says that any combinational logic function can be expanded (or decomposed) in terms of a sum of logical products (normally referred to as "AND-OR networks").

A PLA is based on two arrays of programmable connections (see the example in the following figure). The first array, called the "AND Plane", makes it possible to connect the network inputs (in direct or negated form) to a group of AND gates.

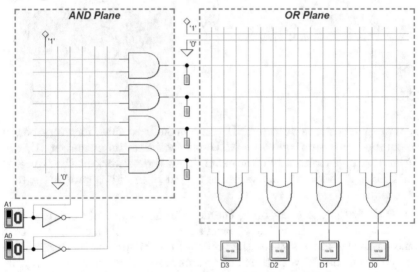

The second array, called the "OR Plane", makes it possible to connect the ANDs to a group of OR gates that then produce the network outputs.

---

[3] This is a virtual conversion. During the simulation, the output value is only shown graphically in the pane internal to the component itself.

[4] PLAs were developed over the second half of the 1960s and were used industrially in integrated circuits about a decade later.

[5] This theorem was actually stated by George Boole in 1854, in his work, "The Laws of Thought". Claude Shannon made mention of this type of expansion in an article in 1948 and applied it to the synthesis of logic networks, which is why this theorem carries his name in the literature.

The example in the figure, which is purposefully less complex, has 2 inputs (A1,A0), 4 AND logic gates, 4 OR logic gates and 4 outputs (D3..D0).

A PLA network makes it possible to create a large number of combinational AND-OR functions obtained by electrically connecting the intersections of the two matrices in an appropriate way. A ROM is a specific kind of PLA where the AND plane is pre-programmed by the manufacturer to obtain the standard logic of a binary decoder (see the figure below). Following this, the constant '1' in the OR plane has also been eliminated.

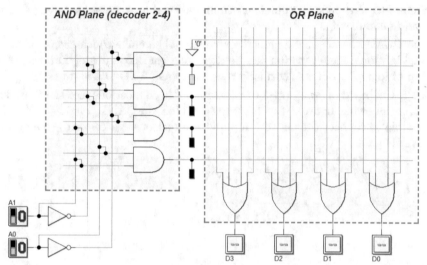

Pre-wired connections on the AND plane make it so that, for every possible combination of inputs, only one AND gate activates its output at '1', while all the others generate a '0'. For example, in the network in the figure, the combination A1A0 = '00' activates the output of the AND gate at the top of the schematic.

In a ROM with N address lines, the decoder activates only one row of the network at a time; among the 2^N available rows, it activates the one corresponding to the address set at the inputs of the component at that moment. If we envision the ROM in terms of a value table, as we did previously, it is easy to see that each row in the table actually corresponds physically to a row in the OR plane.

To program a location at a certain address, we, therefore, take the corresponding selected row into consideration. Note that, formally, the OR gates have as many inputs as the number of array rows[6].

---

[6] As the number of array rows rises (to the thousands, for example), an OR with the same number of inputs may seem unfeasible. In actual fact, it is possible to perform the functions of an OR gate with a specific configuration of circuital elements (transistors) distributed along all the rows. This allows us to use a column made up of a simple wire and obtain the functionality of an OR gate.

In this schematic, all the OR gate inputs are brought up in the array as columns and intersect all the rows[7]. So that the value we want for that location exits from the OR gate output, we need to choose whether to connect the columns to the row of interest.

If we want an OR gate to generate a '1', we'll connect one of its inputs to that row. Otherwise, to generate a '0' that same input will be connected to constant '0'. In so doing, we will find a '1' on the OR gate output under the columns we have connected to the selected row and a '0' will exit the others. The other rows do not disturb the process because they are inactive and only produce '0s' for the other OR inputs.

In the example shown in the figure below, the array of the OR plane is programmed to generate the mathematical function:

$$D = A^2$$

For example, the square of 3 is 9, so if we set $A1A0 = 11_2 = 3_{10}$, the output should generate $D3..D0 = 1001_2 = 9_{10}$. The corresponding row is the lowest and it has been connected as shown in the figure.

Let's look at the input of every OR gate that is farthest to the right in the figure. The OR gates corresponding to the bits that have to be at '1' (D3 and D0) have been connected to the row corresponding to the address $A1A0 = 11_2$ (shown as active) through this input. As regards the other two OR gates, that same input has been connected to '0' (at the top of the figure).

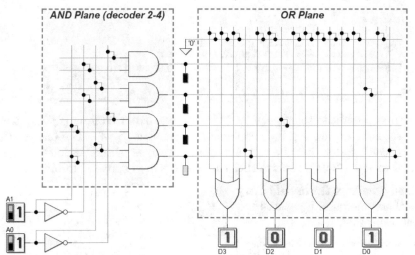

As shown in the simulation in the figure, $A1A0 = 11_2$ and all the rows of the array are at '0', while only the selected row is at '1'. The only outputs at '1' are from the OR gates connected to it.

---

[7] To better organize the structure of a large memory chip, "arrays of arrays" are used and thus, more complex address decoders. In this book, however, we will not delve into these techniques.

There is a similar rationale for the middle rows. However, since the first row has to generate all the outputs at '0', none of the ORs has been connected.

As mentioned before, programming these connections in reality depends on the technology behind the specific component in use. In any case, no designer is required to go into the level of detail of the individual connections, as was done here for the purpose of further study. There is software designed for this purpose. It is generally produced by the manufacturers of the components, and this allows us to start from a formal description of the contents.

Deeds offers the ROM Editor/Programmer, as shown on page 559. This tool allows us to program the memory even by loading the contents of a specific text file with the ".drs" extension ("Deeds ROM Source file"). The file can also be generated in order to export the contents of the memory.

Let's take the ROM of the sine wave generator presented on page 559 as an example. Below, we show how the corresponding file looks inside the ROM/Editor Programmer as it's ready to be exported.

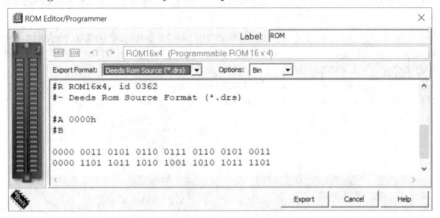

The first line states the type of ROM the file was exported from ("#R"), while the second provides a simple explanatory comment ("#-"). The line starting with "#A" states that the values have been read as of address '0000h', while the next line ("#B") reports that those values are expressed in binary.

Finally, we have the list of binary numbers separated by spaces and read from their locations.

For a complete description of the options offered by the ROM/Editor Programmer, consult the ROM/Editor Programmer guide, available in the context menu highlighted by the yellow arrow in the figure at the right.

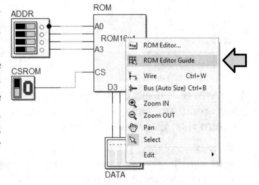

A.2 RAM memory

The term Random Access Memory (RAM) has its origins in history. It was defined as such in opposition to Sequential Access Memory (SAM), which at the time came in the form of perforated paper tape and magnetic tape.

A magnetic tape, for example, does not allow for immediate access to a specific group of information. To access an area of data, one must first fast forward the tape up to the position where the first part of the information is and then read the positions one by one. If the current position of the tape is far from that of the area of the data that needs to be accessed, this operation can take several minutes to execute.

RAM, however, makes it possible to access any data area in a constant amount of time independently of the internal state of the memory. This access is allowed both for read and write operations, which is why it is also called Read Write Memory (RWM).

As mentioned in Chapter 1 on page 75, RAM memory can be considered a group of parallel registers, each of which can be loaded with a number.

Individual registers are called memory locations (or cells) and a RAM can contain a very high number of locations (from a few hundred to a few billion, depending on the component). Each location is identified by an address, as in the case of ROMs.

From a circuital perspective, there are mainly two different types of RAM memory: "static" and "dynamic". Static RAM (SRAM) stores information by using a flip-flop for each bit. Hereinafter, we will always refer to static memory.

The photo on the right shows a historic example of one of the first static RAM memories available on the market, the MM2114, which contained 1024 locations of 4 bits each.

Dynamic memory (DRAM, Dynamic RAM) gets its name because individual bits are stored through capacitors that are charged or discharged according to the desired logic level. Capacitors are actually represented by the capacity of transistors, which are normally defined as "parasites" but here are purposefully used to register individual bits. Since these are not ideal capacitors, they tend to discharge, so they need to be periodically "refreshed" with the last value stored, so that it is not lost.

Although dynamic memory adds to the complexity of refresh circuits, they require many fewer transistors[8] to memorize bits, than static memory, which uses flip-flops.

---

[8] In this book, we will not examine the aspects of microelectronics that allow for this type of storage, or the complex circuits required to execute a refresh.

For this reason and for reasons of cost, they are used where very large memory is needed, in personal computers for example (the photo on the left shows a 4-GByte dynamic memory module).

Aside from these distinctions, there are forms of RAM memory that allow for reading and writing in multiple locations at a time. This is called "multiple port" memory, often used in graphics systems. We, however, will deal with the more common "single port" memory that allows for accessing only one location at a time.

A.2.1 Operating principle

As mentioned before, we will focus on the logical and functional aspects of RAM memory and to do this, the static, single port forms lend themselves better to a linear discussion.

As we saw in Chapter 1 (on page 75), RAM memory can be ideally represented in the form of a table, where each block is identified by an address, as shown in the figure on the right.

When the system is turned on, each location contains an indeterminate value. After that, only the locations where we will write something will store known values.

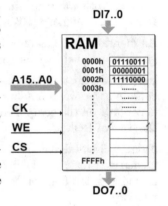

In the example in this figure, the RAM has 16 address wires (A15..A0) and contains 2^{16} locations (64K) of one byte each. Therefore, the location addresses are between 0000h and FFFFh. The RAM also contains 8 input lines (DI7..DI0) for the data to be memorized and 8 output lines (DO7..DO0) for reading the data.

The WE (Write Enable) line makes it possible to command the writing, while the CS (Chip select) input enables the component. The clock CK line will be present if the specific RAM in use is "synchronous" and will not if it is "asynchronous".

Before storing a number in the RAM, we must select the address of the location to use and submit it to the RAM on wires A15..A0. Then we set the number to write in the selected location on inputs DI7..DI0 and activate CS. In asynchronous RAMs, writing is ordered through a pulse on the WE command. In synchronous RAMs, we still need to activate WE but writing is executed on the active edge of CK.

We can re-read the number contained in a location by providing its address on lines A15..A0 and activating CS. The content of the location appears directly at outputs DO7..DO0, if the RAM is asynchronous. For synchronous memory, after providing the address, it is necessary to wait for the next edge of the clock CK to get the number, as with writing.

A.2.2 Internal architecture

Let's examine how we can manage a group of cells inside the memory by gradually building a static, synchronous RAM made up of only 4 locations. With this simplification, the schematic will present the general principles of its architecture without being too difficult to interpret.

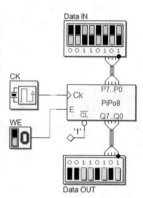

The figure on the right shows an 8-bit parallel register. For the writing operation, the number to store must be presented to inputs P7..P0 and WE, the write enable command must be activated. WE is connected directly to register enable input E.

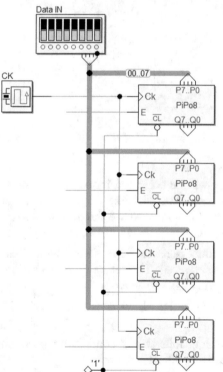

The write operation is done on the rising edge of the clock CK. From then on, the stored number will be available on outputs Q7..Q0.

Note that input $\overline{CL}$ is not connected to a reset line, so to initialize the register to a known value, we need to explicitly write a value in it (this choice is to replicate the behavior of RAMs, which do not have a reset input).

Now, let's take four of these registers (see the figure on the left) and connect all their clock inputs and data input lines in parallel.

When we use the Data IN bus to provide the number to write, that number will be potentially available to all the registers.

However, we will write only in one register at a time. To do this, we will activate the enable line of the chosen register only and keep all the others inactive.

To address 4 locations, we need 2 address lines. Accordingly, we introduce lines A1 and A0, which we will send to a $2 \rightarrow 4$ decoder with enable (see the following figure).

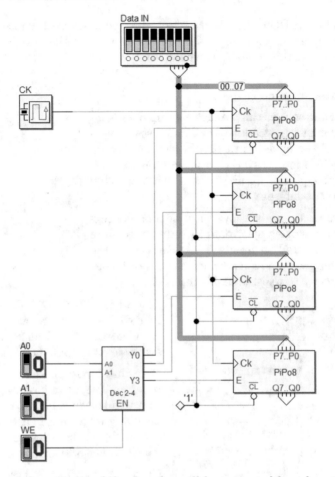

Only one of the outputs of the decoder will be activated based on the corresponding address provided. So, we have connected decoder outputs Y3..Y0 to the registers' E write enable inputs. It will be useful to use the decoder enable (EN) as an overall write enable (WE) line.

When done in this way, enabling individual registers depends on the address set in A1A0 and whether the WE line is activated. The actual write operation, as mentioned before, is done on the rising edge of the clock.

To finish building our simplified example of a RAM, we now need to add the circuitry required to read locations.

To do this, we simply add a 4-channel multiplexer and connect its selection inputs to address lines A1 and A0 (see the following figure).

In the way that it is set, this network allows for an asynchronous read operation. This means that it is enough to provide the address to get the content of the desired location on the Data OUT output, without waiting for the rising edge of the clock.

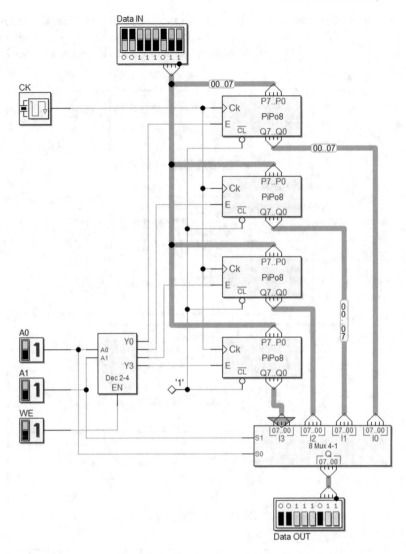

In the simulation screenshot above, a write operation has just concluded. The set address is A1A0 = '11', and WE is active.

Here, the network behavior is captured right after the rising edge of the clock and the number on the Data IN input has been stored in the selected register (the lowest one in the figure). The Data OUT output has the number that has just been written, read from the selected register.

Following the same design criterion, we'll now show a larger synchronous RAM (but a far cry from those of commercial circuits). This RAM has 64 8-bit locations and has external connections similar to the previous RAM (see the following two figures).

Here there are 6 address lines (A5..A0). A chip select CS for general enable is implemented[9]. The schematic is divided into two parts and is shown by way of example, in that it is full of barely legible detail, like the pages of a book[10].

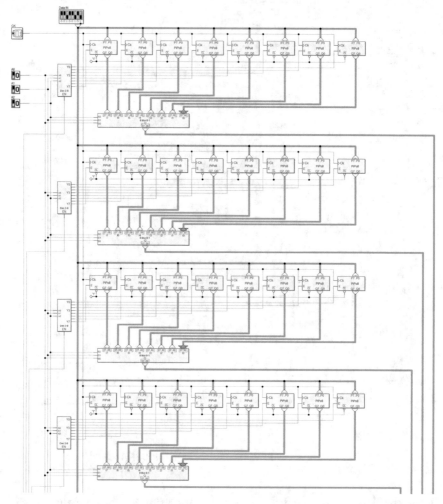

Different from the previous example, here the locations are organized in an 8-row, 8-column rectangular array, similar to real components. This organization optimizes the available space on the chip. The selection of an individual cell is defined by the intersection between the row and column where it is found.

---

[9] In a real device, deactivating the CS also brings the chip to low consumption modality where data are not accessible from the outside but their storage is guaranteed.

[10] For a detailed analysis, this network as well as all the others is available in the digital contents of the book.

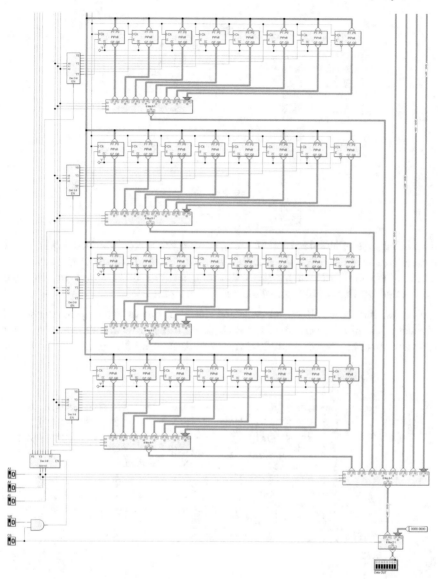

Rather than using one single $6 \to 64$ decoder, here, we use a $(3 \to 8)$ "row decoder" that enables the eight $(3 \to 8)$ "column decoders" one at a time. For the selected row, it is the column decoder that enables the register we need.

The second part of the schematic (the figure above) at the bottom left has the row decoder. As shown (enlarged) in the following figure, the selection of the outputs is given by address lines A5, A4 and A3, and is enabled by inputs WE and CS.

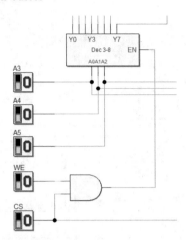

The 8 outputs of the decoder, in turn, enable the 8 column decoders, one for each row, so we can say that address lines A5, A4 and A3 select the row. The column decoders, driven by the remaining address lines, select the register positioned on the column specified by A2, A1 and A0.

The following figure shows an enlarged detail of one of the column decoders. For easy readability, only 4 of the 8 registers in the row are shown, but the 8 decoder outputs are connected to the E enables of all 8 registers.

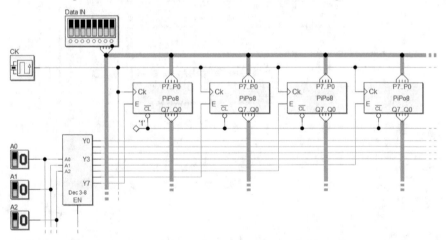

In sum, for a given address A5..A0, only one row will be selected (through lines A5, A4 and A3) and inside of that, only one register (with lines A2, A1 and A0). In the writing phase, the activation of WE will be only routed to the E enable of the addressed register.

For what regards reading the data stored on the register, as shown in the following figure, every row has a column multiplexer whose selection is given by address lines A2, A1 and A0. The content of the desired register is therefore available at the output of the column multiplexer.

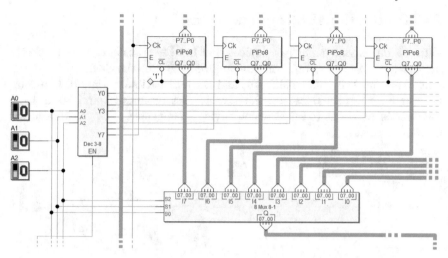

All the column multiplexers converge in the row multiplexer (see the figure below), which selects the content coming from the rows based on lines A5, A4 and A3. The output of the multiplexer is then brought out of the RAM allowing for the data stored in the addressed register to be read.

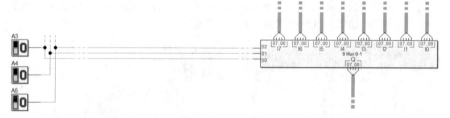

The decoders and multiplexers here have made it possible to describe the RAM architecture in functional terms[11].

The architecture examined in these pages is not the only possible one; there are synchronous RAM memories with different design choices inside. For example, there are memories that have a register to store the address provided from the outside.

Also, memory cells are generally designed with simpler flip-flops than those adopted here, so it is possible to save on the overall number of transistors required. The registers that we have used in these examples employ "PET" (Positive Edge Triggered) flip-flops, while real RAMs have "D-latch" flip-flops, which are much smaller and more economical because they use fewer transistors.

---

[11] In a real component the decoders and multiplexers are replaced by special microelectronic solutions that make it possible to reduce the number of wires and transistors while performing the same functions.

A.3 Bidirectional bus connections

Let's assume we have multiple devices that need to transfer data to a specific destination. In the example in the following figure, the devices are represented by simple inputs (IN3..IN0), and the destination by the OUT output, for simplicity's sake. Based on the selection set by lines S1 and S0, a $4 \rightarrow 1$ multiplexer allows us to select which device is connected to the destination at a given moment.

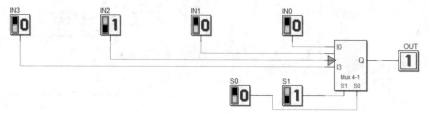

This solution is very general and widely used. Still, when there is a bus connection and a large number of devices, this method becomes costly. This is particularly true if the connections are not inside of a single chip but link physically distinct chips found on different boards.

Let's look at the following example where the devices transfer data through their 8-wire outputs (represented as "busses").

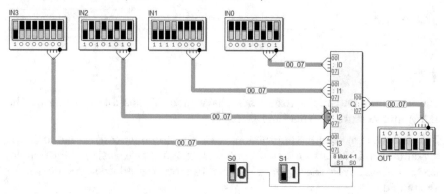

Keep in mind that the multiplexer used here with bus-type connections is built internally with eight $4 \rightarrow 1$ multiplexers (like that of the previous example), one for each of the wires the bus is made of.

Let's determine the number of wires necessary. We have 4 connections with 8 wires each entering the multiplexer and one 8-wire connection in the output plus the two selection wires. Overall, we have 42 wires.

If we wanted to raise the number of devices in the design, we would need to add an 8-wire bus for each additional device. However, we would also need a multiplexer with more channels and this would entail redesigning the board (or boards) the network is made on.

It would be preferable to design the system so as to save on the number of wires. This would reduce costs and keep flexibility and modularity so we would not have to redesign the interconnecting parts every time we need to add elements to the system.

If we use a special type of component, a "tri-state" buffer, we can reduce the number of wires we need to connect the chip to the different boards at the cost of slowing connection speed. With this component, multiple devices can share the same bunch of wires (a bus) to transfer information in both directions where needed, giving us "bidirectional" connections.

A.3.1 Tri-state buffers

A tri-state buffer is a special type of logic device that works like a normal buffer when enabled, and electrically disconnects its outputs when it is not. In the following figure, the left side shows the tri-state buffer symbol while the right shows its functional equivalent.

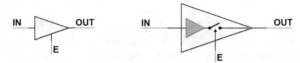

The figure shows an E enable input, normally drawn from the side with respect to the path. When the tri-state buffer output is disabled (E = '0'), this brings it to a state of "high impedance" ('High-Z', or simply 'Z'). It behaves as if there were an open switch at the output of a normal buffer[12]. When E is active, however, output buffering is enabled (the switch is closed).

The truth table of the tri-state buffer (including state 'Z') is shown below.

E	IN	OUT	Function
0	-	'Z'	'High-Z' (output disconnected)
1	0	0	OUT = IN (identity)
1	1	1	

Here below are three examples of tri-state buffer operations, using Deeds and the "animation " simulation.

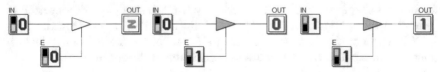

In the example on the left, the E enable is not active, so the output is disconnected and the simulator shows the symbol 'Z'.

---

[12] The switching is done electronically through a control of the transistors that drive the buffer output.

In the center and right examples, E = '1', so the output copies the logic level in the input as in a normal buffer (notice the component shows a different color when the enable is active during the simulation).

A tri-state buffer by itself is not very useful, when employed alone. However, when a certain number of them are used together to reproduce the logical functionality of a multiplexer, their advantages become clear.

The following figure shows the initial example of four devices that transfer their data to a destination that is adapted to tri-state buffers.

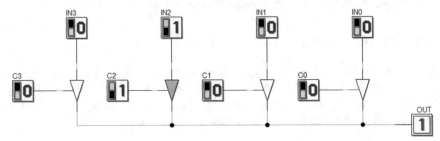

As before, the devices are represented by inputs IN3..IN0, and the destination by the OUT output. Each of the devices can connect with the destination through its own tri-state buffer if it is enabled by the corresponding control (C3..C0).

We will soon see the criteria used to drive these control lines. For now, it should be noted that we will activate them only once, as in the example in the figure. Here, we see that OUT copies IN2 in that control line C2 is active while the others are not.

It would make no sense to activate more than one as it would cause an electrical conflict between the buffer outputs, especially if they tried to set different logical values on the line.

Let's start to look at an important property of this solution that uses tri-state buffers. The output can be connected at any point of the shared connection. While using the multiplexer, the output is generated downstream the component itself.

If we look at the figure above, the information can be produced by the device farthest to the right (IN0) and received by the left side of the connection or generated by the device farthest on the left (IN3) and received by the right side. In other words, this solution allows for a "bidirectional connection"; it makes it possible to transmit data along the line in both directions.

Furthermore, it is clear that this solution saves on the number of wires needed since we use only one common wire rather than one wire for each device. This advantage will become important when the number of devices grows, especially when there is a bus-type data connection.

A.3.2 Tri-state buffers and busses

To connect a bus, we use as many tri-state buffers as the wires it has, and the E enables are connected together (see the left of the following figure). For schematics, it is convenient to use a component with bus-type connections that integrates them into a single symbol (on the right).

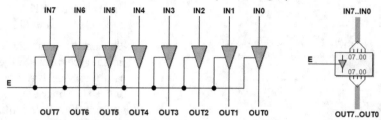

The next figure shows four devices that transfer data to a destination through their 8-wire output busses. This time, however, we employ the tri-state bus-type buffer just introduced.

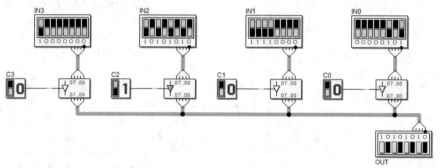

About this example, we can draw the same observations that we have done in the case of the single-wire connection network seen beforehand. The considerations about bi-directional capabilities and the savings in terms of the number of wires in use are noteworthy. As we can see in the figure above, OUT copies IN2 because control line C2 is active while the others are not.

At this point, we have to consider how to manage the selection of the single device ensuring that all the others are disabled.

The example on the following page has 4 units shown inside blue dotted lines. They share a common 8-wire bus to transfer data to the OUT output. Each unit is connected to the bus through a bus-type tri-state buffer. The buffer enables are shown by the C3, C2, C1 and C0 one-bit displays.

Selection lines S1 and S0 are accessible to all the units and control which one will be enabled. Unlike the previous examples, however, each one's tri-state enable is delegated to its own units with the insertion of a private decoder in each one. In every respect, S1 and S0 become the "address" lines of the system. So, every unit has its own, unique "address" attributed to it.

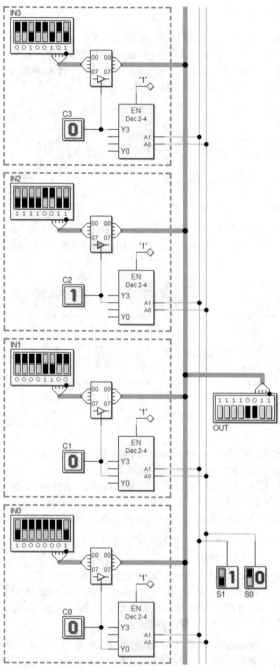

In the simulation screenshot shown in the figure, S1S0 = '10', so unit IN2 is activated. In fact, the decoder in unit IN2 is connected so as to recognize the selection S2S1 = '10' and activate line C2.

The decoders of the other units recognize the other combinations but not this one, so the only enabled tri-state buffer is the one for unit IN2. Therefore, OUT copies the values generated by IN2 to the bus.

If we raise the number S of address lines, we will be able to connect many more units (up to $N = 2^S$) to the bus.

We will then have what we call system "modularity" in the sense that we can add or remove units (e.g. through connectors) without having to change the general architecture, once a standard of bus connections is defined.

The network shown here receives the selection from the outside but in real systems, it is generated by the system itself. To achieve this, one common option used in microcomputers for example, is to assign the special role of "master" of communication to one of the modules.

The other units (typically called "slaves") are controlled by the master and respond to its commands. Generally speaking, each of the modules could be designed not only to send data but also to receive them, thus allowing for a multi-directional exchange of information among the units facing the bus.

Since the bus is a single, shared resource, if two units are exchanging data with each other, the bus will be occupied and it will be impossible to manage another data exchange at the same time. Therefore, data exchanges must be executed one at a time.

Now, let's look at the schematic of a master, regarding the mere control over communication on the bus (see the figure below).

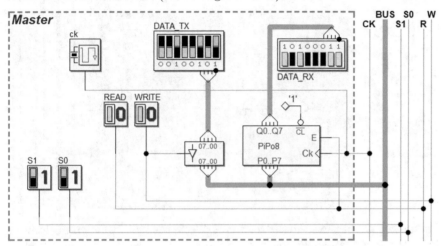

As mentioned, it is the master that generates the selection (the address) of the module to enable from time to time. It controls lines S1S0 that run alongside the bus (on the right side of the figure). On the left, the two switches S1 and S0 symbolize the control of those lines, which in a real application are produced by a logic network that sequences the operations.

At the center, we see the tri-state buffer that makes it possible to transmit data provided by the master (DATA_TX) over the bus. These data are received by the selected unit. To the side, we find the parallel register, introduced to receive and store (DATA_RX) the information from the bus.

From the perspective of the master, an operation that transfers a number to the selected unit is called a "write" operation. Similarly, a "read" operation is the opposite: a number coming from the addressed unit that is acquired by the master.

To assure that the read and write operations are carried out in a mutually exclusive way, we add two more lines to the bus, R (Read) and W (Write), which are also generated by the master. The READ and WRITE buttons shown in the figure represent the lines generated by the sequencing logic in a real system.

Finally, so that all the operations can be carried out synchronously, we add a
CK line to the bus. This will allow all the units to synchronize their operations
with the clock of the master unit.

Now, let's turn our attention to the units that are under the command of the
master. The following figure shows the slave unit designed to be enabled by
address S1S0 = '11'.

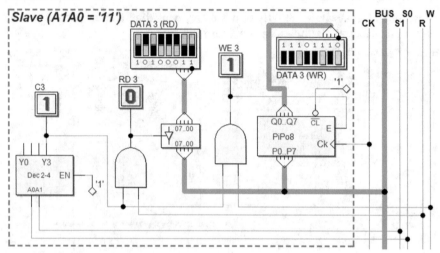

On the left we see the address decoder. It selects the unit by activating internal
line C3 when the master generates S1S0 at '11'. However, this selection by
itself is not enough since the master needs to declare if it wants the read or
the write operation to be carried out (by activating line R or W).

The schematic also shows a tri-state buffer and a parallel register, which have
the same function as those inside the master except that their enable lines are
activated in the presence of the selection of the unit and its respective R or
W signal.

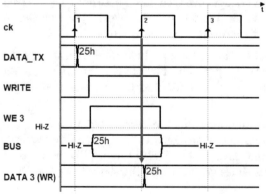

The timing diagram on the
left shows the signal sequence
for the write operation.

Let's assume that the selection generated by the master is S1S0 = '11', maintained
constant along the time interval considered here.

At clock edge 1, the mas-
ter activates WRITE, which
propagates on line W and
reaches all the units.

WRITE also enables the tri-state buffer in the master, so the BUS that was in Hi-Z, is now defined and copies the number set on DATA_TX (25h, in the example).

In response to line W, only the selected receiving unit activates internal line WE3, enabling the write operation in the correct register. So, the number is stored in the destination on clock edge 2.

The following figure shows the signal sequence for the read operation.

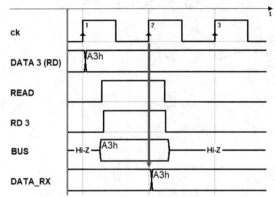

On clock edge 1, the master activates the READ line, which locally enables the write operation in the master's register. This is the selected destination of the data transfer.

The read signal propagates all the units on BUS line R and, of course, reaches the selected unit, which activates the RD3 line internally.

The RD3 line enables its own tri-state buffer and therefore copies the value on DATA3(RD) to the BUS. The master is the only network where the writing on its own register is enabled, so on clock edge 2, the number will be stored there, thus arriving at the desired destination.

On the following page, the schematic represents the complete system in our example with one master (at the top) and the four slave units.

In the timing diagrams studied here, we see that there are time intervals when the BUS is not driven by any of the tri-state buffers since they are all disabled in the Hi-Z state. In these intervals, the logic value of the bus lines is indefinite (indicated as Hi-Z in the figures), because it is not set by any active device[13].

In conclusion, let's make mention of another type of component, the bidirectional tri-state buffer, shown in the figure at the left.

It is made up of two buffers connected from opposite ends and enabled by lines E and DIR.

When E is active, DIR sets the transfer direction. B copies A if DIR = '1', otherwise A copies B. If E = '0', A and B are unconnected (Hi-Z). This buffer is useful when we want to connect two bidirectional buses together.

---

[13] In the time intervals when it is in Hi-Z, the bus of a real system tends to assume values depending on the physical parameters and the design of the bus itself, which should be considered unknown from a logical perspective.

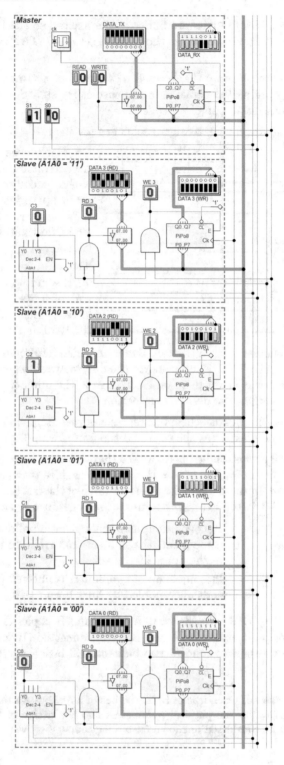

A.3.3 Tri-state memories

A bidirectional bus is often used to connect memory devices. The following figure on the left shows a ROM memory device (4K x 8) that is compatible with a bidirectional-type connection.

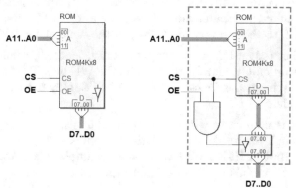

The OE (Output Enable) line commands the activation of data outputs D7..D0, which in this component are tri-state types. The right hand side of the figure shows an equivalent network, in which we use a ROM with non-tri-state outputs, like those examined in Section A.1. A tri-state buffer has been added to the memory in series at the data outputs.

The buffer enable depends on the AND between OE and CS; if the device is enabled (CS = '1'), setting OE = '1' as well, the tri-states of outputs D7..D0 will be enabled, allowing us to read data from the memory. In Chapter 2 we saw an example of how it is used (see page 150).

In the following figure on the left, we show a (4K x 8) RAM memory device. This component is also designed to be used with a bidirectional data bus.

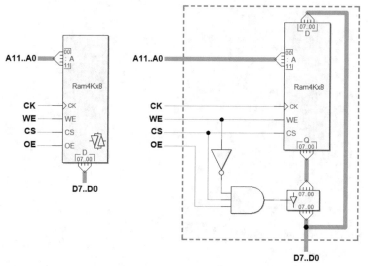

In addition to control lines CS, WE and CK, explained in Section A.2, this component also has the OE enable.

On the right side of the figure is the equivalent schematic of the component, starting from a RAM memory that is the same size but has separate inputs and outputs (as in Section A.2). As with the ROM, we see a tri-state buffer added in series at the data outputs. Nevertheless, we also have a direct connection of lines D7..D0 to the memory input lines of the same name.

Due to the fact that a RAM has the write functionality, output enable does not only depend on OE and CS but also on the WE line. The tri-state buffer is enabled only if write is inactive (WE = '0'), i.e., only if we are in memory data read mode.

For writing, the number to write comes from the bus. To be acquired by the memory, through inputs D7..D0, the tri-state buffer output lines have to be disconnected (put in Hi-Z). We have seen an example of a RAM application with a bidirectional bus in Chapter 2, on page 151.

B

Programmable computing networks: Schematics and tables

B.1 The Mp8A computing network

In this section, we report the table of instructions and the complete schematic of the Mp8A computing network described in Section 1.2.4.

B.1.1 Table of instructions

Mnemonic		Machine Code								
		P1	P0	END	F2	F1	F0	S1	S0	(hex)
IN	A,OP0	0	0	0	1	1	1	1	1	1Fh
IN	A,OP1	0	1	0	1	1	1	1	1	5Fh
IN	A,OP2	1	0	0	1	1	1	1	1	9Fh
IN	A,OP3	1	1	0	1	1	1	1	1	DFh
ADD	A,OP0	0	0	0	0	0	0	1	1	03h
ADD	A,OP1	0	1	0	0	0	0	1	1	43h
ADD	A,OP2	1	0	0	0	0	0	1	1	83h
ADD	A,OP3	1	1	0	0	0	0	1	1	C3h
SUB	A,OP0	0	0	0	0	0	1	1	1	07h
SUB	A,OP1	0	1	0	0	0	1	1	1	47h
SUB	A,OP2	1	0	0	0	0	1	1	1	87h
SUB	A,OP3	1	1	0	0	0	1	1	1	C7h

(cont.)

© The Editor(s) (if applicable) and The Author(s), under exclusive license to Springer Nature Switzerland AG 2022
G. Donzellini et al., *Introduction to Microprocessor-Based Systems Design*,
https://doi.org/10.1007/978-3-030-87344-8

Mnemonic		Machine Code								
		P1	P0	END	F2	F1	F0	S1	S0	(hex)
AND	A,OP0	0	0	0	0	1	0	1	1	0Bh
AND	A,OP1	0	1	0	0	1	0	1	1	4Bh
AND	A,OP2	1	0	0	0	1	0	1	1	8Bh
AND	A,OP3	1	1	0	0	1	0	1	1	CBh
OR	A,OP0	0	0	0	0	1	1	1	1	0Fh
OR	A,OP1	0	1	0	0	1	1	1	1	4Fh
OR	A,OP2	1	0	0	0	1	1	1	1	8Fh
OR	A,OP3	1	1	0	0	1	1	1	1	CFh
NOT	A	0	0	0	1	0	0	1	1	13h
SRL	A	0	0	0	0	0	0	0	1	01h
SLL	A	0	0	0	0	0	0	1	0	02h
NOP		0	0	0	0	0	0	0	0	00h
HALT		0	0	1	0	0	0	0	0	20h

B.1.2 The schematic of the Mp8A computing network

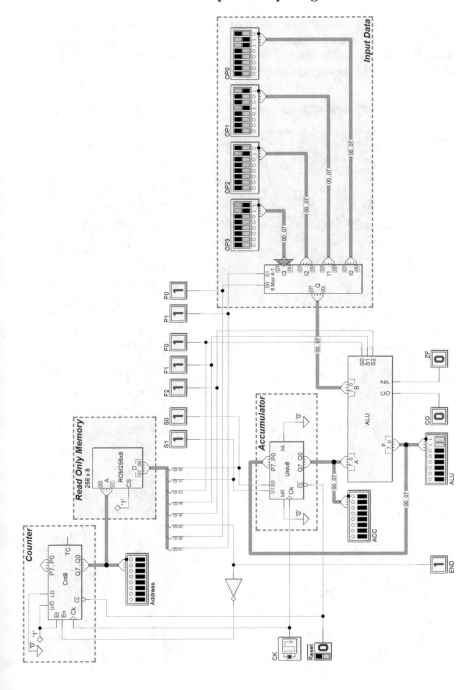

B.2 The Mp8B computing network

In this section, we report the table of instructions and the corresponding microprograms, as well as the complete schematic of the Mp8B computing network described at the start of Section 1.3. This network is the first example of the use of a microprogrammed sequencer.

B.2.1 Table of instructions and the correlated microprograms

Mnemonic		Machine Code	Microprogram
NOP		00h	1100.0000.0000.0000
HALT		01h,01h	1000.0000.0000.0000
ADD	A,OP0	04h	1100.0000.0001.1000
ADD	A,OP1	05h	1100.0000.0001.1010
ADD	A,OP2	06h	1100.0000.0001.1100
ADD	A,OP3	07h	1100.0000.0001.1110
SUB	A,OP0	08h	1100.0000.0011.1000
SUB	A,OP1	09h	1100.0000.0011.1010
SUB	A,OP2	0Ah	1100.0000.0011.1100
SUB	A,OP3	0Bh	1100.0000.0011.1110
AND	A,OP0	0Ch	1100.0000.0101.1000
AND	A,OP1	0Dh	1100.0000.0101.1010
AND	A,OP2	0Eh	1100.0000.0101.1100
AND	A,OP3	0Fh	1100.0000.0101.1110
OR	A,OP0	10h	1100.0000.0111.1000
OR	A,OP1	11h	1100.0000.0111.1010
OR	A,OP2	12h	1100.0000.0111.1100
OR	A,OP3	13h	1100.0000.0111.1110
NOT	A	14h	1100.0000.1001.1000
IN	A,OP0	1Ch	1100.0000.1111.1000
IN	A,OP1	1Dh	1100.0000.1111.1010
IN	A,OP2	1Eh	1100.0000.1111.1100
IN	A,OP3	1Fh	1100.0000.1111.1110
SRL	A	20h	1100.0000.0000.1000
SLL	A	21h	1100.0000.0001.0000

B.2.2 The schematic of the Mp8B computing network

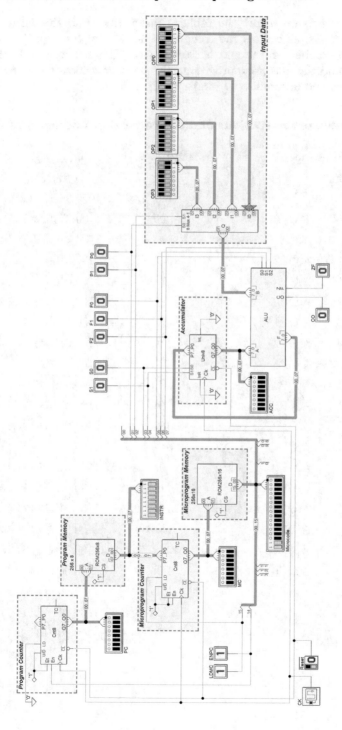

B.3 The Mp8C computing network

In this section, we report the table of instructions and the corresponding microprograms, as well as the complete schematic of the Mp8C computing network described at the end of Section 1.4. This network adds the Flag Register and the new microprogrammed sequencer, which is able to execute conditional and unconditional jumps, to the Mp8B.

B.3.1 Table of instructions and the correlated microprograms

Mnemonic		Machine Code	Microprogram
NOP		00h	1100.0000.0000.0000
HALT		01h,01h	1000.0000.0000.0000
ADD	A,OP0	04h	1100.0100.0001.1000
ADD	A,OP1	05h	1100.0100.0001.1010
ADD	A,OP2	06h	1100.0100.0001.1100
ADD	A,OP3	07h	1100.0100.0001.1110
SUB	A,OP0	08h	1100.0100.0011.1000
SUB	A,OP1	09h	1100.0100.0011.1010
SUB	A,OP2	0Ah	1100.0100.0011.1100
SUB	A,OP3	0Bh	1100.0100.0011.1110
AND	A,OP0	0Ch	1100.0100.0101.1000
AND	A,OP1	0Dh	1100.0100.0101.1010
AND	A,OP2	0Eh	1100.0100.0101.1100
AND	A,OP3	0Fh	1100.0100.0101.1110
OR	A,OP0	10h	1100.0100.0111.1000
OR	A,OP1	11h	1100.0100.0111.1010
OR	A,OP2	12h	1100.0100.0111.1100
OR	A,OP3	13h	1100.0100.0111.1110
NOT	A	14h	1100.0100.1001.1000
IN	A,OP0	1Ch	1100.0000.1111.1000
IN	A,OP1	1Dh	1100.0000.1111.1010
IN	A,OP2	1Eh	1100.0000.1111.1100
IN	A,OP3	1Fh	1100.0000.1111.1110

(cont.)

Mnemonic		Machine Code	Microprogram
SRL	A	20h	1100.0100.1100.1000
SLL	A	21h	1100.0100.1101.0000
JP	\<address\>	22h	0000.1000.0000.0000
			1100.0000.0000.0000
JP	Z, \<address\>	24h	0110.0000.0000.0000
			1100.0000.0000.0000
JP	NZ, \<address\>	26h	0110.1000.0000.0000
			1100.0000.0000.0000
JP	C, \<address\>	28h	0111.0000.0000.0000
			1100.0000.0000.0000
JP	NC, \<address\>	2Ah	0111.1000.0000.0000
			1100.0000.0000.0000
CP	A,OP0	30h	1100.0100.0010.0000
CP	A,OP1	31h	1100.0100.0010.0010
CP	A,OP2	32h	1100.0100.0010.0100
CP	A,OP3	33h	1100.0100.0010.0110

B.3.2 The schematic of the Mp8C computing network

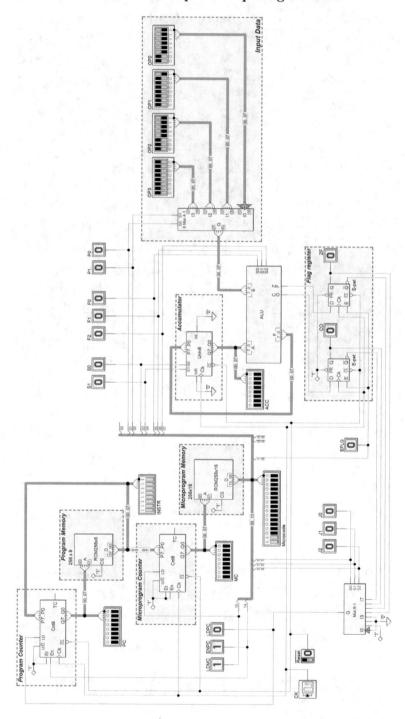

B.4 The Mp8D computing network

In this section, we report the table of instructions and the corresponding microprograms, as well as the complete schematic of the Mp8D computing network described in Section 1.5. This network adds two output ports to the Mp8C.

B.4.1 Table of instructions and the correlated microprograms

Mnemonic		Machine Code	Microprogram
NOP		00h	1100.0000.0000.0000
HALT		01h,01h	1000.0000.0000.0000
ADD	A,OP0	04h	1100.0100.0001.1000
ADD	A,OP1	05h	1100.0100.0001.1010
ADD	A,OP2	06h	1100.0100.0001.1100
ADD	A,OP3	07h	1100.0100.0001.1110
SUB	A,OP0	08h	1100.0100.0011.1000
SUB	A,OP1	09h	1100.0100.0011.1010
SUB	A,OP2	0Ah	1100.0100.0011.1100
SUB	A,OP3	0Bh	1100.0100.0011.1110
AND	A,OP0	0Ch	1100.0100.0101.1000
AND	A,OP1	0Dh	1100.0100.0101.1010
AND	A,OP2	0Eh	1100.0100.0101.1100
AND	A,OP3	0Fh	1100.0100.0101.1110
OR	A,OP0	10h	1100.0100.0111.1000
OR	A,OP1	11h	1100.0100.0111.1010
OR	A,OP2	12h	1100.0100.0111.1100
OR	A,OP3	13h	1100.0100.0111.1110
NOT	A	14h	1100.0100.1001.1000
IN	A,OP0	1Ch	1100.0000.1111.1000
IN	A,OP1	1Dh	1100.0000.1111.1010
IN	A,OP2	1Eh	1100.0000.1111.1100
IN	A,OP3	1Fh	1100.0000.1111.1110
SRL	A	20h	1100.0100.1100.1000
SLL	A	21h	1100.0100.1101.0000

(cont.)

Mnemonic		Machine Code	Microprogram
JP	<address>	22h	0000.1000.0000.0000
			1100.0000.0000.0000
JP	Z, <address>	24h	0110.0000.0000.0000
			1100.0000.0000.0000
JP	NZ, <address>	26h	0110.1000.0000.0000
			1100.0000.0000.0000
JP	C, <address>	28h	0111.0000.0000.0000
			1100.0000.0000.0000
JP	NC, <address>	2Ah	0111.1000.0000.0000
			1100.0000.0000.0000
CP	A,OP0	30h	1100.0100.0010.0000
CP	A,OP1	31h	1100.0100.0010.0010
CP	A,OP2	32h	1100.0100.0010.0100
CP	A,OP3	33h	1100.0100.0010.0110
OUT	PORT0,A	34h	1100.0001.1100.0000
OUT	PORT1,A	35h	1100.0010.1100.0000

B.4.2 The schematic of the Mp8D computing network

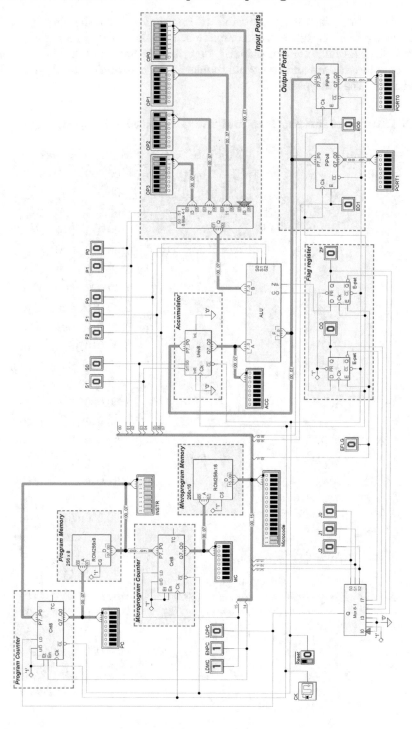

B.5 The Mp8E computing network

In this section, we report the table of instructions and the corresponding microprograms, as well as the complete schematic of the overall computing network in its final version (Mp8E), described in Section 1.6. This network can handle variables and constants thanks to the addition of RAM and to the revision of the data path.

B.5.1 Table of instructions and the correlated microprograms

Mnemonic		Machine Code	Microprogram
NOP		00h	1100.0000.0000.0000
HALT		01h,01h	1000.0000.0000.0000
ADD	A,OP0	04h	1100.0100.0001.1000
ADD	A,OP1	05h	1100.0100.0001.1010
ADD	A,<const>	40h	0100.0100.0001.1110
			1100.0000.0000.0000
ADD	A,(address)	4Ch	0100.0000.0000.0000
			1100.0100.0001.1100
SUB	A,OP0	08h	1100.0100.0011.1000
SUB	A,OP1	09h	1100.0100.0011.1010
SUB	A,<const>	42h	0100.0100.0011.1110
			1100.0000.0000.0000
SUB	A,(address)	4Eh	0100.0000.0000.0000
			1100.0100.0011.1100
AND	A,OP0	0Ch	1100.0100.0101.1000
AND	A,OP1	0Dh	1100.0100.0101.1010
AND	A,<const>	44h	0100.0100.0101.1110
			1100.0000.0000.0000
AND	A,(address)	50h	0100.0000.0000.0000
			1100.0100.0101.1100
OR	A,OP0	10h	1100.0100.0111.1000
OR	A,OP1	11h	1100.0100.0111.1010
OR	A,<const>	46h	0100.0100.0111.1110
			1100.0000.0000.0000
OR	A,(address)	52h	0100.0000.0000.0000
			1100.0100.0111.1100
NOT	A	14h	1100.0100.1001.1000

(cont.)

Mnemonic		Machine Code	Microprogram
IN	A,OP0	1Ch	1100.0000.1111.1000
IN	A,OP1	1Dh	1100.0000.1111.1010
IN	(address),OP0	56h	0100.0000.1110.0001
			1100.0000.0000.0000
IN	(address),OP1	58h	0100.0000.1110.0011
			1100.0000.0000.0000
SRL	A	20h	1100.0100.1100.1000
SLL	A	21h	1100.0100.1101.0000
JP	<address>	22h	0000.1000.0000.0000
			1100.0000.0000.0000
JP	Z, <address>	24h	0110.0000.0000.0000
			1100.0000.0000.0000
JP	NZ, <address>	26h	0110.1000.0000.0000
			1100.0000.0000.0000
JP	C, <address>	28h	0111.0000.0000.0000
			1100.0000.0000.0000
JP	NC, <address>	2Ah	0111.1000.0000.0000
			1100.0000.0000.0000
CP	A,OP0	30h	1100.0100.0010.0000
CP	A,OP1	31h	1100.0100.0010.0010
CP	A,<const>	4Ah	0100.0100.0010.0110
			1100.0000.0000.0000
CP	A,(address)	5Ah	0100.0000.0000.0000
			1100.0100.0010.0100
OUT	PORT0,A	34h	1100.0001.1100.0000
OUT	PORT0,(address)	5Ch	0100.0000.0000.0000
			1100.0001.1110.0100
OUT	PORT1,A	35h	1100.0010.1100.0000
OUT	PORT1,(address)	5Eh	0100.0000.0000.0000
			1100.0010.1110.0100
LD	(<address>),A	38h	0100.0000.1100.0001
			1100.0000.0000.0000
LD	A,(<address>)	3Ah	0100.0000.0000.0000
			1100.0000.1111.1100
LD	A,<const>	3Dh	0100.0000.1111.1110
			1100.0000.0000.0000

B.5.2 The schematic of the Mp8E computing network

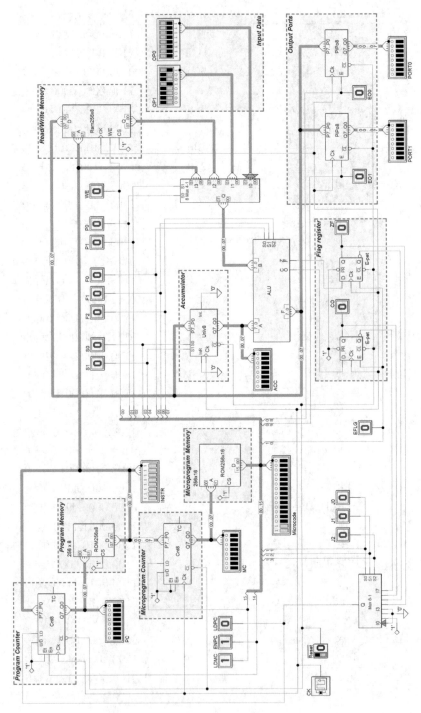

C

DMC8 instruction set tables

This appendix presents, grouped by category, the DMC8 microprocessor's tables of the instruction set.

Meanings of the column headers of the tables (from left to right):

Header	Meaning
Mnemonic	The assembly mnemonic code of the instruction.
Symbolic operation	The operation executed by the instruction, symbolically represented.
Flags	The flags affected according to the operation result (see below the summary of flag operation).
Opcode	The operation code of the instruction, in binary.
Hex	The operation code of the instruction, in hex (when unique).
#B	The number of bytes of the instruction.
#M	The number of machine cycles of the instruction.
#K	The number of clock cycles of the instruction.
Comments	Any comments

Summary of flag operation:

Flag	Name	Operation
S	Sign flag	$S = 1$ if the MSB of the result is one.
Z	Zero flag	$Z = 1$ if the result of the operation is zero.
H	Half-Carry flag	$H = 1$ if the add or subtract operation produced a carry into (or borrow from) bit 4 of the accumulator.
P/V	Parity/Overflow flag	Parity (P) and Overflow (V) share the same flag. For logical operations, $P = 1$ if the result parity is even, $P = 0$ if it is odd. For arithmetic operations, $V = 1$ if the result produces an overflow.
N	Add/Subtract flag	$N = 1$ if the previous operation was a subtract.
C	Carry flag	$C = 1$ if the operation produced a carry from the MSB of the operand or result

G. Donzellini et al., *Introduction to Microprocessor-Based Systems Design*, https://doi.org/10.1007/978-3-030-87344-8

C.1 Data transfer instructions (8-bit)

Mnemonic	Symbolic Operation	Flag S Z H P/v N C	Opcode 76 543 210	Hex	#B	#M	#K	Comments
LD r,r'	r ← r'	• • • • • •	01 r r'		1	1	4	r,r' Reg.
LD r,n	r ← n	• • • • • •	00 r 110 ← n →		2	2	7	000 B
LD r,(HL)	r ← (HL)	• • • • • •	01 r 110		1	2	7	001 C 010 D
LD r,(IX+d)	r ← (IX+d)	• • • • • •	11 011 101 01 r 110 ← d →	DD	3	5	19	011 E 100 H 101 L
LD r,(IY+d)	r ← (IY+d)	• • • • • •	11 111 101 01 r 110 ← d →	FD	3	5	19	111 A
LD (HL),r	(HL) ← r	• • • • • •	01 110 r		1	2	7	
LD (IX+d),r	(IX+d) ← r	• • • • • •	11 011 101 01 110 r ← d →	DD	3	5	19	
LD (IY+d),r	(IY+d) ← r	• • • • • •	11 111 101 01 110 r ← d →	FD	3	5	19	
LD (HL),n	(HL) ← n	• • • • • •	00 110 110 ← n →	36	2	3	10	
LD (IX+d),n	(IX+d) ← n	• • • • • •	11 011 101 00 110 110 ← d → ← n →	DD 36	4	5	19	
LD (IY+d),n	(IY+d) ← n	• • • • • •	11 111 101 00 110 110 ← d → ← n →	FD 36	4	5	19	
LD A,(BC)	A ← (BC)	• • • • • •	00 001 010	0A	1	2	7	
LD A,(DE)	A ← (DE)	• • • • • •	00 011 010	1A	1	2	7	
LD A,(nn)	A ← (nn)	• • • • • •	00 111 010 ← n → ← n →	3A	3	4	13	
LD (BC),A	(BC) ← A	• • • • • •	00 000 010	02	1	2	7	
LD (DE),A	(DE) ← A	• • • • • •	00 010 010	12	1	2	7	
LD (nn),A	(nn) ← A	• • • • • •	00 110 010 ← n → ← n →	32	3	4	13	
Notes:	r, r' = any one of the CPU 8-bit registers A, B, C, D, E, H, L. n = 8-bit value in range 0..255. nn = 16-bit value in range 0..65,535. d = 8-bit signed value in range -128..+127.							
	• = the flag is unchanged by the operation.							

C.2 Data transfer instructions (16-bit)

Mnemonic	Symbolic Operation	Flag S Z H P/v N C	Opcode 76 543 210	Hex	#B	#M	#K	Comments
LD dd,nn	dd ← nn	• • • • • •	00 dd0 001 ← n → ← n →		3	3	10	dd Reg. ─────── 00 BC 01 DE 10 HL 11 SP
LD IX,nn	IX ← nn	• • • • • •	11 011 101 00 100 001 ← n → ← n →	DD 21	4	4	14	
LD IY,nn	IY ← nn	• • • • • •	11 111 101 00 100 001 ← n → ← n →	FD 21	4	4	14	
LD HL,(nn)	L ← (nn) H ← (nn+1)	• • • • • •	00 101 010 ← n → ← n →	2A	3	5	16	
LD dd,(nn)	dd_L ← (nn) dd_H ← (nn+1)	• • • • • •	11 101 101 01 dd1 011 ← n → ← n →	ED	4	6	20	
LD IX,(nn)	IX_L ← (nn) IX_H ← (nn+1)	• • • • • •	11 011 101 00 101 010 ← n → ← n →	DD 2A	4	6	20	
LD IY,(nn)	IY_L ← (nn) IY_H ← (nn+1)	• • • • • •	11 111 101 00 101 010 ← n → ← n →	FD 2A	4	6	20	
LD (nn),HL	(nn) ← L (nn+1) ← H	• • • • • •	00 100 010 ← n → ← n →	22	3	5	16	
LD (nn),dd	(nn) ← dd_L (nn+1) ← dd_H	• • • • • •	11 101 101 01 dd0 011 ← n → ← n →	DD	4	6	20	
LD (nn),IX	(nn) ← IX_L (nn+1) ← IX_H	• • • • • •	11 011 101 00 100 010 ← n → ← n →	DD 22	4	6	20	
LD (nn),IY	(nn) ← IY_L (nn+1) ← IY_H	• • • • • •	11 111 101 00 100 010 ← n → ← n →	FD 22	4	6	20	

(cont.)

Mnemonic	Symbolic Operation	Flag S Z H P/v N C	Opcode 76 543 210	Hex	#B	#M	#K	Comments
LD SP,HL	SP ← HL	• • • • • •	11 111 001	F9	1	1	6	
LD SP,IX	SP ← IX	• • • • • •	11 011 101 11 111 001	DD F9	2	2	10	
LD SP,IY	SP ← IY	• • • • • •	11 111 101 11 111 001	FD F9	2	2	10	
EX DE,HL	DE↔HL	• • • • • •	11 101 011	EB	1	1	4	
EX (SP),HL	H↔(SP+1) L↔(SP)	• • • • • •	11 100 011	E3	1	5	19	
PUSH qq	SP ← SP-1 (SP) ← qq$_H$ SP ← SP-1 (SP) ← qq$_L$	• • • • • •	11 qq0 101		1	3	11	qq Reg. 00 BC 01 DE 10 HL 11 AF
PUSH IX	SP ← SP-1 (SP) ← IX$_H$ SP ← SP-1 (SP) ← IX$_L$	• • • • • •	11 011 101 11 100 101	DD E5	2	4	15	
PUSH IY	SP ← SP-1 (SP) ← IY$_H$ SP ← SP-1 (SP) ← IY$_L$	• • • • • •	11 111 101 11 100 101	FD E5	2	4	15	
POP qq	qq$_L$ ← (SP) SP ← SP+1 qq$_H$ ← (SP) SP ← SP+1	• • • • • •	11 qq0 001		1	3	10	
POP IX	IX$_L$ ← (SP) SP ← SP+1 IX$_H$ ← (SP) SP ← SP+1	• • • • • •	11 011 101 11 100 001	DD E1	2	4	14	
POP IY	IY$_L$ ← (SP) SP ← SP+1 IY$_H$ ← (SP) SP ← SP+1	• • • • • •	11 111 101 11 100 001	FD E1	2	4	14	
Notes:	dd = any one of the CPU 16-bit registers BC, DE, HL, SP. qq = any one of the CPU 16-bit registers BC, DE, HL, AF.							
	n = 8-bit value in range 0..255. nn = 16-bit value in range 0..65,535.							
	• = the flag is unchanged by the operation.							

C.3 Arithmetic and logic instructions (8-bit)

Mnemonic	Symbolic Operation	Flag S Z H P/v N C	Opcode 76 543 210	Hex	#B	#M	#K	Comments
ADD A,r	A ← A+r	⇅ ⇅ ⇅ V 0 ⇅	10 000 r		1	1	4	r Reg.
ADD A,n	A ← A+n	⇅ ⇅ ⇅ V 0 ⇅	11 000 110 ← n →		2	2	7	000 B
ADD A,(HL)	A ← A+(HL)	⇅ ⇅ ⇅ V 0 ⇅	10 000 110		1	2	7	001 C 010 D
ADD A,(IX+d)	A ← A+(IX+d)	⇅ ⇅ ⇅ V 0 ⇅	11 011 101 10 000 110 ← d →	DD	3	5	19	011 E 100 H 101 L
ADD A,(IY+d)	A ← A+(IY+d)	⇅ ⇅ ⇅ V 0 ⇅	11 111 101 10 000 110 ← d →	FD	3	5	19	111 A
ADC A,r	A ← A+r+CY	⇅ ⇅ ⇅ V 0 ⇅	10 001 r		1	1	4	
ADC A,n	A ← A+n+CY	⇅ ⇅ ⇅ V 0 ⇅	11 001 110 ← n →		2	2	7	
ADC A,(HL)	A ← A+(HL)+CY	⇅ ⇅ ⇅ V 0 ⇅	10 001 110		1	2	7	
ADC A,(IX+d)	A ← A+(IX+d)+CY	⇅ ⇅ ⇅ V 0 ⇅	11 011 101 10 001 110 ← d →	DD	3	5	19	
ADC A,(IY+d)	A ← A+(IY+d)+CY	⇅ ⇅ ⇅ V 0 ⇅	11 111 101 10 001 110 ← d →	FD	3	5	19	
SUB r	A ← A-r	⇅ ⇅ ⇅ V 1 ⇅	10 010 r		1	1	4	
SUB n	A ← A-n	⇅ ⇅ ⇅ V 1 ⇅	11 010 110 ← n →		2	2	7	
SUB (HL)	A ← A-(HL)	⇅ ⇅ ⇅ V 1 ⇅	10 010 110		1	2	7	
SUB (IX+d)	A ← A-(IX+d)	⇅ ⇅ ⇅ V 1 ⇅	11 011 101 10 010 110 ← d →	DD	3	5	19	
SUB (IY+d)	A ← A-(IY+d)	⇅ ⇅ ⇅ V 1 ⇅	11 111 101 10 010 110 ← d →	FD	3	5	19	
SBC A,r	A ← A-r-CY	⇅ ⇅ ⇅ V 1 ⇅	10 011 r		1	1	4	
SBC A,n	A ← A-n-CY	⇅ ⇅ ⇅ V 1 ⇅	11 011 110 ← n →		2	2	7	
SBC A,(HL)	A ← A-(HL)-CY	⇅ ⇅ ⇅ V 1 ⇅	10 011 110		1	2	7	
SBC A,(IX+d)	A ← A-(IX+d)-CY	⇅ ⇅ ⇅ V 1 ⇅	11 011 101 10 011 110 ← d →	DD	3	5	19	
SBC A,(IY+d)	A ← A-(IY+d)-CY	⇅ ⇅ ⇅ V 1 ⇅	11 111 101 10 011 110 ← d →	FD	3	5	19	

(cont.)

Mnemonic	Symbolic Operation	Flag S Z H P/v N C	Opcode 76 543 210	Hex	#B	#M	#K	Comments
CP r	A-r	↕ ↕ ↕ V 1 ↕	10 111 r		1	1	4	
CP n	A-n	↕ ↕ ↕ V 1 ↕	11 111 110 ← n →		2	2	7	
CP (HL)	A-(HL)	↕ ↕ ↕ V 1 ↕	10 111 110		1	2	7	
CP (IX+d)	A-(IX+d)	↕ ↕ ↕ V 1 ↕	11 011 101 10 111 110 ← d →	DD	3	5	19	
CP (IY+d)	A-(IY+d)	↕ ↕ ↕ V 1 ↕	11 111 101 10 111 110 ← d →	FD	3	5	19	
AND r	A ← A and r	↕ ↕ 1 P 0 0	10 100 r		1	1	4	
AND n	A ← A and n	↕ ↕ 1 P 0 0	11 100 110 ← n →		2	2	7	
AND (HL)	A ← A and (HL)	↕ ↕ 1 P 0 0	10 100 110		1	2	7	
AND (IX+d)	A ← A and (IX+d)	↕ ↕ 1 P 0 0	11 011 101 10 100 110 ← d →	DD	3	5	19	
AND (IY+d)	A ← A and (IY+d)	↕ ↕ 1 P 0 0	11 111 101 10 100 110 ← d →	FD	3	5	19	
OR r	A ← A or r	↕ ↕ 1 P 0 0	10 110 r		1	1	4	
OR n	A ← A or n	↕ ↕ 1 P 0 0	11 110 110 ← n →		2	2	7	
OR (HL)	A ← A or (HL)	↕ ↕ 1 P 0 0	10 110 110		1	2	7	
OR (IX+d)	A ← A or (IX+d)	↕ ↕ 1 P 0 0	11 011 101 10 110 110 ← d →	DD	3	5	19	
OR (IY+d)	A ← A or (IY+d)	↕ ↕ 1 P 0 0	11 111 101 10 110 110 ← d →	FD	3	5	19	
XOR r	A ← A xor r	↕ ↕ 1 P 0 0	10 101 r		1	1	4	
XOR n	A ← A xor n	↕ ↕ 1 P 0 0	11 101 110 ← n →		2	2	7	
XOR (HL)	A ← A xor (HL)	↕ ↕ 1 P 0 0	10 101 110		1	2	7	
XOR (IX+d)	A ← A xor (IX+d)	↕ ↕ 1 P 0 0	11 011 101 10 101 110 ← d →	DD	3	5	19	
XOR (IY+d)	A ← A xor (IY+d)	↕ ↕ 1 P 0 0	11 111 101 10 101 110 ← d →	FD	3	5	19	

(cont.)

Mnemonic	Symbolic Operation	Flag S Z H P/v N C	Opcode 76 543 210	Hex	#B	#M	#K	Comments
INC r	r ← r+1	↕ ↕ ↕ V 0 ●	00 r 100		1	1	4	
INC (HL)	(HL) ← (HL)+1	↕ ↕ ↕ V 0 ●	00 110 100		1	3	11	
INC (IX+d)	(IX+d) ← (IX+d)+1	↕ ↕ ↕ V 0 ●	11 011 101 00 110 100 ← d →	DD	3	6	23	
INC (IY+d)	(IY+d) ← (IY+d)+1	↕ ↕ ↕ V 0 ●	11 111 101 00 110 100 ← d →	FD	3	6	23	
DEC r	r ← r-1	↕ ↕ ↕ V 1 ●	00 r 101		1	1	4	
DEC (HL)	(HL) ← (HL)-1	↕ ↕ ↕ V 1 ●	00 110 101		1	3	11	
DEC (IX+d)	(IX+d) ← (IX+d) -1	↕ ↕ ↕ V 1 ●	11 011 101 00 110 101 ← d →	DD	3	6	23	
DEC (IY+d)	(IY+d) ← (IY+d)-1	↕ ↕ ↕ V 1 ●	11 111 101 00 110 101 ← d →	FD	3	6	23	
DAA	Convert A content into "packed BCD", following add or subtract with packed BCD operands	↕ ↕ ↕ P ● ↕	00 100 111	27	1	1	4	
CPL	A ← $\overline{A}$	● ● 1 ● 1 ●	00 101 111	2F	1	1	4	One's compl.
NEG	A ← $\overline{A}$+1	↕ ↕ ↕ V 1 ↕	11 101 101 01 000 100	ED 44	2	2	8	Two's compl.
Notes:	The V symbol in the P/V column of flags indicates that this flag reports the operation overflow (V = 1). Similarly, the symbol P indicates parity even (P = 1) or odd (P = 0).							
	● = the flag is unchanged by the operation. 0 = the flag is reset by the operation. 1 = the flag is set by the operation. ↕ = the flag is affected according to the result of the operation.							
	r = any one of the CPU 8-bit registers A, B, C, D, E, H, L. n = 8-bit value in range 0..255. d = 8-bit signed value in range -128..+127. CY = the Carry flag. compl. = complement.							

C.4 Arithmetic instructions (16-bit)

Mnemonic	Symbolic Operation	S	Z	H	P/v	N	C	Opcode 76 543 210	Hex	#B	#M	#K	Comments
ADD HL,ss	HL ← HL+ss	•	•	$\updownarrow^2$	•	0	$\updownarrow^1$	00 ss1 001		1	3	11	ss Reg.
ADC HL,ss	HL ← HL+ss+CY	$\updownarrow^1$	$\updownarrow^1$	$\updownarrow^2$	V^1	0	$\updownarrow^1$	11 101 101 01 ss1 010	ED	2	4	15	00 BC 01 DE
SBC HL,ss	HL ← HL-ss-CY	$\updownarrow^1$	$\updownarrow^1$	$\updownarrow^2$	V^1	1	$\updownarrow^1$	11 101 101 01 ss0 010	ED	2	4	15	10 HL 11 SP
ADD IX,pp	IX ← IX+pp	•	•	$\updownarrow^2$	•	0	$\updownarrow^1$	11 011 101 00 pp1 001	DD	2	4	15	pp Reg.
ADD IY,rr	IY ← IY+rr	•	•	$\updownarrow^2$	•	0	$\updownarrow^1$	11 111 101 00 rr1 001	FD	2	4	15	00 BC 01 DE 10 IX
INC ss	ss ← ss+1	•	•	•	•	•	•	00 ss0 011		1	1	6	11 SP
INC IX	IX ← IX+1	•	•	•	•	•	•	11 011 101 00 100 011	DD 23	2	2	10	
INC IY	IY ← IY+1	•	•	•	•	•	•	11 111 101 00 100 011	FD 23	2	2	10	rr Reg.
DEC ss	ss ← ss-1	•	•	•	•	•	•	00 ss1 011		1	1	6	00 BC 01 DE
DEC IX	IX ← IX-1	•	•	•	•	•	•	11 011 101 00 101 011	DD 2B	2	2	10	10 IY 11 SP
DEC IY	IY ← IY-1	•	•	•	•	•	•	11 111 101 00 101 011	FD 2B	2	2	10	

Notes:	
	The V symbol in the P/V column of flags indicates that this flag reports the operation overflow ($V = 1$). ^{1} The flag is affected by the 16-bit result of the operation. ^{2} The flag is affected by the higher 8-bit result of the operation.
	• = the flag is unchanged by the operation. 0 = the flag is reset by the operation. 1 = the flag is set by the operation. $\updownarrow$ = the flag is affected according to the result of the operation.
	ss = any one of the CPU 16-bit registers BC, DE, HL, SP. pp = any one of the CPU 16-bit registers BC, DE, IX, SP. rr = any one of the CPU 16-bit registers BC, DE, IY, SP.
	The 16-bit additions are executed by adding the two least significant bytes first, and then the two most significant bytes. CY = the Carry flag.

C.5 Rotate and shift instructions

Mnemonic	Symbolic Operation	Flag S Z H P/v N C	Opcode 76 543 210	Hex	#B	#M	#K	Comments
RLCA	CY←[7←0]← A	• • 0 • 0 ↕	00 000 111	07	1	1	4	
RLA	CY←[7←0]← A	• • 0 • 0 ↕	00 010 111	17	1	1	4	
RRCA	[7→0]→CY A	• • 0 • 0 ↕	00 001 111	0F	1	1	4	
RRA	[7→0]→CY A	• • 0 • 0 ↕	00 011 111	1F	1	1	4	
RLC r	CY←[7←0]← r	↕ ↕ 0 P 0 ↕	11 001 011 00 000 r	CB	2	2	8	r Reg. 000 B
RLC (HL)	CY←[7←0]← (HL)	↕ ↕ 0 P 0 ↕	11 001 011 00 000 110	CB	2	4	15	001 C 010 D 011 E
RLC (IX + d)	CY←[7←0]← (IX+d)	↕ ↕ 0 P 0 ↕	11 011 101 11 001 011 ← d → 00 000 110	DD CB	4	6	23	100 H 101 L 111 A
RLC (IY + d)	CY←[7←0]← (IY+d)	↕ ↕ 0 P 0 ↕	11 111 101 11 001 011 ← d → 00 000 110	FD CB	4	6	23	
RL r	CY←[7←0]← r	↕ ↕ 0 P 0 ↕	11 001 011 00 010 r	CB	2	2	8	
RL (HL)	CY←[7←0]← (HL)	↕ ↕ 0 P 0 ↕	11 001 011 00 010 110	CB	2	4	15	
RL (IX + d)	CY←[7←0]← (IX+d)	↕ ↕ 0 P 0 ↕	11 011 101 11 001 011 ← d → 00 010 110	DD CB	4	6	23	
RL (IY + d)	CY←[7←0]← (IY+d)	↕ ↕ 0 P 0 ↕	11 111 101 11 001 011 ← d → 00 010 110	FD CB	4	6	23	
RRC r	[7→0]→CY r	↕ ↕ 0 P 0 ↕	11 001 011 00 001 r	CB	2	2	8	
RRC (HL)	[7→0]→CY (HL)	↕ ↕ 0 P 0 ↕	11 001 011 00 001 110	CB	2	4	15	

(cont.)

Mnemonic	Symbolic Operation	Flag S Z H P/v N C	Opcode 76 543 210	Hex	#B	#M	#K	Comments
RRC (IX + d)	⌐7→0→CY (IX+d)	↕ ↕ 0 P 0 ↕	11 011 101	DD	4	6	23	
			11 001 011	CB				
			← d →					
			00 001 110					
RRC (IY + d)	⌐7→0→CY (IY+d)	↕ ↕ 0 P 0 ↕	11 111 101	FD	4	6	23	
			11 001 011	CB				
			← d →					
			00 001 110					
RR r	⌐7→0→CY⌐ r	↕ ↕ 0 P 0 ↕	11 001 011	CB	2	2	8	
			00 011 r					
RR (HL)	⌐7→0→CY⌐ (HL)	↕ ↕ 0 P 0 ↕	11 001 011	CB	2	4	15	
			00 011 110					
RR (IX + d)	⌐7→0→CY⌐ (IX+d)	↕ ↕ 0 P 0 ↕	11 011 101	DD	4	6	23	
			11 001 011	CB				
			← d →					
			00 011 110					
RR (IY + d)	⌐7→0→CY⌐ (IY+d)	↕ ↕ 0 P 0 ↕	11 111 101	FD	4	6	23	
			11 001 011	CB				
			← d →					
			00 011 110					
RLD	03|47 03|47 A (HL)	↕ ↕ 0 P 0 ●	11 101 101	ED	2	5	18	
			01 101 111	6F				
RRD	03|47 03|47 A (HL)	↕ ↕ 0 P 0 ●	11 101 101	ED	2	5	18	
			01 100 111	67				
SLA r	CY←7←0←0 r	↕ ↕ 0 P 0 ↕	11 001 011	CB	2	2	8	r Reg.
			00 100 r					000 B
SLA (HL)	CY←7←0←0 (HL)	↕ ↕ 0 P 0 ↕	11 001 011	CB	2	4	15	001 C 010 D
			00 100 110					011 E
SLA (IX + d)	CY←7←0←0 (IX+d)	↕ ↕ 0 P 0 ↕	11 011 101	DD	4	6	23	100 H 101 L
			11 001 011	CB				111 A
			← d →					
			00 100 110					
SLA (IY + d)	CY←7←0←0 (IY+d)	↕ ↕ 0 P 0 ↕	11 111 101	FD	4	6	23	
			11 001 011	CB				
			← d →					
			00 100 110					

(cont.)

Mnemonic	Symbolic Operation	Flag S Z H P/v N C	Opcode 76 543 210	Hex	#B	#M	#K	Comments
SRA r	$\uparrow\boxed{7\rightarrow0}\rightarrow\boxed{CY}$ r	↕ ↕ 0 P 0 ↕	11 001 011 00 101 r	CB	2	2	8	
SRA (HL)	$\uparrow\boxed{7\rightarrow0}\rightarrow\boxed{CY}$ (HL)	↕ ↕ 0 P 0 ↕	11 001 011 00 101 110	CB	2	4	15	
SRA (IX + d)	$\uparrow\boxed{7\rightarrow0}\rightarrow\boxed{CY}$ (IX+d)	↕ ↕ 0 P 0 ↕	11 011 101 11 001 011 ← d → 00 101 110	DD CB	4	6	23	
SRA (IY + d)	$\uparrow\boxed{7\rightarrow0}\rightarrow\boxed{CY}$ (IY+d)	↕ ↕ 0 P 0 ↕	11 111 101 11 001 011 ← d → 00 101 110	FD CB	4	6	23	
SRL r	$0\rightarrow\boxed{7\rightarrow0}\rightarrow\boxed{CY}$ r	↕ ↕ 0 P 0 ↕	11 001 011 00 111 r	CB	2	2	8	
SRL (HL)	$0\rightarrow\boxed{7\rightarrow0}\rightarrow\boxed{CY}$ (HL)	↕ ↕ 0 P 0 ↕	11 001 011 00 111 110	CB	2	4	15	
SRL (IX + d)	$0\rightarrow\boxed{7\rightarrow0}\rightarrow\boxed{CY}$ (IX+d)	↕ ↕ 0 P 0 ↕	11 011 101 11 001 011 ← d → 00 111 110	DD CB	4	6	23	
SRL (IY + d)	$0\rightarrow\boxed{7\rightarrow0}\rightarrow\boxed{CY}$ (IY+d)	↕ ↕ 0 P 0 ↕	11 111 101 11 001 011 ← d → 00 111 110	FD CB	4	6	23	
Notes:	The P symbol in the P/V column of flags indicates that this flag reports the parity even (P = 1) or odd (P = 0) of the result.							
	• = the flag is unchanged by the operation. 0 = the flag is reset by the operation. 1 = the flag is set by the operation. ↕ = the flag is affected according to the result of the operation.							
	CY = the Carry flag. r = any one of the CPU 8-bit registers A, B, C, D, E, H, L. d = 8-bit signed value in range -128..+127.							

C.6 Bit manipulation instructions

Mnemonic	Symbolic Operation	Flag S Z H P/v N C	Opcode 76 543 210	Hex	#B	#M	#K	Comments
BIT b,r	$Z \leftarrow \overline{r_b}$	X ↕ 1 X 0 ●	11 001 011 01 b r	CB	2	2	8	**r Reg**
BIT b,(HL)	$Z \leftarrow \overline{(HL)_b}$	X ↕ 1 X 0 ●	11 001 011 01 b 110	CB	2	3	12	000 B 001 C 010 D
BIT b,(IX+d)	$Z \leftarrow \overline{(IX+d)_b}$	X ↕ 1 X 0 ●	11 011 101 11 001 011 ← d → 01 b 110	DD CB	4	5	20	011 E 100 H 101 L 111 A
BIT b,(IY+d)	$Z \leftarrow \overline{(IY+d)_b}$	X ↕ 1 X 0 ●	11 111 101 11 001 011 ← d → 01 b 110	CB FD	4	5	20	**b Bit** 000 0 001 1
SET b,r	$r_b \leftarrow 1$	● ● ● ● ● ●	11 001 011 11 b r	CB	2	2	8	010 2 011 3 100 4
SET b,(HL)	$(HL)_b \leftarrow 1$	● ● ● ● ● ●	11 001 011 11 b 110	CB	2	4	15	101 5 110 6 111 7
SET b,(IX+d)	$(IX+d)_b \leftarrow 1$	● ● ● ● ● ●	11 011 101 11 001 011 ← d → 11 b 110	DD CB	4	6	23	
SET b,(IY+d)	$(IY+ d)_b \leftarrow 1$	● ● ● ● ● ●	11 111 101 11 001 011 ← d → 11 b 110	FD CB	4	6	23	
RES b,r	$r_b \leftarrow 0$	● ● ● ● ● ●	11 001 011 10 b r	CB	2	2	8	
RES b,(HL)	$(HL)_b \leftarrow 0$	● ● ● ● ● ●	11 001 011 10 b 110	CB	2	4	15	
RES b,(IX+d)	$(IX+ d)_b \leftarrow 0$	● ● ● ● ● ●	11 011 101 11 001 011 ← d → 10 b 110	DD CB	4	6	23	
RES b,(IY+d)	$(IY+ d)_b \leftarrow 0$	● ● ● ● ● ●	11 111 101 11 001 011 ← d → 10 b 110	FD CB	4	6	23	

Notes:	The notation r_b indicates the bit b (0..7) of register r. The notations $(HL)_b$, $(IX+d)_b$ and $(IY+d)_b$ indicate the bit b (0..7) of the memory location referred by the specified addressing mode. BIT instructions are executed using a bitwise AND operation.
	● = the flag is unchanged by the operation. 0 = the flag is reset by the operation. 1 = the flag is set by the operation. X = the flag is a "don't care" and can assume any value. ↕ = the flag is affected according to the result of the operation.
	r = any one of the CPU 8-bit registers A, B, C, D, E, H, L. d = 8-bit signed value in range -128..+127.

C.7 Jump instructions

Mnemonic	Symbolic Operation	Flag S Z H P/v N C	Opcode 76 543 210	Hex	#B	#M	#K	Comments
JP nn	PC ← nn	• • • • • •	11 000 011 ← n → ← n →	C3	3	3	10	
JP cc,nn	if cc is true, PC ← nn, otherwise continue	• • • • • •	11 cc 010 ← n → ← n →		3	3	10	cc Cond. ─────── 000 NZ 001 Z 010 NC 011 C 100 PO 101 PE 110 P 111 M
JP (HL)	PC ← HL	• • • • • •	11 101 001	E9	1	1	4	
JP (IX)	PC ← IX	• • • • • •	11 011 101 11 101 001	DD E9	2	2	8	
JP (IY)	PC ← IY	• • • • • •	11 111 101 11 101 001	FD E9	2	2	8	
Notes:	Cond. = condition (one among the following): NZ = non zero; Z = zero; NC = non carry; C = carry; PO = parity odd; PE = parity even; P = sign positive; M = sign negative.							
	n = 8-bit value in range 0..255. nn = 16-bit value in range 0..65,535.							
	• = the flag is unchanged by the operation.							

C.8 Subprogram call and return instructions

Mnemonic	Symbolic Operation	Flag S Z H P/v N C	Opcode 76 543 210	Hex	#B	#M	#K	Comments
CALL nn	SP ← SP-1 (SP) ← PC$_H$ SP ← SP-1 (SP) ← PC$_L$ PC ← nn	• • • • • •	11 001 101 ← n → ← n →	CD	3	5	17	
CALL cc,nn	if cc is true, SP ← SP-1 (SP) ← PC$_H$ SP ← SP-1 (SP) ← PC$_L$ PC ← nn, otherwise continue	• • • • • •	11 cc 100 ← n → ← n →		3 3	3 5	10 17	if cc is false if cc is true cc Cond. 000 NZ 001 Z 010 NC 011 C 100 PO 101 PE 110 P 111 M
RET	PC$_L$ ← (SP) SP ← SP+1 PC$_H$ ← (SP) SP ← SP+1	• • • • • •	11 001 001	C9	1	3	10	
RET cc	if cc is true, PC$_L$ ← (SP) SP ← SP+1 PC$_H$ ← (SP) SP ← SP+1, otherwise continue	• • • • • •	11 cc 000		1 1	1 3	5 11	if cc is false if cc is true
RST p	SP ← SP-1 (SP) ← PC$_H$ SP ← SP-1 (SP) ← PC$_L$ PC ← p	• • • • • •	11 t 111		1	3	11	t p 000 0000h 001 0008h 010 0010h 011 0018h 100 0020h 101 0028h 110 0030h 111 0038h
Notes:	Cond. = condition (one among the following): NZ = non zero; Z = zero; NC = non carry; C = carry; PO = parity odd; PE = parity even; P = sign positive; M = sign negative.							
	n = 8-bit value in range 0..255. nn = 16-bit value in range 0..65,535.							
	• = the flag is unchanged by the operation.							

C.9 Input/output instructions

Mnemonic	Symbolic Operation	Flag S Z H P/v N C	Opcode 76 543 210	Hex	#B	#M	#K	Comments
IN A,(n)	A ← (n)	• • • • • •	11 011 011 ← n →	DB	2	3	11	r Reg.
IN r,(C)	r ← (C)	↕ ↕ 0 P 0 •	11 101 101 01 r 000	ED	2	3	12	000 B / 001 C / 010 D
OUT (n),A	(n) ← A	• • • • • •	11 010 011 ← n →	D3	2	3	11	011 E / 100 H
OUT (C),r	(C) ← r	• • • • • •	11 101 101 01 r 001	ED	2	3	12	101 L / 111 A

Comments column register table:

r	Reg.
000	B
001	C
010	D
011	E
100	H
101	L
111	A

Notes:	The P symbol in the P/V column of flags indicates that this flag reports the parity even (P = 1) or odd (P = 0) of the result.
	• = the flag is unchanged by the operation.
	0 = the flag is reset by the operation.
	1 = the flag is set by the operation.
	↕ = the flag is affected according to the result of the operation.
	n = 8-bit value in range 0..255.
	r = any one of the CPU 8-bit registers A, B, C, D, E, H, L.

C.10 CPU control instructions

Mnemonic	Symbolic Operation	Flag S Z H P/v N C	Opcode 76 543 210	Hex	#B	#M	#K	Comments
CCF	CY ← $\overline{CY}$	• • X • 0 ↕	00 111 111	3F	1	1	4	Complement carry flag
SCF	CY ← 1	• • 0 • 0 1	00 110 111	37	1	1	4	
NOP	No Operation	• • • • • •	00 000 000	00	1	1	4	
HALT	CPU halted	• • • • • •	01 110 110	76	1	1	4	
DI[1]	IFF ← 0	• • • • • •	11 110 011	F3	1	1	4	
EI[1]	IFF ← 1	• • • • • •	11 111 011	FB	1	1	4	
Notes:	CY = the Carry flag.							
	• = the flag is unchanged by the operation. 0 = the flag is reset by the operation. 1 = the flag is set by the operation. X = the flag is a "don't care" and can assume any value. ↕ = the flag is affected according to the result of the operation.							
	[1] No interrupt is generated due to the execution of DI or EI instructions.							

Printed in the United States
by Baker & Taylor Publisher Services